Visualization
of Categorical Data

Visualization
of Categorical Data

Edited by

JÖRG BLASIUS
Zentralarchiv für Empirische Sozialforschung
University of Cologne
Cologne, Germany

MICHAEL GREENACRE
Departament d'Economia i Empresa
Pompeu Fabra University
Barcelona, Spain

 **ACADEMIC PRESS**

San Diego   London   Boston
New York   Sydney   Tokyo   Toronto

Find Us on the Web!   http://www.apnet.com

This book is printed on acid-free paper. ∞

The cover design is an inverted part of a time series of weekly amounts of rye
sold in Cologne, 1542–1648, showing a long-term trend line. The time series is
reminiscent of the Cologne cathedral.

Academic Press
525 B St. Suite 1900, San Diego, CA 92101-4495, USA
1300 Boylston Street, Chestnut Hill, MA 02167, USA

Academic Press Limited
24-28 Oval Road, London NW1 7DX, UK
http://www.hbuk.co.uk/ap/

Library of Congress Cataloging-in-Publication Data

Visualization of categorical data / [edited by] Jörg Blasius, Michael
    Greenacre.
        p.   cm.
    Includes bibliographical references and index.
    ISBN 0-12-299045-5 (alk. paper)
    1. Multivariate analysis--Graphic methods.   I. Blasius, Jörg,
    1957-   . II. Greenacre, Michael L.
    QA278.V57   1997
001.4'226--dc21
                                                            97-35782
                                                            CIP

Printed in the United States of America
97 98 99 00 EB 9 8 7 6 5 4 3 2 1

For Clifford C. Clogg
In Memoriam

Contents

Part 2   Correspondence Analysis   107

Preface

The story starts in Cologne with the conference on Recent Developments and Applications of Correspondence Analysis and the subsequent publication of the book *Correspondence Analysis in the Social Sciences* by Academic Press, London, in 1994. This book contained 16 chapters written by a total of 24 authors and edited by ourselves.

The idea was to bring together social science researchers and statisticians in a collaborative project to bridge the gap between theory and practice in social science methodology. We are happy to report that the exercise proved to be so successful that we acquired enough energy to try it once again. This time, however, the subject would be put into a wider context, defined by three keywords: data, categorical, and visualization.

The keyword *data* stresses the central importance of application, the need for methodology to be illustrated in a particular data context and not isolated as theory for theory's sake. The method is judged essentially by its applicability to data.

The data context is *categorical*, that is, classified, grouped, categorized, or ranked. Wherever observations can be put into boxes and counted, this is what we are interested in. This puts us into the realm of social science observation without the need to specify it.

The context of the categorical data is *visualization*. We are interested in exploring categorical data through graphical displays, be they maps, trees, custom-designed computer graphics, or geometric shapes. We are also interested in models for categorical data and especially in the potential to use visual tools to aid in the interpretation of models and the modeling process. Hence the title: *Visualization of Categorical Data*.

In 1995 the Zentralarchiv für Empirische Sozialforschung (Central Archive for Empirical Social Research) in Cologne again hosted an international conference, this time on the visualization of categorical data. The response was double that of the previous conference and 21 countries were represented—clearly the topic was of wide interest. Selected papers as well as specially invited contributions, totaling 35 chapters by 63 authors, have been refereed, edited, and—to be sure—categorized to bring to you a state-of-the-art collection on how to achieve visual summaries and presentations of categorical data. In our editing we have tried to reduce the material as much as possible to accommodate as many contributions as possible. We have also introduced cross-referencing between the chapters, tried to unify the notation as much as possible, and established a common reference list and index. We hope that the book forms a coherent whole and that the collection proves worthy of the patient and long-suffering work of all the authors who have contributed to it.

As a prologue we are pleased to have a "keynote chapter" written by Jan de Leeuw, titled "Here's Looking at Multivariables," with his personal view on data visualization. After this initial chapter, we have divided the book into four parts: Graphics for Visualization, Correspondence Analysis, Multidimensional Scaling and Biplot, and finally Visualization and Modeling.

The first part concentrates on a variety of graphical methods, ranging from very simple graphical displays to sophisticated computer graphics. This part will give you a flavor of the wide variety of visualization approaches inspired by a diversity of data contexts, such as contingency tables, questionnaire responses, and event histories.

The second part deals with correspondence analysis, some "classic" applications to social science data, and some interesting new developments, especially in the interpretation of correspondence analysis of multivariate data. This part can be considered the continuation of the story started in the 1994 book mentioned earlier.

The third part is devoted to multidimensional scaling and biplot methods, which visualize data in the form of distances and scalar products, respectively. Here there are various chapters dealing with issues of interpretation, diagnosis of structure, applications, and new methods.

Finally, the fourth part aims to demonstrate the visualization possibilities in categorical data modeling such as latent class analysis, ideal point discriminant analysis, latent budget analysis, and general log-bilinear models. In many cases, the complementary nature of the modeling and exploratory visualization approaches is illustrated.

Many of the methods in this book are associated with what are known as "exploratory" methods as opposed to "confirmatory" methods. We believe that both these approaches have a place in our understanding of social science phenomena and that visualization techniques have an important role to play in every approach to data analysis.

Exploratory methods are often criticized as having no "traditional" social science hypothesis such as "income is dependent on sex and education." Researchers working with correspondence analysis or other exploratory methods often have no strong conclusions such as "females earn significantly less money than males" or "there is a significant relationship between education and social status." But solutions like these are usually not the aim of the researchers who prefer exploratory techniques; they would very often argue that they like to describe the structure of the data only and that they have no need for modeling the relationships between the variables.

A typical research question for these researchers would be to describe movements in the "social space." In the theoretical part of a work one could express assumptions about the closeness of variables to each other in this social space. In the empirical part of such a work, using correspondence analysis, for example, one could see whether or not the variables that should be close to each other belong to a common cluster.

Visualization can considerably enhance the modeling process. Before running a bivariate regression analysis, for example, a simple scatterplot between the variables would tell us whether the relation between both variables is linear. On the basis of this solution, one could decide whether a linear model is appropriate or what other relationship holds. The same concept holds for categorical data—before the modeling process starts, visualization techniques will help to understand the structure of the data, which categories can be combined without affecting the analysis, for example, or what interactions are present. Visualization can also be used to give further information after the data are modeled—for example, to search for possible structure in the residuals from the model or to investigate which additional effects have to be considered in the model.

A novel aspect of this book is the section with color illustrations. We felt that it was necessary to see some figures in their original colors rather than shades of gray. Color adds a tremendous benefit to the potential of any visualization, as is clearly demonstrated in these examples. We feel that use of color should be encouraged as an essential component of visualization methods.

We wish to thank all our authors for participating in this project and for repeatedly revising their papers with patience, dedication, and timeliness. We would also like to thank all the authors who helped to review their peers' contributions. In addition, there were many other colleagues who assisted us in the anonymous reviewing process: many thanks to Hans-Jürgen Andreß (University of Bielefeld, Germany), Phipps Arabie (Rutgers University, New Jersey), Gerhard Arminger (University of Wuppertal, Germany), Johann Bacher (University of Linz, Austria), Hans Hermann Bock (University of Aachen, Germany), Jürgen Friedrichs (University of Cologne), Helmut Giegler (University of Augsburg, Germany), Werner Georg (University of Konstanz, Germany), Jacques Hagenaars (Tilburg University, The Netherlands), Wolfgang Jagodzinski (Zentralarchiv für Empirische Sozialforschung, Cologne), Walter Kristof (University of Hamburg), Ulrich Kockelkorn (University of Berlin), Steffen Kühnel (University of Gießen, Germany), Warren Kuhfeld (SAS Institute, Cary, North Carolina), Rolf Langeheine (Institute for Pedagogics in the Natural Sciences, Kiel, Germany), Herbert Matschinger (University of Leipzig, Germany), Ekkehard Mochmann (Zentralarchiv für Empirische Sozialforschung, Cologne), Jost Reinecke (University of Münster, Germany), Götz Rohwer (Max Planck Institute for Educational Research, Berlin), Jože Rovan (University of Ljubljana, Slovenia), Cajo ter Braak (University of Wageningen, The Netherlands), Karl van Meter (LASMAS/IRESCO-CNRS, Paris), Wijbrandt van Schuur (University of Groningen, The Netherlands) and Ken Warwick (Ken Warwick Associates, New York).

This project could never have been undertaken without the continual encouragement and financial support of the Zentralarchiv in Cologne, and we would like to thank the executive manager, Ekkehard Mochmann, for his and his organization's cooperation in every aspect of the venture. We thank the secretaries at the Zentralarchiv, Friederika Priemer and Angelika Ruf, as well as Hanni Busse and Bernd Reutershan from the administrative staff, and our students Udo Dillmann, Gabriele Franzmann, Ulla Läser, and Rainer Mauer for their assistance. We would like to make a special mention of Friederika Priemer's invaluable contribution in editing the reference list and the section "About the Authors," as well as all her proofreading and handling an endless stream of inquiries and problems on our behalf.

To finalize the manuscript, several trips were made between Barcelona and Cologne. For generously supporting Michael's travel to Cologne and Jörg's to Barcelona, we thank the Zentralarchiv für Empirische Sozialforschung as well as the Universitat Pompeu Fabra and appreciate partial support from Spanish DGICYT grant PB93-0403. The Christmas "break" was spent working in the German spa town of Bad Orb, with walks in temperatures of $-15°C$ to clear the mind for another three-hour session working on a chapter. And we both agree that it is worth mentioning the small 48-seater Lufthansa jet operating between Barcelona and Cologne, with visits to the flight cabin to witness spectacular views of the Alps and excellent meals and personal service.

Our final thanks go to Karen Wachs, Julie Champagne, and all the Academic Press staff involved in this project.

It is with sadness that we have learned of the recent death of one of the contributors to this book, Frans Symons, of Leuven University in Belgium. A few lines about Frans' work, written by his friend and colleague Jaak Billiet, have been included in the "About the Authors" section at the end of the book.

This book is dedicated to Cliff Clogg, who was to have been one of the keynote speakers at our conference but who died unexpectedly ten days before the meeting began. Cliff was one of the most prominent researchers in both statistics and sociology and embodied the spirit of this project in bridging the gap between these two disciplines. His contribution to this book, finished by his friend Tamás Rudas, is one of the last he worked on. For this reason and as a tribute to his work in the areas of statistics and sociology, we and all the authors join together in dedicating this book to his memory as a colleague and friend.

Jörg Blasius and Michael Greenacre
Cologne and Barcelona
October 1997

Chapter 1

Here's Looking at Multivariables

Jan de Leeuw

1 Introduction

I don't really understand what "visualization of categorical data" is about. This is a problem, especially when one is supposed to write the opening chapter for a book on this topic. One way to solve this problem is to look at the other chapters in the book. This empirical, data analysis–oriented, approach is based on the idea that the union of all published chapters defines the topic of the book.

For this book, the approach of looking at the titles of the chapters produced somewhat disappointing results. Whatever "visualization of categorical data" is, it consists of about 50% correspondence analysis, about 10% multidimensional scaling, about 10% cluster analysis, about 20% contingency table techniques, and the remaining 10% other ingredients. It is unclear from the titles of the chapters what they have in common. When writing this introduction I assumed, and I have since then verified, that every author in this book shows at least one graph or one plot. But this is a very weak common component. Not enough to base an opening chapter on.

Thus the empirical approach fails. Alternatively, I can try to define my way out of the problem. This is intellectually a more satisfying approach.

We start with *data*, more precisely *categorical data*, and these data are analyzed by using a (data-analytical or statistical) *technique*.

Such a technique produces an *image* or a *smooth* or a *representation* of the data. The title of the book indicates that we are particularly interested in *visual* smooths or representations of the data.

1

Thus our course is clear; we have to define *data*, *technique*, and *representation* and then single out categorical data and visual representations. After that exercise, we can go back and see if and how the contents of the conference fit in.

This paper can be seen as the next member of the sequence of de Leeuw (1984, 1988, 1990, 1994). The general approach, in an early form, can also be found in the first chapter of Gifi (1990).

2   Data

The data D are an element of the *data space* $\mathcal{D}$. The design of the experiment, where both "design" and "experiment" are used in a very general sense, defines the data space.

If you distribute a particular questionnaire to 1000 persons and your questionnaire has 50 items with 7 alternatives each, then the data space has $1000^{7^{50}}$ possible elements, and if you allow for missing data and nonresponse, it has even more. In general, these data spaces, which are the sets of all possible outcomes of our experiment, tend to be very large.

The same thing is true if you make measurements in a particular physical experiment, or if you plant a number of seeds in a number of pots, or if you watch a number of infants grow up. Even with a limited number of variables, the possible number of outcomes is very, very large.

In all these cases the data space is defined before the observations are actually made, and the possible outcomes of the experiment are known beforehand as well. Is it possible to be surprised? I guess it is, but that is a flaw in the design.

2.1   Coding

We do not find data on the street. Data are not sense impressions, which are simply recorded. Data are *coded*, by which we mean that they are entered into a preformatted database. This is not necessarily a computerized database; it could simply be the codebook given to interviewers or to data-entry persons, or it could be an experimental protocol.

The important notion is that data are already categorized and cleaned and that the protocol tells us how to reduce data from a data space of quadri-zillions of elements to one of trillions of elements. We know, for instance, that we can ignore the look on the face of the person filling in the questionnaire, and the doodles on the student's examination forms are not counted toward the grade.

Another key point is that usually the greatest amount of data reduction goes on in this coding stage. The really important scientific decisions, and the places where the prior knowledge has the greatest impact, are not necessarily the choice between the normal distribution and Student's t or between frequentist and Bayesian procedures.

2.2 Example

This could perhaps be illustrated by an actual example. One of the clients of University of California Los Angeles (UCLA) Statistical Consulting is the California Department of Corrections. There is a gigantic project set up to study whether the classification of prisoners into four security categories actually reduces within-prison violence. The data for answering this study are the prison careers of all individuals who were in one of the state prisons in California in the last 10 years. It will not surprise you to hear that these are hundreds of thousands of individuals. Many of them have been in and out of prison for 20 or more years. They have been shifted between security levels many times, often on the basis of forms that are filled in and that have objective cutoffs, but often also on the basis of "administrative overrides" of these objective results.

It is generally known that propensity to violence in prison is related to age, to previous prison career, and to gang membership. Clearly, there are potentially thousands of variables that could be coded because they might be relevant. Wardens and other prison personnel observe prisoners and make statements and judgments about their behavior and their likelihood to commit violent acts while in prison. Presumably, many of these judgments and observations change over time for a given prisoner and maybe even for a given prison guard.

The observations and classifications of the prison personnel, however, are not data and not variables. They become data as soon as they are organized and standardized, as soon as variables are selected and it is decided that comparable information should be collected on each prisoner, over time, over changing security levels, and perhaps over changing institutions. It is decided that a study will be done, a database will be constructed, and the integrity and completeness of the database become so important that observations in different institutions and time periods by different observers on the same individual are actually coded uniformly and combined. Without reducing the chaos of impressions and judgments to a uniform standard and format, there really are no data.

2.3 Categorical Data

Classically, data are called categorical when the data space is discrete. I think it is useful to repeat here that all data are categorical. As soon as we have set the precision of our measurements, the grid on which we measure, and the mesh of our classifications, then we have defined a discrete and finite data space.

Statistics has been dominated by mathematics for such a long time that some people have begun to act as if "continuous" data is the rule. Continuous data is a contradiction. Continuity is always part of the mathematics, that is, of the *model* for the data. The question whether continuity "really" occurs in nature is clearly a metaphysical one, which need not concern us here. We merely emphasize that continuity is used mostly to simplify computations, in the same way as the normal distribution was first used to simplify binomial calculations.

The codebook, or the rules for entry into the database, also contains rules for coding numerical information. It has to be categorized (or rounded), because our data entry persons and our computers cannot deal with infinite data spaces.

Thus:

All Data Are Categorical

although perhaps some data are more categorical than others. This suggests that, in a strict sense, it is impossible to distinguish "categorical data" from "other data." In actual practice, however, we continue to use the distinction and speak about categorical data when our variables are non-numerical and/or have only a small number of discrete values.

2.4 Multivariables

The most important type of data that science has been able to isolate is the variable or, if you like, *multivariable*. This is closely related to the "fluents" in Newtonian physics, the random variables of statistics, and the variables in mathematical expressions. For some useful philosophical discussion of these concepts we refer to Menger (1954, 1955, 1961) and quite a few other publications by the same illustrious author.

In the case of a multivariable, the data space is the product of a number of functions defined on a common domain, with different images. Table 1 shows a simple example of a bivariable, describing the nine faculty members in the new UCLA statistics department. Two variables are used: department of origin and country of origin.

If we look at the types of data spaces most frequently discussed at this conference, we find the multivariable in various disguises. In formal concept analysis multivariables are called *many-valued contexts (mehrwertige Kontexte)*, the variables are *attributes (Merkmale)*, and the domain of the variables is the *objects (Gegenstände)*—see Wolff and Gabler (Chapter 7) and Frick *et al.* (Chapter 6).

In cluster analysis, multidimensional scaling, contingency table analysis, and multivariate analysis, the multivariable is often preprocessed to form a *distance*

Table 1: A Multivariable

	Department	Born in the United States
Ferguson	Mathematics	Yes
Li	Mathematics	No
Ylvisaker	Mathematics	Yes
Berk	Sociology	Yes
DeLeeuw	Statistics	No
Mason	Sociology	Yes
Bentler	Psychology	No
Muthén	Education	No
Jennrich	Mathematics	Yes

matrix, a *covariance matrix*, or a *cross-table*. This often involves data reduction, although sometimes the map is one-to-one. This particular step in the data reduction process can be thought of as either the last step of coding or the first step of the statistical analysis.

Also, in some cases, we observe dissimilarities or measure distances directly. This can be coded as a single real variable on $I \otimes I$ or as three variables, the first two being labels.

3   Representation

The process of coding maps the possible outcomes of an experiment into the data space, which is defined by the design. Although in some experiments coding may be relatively straightforward, in others it involves many decisions.

The mappings used in coding are not often studied in statistics, although perhaps they should be analyzed more. Design in the narrow sense is generally seen to be a part of statistics, but codebooks and experimental protocols are usually assumed to be part of the actual science.

What definitely is a part of statistics is the next phase, the mapping of data into representations. We take the data, an element of the data space, and we compute the corresponding element of the representation space. This mapping is called a *statistical technique* (see Figure 1).

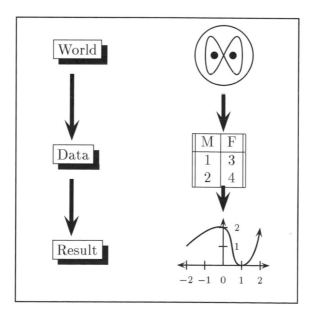

Figure 1

Not surprisingly, many types of representations are used in statistics. In formal concept analysis, data are represented as lattices or graphs; in cluster analysis as trees, hierarchies, or partitionings; in correspondence analysis (CA), multidimensional scaling (MDS), and biplots as maps in low-dimensional spaces. In regression analysis and generalized linear modeling we find many types of scatterplots, either to picture the structural relationships (added variable plots, smooths) or to portray the residuals and other diagnostics. There are very many variations of these mapped plots, and new ones are continually being invented.

In contingency table analysis we now also have graphical models, familiar from path analysis and structural equation modeling. Residuals from contingency table analysis are modeled with Euclidean techniques. We should remember, however, that 500 pages of computer output also defines a representation space and that people look at the tables in CROSSTABS output from SPSS as primitive visualizations as well.

4 Techniques

We have seen that techniques map data space into representation space. What are the desirable properties of the techniques? We mention the most important ones.

- A technique has to be as *into* as possible; that is, it should be maximally data reducing.
- A technique should incorporate as much prior knowledge from the science as possible (this could, however, be prejudice or fashion).
- A technique should separate the stable and interesting effects from the background or noise.
- A technique should show the most important aspects of the data.
- A technique should be stable, that is, continuous and/or smooth.

Some qualifying remarks are in order here. Data reduction cannot be the only criterion, because otherwise we could replace any data set with the number zero, and this would be a perfect technique. In the same way, stability cannot be the only criterion either (same example).

We also need some notion of fit, and this is embedded in what we think is interesting (i.e., in our prior knowledge). In homogeneity analysis (or multiple correspondence analysis) we apply a singular value decomposition to a binary matrix of indicators (also called dummy variables). In analyses using the ordinary singular value decomposition, fit is defined as least-squares approximation to the observed matrix by a matrix of low rank. But in homogeneity analysis we do not want to approximate the zeros and ones, we want to make a picture of the qualitative relations in the data. Thus we look at partitionings, coded as star plots (Hoffman and de Leeuw, 1992).

Also observe that continuity of a technique requires a topology on $\mathcal{D}$ and $\mathcal{R}$, and smoothness in the sense of differentiability even requires a linear structure. This

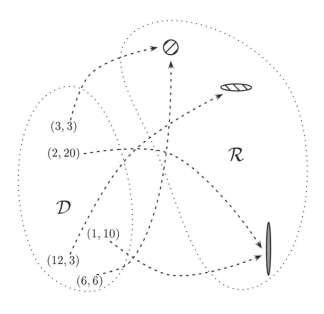

Figure 2

already provides so much mathematical, in fact geometrical, structure that we can almost say we are visualizing the data.

Tentatively, we could maintain the working hypothesis:

All Statistical Techniques Visualize Data

There seems to be some idea that visualization takes place by using representations that are geometrical and that maybe even have the geometry of Euclidean space. This is certainly suggested by the contents of this book, given the predominance of CA and MDS.

But this point of view is certainly much too narrow, because the notions of geometry pervade much of analysis, algebra, discrete mathematics, topology, and so on. Even the real numbers (the real line) are geometrical, and computing a one-dimensional statistic means mapping the data space into the real line (think of confidence intervals, for instance). Again, as with the notion of categorical data, all analysis is visualization, but some analyses are more visual than others.

As we mentioned earlier, it is difficult to draw the line between coding and analysis. Both involve data reduction, and both involve following certain rules. But usually there is a decision to ignore a part of the process and to consider the outcome of this ignored part of the data, which will be fed into the technique.

Very often the technique has multiple stages. We start by reducing the data to a contingency table, or a covariance matrix, or a set of moments, or an empirical distribution function. This stage is often formalized by the optimistic concept of sufficient statistics, which gives conditions under which we do not lose information.

Only after this stage of preliminary reduction, the serious data analysis starts. Such a serious analysis is often based on a model.

5   Additional Tools

We have discussed data and the forms they take, emphasizing multivariables. We have also discussed the most common types of representations, including simple statistics, tables, graphs, plots, lattices and other ordered structures, partitions, and Euclidean representations. And finally, we have discussed the maps of the data space into the presentation space, which associate the outcome of the statistical analysis with the data in the study.

There are some ideas that can guide us in the construction of visualizations. If the data themselves are spatial, we do not need such guides (and this is recognized more and more often in the use of geographical information systems, or GISs). But otherwise we can use models, and we can try to represent properties of the data as well as possible in our visualizations (using some notion of fit).

5.1   Role of Models

Figure 3 illustrates the use of a model. In this particular case, the model is that gender and size of the ego are independent. The data P are in the upper left-hand corner;

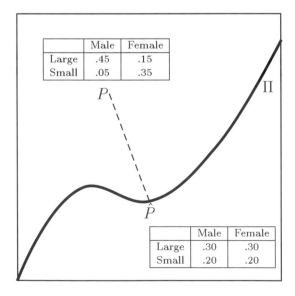

Figure 3

they are proportions in a 2×2 table. The model is the set of all 2×2 tables with independence, a curved surface in three-dimensional space, represented in the figure by the curved line Π. To find out whether the model fits the data, we look at the distance between the model and the data. The statistical technique actually projects the data P on the model Π and comes up with the fitted (or reduced) data $\hat{P}$.

Models are convenient tools with which to capture prior information and to construct statistical techniques. The idea is that a model is some subset of the representation space $\mathcal{R}$ and that prior information tells us that the data, suitably reduced perhaps to a table or covariance matrix, should be close to the model.

This discussion of the role of models covers maximum likelihood methods, the linear model, the t-test, and much of nonparametric statistics as well. It works, provided we are willing to specify a model in the representation space, that is, a subset of that space that we are particularly comfortable with (for scientific reasons, but often only for aesthetic reasons).

5.2 Fit

Figure 3 illustrates one notion, the distance between suitably reduced data and the model. More generally, we may want to know how good or faithful a visualization of the data is. Sometimes representations are very faithful, in fact one-to-one.

Some of this is also illustrated in the pictures that follow, where we first make a graph of the data (Figure 4) and then modify the graph by using multiple correspon-

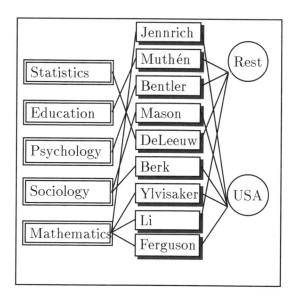

Figure 4

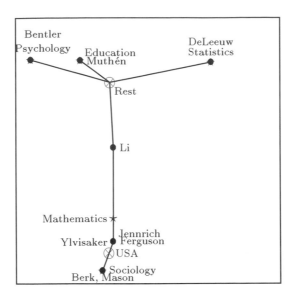

Figure 5

dence analysis (to "make the lines shorter"), shown in Figure 5. As long as the lines are still there, we do not lose information. If we leave them out and interpret on the basis of proximity, we have to guess, and we'll sometimes guess wrong.

In Figure 5 the distances between the statisticians approximate the "chi-squared distances," while the categories of the variables are plotted using the "centroid principle."

6 Visualization of Statistics

Due to the fast personal computer and the bitmapped screen, our day-to-day use of statistics is changing. We can replace assumptions by computations and long lists of tables by graphs and plots.

But, even more profoundly, our interpretation of statistics has been changing too. Moving away from specific calculation-oriented formulas has led to a much more geometrical approach to the discipline (most clearly illustrated in the differential geometric approach, but also in the more applied graphical models approach and of course in the use of GISs).

In a sense, this is nothing new, because modern analysis has been thoroughly geometrized as well. And even in the area of the greatest rigor, that of proofs, a picture is sometimes worth a thousand numbers.

To close with an example of this, an illustration is shown in Figure 6. This is a familiar picture, and it can be used to illustrate many of the basic regression principles.

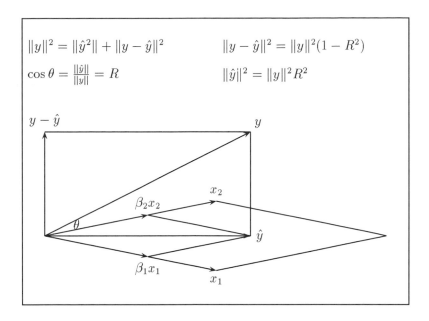

$$\|y\|^2 = \|\hat{y}^2\| + \|y - \hat{y}\|^2 \qquad\qquad \|y - \hat{y}\|^2 = \|y\|^2(1 - R^2)$$

$$\cos\theta = \frac{\|\hat{y}\|}{\|y\|} = R \qquad\qquad\qquad \|\hat{y}\|^2 = \|y\|^2 R^2$$

Figure 6

In my regression course, I use it perhaps in 10 of the 30 lectures. It portrays projection, orthogonality *à la* Pythagoras, the regression coefficients, the residuals, the predicted values, the multiple correlation coefficient, and the residual sum of squares. Thus it provides a picture of the basic regression statistics that are printed out by all packages, in a form in which we can use quite a few lectures, of 1000 words each, to explain to the students what is actually going on.

PART I

Graphics for Visualization

Visualization of data is a vast subject with a long tradition in the social sciences. Pictures are easier to understand than numbers, especially when there are many numbers to understand.

Different visualization techniques are appropriate to the measurement level of the data, and special methods have been developed to handle univariate, bivariate, and multivariate data. Even in the simplest case in which we have only one variable, there is a wide range of techniques: various types of histograms, bar charts, stem-and-leaf displays, box plots, and pie charts (for an overview, see Tukey, 1977). Most of these techniques are available in widely used spreadsheet packages and thus have found their way into the popular media. We often see graphics of a single set of observations in newspapers and on television, for example, a pie chart of the number of seats for each political party in a parliament after an election, or a plot of a sequence of values over time of a continuously scaled variable such as the interest rate. Sometimes one method of visualization is clearly better than another. From perception psychology one knows, for example, that it is easier to recognize differences in the data when using bar charts than using pie charts—in bar charts differences are visualized in one dimension, and small differences in the data are easy to discern. Pie charts include two dimensions, and one has the more difficult task of judging arc lengths or areas. Another example is the use of a Q–Q (quantile–quantile) probability plot to diagnose normality in a set of data: it is easier to see whether a set of points lies in a straight line than to judge whether a histogram of the observations looks like a normal density.

In the bivariate case the number of possibilities for visualizing the data is even larger as we consider the different cases: both variables continuous, one continuous and one categorical, or both categorical. When one is categorical with only a few categories we can juxtapose univariate displays, for example, two histograms "back to back" for a comparison of age distributions of males and females, often used in demographic studies, or different pie charts showing the state of the political parties before and after the election. Scatterplots are used when both variables are continuous, for example, income and age, and allow one to diagnose the nature of the relationship between the two variables. In the case of two categorical variables with many categories we may wish to compare observed frequencies of co-occurrence with the marginal frequencies of the cross-tabulation. We shall see many original ways of exploring such cross-tablulations in the intial chapters after this introduction.

For three or more dimensions, visualization demands more sophistication and even more originality. In the case of three continuous variables, there are several praiseworthy attempts to see data in three-dimensional space involving real-time computer graphics. For many variables some methods try to depict all the data exactly in a way that allows interpretation of all variables simultaneously, for example, Chernoff faces and star plots. The advantage of visualizing all information in the data is often outweighed by the disadvantage of the complexity of the displays and their interpretation. However, once we recognize that there is a certain level of redundancy in multivariate data, it is possible to exploit this surfeit of information to simplify the problem to one of few dimensions, where we can again resort to the existing graphical tools. This aspect of "dimension reduction" is one of the main themes of this book.

In the first part of the book we will start with a range of examples of what we might call "straight" visualization of bivariate and multivariate data. In most cases these will be innovative displays of the original data or transformed versions of the data, specially developed for a particular context. Some methods are especially useful for the analysis of small data sets. Formal concept analysis (Chapters 6 and 7) permits the visualization of every item of information in the data; the solution is a display considering all connections between all variables and between all subjects as well as between subjects and variables. In event history analysis, which has strongly increased in the social sciences in the nineties, Lexis pencils can be used to visualize the information for each subject (Chapter 4). In other examples of visualization, the type of picture will be used to diagnose a model or the picture will be the model itself.

Visualization is often improved using the tools used by graphical artists, for example, shading and colors. We have included color graphics in some cases to illustrate the use of this important aspect of visualization. As color printing becomes cheaper and more accessible, we expect to see more widespread and routine use of color graphics.

Chapter 2, by Michael Friendly, gives an overview of different ways of visualizing a contingency table. So-called sieve and mosaic displays rely on displaying observed frequencies or expected frequencies under independence as areas of rectangles drawn in the same row–column positions as in the original table. Differences between observed and expected frequencies can be visualized using shading and color, and it also helps to reorder the rows or columns if they represent unordered (nominal) categories. Friendly shows how these ideas extend to multiway tables and uses interesting analogies with physical concepts such as pressure and energy to make the visualizations even more interpretable. Another type of diagram, the fourfold display, is used for comparing sets of 2×2 tables and also relies on depicting cell frequencies by areas, but in such a way that the odds ratios are displayed.

In Chapter 3, Jean-Hugues Chauchat and Alban Risson demonstrate the application of the ideas of Jacques Bertin, the French graphical "semiologist," to visualizing the rows and the columns of a contingency table. Bertin's graphics remain true to the original data, using only permutations of the rows and the columns of the data matrix. The authors use different data sets from the social sciences and also show how the same ideas can be used to visualize solutions obtained by cluster analysis and correspondence analysis.

Chapter 4, by Brian Francis, Mark Fuller, and John Pritchard, is dedicated to the visualization of event history data. Event history studies usually involve the collection of large and complex amounts of information on a set of individuals over time. A typical event history study consists of records of individual job careers over a number of years, including other information such as income and family status. The aim of event history analysis is to find common events in time, for example, getting married and having the first child a specific number of months later. Using the Lexis diagram, the authors introduce a visualization technique that allows the representation of both duration and state transitions in all variables relevant to an analysis. In their empirical example, the authors visualize data from the British Social Change and Economic Life Surveys using a three-dimensional Lexis diagram with the duration variables date of marriage, female age, and time since marriage. The Lexis diagram includes a "pencil" for every subject indicating the time duration. Coloring the pencils shows the individual status at every time point, for example, indicating the work status or number of children.

In Chapter 5, Tomàs Aluja-Banet and Eduard Nafría discuss generalized impurity measures and data diagnostics in decision trees. Decision trees are a specific means of displaying a set of multivariate categorial data in which one variable is of special interest and is regarded as a response. The decision tree is the visualization of a simple rule for predicting the response from the other variables, which are regarded as predictors. Using a survey of mobility preferences of the inhabitants of Barcelona, the authors show in descending order the different importance of the predictors for the decision if one prefers to have "no change" or to move to "another district," to the "surroundings," to the "rest of Catalonia," or to the "rest of Spain." It turns out that the most important variable for this decision is the district of residence. Other variables that define the nodes of the decision tree are socioeconomic status of the household, age of head of the household, and years living in the neighborhood. The authors also discuss the stability of the results according to split and stop criteria.

Chapter 6, by Ulrich Frick, Jürgen Rehm, Karl Erich Wolff, and Michael Laschat, describes obstetricians' attitudes on prenatal risks by using formal concept analysis (FCA). FCA simultaneously groups both objects and attributes and reveals dependencies between attributes and between objects. In one of their empirical examples, the objects are heads of obstetrics departments in Vienna and the attributes are fetal risks. The data are binary according to whether or not the risks are accepted by the departments. By applying FCA to these binary data, the authors arrive at line diagrams from which one can read the position of each of the departments and which fetal risks they accept. The solution provides both an ordering of the acceptance of fetal risks and an ordering of the departments according to the risks they accept. The display is exact: at every position of the line diagram one gets full information on the departments (which risks they accept) as well as on the fetal risks (in which departments they are accepted).

In Chapter 7, Karl Erich Wolff and Siegfried Gabler also discuss FCA, here in comparison with correspondence analysis. Whereas FCA displays all information in the data, correspondence analysis is a data reduction technique in which loss of information is incurred. Applying both methods to the same data, the authors show that the solutions are quite similar. Discussing the advantages and the disadvantages of both methods, Wolff and Gabler suggest that FCA should be used only when there is a relatively small number of attributes. For the visualization of a higher number of attributes they suggest applying correspondence analysis to find some interesting attribute clusters that might serve as a starting point for the data analysis with FCA. In addition to the comparison of both methods, the chapter gives some background information on FCA as well as some rules for reading a correspondence analysis map.

Chapter 8, by Vartan Choulakian and Jacques Allard, describes the Z-plot as a graphical procedure for contingency tables with an ordered response variable. For such contingency tables, the proportional odds models of McCullagh and Goodman's R or C model are often used. The authors discuss the use of the Z-plot as a preliminary aid to screen the data before applying formal statistical analysis. If the Z-plot does not reflect the ordinal order of the response variable, then the preceding relatively simple models do not describe the data well, and a more complex model such as the RC association model should be applied (the latter model is also known as the log-bilinear model; for a description of this model see de Falguerolles, Chapter 35).

Chapter 2

Conceptual Models for Visualizing Contingency Table Data

Michael Friendly

1 Introduction

For some time I have wondered why graphical methods for categorical data are so poorly developed and little used compared with methods for quantitative data. For quantitative data, graphical methods are commonplace adjuncts to all aspects of statistical analysis, from the basic display of data in a scatterplot, to diagnostic methods for assessing assumptions and finding transformations, to the final presentation of results. In contrast, graphical methods for categorical data are still in their infancy. There are not many methods, and those that are discussed in the literature are not available in common statistical software; consequently, they are not widely used.

What has made this contrast puzzling is the fact that the statistical methods for categorical data are in many respects discrete analogues of corresponding methods for quantitative data: log-linear models and logistic regression, for example, are such close parallels of analysis of variance and regression models that they can all be seen as special cases of generalized linear models.

Several possible explanations for this apparent puzzle may be suggested. First, it may be that those who have worked with and developed methods for categorical data are just more comfortable with tabular data, or that frequency tables, representing sums over all cases in a data set, are more easily apprehended in tables than quantitative data. Second, it may be argued that graphical methods for quantitative

17

data are easily generalized so, for example, the scatterplot for two variables provides the basis for visualizing any number of variables in a scatterplot matrix; available graphical methods for categorical data tend to be more specialized. However, a more fundamental reason may be that quantitative data display relies on a well-known natural visual mapping in which a magnitude is depicted by length or position along a scale; for categorical data, we shall show that a count is more naturally displayed by an area or by the visual density of an area.

2   Some Graphical Methods for Contingency Tables

Several schemes for representing contingency tables graphically are based on the fact that when the row and column variables are independent, the expected frequencies, m_{ij}, are products of the row and column totals, divided by the grand total. Then each cell can be represented by a rectangle whose area shows the cell frequency, n_{ij}, or deviation from independence.

2.1   Sieve Diagrams

Table 1 shows data on the relation between hair color and eye color among 592 subjects (students in a statistics course) collected by Snee (1974). The Pearson χ^2 for these data is 138.3 with nine degrees of freedom, indicating substantial departure from independence. The question is how to understand the *nature* of the association between hair and eye color.

For any two-way table, the expected frequencies m_{ij} under independence can be represented by rectangles whose widths are proportional to the total frequency in each column, $n_{.j}$, and whose heights are proportional to the total frequency in each row, $n_{i.}$; the area of each rectangle is then proportional to m_{ij}. Figure 1 shows the expected frequencies for the hair and eye color data.

Riedwyl and Schüpbach (1983, 1994) proposed a *sieve diagram* (later called a *parquet diagram*) based on this principle. In this display the area of each rectangle

Table 1: Hair color, eye color data

Eye color	Hair color				Total
	Black	**Brown**	**Red**	**Blond**	**Total**
Green	5	29	14	16	64
Hazel	15	54	14	10	93
Blue	20	84	17	94	215
Brown	68	119	26	7	220
Total	**108**	**286**	**71**	**127**	**592**

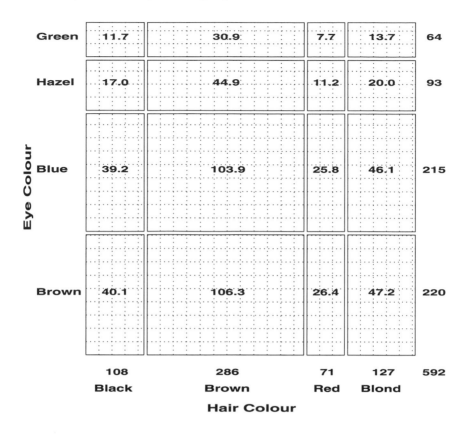

Figure 1: Expected frequencies under independence

is proportional to the expected frequency and the observed frequency is shown by the number of squares in each rectangle. Hence, the difference between observed and expected frequencies appears as the density of shading, using color to indicate whether the deviation from independence is positive or negative. (In monochrome versions, positive residuals are shown by solid lines, negative by broken lines.) The sieve diagram for hair color and eye color is shown in Figure 2.

2.2   Mosaic Displays for *n*-way Tables

The *mosaic display*, proposed by Hartigan and Kleiner (1981) and extended by Friendly (1994a), represents the counts in a contingency table directly by tiles whose area is proportional to the cell frequency. This display generalizes readily to *n*-way tables and can be used to display the residuals from various log-linear models.

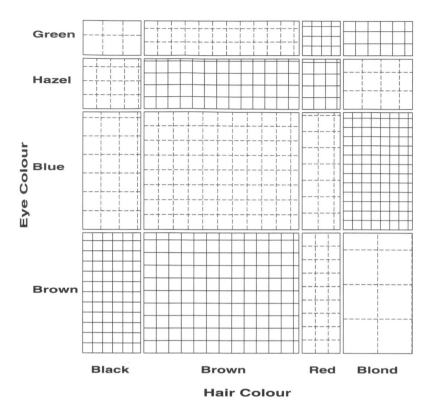

Figure 2: Sieve diagram for hair color, eye color data

Condensed Mosaic Displays   One form of this plot, called the *condensed mosaic display*, is similar to a divided bar chart. The width of each column of tiles in Figure 3 is proportional to the marginal frequency of hair colors; the height of each tile is determined by the conditional probabilities of eye color in each column. Again, the area of each box is proportional to the cell frequency, and complete independence is shown when the tiles in each row all have the same height.

Enhanced Mosaics   The *enhanced mosaic display* (Friendly, 1992b, 1994a) achieves greater visual impact by using color and shading to reflect the size of the residuals from independence and by reordering rows and columns to make the pattern more coherent. The resulting display shows both the observed frequencies and the pattern of deviations from a specified model.

Plate 1 shows the extended mosaic plot, in which the standardized (Pearson) residual from independence, $d_{ij} = (n_{ij} - m_{ij})/\sqrt{m_{ij}}$, is shown by the color and shading of each rectangle: cells with positive residuals are outlined with solid lines and filled with slanted lines; negative residuals are outlined with broken lines and filled with gray scale. The absolute value of the residual is portrayed by shading

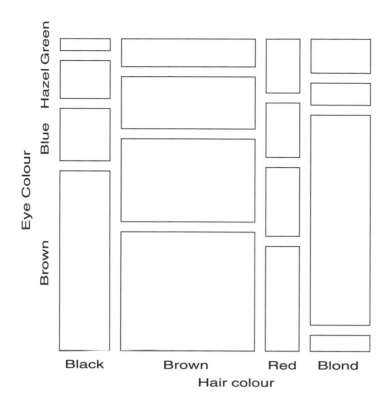

Figure 3: Condensed mosaic for hair color, eye color data

density: cells with absolute values less than 2 are empty; cells with $|d_{ij}| \geq 2$ are filled; those with $|d_{ij}| \geq 4$ are filled with a darker pattern. Color versions use blue and red with varying lightness to portray both sign and magnitude of residuals. Under the assumption of independence, these values roughly correspond to two-tailed probabilities $p < .05$ and $p < .0001$ that a given value of $|d_{ij}|$ exceeds 2 or 4. For exploratory purposes, we do not usually make adjustments for multiple tests (for example, using the Bonferroni inequality) because the goal is to display the pattern of residuals in the table as a whole. However, the number and values of these cutoffs can be easily set by the user.

When the row or column variables are unordered, we are also free to rearrange the corresponding categories in the plot to help show the nature of association. For example, in Plate 1, the eye color categories have been permuted so that the residuals from independence have an opposite-corner pattern, with positive values running from the bottom left to the top right corner and negative values along the opposite diagonal. Coupled with size and shading of the tiles, the excess in the black-brown and blond-blue cells, together with the underrepresentation of brown-eyed blonds and people with black hair and blue eyes, is now quite apparent. Although the table was reordered on the basis of the d_{ij} values, both dimensions in Plate 1 are ordered from dark to light, suggesting an explanation for the association. In this example the eye color categories could be reordered by inspection. A general method (Friendly, 1994a) uses category scores on the first principal axis of a correspondence analysis.

Multiway Tables   Like the scatterplot matrix for quantitative data, the mosaic plot generalizes readily to the display of multidimensional contingency tables. Imagine that each cell of the two-way table for hair and eye color is further classified by one or more additional variables: sex and level of education, for example. Then each rectangle can be subdivided horizontally to show the proportion of males and females in that cell, and each of those horizontal portions can be subdivided vertically to show the proportions of people at each educational level in the hair–eye–sex group.

Fitting Models   When three or more variables are represented in the mosaic, we can fit several different models of independence and display the residuals from each model. We treat these models as null or baseline models, which may not fit the data particularly well. The deviations of observed frequencies from expected ones, displayed by shading, will often suggest terms to be added to an explanatory model that achieves a better fit.

- Complete independence: The model of complete independence asserts that all joint probabilities are products of the one-way marginal probabilities:

$$\pi_{ijk} = \pi_{i\cdot\cdot}\ \pi_{\cdot j\cdot}\ \pi_{\cdot\cdot k} \tag{1}$$

for all i, j, k in a three-way table. This corresponds to the log-linear model $[A][B][C]$. Fitting this model puts all higher terms, and hence all association among the variables, into the residuals.

- Joint independence: Another possibility is to fit the model in which variable C is jointly independent of variables A and B,

$$\pi_{ijk} = \pi_{ij\cdot}\ \pi_{\cdot\cdot k}. \tag{2}$$

This corresponds to the log-linear model $[AB][C]$. Residuals from this model show the extent to which variable C is related to the combinations of variables A and B, but they do not show any association between A and B.

For example, with the data from Table 1 broken down by sex, fitting the model [HairEye][Sex] allows us to see the extent to which the joint distribution of hair color and eye color is associated with sex. For this model, the likelihood ratio G^2 is 19.86 with $df = 15$ ($p = .178$), indicating an acceptable overall fit. The three-way mosaic, shown in Plate 2, highlights two cells: among blue-eyed blonds, there are more females (and fewer males) than would be the case if hair color and eye color were jointly independent of sex. Except for these cells, hair color and eye color appear unassociated with sex.

2.3 Fourfold Display

A third graphical method based on the use of area as the visual mapping of cell frequency is the "fourfold display" (Friendly, 1994b, 1994c) designed for the display of 2×2 (or $2 \times 2 \times k$) tables. In this display the frequency n_{ij} in each cell of a fourfold table is shown by a quarter-circle, whose radius is proportional to $\sqrt{n_{ij}}$, so the area is proportional to the cell count.

For a single 2×2 table the fourfold display described here also shows the frequencies by area, but scaled in a way that depicts the sample odds ratio, $\hat{\theta} = (n_{11}/n_{12}) \div (n_{21}/n_{22})$. An association between the variables ($\theta \neq 1$) is shown by the tendency of diagonally opposite cells in one direction to differ in size from those in the opposite direction, and the display uses color or shading to show this direction. Confidence rings for the observed θ allow a visual test of the hypothesis $H_0 : \theta = 1$. They have the property that the rings for adjacent quadrants overlap if and only if the observed counts are consistent with the null hypothesis.

As an example, Figure 4 shows aggregate data on applicants to graduate school at Berkeley for the six largest departments in 1973 classified by admission and sex.

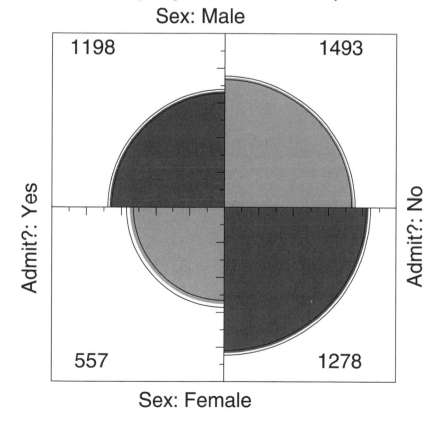

Figure 4: Fourfold display for Berkeley admissions

At issue is whether the data show evidence of sex bias in admission practices (Bickel *et al.*, 1975). The figure shows the cell frequencies numerically in the corners of the display. Thus there were 2691 male applicants, of whom 1198 (44.4%) were admitted, compared with 1855 female applicants, of whom 557 (30.0%) were admitted. Hence the sample odds ratio, Odds (Admit | Male)/(Admit | Female) is 1.84, indicating that males were almost twice as likely to be admitted.

The frequencies displayed graphically by shaded quadrants in Figure 4 are not the raw frequencies. Instead, the frequencies have been standardized by iterative proportional fitting so that all table margins are equal, while preserving the odds ratio. Each quarter-circle is then drawn to have an area proportional to this standardized cell frequency. This makes it easier to see the association between admission and sex without being influenced by the overall admission rate or the differential tendency of males and females to apply. With this standardization the four quadrants will align when the odds ratio is 1, regardless of the marginal frequencies.

The shaded quadrants in Figure 4 do not align and the 99% confidence rings around each quadrant do not overlap, indicating that the odds ratio differs significantly from 1. The width of the confidence rings gives a visual indication of the precision of the data.

Multiple Strata   In the case of a $2 \times 2 \times k$ table, the last dimension typically corresponds to "strata" or populations, and we would like to see if the association between the first two variables is homogeneous across strata. The fourfold display allows easy visual comparison of the pattern of association between two dichotomous variables across two or more populations.

For example, the admissions data shown in Figure 4 were obtained from a sample of six departments; Figure 5 displays the data for each department. The departments are labeled so that the overall acceptance rate is highest for Department A and decreases steadily to Department F. Again, each panel is standardized to equate the marginals for sex and admission. This standardization also equates for the differential total applicants across departments, facilitating visual comparison.

Figure 5 shows that, for five of the six departments, the odds of admission are approximately the same for both men and women applicants. Department A appears to differ from the others, with women approximately 2.86 [$= (313/19)/(512/89)$] times as likely to gain admission. This appearance is confirmed by the confidence rings, which in Figure 5 are *joint* 99% intervals for θ_c, $c = 1, \ldots, k$.

This result, which contradicts the display for the aggregate data in Figure 4, is a nice example of Simpson's paradox. The resolution of this contradiction can be found in the large differences in admission rates among departments. Men and women apply to different departments differentially, and in these data women apply in larger numbers to departments that have a low acceptance rate. The aggregate results are misleading because they falsely assume men and women are equally likely to apply in each field. (This explanation ignores the possibility of structural bias against women, e.g., lack of resources allocated to departments that attract women applicants.)

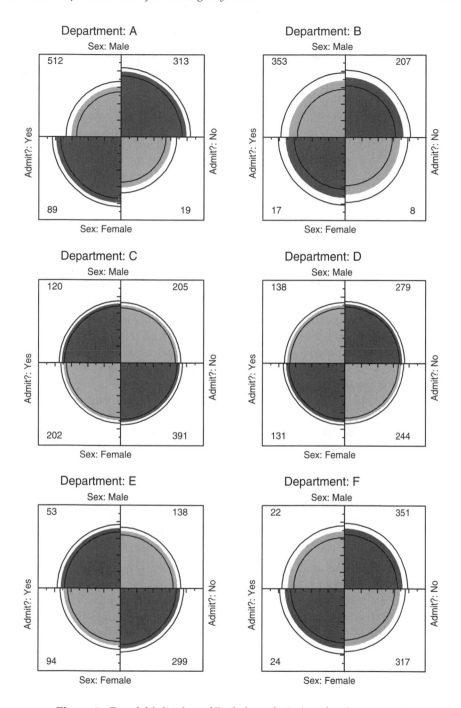

Figure 5: Fourfold display of Berkeley admissions by department

3   Conceptual Models for Visual Displays

Visual representation of data depends fundamentally on an appropriate visual scheme for mapping numbers into graphic patterns (Bertin, 1983). The widespread use of graphical methods for quantitative data relies on the availability of a natural visual mapping: magnitude can be represented by length, as in a bar chart, or by position along a scale, as in dot charts and scatterplots. One reason for the relative paucity of graphical methods for categorical data may be that a natural visual mapping for frequency data is not so apparent. And, as I have just shown, the mapping of frequency to area appears to work well for categorical data.

Closely associated with the idea of a visual metaphor is a conceptual model that helps you interpret what is shown in a graph. A good conceptual model for a graphical display will have deeper connections with underlying statistical ideas as well. In this section we consider conceptual models for both quantitative and frequency data that have these properties.

3.1   Quantitative Data

The simplest conceptual model for quantitative data is the balance beam, often used in introductory statistics texts to illustrate the sample mean as the point along an axis where the positive and negative deviations balance.

A more powerful model (Sall, 1991a) likens observations to fixed points connected to a movable junction by springs of equal spring constant, $k \sim 1/\sigma$. For example, least-squares regression can be represented as shown in Figure 6, where the points are again fixed and attached to a movable rod by unit length, equally stiff springs. If the springs are constrained to be kept vertical, the rod, when released, moves to the position of balance and minimum potential energy, the least-squares solution. The normal equations,

$$\sum e_i = \sum_{i=1}^{n} (y_i - a - bx_i) = 0 \qquad (3)$$

$$\sum x_i e_i = \sum_{i=1}^{n} (y_i - a - bx_i)\, x_i = 0 \qquad (4)$$

are seen, respectively, as conditions that the vertical forces balance and the rotational moments about the intercept $(0, a)$ balance.

The appeal of the spring model lies in the intuitive explanations it provides for many statistical phenomena and the understanding it can bring to our perception of graphical displays—see Sall (1991a) and Farebrother (1987) for more details.

3.2   Categorical Data

For categorical data, we need a visual analogue for the sample frequency in k mutually exclusive and exhaustive categories. Consider first the one-way marginal frequencies of hair color from Table 1.

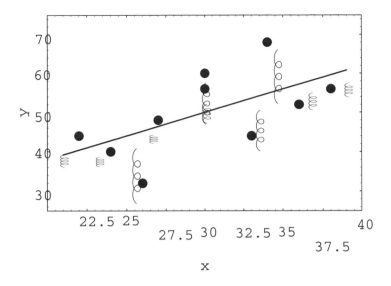

Figure 6: Spring model for least-squares regression

Urn Model The simplest physical model represents the hair color categories by urns containing marbles representing the observations (Figure 7). This model is sometimes used in texts to describe multinomial sampling and provides a visual representation that equates the count n_i with the area filled in each urn. In Figure 7 the urns are of equal width, so the count is also reflected by height, as in the familiar bar chart. However, the urn model is a static one and provides no further insights. It does not relate to the concept of likelihood or to the constraint that the probabilities sum to 1.

Pressure and Energy A dynamic model gives each observation a force (Figure 8). Consider the observations in a given category (red hair, say) as molecules of an ideal gas confined to a cylinder whose volume can be varied with a movable piston (Sall, 1991b), set up so that a probability of 1 corresponds to ambient pressure, with no force exerted on the piston. An actual probability of red hair equal to p means that the same number of observations are squeezed down to a chamber of height p. By Boyle's law, which states that pressure $\times$ volume is a constant, the pressure is proportional to $1/p$. In the figure, pressure is shown by *observation density*, the number of observations per unit area. Hence, the graphical metaphor is that a count can be represented visually by observation density when the count is fixed and area is varied (or by area when the observation density is fixed as in Figure 7).

The work done on the gas (or potential energy imparted to it) by compressing a small distance δy is the force on the piston times δy, which equals the pressure times the change in volume. Hence, the potential energy of a gas at a height of p is $\int_p^1 (1/y)\, dy$, which is $-\log(p)$, so the energy in this model corresponds to negative log-likelihood.

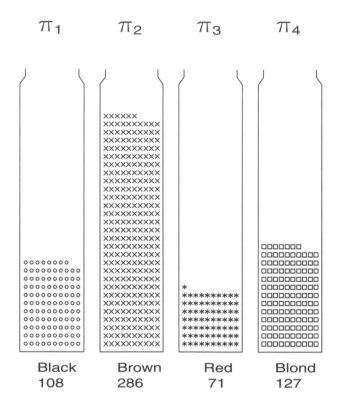

Figure 7: Urn model for multinomial sampling

Fitting Probabilities: Minimum Energy, Balanced Forces   Maximum likelihood estimation means literally finding the values, $\hat{\pi}_i$, of the parameters under which the observed data would have the highest probability of occurrence. We take derivatives of the (log-) likelihood function with respect to the parameters, set these to zero, and solve:

$$\frac{\partial \log L}{\partial \pi_i} = 0 \quad \longrightarrow \quad \frac{n_1}{\pi_1} = \frac{n_2}{\pi_2} = \cdots = \frac{n_c}{\pi_c} \quad \longrightarrow \quad \hat{\pi}_i = \frac{n_i}{n} = p_i$$

Setting derivatives to zero means minimizing the potential energy; the maximum likelihood estimates (MLEs) are obtained by setting parameter values equal to corresponding sample quantities, where the forces are balanced.

In the mechanical model (Figure 9) this corresponds to stacking the gas containers with movable partitions between them, with one end of the bottom and top containers fixed at 0 and 1. The observations exert pressure on the partitions, the likelihood equations are precisely the conditions for the forces to balance, and the partitions move so that each chamber is of size $p_i = n_i/n$. Each chamber has potential energy $-\log p_i$, and the total energy, $-\sum_i^c n_i \log p_i$, is minimized. The constrained top and

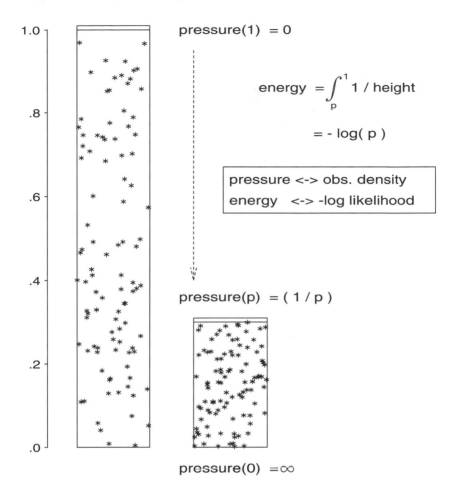

Figure 8: Pressure model for categorical data

bottom force the probability estimates to sum to 1, and the number of movable partitions is literally and statistically the degrees of freedom of the system.

Testing a Hypothesis   This mechanical model also explains how we test hypotheses about the true probabilities (Figure 10). To test the hypothesis that the four hair color categories are equally probable, $H_0 : \pi_1 = \pi_2 = \pi_3 = \pi_4 = \frac{1}{4}$, simply force the partitions to move to the hypothesized values and measure how much energy is required to force the constraint. Some of the chambers will then exert more pressure, some less than when the forces are allowed to balance without these additional restraints. The change in energy in each compartment is then $-(\log p_i - \log \pi_i) =$

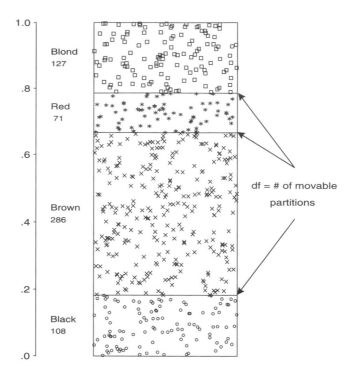

Figure 9: Fitting probabilities for a one-way table

$-\log(p_i/\pi_i)$, the change in negative log-likelihood. Sum these up and multiply by 2 to get the likelihood ratio G^2.

The pressure model also provides simple explanations of other results. For example, increased sample size increases power, because more observations mean more pressure in each compartment, so it takes more energy to move the partitions and the test is sensitive to smaller differences between observed and hypothesized probabilities.

Multiway Tables   The dynamic pressure model extends readily to multiway tables. For a two-way table of hair color and eye color, partition the sample space according to the marginal proportions of eye color, and then partition the observations for each eye color according to hair color as before (Figure 11). Within each column the forces balance as before, so that the height of each chamber is $n_{ij}/n_{i\cdot}$. Then the area of each cell is proportional to the MLE of the cell probabilities, $(n_{i\cdot}/n)\,(n_{ij}/n_{i\cdot}) = n_{ij}/n = p_{ij}$, which again is the sample cell proportion.

For a three-way table, the physical model is a cube with its third dimension partitioned according to conditional frequencies of the third variable, given the first

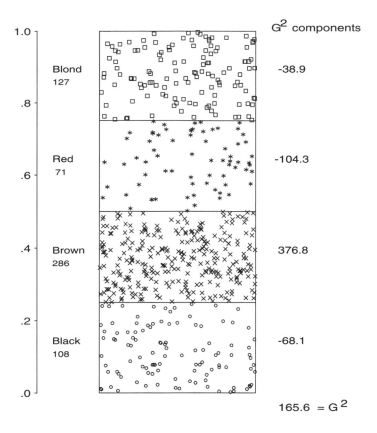

Figure 10: Testing a hypothesis

two. If the third dimension is represented instead by partitioning a two-dimensional graph, the result is the mosaic display.

Testing Independence   For a two-way table of size $I \times J$, independence is formally the same as the hypothesis that conditional probabilities (of hair color) are the same in all strata (eye colors). To test this hypothesis, force the partitions to align and measure the total additional energy required to effect the change (Figure 12). The degrees of freedom for the test is again the number of movable partitions, $(I - 1)(J - 1)$.

Each log-linear model for three-way tables can be interpreted analogously. For example, the log-linear model $[A][B][C]$ (complete independence) corresponds to the cube in which all chambers are forced to conform to the one-way marginals, $\pi_{ijk} = \pi_{i..}\,\pi_{.j.}\,\pi_{..k}$ for all i, j, k. G^2 is again the total additional energy required to move the partitions from their positions in the saturated model in which the volume of each cell is $p_{ijk} = n_{ijk}/n$ (so the pressures balance) to the positions where each

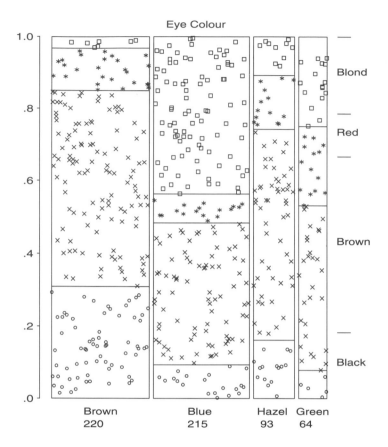

Figure 11: Two-way tables

cell is a cube of size $\pi_{i..} \times \pi_{.j.} \times \pi_{..k}$. Other models have a similar representation in the pressure model.

Iterative Proportional Fitting   For three-way (and higher) tables some log-linear models have closed-form solutions for expected cell frequencies. The cases in which direct estimates exist are analogous to the two-way case, in which the estimates under the hypothesized model are products of the sufficient marginals. Here we see that the partitions in the observation space can be moved directly in planar slices to their positions under the hypothesis, so that iteration is unnecessary.

When direct estimates do not exist, the MLEs can be estimated by iterative proportional fitting (IPF). This process simply matches the partitions corresponding to each of the sufficient marginals of the fitted frequencies to the same marginals of the data. For example, for the log-linear model $[AB][BC][AC]$, the sufficient statistics are $n_{ij\cdot}$, $n_{i\cdot k}$, and $n_{\cdot jk}$. The conditions that the fitted margins must equal

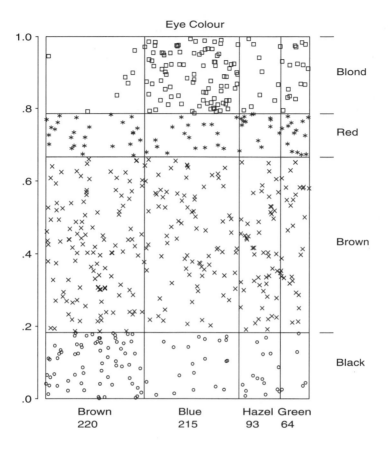

Figure 12: Testing independence

these observed margins are

$$\frac{n_{ij\cdot}}{\hat{m}_{ij\cdot}} = \frac{n_{i\cdot k}}{\hat{m}_{i\cdot k}} = \frac{n_{\cdot jk}}{\hat{m}_{\cdot jk}} = 1 \tag{5}$$

which is equivalent to balancing the forces in each fitted marginal. The steps in IPF follow directly from equation (5). For example, the first step in cycle $t + 1$ of IPF matches the frequencies in the $[AB]$ marginal table,

$$\hat{m}_{ijk}^{(t+1)} = \hat{m}_{ijk}^{(t)} \left(\frac{n_{ij\cdot}}{\hat{m}_{ij\cdot}^{(t)}}\right) \tag{6}$$

which makes the forces balance when equation (6) is summed over variable C: $\hat{m}_{ij\cdot}^{(t+1)} = n_{ij\cdot}$. The other steps in each cycle make the forces balance in the $[BC]$ and $[AC]$ margins.

The iterative process can be shown visually (Friendly, 1995), in a way that is graphically exact, by drawing chambers whose area is proportional to the fitted frequencies, $\hat{m}_{ijk}$, and which are filled with a number of points equal to the observed n_{ijk}. Such a figure will then show equal densities of points in cells that are fitted well but relatively high or low densities where $n_{ijk} > \hat{m}_{ijk}$ or $n_{ijk} < \hat{m}_{ijk}$, respectively. The IPF algorithm can in fact be animated, by drawing one such frame for each step in the iterative process. When this is done, it is remarkable how quickly IPF converges, at least for small tables.

Likewise, numerical methods for minimizing the negative log-likelihood directly can also be interpreted in terms of the dynamic model (Farebrother, 1988; Friendly, 1995). For example, in steepest descent and Newton–Raphson iteration, the update step changes the estimated model parameters $\boldsymbol{\beta}^{(t+1)}$ in proportion to the score vector $\mathbf{f}^{(t)}$ of derivatives of the likelihood function, $\mathbf{f}^{(t)} = \partial \log L / \partial \boldsymbol{\beta} = \mathbf{X}^{\mathsf{T}}(\mathbf{n} - \mathbf{m}^{(t)})$ to give $\boldsymbol{\beta}^{(t+1)} = \boldsymbol{\beta}^{(t)} + \lambda \mathbf{f}^{(t)}$. But $\mathbf{f}^{(t)}$ is just the vector of forces in the mechanical model attributed to the differences between $\mathbf{n}$ and $\mathbf{m}^{(t)}$ as a function of the model parameters.

4 Conclusion

This chapter started with the puzzling contrast in use and generality between graphical methods for quantitative data and those for categorical data, despite strong formal similarities in their underlying methods. In this chapter we have seen that categorical data require a different graphical metaphor and hence a different visual representation (count $\leftrightarrow$ area) from that which has been useful for quantitative data (magnitude $\leftrightarrow$ position on a scale). The sieve diagram, mosaic, and the fourfold display all show frequencies in this way and are valuable tools for both the analysis and presentation of categorical data.

We then showed that physical models for both quantitative and categorical data and their graphic representation can yield a wide range of interpretations for statistical principles and phenomena. Although the spring and pressure models differ fundamentally in their mechanics, both can be understood in terms of balancing of forces and the minimization of energy. The recognition of these conceptual models can make a graphical display a tool for thinking, as well as a tool for data summarization and exposure.

Finally, we can see two areas needing improvement in the future development of graphical methods for categorical data. First, much of the power of graphical methods for quantitative data stems from the availability of tools that generalize readily to multivariable data and can make important contributions to model building, model criticism, and model interpretation. The mosaic display possesses some of these properties, and other chapters in this book attest to the widespread utility of biplots and correspondence analysis. However, I believe there is need for further development of such methods, particularly as tools for constructing models and communicating their import. Second, I am reminded of the statement (Tukey, 1959, attributed to

Churchill Eisenhart) that the *practical power* of any statistical tool is the product of its statistical power and its probability of use. It follows that statistical and graphical methods are of practical value to the extent that they are implemented in standard software, available, and easy to use. Statistical methods for categorical data analysis have nearly reached that point. Graphical methods still have some way to go.

Chapter 3

Bertin's Graphics and Multidimensional Data Analysis

Jean-Hugues Chauchat and Alban Risson

1 Introduction

The objective of this chapter is to show how Bertin's graphics are a straightforward and accurate method for communicating the results of some multidimensional statistical methods such as principal components analysis, correspondence analysis, and cluster analysis. These graphics remain true to the original data, using only permutations of rows and columns of the data matrix.

The idea of permuting the rows and columns of a matrix for the purpose of revealing hidden structure in a data matrix is an old one: the pioneering work was done by Sir W. M. Flinders Petrie almost a century ago. He was looking for a "sequence in prehistoric remains," that is, a chronological "seriation." As noticed by Arabie *et al.* (1978), Caraux (1984), and Marcotorchino (1987), this idea is having an increasing influence in applied mathematics, especially in the behavioral sciences. Bertin (1967, 1981) laid histograms side by side, using an appropriate scale, and permuted the elements to reveal underlying structures in the data.

We consider two types of statistical methods that can help us to discover rapidly the best pair of permutations of the rows and columns of the table among the $n! \times p!$ possible solutions: (1) identification of a diagonal pattern when it exists, for example, a predominant factor in correspondence analysis or principal components analysis, and (2) classification of rows and columns by cluster analysis.

The first type of solution (diagonal pattern) is known as "seriation" or "ordination," the second type as "block seriation" or "cliques." In both cases, Bertin's graphics gives an easily understandable visual representation of the results of the statistical data analysis; each bit of information, each entry of the table is presented in its original form, with only the order of the rows and columns changed.

We present a new exploratory method that integrates multidimensional data analysis and graphical methods and is implemented in the software program AMADO (Risson *et al.*, 1994). This methodology can be applied to any matrix consisting of positive values: contingency tables, logical tables representing a response pattern or a graph, symmetric tables of similarities, and so on.

2   Bertin's Rules of Graphical Syntax

Contrary to a table, with which the aim is to make every cell available to the reader, a graph should be read in an instant—similarities and differences should be immediately apparent. Let the rows of a table be the horizontal dimension (say X) and the columns of the table be the vertical dimension (say Y). A color variation or shading in light intensity can induce a third visible dimension (say Z); this third dimension is used to represent the numerical values of the data table.

During the 1960s, Bertin and his team worked with groups of wooden cubes covered with paper on which were drawn rectangles from histograms; rows (or columns) were then moved by hand until a diagonal structure, or a "block model," was obtained. Later, the use of numerical multivariate descriptive statistical analysis methods (Lebart *et al.*, 1984) replaced this purely visual approach.

Looking for a unidimensional ordering, one may use correspondence analysis (CA) to find the optimal ranking of the row and column variables. The first axis of the CA solution gives the numerical scale for the rows and columns so that each individual may be characterized in a scatterplot by the coordinates of the individual's

Table 1: The (0/1) matrix, logic table that represents Jan De Leeuw's UCLA statistics program graph

	Mathematics	Sociology	Statistics	Psychology	Education	Born in the U.S.A.	Born out of the U.S.A.
Ferguson	1	0	0	0	0	1	0
Li	1	0	0	0	0	0	1
Ylvisaker	1	0	0	0	0	1	0
Berk	0	1	0	0	0	1	0
De Leeuw	0	0	1	0	0	0	1
Mason	0	1	0	0	0	1	0
Bentler	0	0	0	1	0	0	1
Muthén	0	0	0	0	1	0	1
Jennrich	1	0	0	0	0	1	0

categories i and j. There are n_{ij} individuals at the same position, so the number n_{ij} can be used as the third dimension Z. The correlation coefficient between the two scaled row and column variables is the square root of the eigenvalue associated with this first principal axis (Nishisato, 1980, chap. 3; Tenenhaus and Young, 1985), also called the canonical correlation (see, for example, Greenacre, 1993a, chap. 7).

Bertin's graphics (see Figure 3) can be seen as a type of scatterplot: coordinates from CA become ranks, and the area of each rectangle is proportional to the number of observations/cases with those ranks. With this interpretation, the best permutation of rows and columns would maximize the Spearman rank correlation coefficient. Looking for a block seriation, one may use any appropriate cluster analysis method on rows and/or columns.

3   A Simple Example of Bertin's Graphics

In Chapter 1 of this book, de Leeuw presents a small data set on the UCLA statistics department. The data are given in Table 1; Figure 1 shows the corresponding display using Bertin's graphics. From this display it is easy to see that both sociologists as

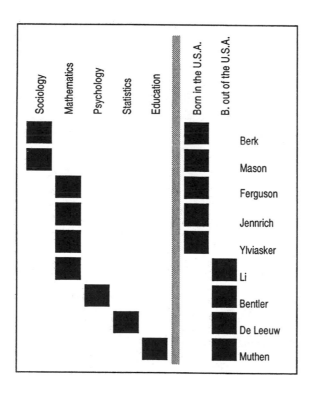

Figure 1: Bertin's graphic from de Leeuw's graph.

well as three of the four mathematicians were born in the United States, whereas the fourth mathematician, the psychologist, the statistician, and the educator were born outside of the United States.

4   Livestock Slaughtered in the European Community in 1995

The data in Table 2 were obtained from the European Community Statistical Office (EUROSTAT) in Luxembourg. Here, we consider the number of livestock (in thousands) slaughtered in 1995 in EEC countries.

4.1   Correspondence Analysis and Bertin's Graphics

Such a contingency table can be displayed via correspondence analysis. The first factorial plane is shown in Figure 2. This map might be hard to read: many people will read that "Adult Bovines" are mostly found in Italy, because these two points appear near one another on the plot, or that there are more pigs in Finland than in Denmark because the former is closer to "Pigs" than the latter. These erroneous conclusions are quite common.

Table 2: Livestock slaughtered in the European Community in 1995 (1000 animals)

	Austria	Belgium	Denmark	Finland	France	Germany	Greece
Heifers	69	67	57	51	577	674	31
Adult bovines	533	711	703	382	3968	4251	235
Calves	130	336	55	10	2042	501	80
Pigs	4954	11294	19873	2066	24859	39353	2268
Sheep	280	22	69	74	7696	2057	7712
Caprines	0	0	0	1	1058	12	4819

	Ireland	Italy	NL	Portugal	Spain	Sweden	UK
Heifers	487	558	48	53	591	52	940
Adult bovines	1514	3411	1181	325	1965	501	3266
Calves	0	1321	1198	71	25	30	26
Pigs	3002	11992	18616	4209	27539	3743	14376
Sheep	4298	7960	626	1083	20085	189	19311
Caprines	0	483	17	205	1891	0	30

Source: EUROSTAT.

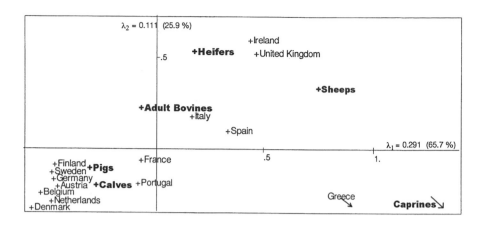

Figure 2: The European Community's livestock correspondence analysis.

Such errors are impossible with graphics in Figures 3 and 4. Figure 3 depicts the raw data after rows and columns have been permuted with respect to their order on the first CA axis. The seven northern and eastern countries (Denmark, Belgium, Netherlands, Germany, Finland, Sweden, and Austria), where pigs and beef are the most important products, are opposed to the western and southern countries (Portugal, France, Italy, Spain, Ireland, United Kingdom, and Greece), which produce sheep and goats rather than pigs. In Figure 4 the row and column reordering is maintained but the *conditional* distributions are shown: first of countries, given species (i.e., row profiles), and second of species, given countries (i.e., column profiles). Figure 4a shows that the larger part of pigs produced in the European Community comes from Germany and then from Spain, France, The Netherlands, and so on, whereas Figure 4b shows that the countries in which pigs are the main product are Denmark, Belgium, The Netherlands, Germany, and so on.

These Bertin's graphics represent the original data perfectly, but contrary to correspondence analysis, they are limited in their ability to display more than one factor at a time. Figure 5 is similar to Figure 4 after permutation of rows and columns with respect to the second principal axis; beef- and sheep-producing countries are opposed to those producing pigs or goats. These graphics show the additional information carried by the second axis, as well as the peculiar position of "Greece" and "goats."

4.2 Cluster Analysis and Bertin's Graphics

Usually, results from hierarchical clustering are depicted by a "tree"; the tree shows how the clusters were formed but it distorts the distances between the clustered rows or columns into "ultrametric" distances. Moreover, the tree does not give information on why two rows, or two columns, were found to be "close" or "distant." Bertin's

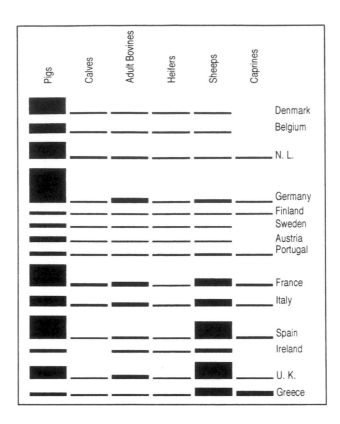

Figure 3: The European Community's livestock Bertin's graphic after reclassification according to the ranking of values on the first axis of correspondence analysis.

graphics can be applied to the original data, where rows or columns are ranked according to their location in the tree. Clustering of the countries is performed by Ward's method (Ward, 1963) using the chi-squared distance (Greenacre, 1988a; Jambu, 1989). Figure 6 shows the hierarchical clustering tree, revealing the main geographic and cultural ensembles of Europe: the British Isles, the Roman world (Italy, France, Spain, Portugal), Greece on its own, and northeastern Europe around Germany, where the three countries Belgium, Netherlands, and Denmark stand out.

Figure 6b shows the profile for each country, and it is now apparent what links countries of the same cluster and what separates those in different clusters. Northeastern countries produce pigs, no goats, and hardly any sheep; large bovines and sheep are produced in the British Isles, but no goats or calves, and so on. One sees

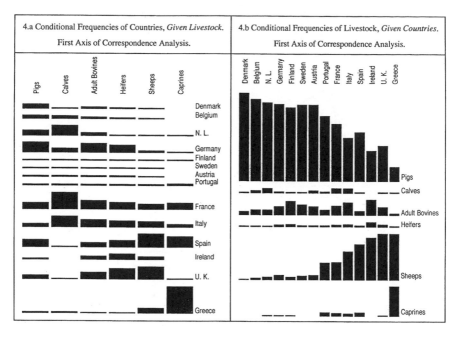

Figure 4: Conditional frequencies Bertin's graphic after reclassification according to the ranking of values on the first axis of correspondence analysis.

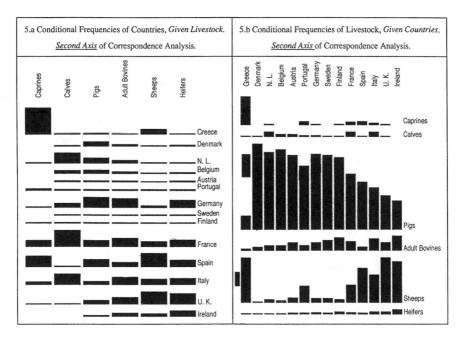

Figure 5: Conditional frequencies Bertin's graphic after reclassification according to the ranking of values on the second axis of correspondence analysis.

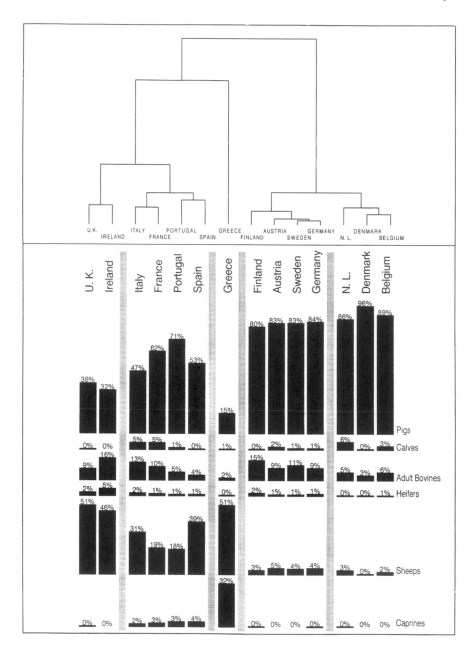

Figure 6: (a) Hierarchical tree for European Community countries. (b) Cluster Bertin's graphic.

how the two methods complement each other: clusters are found "automatically" and give a good classification of countries, and Bertin's graphics assist in the interpretation of the clusters by returning to the original data.

5 Conclusion

Bertin's graphics provide a visual complement to the solutions of correspondence and cluster analyses. Data matrices are represented by a matrix of histograms, all on the same scale, where rows and columns are optimally permuted. This permutation is defined in terms of either progressive variation or seriation, or by homogeneous groups distinct from one another, or block modeling. These permutation criteria, which Bertin defined empirically, are the very criteria of the multivariate statistical methods: the diagonal seriation corresponds to the maximum correlation permutation of rows and columns in CA, and the block criteria correpond to the homogeneity of groups in cluster analysis, for example, Ward's minimization of intracluster variance.

Here, we can quote Arabie *et al.*, (1978):

> "It is intuitively convincing that row-column permutations of a matrix leave the raw data far more chaste than do data analysis techniques requiring a priori replacement or aggregation, e.g. taking ranks, or replacing subsets of the data by various summary statistics (...). For this reason, permutation methods are an important member of the small but growing family of data analysis methods following the philosophy that aggregation is to be inferred at the end of the analysis, not imposed at the beginning."

Software Note: AMADO

The program AMADO is an implementation of Bertin's method. AMADO is distributed in Windows and Macintosh versions by CISIA (1 av. Herbillon, 94160 Saint-Mandé, France). A user's guide (Risson *et al.*, 1994) is available in French, and Italian and English versions will become available in 1998.

Acknowledgments

We would like to thank Jean Dumais, visiting professor in Université Lumière, for his assistance.

Chapter 4

The Use of Visualization in the Examination of Categorical Event Histories

Brian Francis, Mark Fuller, and John Pritchard

1 Introduction

Graphical displays of multivariate data provide much insight into the nature of a data set before statistical analysis. Such displays identify unusual observations, potential clusters of observations, and possible relationships between variables and thus can suggest appropriate statistical models in later analysis. Everitt (1978), for example, described a collection of graphical techniques for certain types of multivariate data. The primary aim in producing such displays was well summarized by Andrews (1972), who described exploratory data analysis as "the manipulation, summarisation, and display of data to make them more comprehensible to human minds, thus uncovering structure in the data and detecting important departures from that structure."

Event history studies usually involve the collection of large and complex amounts of information on a set of individuals. A typical work history for an individual consists of records of that individual's employment state containing the start and end dates of each period of employment or unemployment, social class and industrial classification, number of hours worked, and so on. Life history data contain further records of other life events such as the individual's marital history, residential history, educational history, criminal history, medical history, and other demographic information such as the dates of birth of children and the size and composition of the individual's household over time. Life histories are special cases of event histories, which could also include shorter term studies such as the medical history of an individual since the first onset of an illness such as Acquired Immunodeficiency Syndrome (AIDS).

Although in many short-term studies, event history data can be collected prospectively, it is common when assembling life histories for information to be collected retrospectively through questionnaire or interview, leading to problems of recall for both dates of state changes and associated covariate information. Such data are therefore characterized for an individual by a set of multiple durations in each of a number of states, with additional complex covariate information varying over time. Furthermore, censoring may be present for some durations in some states, and data may be missing.

Our aim in developing visual techniques for event history data is therefore to allow the representation of both durations and state transitions in all variables relevant to an analysis. Such displays should allow both the examination of a single event history and the comparison of multiple event histories.

Scientific data visualization (McCormick *et al.*, 1987) has developed over the past decade and is characterized by highly interactive computer software with a comprehensive set of tools for viewing scientific data. Scientific visualization has traditionally been used to display data in such areas as engineering (computational fluid dynamics), medicine (computed tomography displays), and meteorology (pressure, wind, and cloud systems). These applications typically represent the coordinates of real physical three-dimensional (3D) objects. Further information can be added to the 3D representation by using color and superimposed symbols. Applications in which there is no underlying 3D physical representation are rare, although visualization has been used, for example, in geography (Hearnshaw and Unwin, 1994), where pollution measures supply the third dimension on a 2D map, and for the examination of stock exchange data (Koh, 1993), where the graphical representation has no underlying physical model.

The use of the term visualization in this chapter is perhaps different from that in other chapters in this book. "Visualization" in statistics is often taken to be simply a static representation of a set of data. We prefer to reserve this term for the highly interactive displays just described, using the term "graphical displays" for static representations of data. It is clear that visualization software can aid the statistical practitioner both in exploring complex data before analysis and in the presentation of the results of a fitted model. The problem with event history data is that there is no unique 3D representation of such data, and we confront this issue in the following sections.

2   Graphical Representations of Event Data

Francis and Fuller (1996) reviewed existing methods for graphically representing event history data. The methods fall into two categories: those summarizing a collection of event histories by defining a set of ranked key events present in a large proportion of the histories (Blossfeld *et al.*, 1989) and those attempting to graph a single, complex event history in its full detail. Methods in the first category cannot usually represent the full complexity of a data set, and a proportion of the data needs to be excluded from the display. Although useful as presentational displays illustrating the progress of individuals through a sequence of states, they are not considered further here.

Methods that take the first route have commonly used a straight line to represent a single history, with various forms of textual annotation or shading to represent events. In his book on the principles of graph construction, Cleveland (1994) considers one example of an event history chart showing the activities of a woman and her baby from the African !Kung tribe over a 12-hour period, moving into and out of four different activities: sleep, nursing, fretting, and holding. As more than one activity can occur at the same time, the single-line model suggested by Konner and Worthman (1980), with different forms of shading on a line to represent different variables, is rejected. Instead, Cleveland is in favor of a simple diagram with time on the horizontal axis and with four horizontal bands stacked one above the other representing the four activity states. Where a particular activity is present, the band is shaded, giving a "block diagram." Cleveland's graph is essentially similar to a tulip plot (Barry *et al.*, 1989), in which a circular rather than a linear block diagram is used, with concentric rings representing the variables.

The foregoing ideas are suitable for examination of a single or a small number of event histories. However, we are concerned here with displays that might be applied to larger collections of histories and also with methods that can deal with all types of variables, which extend from simple binary state variables to cover all types of variables encountered in event history analysis. Variables can belong to one of five possible types:

1. Time variables: variables directly related to time, which measure the progress of an individual in time, such as age or calendar year.
2. Time-varying variables: variables that vary within histories as well as between histories. Examples are the number of hours worked per week (continuous), highest academic qualification (ordinal), and marital status (nominal).
3. Time-constant variables: as above, but constant within a history. Examples include sex, ethnicity, and place of birth.
4. Internal events: events directly related to the individual, which will vary from individual to individual, such as the date of death of a parent and the date a driving licence was gained.
5. External events: these affect the whole sample under study at the same calendar time. Examples include a change of government or the closure of a major factory in a locality.

Naturally, many variables can belong to more than one type. For example, age can be thought of as a time-constant variable if treated as "age at entry to study," and number of children can be represented as time-varying continuous, ordinal, or nominal.

3   Lexis Pencils for Event Histories

We now consider possible objects or glyphs that can be used as a model for the display of event history information. The object needs to be compact, to allow many such objects to be displayed simultaneously, and also needs to have many faces, to allow a suitable selection of event history variables to be displayed. A linear object seems more suitable than a circular object, as variables on the outside of a circle carry a different visual impact compared to those on the inside—an undesirable characteristic. A linear pencil-like object therefore seems appropriate as a means of representing an event history. A suitable time variable such as calendar time or age would be measured along the length of the pencil. Each time-varying variable required would be represented by a different face of the pencil. Continuous or ordinal time-varying variables can be represented either by continuous changes in color or alternatively by protuberances from faces of the object, with the height of the protuberances representing the values of the variable. Categorical time-varying variables can be represented by changes in color or texture. Events can be marked by solid rings around the object, with different types or colors of rings for different event types. Time-constant variables can be represented by different colors or glyphs at the end of each pencil or by additional faces to the object.

These pencils can be displayed side by side in case number or other order (perhaps sorted by the length of the history or by a suitable time-constant variable), thus providing a full, informative display for viewing the data. However, it is possible to go further, and to do this, we turn to the demographic literature.

The Lexis diagram (Lexis, 1875) provides a graphical method of displaying demographic data. The modified form of the diagram used today is based on work by Pressat (1961) and is shown in Figure 1. The x-axis represents calendar time t, and the y-axis represents age a. Each individual is represented by a distinct line on the diagram. An individual born at calendar time T and dying at age A will die at time $T + A$. The individual will therefore be represented in the diagram by a 45° line joining the time and age at birth $(T,0)$ to the time and age at death $(T + A, A)$. The Lexis diagram is also commonly used in survival analysis studies to represent the progress of individuals through a study. In these diagrams, the x-axis still represents calendar time but the y-axis represents the time in the study. An individual entering a study at calendar time T will stay in the study for a period of length A, either until an event of interest occurs (uncensored) or until the end of the study (censored). Symbols placed at the end of the lines are used to indicate the presence or absence of censoring.

Keiding (1990) described some statistical properties of the Lexis diagram. For example, if death intensities $\mu(t, a)$ are assumed to be constant within some principal set in the Lexis diagram and varying between principal sets, with a multiplicative age-period model for the death intensities $\mu(t, a) = \alpha_t \beta_a$, then this gives a piecewise constant intensity model. A Poisson-type likelihood can be derived, which can be fitted as a special case of a generalized linear model. The diagram thus has a good statistical rationale. Keiding also describes a continuous form of this model.

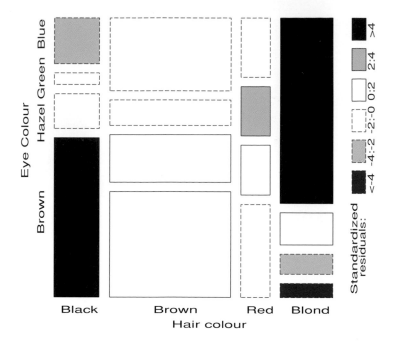

Plate 1: Condensed mosaic, reordered and shaded.

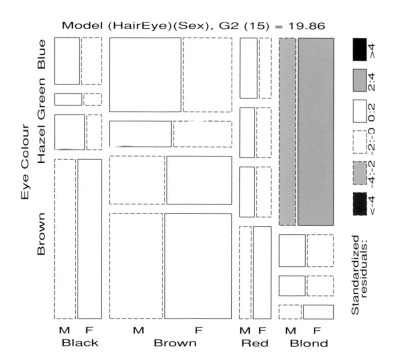

Plate 2: Three-way mosaic, joint independence.

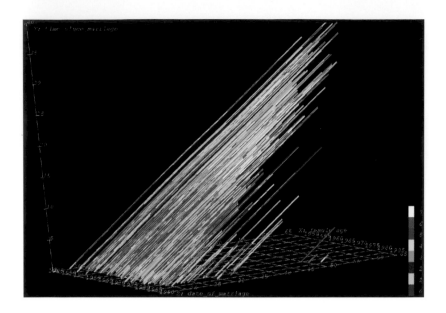

Plate 3: A general view of the 188 life histories of married couples in Kirkcaldy.

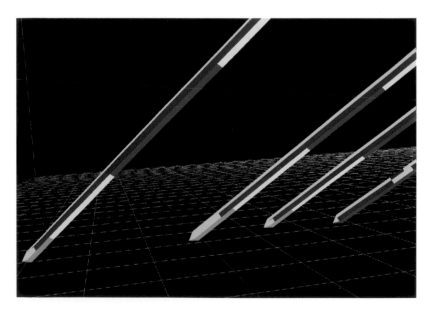

Plate 4: A close-up view of the Lexis pencil representation of couples marrying in 1967 and 1968. The bottom face of each pencil represents the age of the youngest child (green, no child; yellow, child under 1; red, child under 5), the middle face the employment history of the women (light blue, working; dark blue, not working), and the top face the employment history of the man. The woman stops work before the birth of a child; this pregnancy effect can clearly be seen.

Plate 5: A close-up view of the Lexis pencil representation of couples marrying before 1955. Couples marrying in 1951 and 1952 are represented by solid pencils; other histories are represented by ghosted pencils. Women in this cohort tend not to work at all or tend not to return to work after the birth of a child.

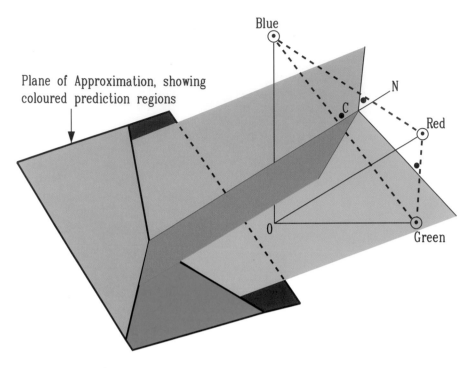

Plate 6: The geometry of neighbor regions and prediction regions for a categorical variable ("color") with three categories ("blue," "green," "red"). The CLPs are denoted by the three small circle-enclosed dots on three rectangular axes as in Figure 3 and these define the dotted triangle. The perpendicular bisectors of the sides of the dashed triangle meet at the circumcenter C and are the boundaries of the neighbor regions in the plane of the triangle; CN is normal to the plane of the triangle. The full neighbour regions are obtained by sliding the triangular neighbor regions along CN, giving the separators between neighbor regions that are shown. The plane of approximation represents a sheet of paper or a computer screen showing the best approximation of the samples. This plane intersects the neighbor regions, giving the prediction regions shown. The prediction region for "red" is largely hidden behind the planes separating the blue/red and red/green regions and hence only two ends are shown, the remainder being indicated by the dashed line.

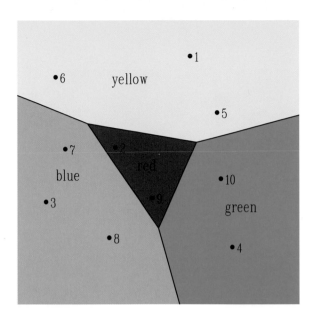

Plate 7: The prediction regions for a categorical variable "color" with four levels: "blue," "green," "red," "yellow." Also shown are the positions of 10 numbered sample points.

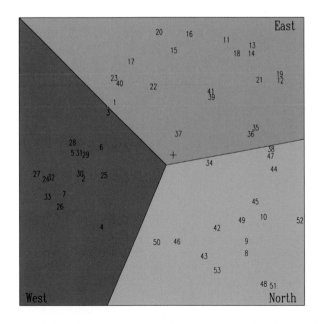

Plate 8: Analysis of the EMC showing prediction regions for the variable "region." The numbers refer to the 53 farms.

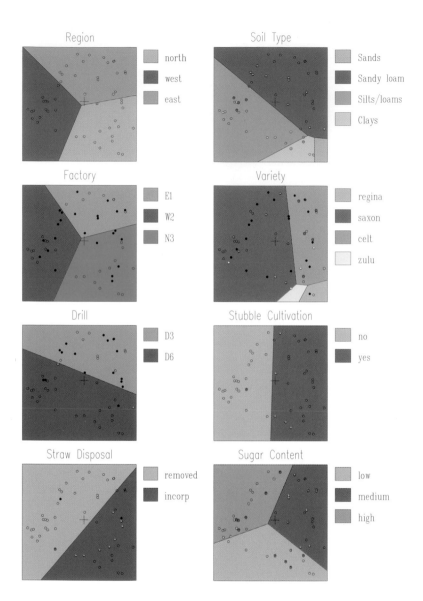

Plate 9: EMC prediction regions for each variable. The farm numbers are omitted but may be found in Plate 8. Farms falling into the correct regions are indicated by open circles and those in incorrect regions are indicated either by circles colored according to the key or, for category-levels absent from the key, by black circles.

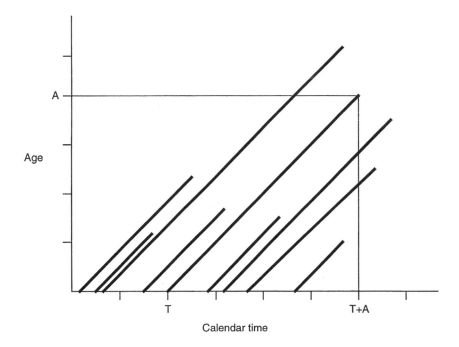

Figure 1: A Lexis diagram.

One method of analyzing event history data is through multiple duration survival analysis models, and the standard Lexis diagram could straightforwardly be adapted for complex event history data, with pencil objects replacing the Lexis lines. However, the complexity of the resulting diagrams with more than one time-varying variable and the strong likelihood of pencils overlapping suggest that this 2D approach would be suitable only for simple data sets. If the pencils could instead be placed in 3D space, this approach would be more attractive. A 3D version of the Lexis diagram was also suggested by Lexis, to represent irreversible changes of state such as termination of marriage during the lifetime of an individual. He noted that such data from studies usually have an extra time dimension, with individuals entering this new state at varying ages. This led him to suggest a 3D extension to his diagram, where the x–z base plane had two dimensions, namely year of birth and age upon entering the new state. The vertical y-axis would then represent time spent in the new state. In effect, this approach would generate 3D age–period–cohorts displays of the raw data, and it provides us with a suitable spatial framework for visualization, with pencils again replacing the Lexis lines.

How should the pencils be angled? In the original work by Lexis, the Lexis lines arose vertically from and perpendicular to the x–z base plane. However, we can also angle the lines at 45° to the x-axis, to the z-axis, or to both the x-axis and z-axis. The choice of display depends on the method chosen for representing the time variables. For example, age can be represented either as a constant for each individual (e.g., age of the individual when entering the study) or as a continuously changing variable over the individual's time in the study. Similarly, calendar time can be represented as constant for an individual (time of entry to study; date of marriage) or as a continuously changing variable. There are four possible ways to represent the two age options and the two calendar time options, giving four possible displays that can be constructed, with suitable orientations of the pencils for each display. For example, when both age and calendar time are continuously varying, the resulting display will consist of angled pencils, at 45° to both the age and time axes. Choosing an appropriate orientation will be guided to some extent by the data analyst's proposed statistical model and the representation of time and age in that model. Alternatively, displaying the data in more than one orientation can often be useful. If there is no suitable secondary time axis, then a variable indexing the individuals may be used as a substitute. This will space the histories equally along the z-axis. Variations on this display would sort the individuals into date, age, or other order before construction of the index variable.

Once the display has been constructed, standard visualization tools such as rotation, panning, zooming, and slicing can be used to explore the data. Both the location of the pencils and the interrelationship of the faces of the pencil are of interest. Dynamic statistical graphics (see, for example, Cleveland and McGill, 1988) provide flexible graphical tools, for example:

1. Identification. The ability to identify case number or the values of other displayed variables on screen. Some visualization systems such as AVS (see Software Notes) provide a method of identifying case number by defining each pencil as a separate object labeled by the case number—the pencil can be identified by clicking with a mouse.

2. Selection and brushing. Case selection in dynamic statistical graphical systems is highly interactive and general, using a combination of simple mouse operations and a graphical toolbar. Selection is also available in visualization systems but would usually be available by specifying ranges of the data through sliders or interactive gauges.

3. Linked displays. The linking of two or more displays, with highlighted objects in one display also highlighted in the remaining displays, is desirable. Visualization software can offer several different views of the same object and also linkage between different geometrical views of the same object, but this requires extra programming effort by the user.

4   An Example

Davies *et al.* (1992) used data from the Social Change and Economic Life Surveys to reexamine the "employment shortfall" effect noted in cross-sectional studies, in which the wives of unemployed men are less likely to be working than the wives of employed men. The data, collected in six UK localities, consisted of work and life histories collected retrospectively for 1171 partnerships existing in 1987. The partnership histories started at the month of last marriage and continued until the date of the survey. Various hypotheses have been suggested to explain this shortfall. Dilnot and Kell (1987) argue that it may be a financial effect related to the payment of benefit. Alternatively, Barrere-Maurisson *et al.* (1985) suggest that wives may be reluctant to work because this could damage their unemployed husband's self-esteem. A further hypothesis suggested by Dilnot and Kell is that women married to unemployed men may have personal characteristics that make them less likely to find work or may live in areas with few jobs available.

A binary logistic regression was carried out by Davies *et al.* (1992) on the monthly employment state of each wife (1 = employed, 0 = unemployed) separately for each of the six localities. Heterogeneity was allowed for by including an individual-specific error term, which was assumed to have a normal distribution, and by the incorporation of two end points at infinity and minus infinity, to allow for women with very low and very high probabilities of taking paid employment. The authors fitted a main-effects model with number of children, husband's length of unemployment (grouped into four duration categories), and a set of dummy variables measuring the age of the youngest child as covariates. Also included in the model were linear and quadratic effects of the husband's age, wife's age, and calendar time. The first column of Table 1 contains the parameter estimates for Kirkcaldy, one of the six localities. From 188 partnership histories and 40,960 partnership months, this model gives a value of minus twice the log-likelihood of 30,987.2 with 40,942 degrees of freedom. Note that the degrees of freedom are used solely to determine the change in the number of parameters between competing models and are not used to assess goodness of fit.

Examination of the parameter estimates and their associated standard errors shows that there is little effect of a husband's unemployment on his wife's employment status for the first 12 months, then a highly significant effect thereafter. The effect of the age of the youngest child is as expected, with strongly significant negative effects for all age groups. The effect of a child under 1 year old is particularly strong (-4.48), formed from the sum of the estimates for <1, <5, and <11 years old. However, combined with this is a positive effect for the number of children. All quadratic and linear terms representing age and calendar time were also significant.

We can now reexamine these data using the visualization ideas described earlier. The Lexis pencils were chosen to have three faces, representing, in clockwise order, the variables of husband's and wife's employment state and a composite variable representing the age of the youngest child in the family. The angle between faces of the pencil is set to be $45°$. A simple color representation of state changes was chosen. For employment state, light blue is used for employed and dark blue for unemployed for both the male and female in the partnership. Similarly, color is again used for the age of the youngest child, with green representing no children in the household, yellow representing under 1 year, red under 5 years, magenta under 11 years, and white under 16 years. We choose the x-axis to be age at marriage, the z-axis to be calendar year, and the vertical y-axis time since marriage.

Table 1: Parameter estimates (with standard errors in parentheses) and log-likelihoods for the Kirkcaldy female unemployment data fitting the logistic-normal model with end points

	A. Davies *et al.* model	B. Model A with additional pregnancy factor included	C. Model B for pre-1962 marriage cohort	D. Model B for 1962–1987 marriage cohort
Intercept	−0.25 (0.02)	−0.004 (0.02)	−0.12 (0.05)	0.67 (0.04)
Husband's unemployment duration:				
1–6 months	−0.05 (0.06)	0.03 (0.06)	−0.70 (0.17)	0.37 (0.08)
7–11 months	−0.19 (0.12)	−0.02 (0.13)	0.57 (0.71)	−0.29 (0.17)
1–2 years	−0.81 (0.10)	−0.67 (0.10)	0.41 (0.28)	−0.85 (0.17)
>2 years	−2.71 (0.03)	−2.65 (0.03)	−2.76 (0.11)	−0.93 (0.12)
Age of youngest child:				
< 1 year	−1.08 (0.03)	−1.32 (0.03)	−0.51 (0.06)	−1.66 (0.04)
< 5 years	−2.09 (0.01)	−2.18 (0.01)	−1.95 (0.02)	−2.23 (0.02)
<11 years	−1.31 (0.01)	−1.45 (0.01)	−1.02 (0.02)	−1.87 (0.02)
Number of children	0.63 (0.01)	0.52 (0.01)	0.50 (0.01)	0.17 (0.01)
Age:				
Husband's age	0.01 (0.01)	0.03 (0.01)	0.09 (0.01)	−0.01 (0.01)
(Husband's age)$^2 \times 10^{-2}$	−0.05 (0.01)	−0.06 (0.01)	−0.09 (0.01)	0.07 (0.01)
Wife's age	−0.10 (0.01)	−0.10 (0.01)	−0.29 (0.01)	0.36 (0.01)
(Wife's age)$^2 \times 10^{-2}$	0.14 (0.01)	0.10 (0.01)	0.32 (0.01)	−0.73 (0.02)
Time:				
Calendar year $\times 10^{-1}$	6.08 (0.06)	6.31 (0.06)	7.56 (0.12)	3.30 (0.19)
(Calendar year)$^2 \times 10^{-3}$	−3.60 (0.04)	−3.74 (0.04)	−4.60 (0.08)	−1.59 (0.13)
Pregnancy:				
2nd trimester		−1.25 (0.07)	−1.03 (0.14)	−1.66 (0.10)
3rd trimester		−3.29 (0.08)	−2.15 (0.11)	−4.29 (0.12)
Scale parameter for normal	1.03	1.05	1.29	1.41
End point probabilities:				
p_0 (at minus infinity)	0.062	0.060	0.058	0.056
p_1 (at infinity)	0.034	0.030	0.019	0.010
−2 log-likelihood	30987.2	30127.7	13842.2	15819.7

Plate 3 shows the resulting rendered display of all 188 histories. At this viewpoint distance, it is difficult to see changes in employment and family state within event histories, but the structure of the sample becomes clear. For example, women who married in the 1950s were over 50 at the time of the survey and contribute long Lexis pencils to the display. In addition, women from all marriage cohorts contribute partial histories to the 1980s. It is possible to use this plot to search for influential histories that might have an influential effect on the regression parameter estimates of the calendar time and age variables. As each history is represented by a different graphical object, it is easy to identify the case number by clicking on the pencil of interest. In this data set, there is no obviously influential history, and we proceed to examination of the display in greater detail.

By zooming into the histories, further features of the data become apparent. Plate 4 illustrates some typical histories for women marrying at the age of 25 in 1967 and 1968. Following the histories through time, from the bottom to the top of each pencil, changes of state in the middle face of the pencil from light blue (employed) to dark blue (unemployed) on the female employment history usually occur before changes of state from green (no children) into yellow bars (child under 1 year) on the child history. In other words, the female partner usually stops work before, not at, the introduction of a child under 1 year old into the household. Further examination of the histories also shows evidence of differences between women marrying earlier (before 1960) and those marrying later (after 1970). Plate 5 shows a close-up view of the histories where marriage occurred before 1955. Using the transparency and selection tools, histories with a date of marriage in 1951 or 1952 are highlighted— all other histories are ghosted, making them visually less important. Women in this cohort appear to have fewer state changes in female employment (the center face of the displayed pencils), either working until the first child or not working at all after marriage, and also appear to be less likely to reenter work than those in later marriage cohorts.

Two features of interest have therefore been found in the event histories. We have observed that women in work often stop work a number of months before the "arrival of a child under one" in the partnership. Therefore, we can assume that the wives stop work because of their pregnancies. This effect is expected but was omitted from the original analysis. The second observation is that women in the survey who married prior to 1960 seem to have a different pattern of female employment history, either not working at all or stopping work when the first child arrives in the household.

We investigate these features of the display by statistically reanalyzing the Kirkcaldy data set. Three further analyses are performed. The first introduces a new three-level factor representing pregnancy: not pregnant or in first trimester of pregnancy, in second trimester, and in third trimester. There is a dramatic increase in twice the log-likelihood of 859.5, with 2 degrees of freedom. The parameter estimates from this model are shown in Table 1, column B. Most parameter estimates remain close to the Davies *et al.* estimates, but those for the effect of husband's unemployment show a change, with the effect of unemployment duration between 1 and 2 years less strong than before. The effects of pregnancy are dramatic, with the strongest effect, as expected, in the third trimester. The estimate of -3.29 in this category is nearly equivalent to the effect (-3.63) of the youngest child being under 5 (the sum of the estimates for <11 and <5).

The second and third analyses repeat the first analysis, but on subsets of the data. We divide the histories into two marriage cohorts—those marrying before 1962 (52 individuals) and those marrying in 1962 or later (136 individuals). In terms of partnership-months, however, the data are divided approximately equally. The results are shown in columns C and D of Table 1. For the whole data set, there is a further decrease in minus twice the log-likelihood of 465.8, with 18 degrees of freedom. More important, the parameter estimates differ substantially between the two groups. The effect of male unemployment on female unemployment in the pre-1962 marriage cohort can be neglected for all categories except for male long-term unemployment of over 2 years. In contrast, the later marriage cohort shows the "12 month" effect discussed earlier. There are also substantial differences between the two cohorts in the effect of the youngest child and number of children on female unemployment.

5   Conclusions

We have illustrated that graphical techniques for exploring event history data can lead to important model improvements, changing the substantive conclusions of a data analysis. Scientific data visualization seems to offer a suitable environment in which to explore such data, allowing the user to examine the whole data set, as well as to explore small subsets of histories in greater detail. The use of Lexis pencils allows the researcher to examine both the changes of state and the relationship between selected time-varying variables of interest, and the pencils can be positioned in 3D space using axes appropriate to the study. There are some difficulties that need to be addressed. One is the question of user perception, such as the best choice of colors to represent changes of state; an initial investigation has been made by Travis (1991). Another issue is the number of histories and variables that can usefully be visualized in a single display. Displays with more than 250 histories appear very cluttered, and our experience suggests that with larger numbers of histories, disjoint subsets of histories can be examined in sequence. In addition, we would recommend that users not attempt to display all potential variables in a Lexis pencil but choose several smaller selections, keeping the number of faces of the pencil small.

Acknowledgments

This work was funded under the ESRC's Analysis of Large and Complex Datasets initiative (H519255029). We are grateful to Professor Fred Smith for useful discussions on the Lexis diagram.

Software Notes

Modern visualization computer systems such as AVS (Advanced Visual Systems, 1992) and Explorer (Silicon Graphics, 1993) provide a set of highly interactive graphical tools, such as rotation, panning, and zooming; clipping (allowing portions of the display either above, below, or intersecting a clipping plane to be deleted); and color redefinition, transparency, texture, and lighting. Most systems also have a modular graphical programming environment. Modules are assembled on the screen in a graphical environment, with graphical links added interactively to define the data flows from one module to another; this forms a network or map. In this way, new applications and displays can generally be assembled without the need for extensive programming.

Francis and Fuller (1996) considered the implementation of their ideas in Explorer. The work reported here was done with AVS, which we have found offers better tools and functionality for the investigation of complex statistical data. In this application, the need for the user to specify and change interactively certain features of the display led to the development of a customized module called *Lexis plot*. This gives the user control over the variables to be assigned to the axes and whether these are time varying or not; the pencil geometry, such as the number of faces of the pencils, the variables to be assigned to them, the pencil thickness, and the angle subtended between adjacent faces; and the color map to be used and whether any selection and clipping required.

AVS is available in the UK from AVS / UNIRAS Ltd. Montrose House, Chertsey Boulevard, Hanworth Lane, Chertsey, Surrey KT16 9JX. Telephone: +44 (0)1932 566608. E-mail: sales@avsuk.com.

Chapter 5

Generalized Impurity Measures and Data Diagnostics in Decision Trees

Tomàs Aluja-Banet and Eduard Nafría

1   Introduction

The objective of tree-based methods is to provide a simple rule for predicting a response variable from a set of predictors. The response variable can be either continuous or categorical, leading to what are called "regression trees" or "classification trees," respectively. The well-known Classification and Regression Trees (CART) method (Breiman *et al.*, 1984) and associated computer program perform both types of tree construction. In this chapter we concentrate on classification trees and essentially follow the CART methodology. Classification trees have the same objective as such multivariate methods as discriminant analysis and logistic regression or more recent techniques such as neural networks, which are being used increasingly in decision making in financial institutions. The main advantage of tree-based classification is the simplicity of the results, given visually in the form of a decision tree. The branchings of the tree follow the human process for decision making very closely.

The heart of the tree-growing process is the splitting criterion used at each node of the tree. The general idea is to split the cases into two or more subgroups at each node so that the heterogeneity between the subgroups is maximized each time in a certain predefined sense. In Chapter 22 Siciliano and Mola use a criterion related to the predictability index τ of Goodman and Kruskal to measure the heterogeneity. In CART, binary splits are made at each node to minimize a so-called impurity index. We present a general formulation for the impurity of a node as a function of the proximity of the individuals in the node to its "representative." We also show how this impurity, or heterogeneity, measure can be decomposed into contributions that can be used to assess the stability of the split at each node.

Although the results of trees are in general attractive and clearly meaningful, a major problem is the stability of the results obtained. Small fluctuations in data may cause a major change in the tree-growing process, although the predictive power may remain the same. We distinguish internal stability from external stability, in the same sense as described by Greenacre (1984, sec. 8.1). External stability refers to the tree sensitivity with respect to independent random samples and can be assessed by means of a test sample, cross-validation, or a bootstrap technique, whereas internal stability refers to the influence of each observation of the learning sample on the tree construction. Our use of diagnostics at each node enhances internal stability in the tree-growing process and hence increases the predictive power.

We apply our methodology to a survey of the mobility preferences of the inhabitants of Barcelona. From past censuses it has been detected that Barcelona's population is decreasing and the objective is to explain this behavior in terms of several socioeconomic, biographic, and living status variables. We take as a response variable the question "If you could change your residence, where would you like to move?" The labels of the possible responses were:

1. No change: I want to stay in the same place or neighborhood.
2. Other district: I want to move to another district of the city.
3. Surroundings: I want to move to the surroundings of Barcelona.
4. Rest of Catalonia: I want to move to the rest of Catalonia.
5. Rest of Spain: I want to move to the rest of Spain.

The list of predictors, all of which are categorical variables (we indicate the number of categories in each case and whether the variable is nominal or ordinal), is the following:

Nominal		**Ordinal**	
district of residence	10	age of head of household	7
region of origin	13	level of studies	10
socioeconomic status		years living in the neighborhood	6
of household	13		
tenancy of the house	3	m^2 of apartment	6
job stability	4	family income per capita	8

The classification tree procedure will identify which variables and which combinations of categories of the variables are related to predicting each of the five response categories.

2   The CART Methodology

Originating in the pioneering work of Automatic Interaction Detection, or AID (Sonquist and Morgan, 1964), tree-growing methodology consists of a recursive splitting of each group of individuals into two subgroups, starting from the total sample and continuing until the subgroups contain mostly individuals giving similar responses, without allowing these subgroups to become too small. The core of the procedure is the identification of the variable that optimally splits a group into two at each node of the tree. The following steps are taken at each node:

- Defining the set of possible splits
- Selecting the best split according to a statistical criterion
- Verifying a stopping criterion, based on a statistical threshold

Most research has been concentrated on the splitting and stopping criteria. Kass (1980) developed tree methodology for a categorical response using a chi-squared criterion, leading to the technique called CHAID (chi-square automatic interaction detection). Celeux and Lechevallier (1982) proposed a splitting criterion based on a measure of distance between distribution functions. Ciampi (1991) proposed instead using the measure of deviance for a generalized linear model. Although the results obtained were an improvement, they still suffered from the criticism of the optimality of the tree and its dependence on the actual data.

The main innovations of CART are:

- Unification of the case of the categorical response variable (classification tree) and that of the quantitative response variable (regression tree) within a similar framework
- Use of an impurity index to measure the heterogeneity of each node
- Pruning from a maximal tree instead of using a stop criterion
- Giving honest estimates of the misclassification error

2.1   Notation

Let t represent a node of the tree. Let n_t be the number of individuals associated with this node and let n_{tj} be the number of individuals of node t with response category j, where $j = 1, \ldots, J$. For a particular node t, called the *parent* node, with n_t individuals, we distinguish its *descendants* t_l and t_r, left and right nodes with n_{tl} and n_{tr} individuals, respectively. A terminal node is a *leaf* of the tree, whereas the initial node, consisting of all the individuals, is the *root* of the tree. A *branch* is formed by a particular path from the root node to a leaf.

For a particular parent node t and for each predictor there exists a set of admissible splits, depending on the nature and coding of the predictor: a binary predictor, with one split; a nominal predictor with k categories, having $2^{k-1} - 1$ admissible splits; an ordinal predictor with k categories, having $k - 1$ splits; and a continuous predictor with r distinct values, having $r - 1$ admissible splits. For example, for the job stability variable with four categories, we can define up to seven possible splits by different groupings of these categories into two sets: four splits in which one category is split off from the other three, and three splits of two categories each. For an ordinal variable, we can make only splits consistent with the ordering of the categories. For example, for age of the head of the household, with age groups 18–25, 26–35, 36–45, and so on, we can just split individuals into groups younger and older than a specified age, up to age 25 and older than 25, up to age 35 and older than 35, and so on.

2.2   The Impurity Index

CART uses an impurity index to assess the split at a node. For a categorical response variable, the impurity index can be written as a function of the probabilities of the response classes:

$$i(t) = F(p(j|t)) \qquad \text{for } j = 1, \ldots, J \tag{1}$$

where $p(j \mid t)$ is the relative frequency of class j at node t.

A node is pure when it contains individuals of just one class, in which case $i(t) = 0$. At the other extreme, when a node contains all classes with equal relative frequency, then $i(t) = \text{MAX}$. An essential property of the impurity measure is that it decreases across the splitting process, that is,

$$i(t) \geq \alpha i(t_r) + (1 - \alpha)i(t_l), \qquad 0 \leq \alpha \leq 1 \tag{2}$$

Any measure that follows properties (1) and (2) can be considered an acceptable impurity index. The indices available in CART are:

Gini $\qquad\qquad\qquad i(t) = \sum_{i \neq j} \sum p(j \mid t)p(i \mid t)$

Misclassification $\qquad i(t) = 1 - \max_j p(j \mid t)$

Entropy $\qquad\qquad i(t) = - \sum_j p(j \mid t) \log(p(j \mid t))$

Twoing $\qquad\qquad i(t) = \dfrac{p_l p_r}{4} \left(\sum_j |p(j \mid t_l) - p(j \mid t_r)| \right)^2$

In practice, the most commonly used index is the Gini index, which may be written equivalently as

$$i(t) = 1 - \sum_j p(j \mid t)^2$$

2.3   Splitting Criterion

For a given node and for all predictors and admissible splits, the predictor and split are chosen that maximize the impurity reduction between the parent node and its descendants:

$$\Delta i(t) = i(t) - \frac{n_{tl}}{n_t} i(t_l) - \frac{n_{tr}}{n_t} i(t_r) \tag{3}$$

This criterion is applied recursively to the descendants, which become the parents of successive splits, and so on, until we arrive at the final nodes, or leaves.

Applying the CART methodology with the Gini index of impurity to the sample of 2492 Barcelona inhabitants in our study, we obtained the tree shown in Figure 1. We can see that the variable that explains most of the geographical mobility is the district of residence, which is, of course, correlated with social class. District of residence splits the total sample of 2492 individuals into two groups, one of 1688 individuals and the other of 804 individuals. Below the number of individuals at each node, the proportion of the five response classes is indicated. Thus, district of residence defines two groups, one aiming to change their actual residence and the other (the minority) preferring to stay in the same place. Then the former group is split according to socioeconomic status into a group of 814 people (the upper classes), who want to move mainly to another district of the city, the surroundings, or the rest of Catalonia, and another group of 874 who prefer to stay in the same place or move to the rest of Spain (the lower classes). In this way we can continue explaining all the

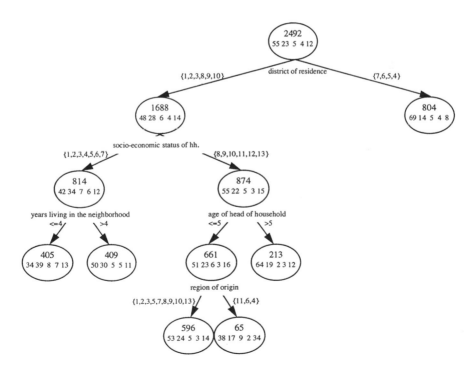

Figure 1: Mobility tree of citizens of Barcelona.

splits of the tree; for instance, the latter group of 874 individuals is split according to the age of the head of the household, those younger than 65 years old wanting to move out of their actual residence, whereas the older ones prefer to stay in it. Finally, there is a split according to their region of origin, and we identify a small group with a very high percentage wanting to return to the rest of Spain. A question that arises is when to stop the splitting process. This is intimately connected with the error rate of the tree. Every leaf of the tree is assigned to one response class, currently the predominant class of the leaf; then the error rate is the proportion of individuals misclassified by the tree. In general, we can have costs of misclassification, in which case we assign the leaves to the class with minimum risk.

2.4   Defining the Right-Sized Tree

The problem of when to stop the splitting process is solved, rather than by having a threshold defined upon the splitting criterion, as in other techniques, simply by growing a maximal tree (a tree with every terminal node pure or with, say, five individuals or fewer) and from this maximal tree defining a sequence of optimal trees by successively removing noninformative subtrees of the maximal tree, minimizing an error complexity measure, and thus obtaining a sequence of nested trees with a decreasing number of terminal nodes. This is done by measuring for each subtree of the maximal tree, its worth, that is, the relative decrease of the error rate relative to the size of the subtree. Then the problem is transformed into one of choosing the right tree of the sequence. This problem is connected with giving honest estimates of the error rate.

In fact, the error rate $C(T)$ decreases with the size of the tree. Any split produces a monotone reduction of the error rate, thus giving an optimistic measure of the goodness of the tree. For that reason, CART proposes to use a test sample or a cross-validation technique to evaluate the error rate of every tree of the sequence.

Then the honest criterion is to select the smallest tree of the sequence with minimum error rate in the test sample or, alternatively, in the cross-validation procedure. Notice that the criterion we use for selecting the tree, that is, the error rate, which is usually the percentage of misclassification, is different from that for growing it (the impurity measure adopted).

Although this approach is neutral in the sense that it gives the right-sized tree with an honest estimate of the error rate, the growing process is still very dependent on data. In other words, the split criterion can be very dependent on data, which implies that the overall tree is also unstable. This is particularly true when using the Gini impurity index, because the Gini index attempts to favor small but very pure nodes rather than equal-sized but less pure ones. To tackle this problem, we have studied the relationship of the splitting criterion with the contribution of each observation to the reduction of impurity, that is, the internal stability of a split. In order to do this, we first present a general formulation of the impurity, from which we will compute the contribution per individual to the reduction of impurity.

3   General Formulation of the Impurity

For a given node t, with n_t individuals to classify according to J classes of the response variable, we generalize the notion of impurity, using a geometric approach in which each response class defines a J-dimensional point, for example, $(1, 0, 0, \ldots)$ for the first response class. Hence, a node is identified with the set of unit points of R^J, at which the individuals are located depending on their response class, for example, n_{t1} individuals at unit point $(1, 0, 0, \ldots)$, n_{t2} at unit point $(0, 1, 0, \ldots)$, and so on.

For a particular node we take its representative point (we represent it by m_t), defined as the point of the convex polygon of R^J, with vertices on the stated unit points, that minimizes the impurity measure $i(t)$.

Then we define the impurity as a function of the squared distance between each individual in the node and its representative.

$$i(t) = \frac{\sum_{j=1}^{J} n_{tj} d^2(j, m_t)}{n_t} \tag{4}$$

where $d(j, m_t)$ is the distance of an individual of class j and m_t (obviously, all individuals of the same response class share the same distance). For example, for three response classes, the geometrical picture would be as shown in Figure 2. In formula (4) we can define the distances $d(j, m_t)$ in several different ways. In particular, we could use the L_2 norm or the L_1 norm.

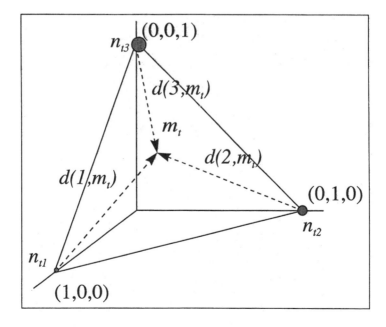

Figure 2: Geometrical representation of a node.

3.1   The Impurity Measure with L_2 Norm

Then the proximity $d(j,m)$ is equal to the Euclidean norm between two points. It can be shown that the representative of the node that minimizes expression (4) is the multinomial vector of probabilities for a given node $\{p(j \mid t), j = 1,\ldots,J\}$ (or the mean of the response variable for that node, in the quantitative case).

$$i(t) = \frac{\sum_{j=1}^{J} n_{jt} d^2(j,m_t)}{n_t} = \frac{\sum_{j=1}^{J} n_{jt} \left(1 - 2m_{jt} + \sum_{l=1}^{J} m_{lt}^2\right)}{n_t}$$

$$\frac{\partial i(t)}{\partial m_{jt}} = \frac{2}{n_t}(-n_{jt} + m_{jt}n_t) = 0 \Rightarrow m_{jt} = \frac{n_{jt}}{n_t}$$

Then the impurity index (4) is equal to the Gini index defined previously:

$$i(t) = \frac{\sum_{j=1}^{J} n_{jt} d^2(j,m_t)}{n_t} = \frac{\sum_{j=1}^{J} n_{jt} \left((1 - p_{(j|t)})^2 + \sum_{l \neq j} p_{(l|t)}^2\right)}{n_t}$$

$$= 1 - \sum_{j=1}^{J} p_{jt}^2 = \sum_{i \neq j} \sum p_{it} p_{jt}$$

or the variance index for the quantitative case.

3.2   The Impurity Measure with L_1 Norm

If we consider the L_1 norm to measure the proximities, then the representative of the node is the unit point e_j with maximum probability $p(j \mid t)$ in node t (or the median for the quantitative case).

$$i(t) = \frac{\sum_{j=1}^{J} n_{tj} |e_j - m_{tj}|}{n_t} = 1 - 2\sum_{j=1}^{J} m_{tj} p(j \mid t) + \sum_{j=1}^{J} m_{tj}$$

Then minimizing this expression leads to

$$\min\left\{1 - 2\sum_{j=1}^{J} m_{jt} p(j \mid t) + \sum_{j=1}^{J} m_{jt}\right\} \Rightarrow \max\left\{\sum_{j=1}^{J} m_{jt} p(j \mid t)\right\}$$

$$\sum_{j=1}^{J} m_{jt} = 1, \qquad m_{jt} \geq 0, \qquad j = 1,\ldots,k$$

The solution is all m_{tj} equal zero, except for the class where $p(j \mid t)$ is maximum.

Then it can be shown that the impurity index is equal to twice the misclassification index (or the absolute deviation for the quantitative case).

$$i(t) = \frac{\sum_{j=1}^{J} n_{jt}\,|e_j - m_{jt}|}{n_t} = \sum_{j=1}^{J} p(j\,|\,t) \times \left(1 - m_{jt} + \sum_{l \neq j} m_{lt}\right)$$

$$= 2 \sum_{j \neq \max} p(j\,|\,t)$$

4   Contributions to the Impurity Reduction

The contribution of each observation to the reduction of impurity depends on the metric used to evaluate the proximities. Developing expression (3),

$$\Delta i(t) = \frac{\sum_{i \in t} \delta(i, m_t)}{n_t} - \frac{\sum_{i \in t_l} \delta(i, m_{t_l})}{n_t} - \frac{\sum_{i \in t_r} \delta(i, m_{t_r})}{n_t} \tag{5}$$

Then we can express the impurity reduction as a sum over all the individuals of the parent node.

$$n_t \Delta i(t) = \sum_{i \in t} \left[\left(\delta(i, m_t) - \delta(i, m_{t_l}) \right) \times 1_{t_l} + \left(\delta(i, m_t) - \delta(i, m_{t_r}) \right) \times 1_{t_r} \right]$$

where 1_{t*} is the Kronecker delta to express membership in one of the successors of the parent node. Finally, the contribution of any individual to the reduction of impurity is simply the difference between its distance from the representative of the parent node and the distance from the representative of its successor.

$$c_i = \delta(i, m_t) - \delta(i, m_{t*}) \tag{6}$$

This quantity is positive or negative when the individual increases its distance from the representative of the offspring. We can compare this measurement to the average contribution. We find that, taking into account formula (5), this average coincides with the impurity reduction. This gives us another interpretation of the impurity reduction.

$$\bar{c}_t = \frac{\sum_{i \in t} c_i}{n_t} = \Delta i(t)$$

Then the ratio

$$\frac{c_i}{\Delta i(t)} \tag{7}$$

allows the control and diagnosis of splits with high dependence on data. This ratio is easy to compute and hence provides a measure of the internal stability for a given split. Then between two splits with similar reduction of stability, we can choose the split with more homogeneous contributions to the impurity reduction. That is, we think that the best strategy is not to follow the best split but to take into account its stability as well.

Table 1: Split 1 (region of origin), $\Delta i(t) = 0.005817$

	Proportions m_t			Distances $\delta(j, m_t)$			Contributions to reduction of impurity		Contributions over $\Delta i(t)$	
	Parent node	Left node	Right node	Parent node	Left node	Right node	Left node	Right node	Left node	Right node
No change	0.51	0.53	0.38	0.32	0.31	0.53	0.02	−0.21	2.81	−35.93
Other dist.	0.23	0.24	0.17	0.88	0.88	0.96	0.00	−0.08	0.39	−13.75
Surround.	0.06	0.05	0.09	1.24	1.26	1.12	−0.02	0.12	−3.42	21.18
Rest Catal.	0.03	0.03	0.02	1.29	1.30	1.27	−0.01	0.02	−1.44	3.05
Rest Spain	0.16	0.14	0.34	1.02	1.07	0.62	−0.05	0.40	−8.62	68.88
Size	661	596	65							

For example, in the mobility tree for the citizens of Barcelona, we found a split at the fourth level of 661 individuals into two groups of 596 and 65 individuals, according to their region of origin (Table 1). The reduction of impurity of this split is 0.005817, very low. Thus, computing the contributions to the impurity reduction in every split, we find for this one a concurrent split, dividing the 661 individuals into groups of 305 and 356 depending on their level of studies (Figure 3, Table 2). For this second split, the reduction of impurity is very similar, 0.00538. Thus, we compute the contribution of individuals of both splits to its reduction of impurity.

We can see that the contribution to the reduction of impurity is clearly more homogeneous in the second split, whereas for the first split it is highly dependent on the 22 individuals of the right child, moving to the rest of Spain, with a contribution to the reduction of impurity 68.88 times higher than the average. The first split is a

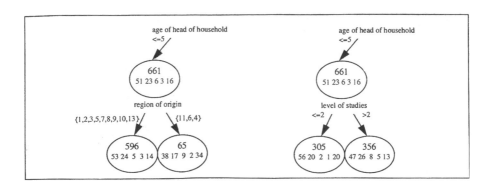

Figure 3: (a) Split 1. (b) Split 2.

Table 2: Split 2 (level of studies), $\Delta i(t) = 0.00538$

	Proportions m_t			Distances $\delta(j, m_t)$			Contributions to reduction of impurity		Contributions over $\Delta i(t)$	
	Parent node	Left node	Right node	Parent node	Left node	Right node	Left node	Right node	Left node	Right node
No change	0.51	0.56	0.47	0.32	0.28	0.37	0.05	−0.05	8.43	−9.08
Other dist.	0.23	0.20	0.26	0.88	0.99	0.80	−0.11	0.08	−20.38	15.60
Surround.	0.06	0.02	0.08	1.24	1.35	1.15	−0.11	0.09	−21.05	16.18
Rest Catal.	0.03	0.01	0.05	1.29	1.38	1.22	−0.09	0.07	−16.93	12.65
Rest Spain	0.16	0.20	0.13	1.02	0.99	1.06	0.03	−0.04	6.06	−7.05
Size	661	305	356							

high candidate to be eliminated in the pruning phase. Thus, it is advisable for these types of splits to consider more stable ones.

Of course, one way to prevent unstable splits is to consider other impurity measures, different from the Gini index, in particular those based on L_1, which are more robust and tend to present more homogeneous contributions to the reduction of impurity. Anyway, the misclassification impurity index needs a local weighting of individuals at each node to give the same probability to every response class, that is, to put all possible splits in a neutral context. Otherwise, if one class is clearly predominant, it can be very unlikely for a split to change the final assignment of a node and hence to decrease the impurity measure.

Chapter 6

Obstetricians' Attitudes on Perinatal Risk: The Role of Quantitative and Conceptual Scaling Procedures

Ulrich Frick, Jürgen Rehm, Karl Erich Wolff,
and Michael Laschat

1 Introduction

Perinatal health outcome of newborns is dependent primarily on the medical condition of their mothers during pregnancy and birth and on their own physical constitution. Especially in cases of medical complications during pregnancy or birth, the outcome is also dependent on the level of health care facilities available at the time of birth (LeFevre *et al.*, 1992). Contemporary concepts of perinatal medicine try to minimize this secondary risk by allocating mothers to different levels of perinatal care before delivery according to prenatal diagnosis. For this purpose, screening programs for pregnant women have been established in nearly all developed health care systems.

A pivotal position in the process of allocation of a mother–child dyad at risk to an appropriate health care facility belongs to physicians in obstetrics departments. Of special interest is the risk acceptance of departments offering all usual medical facilities for delivering mothers (e.g., obstetrical surgery) but offering few or only intermediate possibilities for treating ill newborns (Shenai, 1993). Many authors (for example, Modanlou *et al.*, 1980; Obladen *et al.*, 1994) have shown that antenatal referral of high-risk neonates from departments with low-level pediatrics to high-level perinatal centers decreases mortality, length of hospitalization, and morbidity of transferred neonates. In any case, it is of great interest to examine physicians' perinatal referral decisions in order to obtain background information for quality assurance measures.

Two components of prenatally recognizable risk for the fetus can be distinguished (Renwick, 1992; D'Alton and DeCherney, 1993). On the one hand, a pregnant mother may suffer from one or more conditions of a series of pathological processes during pregnancy. On the other hand, the fetus may develop irregularly due to genetic disposition or due to external noxes. Whereas for the former quantifications of the mortality risk associated with different levels of perinatal care have been investigated by a series of studies (Rudolph and Borker, 1987; Miller *et al.*, 1983), no comparable risk evaluations taking into consideration intensity of perinatal care can be given for fetal anomalies.

One aim of this chapter is to demonstrate the merits of a scaling method in a situation in which unidimensionality of the theoretical concept, which in our case is subjective risk perception (see Johnson and Tversky, 1984), cannot be expected. Whereas "objective" epidemiological risk in the case of mothers' complications constitutes a homogeneous (in the sense of Wottawa, 1980) risk scale, onto which a physician can be placed according to his or her acceptance of a list of known risks, in the case of fetal risk acceptance one exclusively has to deal with subjective risk perceptions of individuals. No method for a unidimensional representation of physicians' decisions on an "objective" scale can be given in the latter case. Nevertheless, the referral decisions of different obstetrics departments should be compared in a manner that simultaneously analyzes possible disagreement of risk policies and ranks perceived health risks of the abnormalities in question.

2 Data and Methods

The heads of all obstetrics departments in the city of Vienna (18 wards in eight public and nine private hospitals) were asked to respond to a self-administered questionnaire containing items concerning their technical equipment and the level of experience of medical and nursing staff. The survey took place in November 1994 and was part of a statewide health planning project of the local government. For various reasons two wards could not complete the questionnaire, so our study is concerned with the responses of 16 wards. The questionnaire consisted of two lists reflecting the two categories of risk just described: maternal risks and fetal risks.

2.1 Maternal Risks

The Maternal Transport Index (MTI) of Strobino *et al.* (1993) was used to measure decisions about risk with regard to medical conditions of pregnant women. This instrument consists of 32 detailed descriptions of hazardous situations with potentially lethal outcomes for the unborn baby. For each situation the responsible departmental medical director was asked to decide whether or not she or he would accept the respective mother for delivery in her or his own ward or would transfer her antenatally to another hospital of higher service level.

The MTI is thus based on an existing classification of pediatric facilities into three levels according to technical equipment and skills of medical staff. Each level can be regarded as optimal for coping with certain maternal risk factors. Referral of high-risk mothers to insufficient service levels increases all kinds of complications including mortality. For example, if a level II facility accepts a mother with heavy vaginal bleeding during the 27th week of pregnancy instead of transferring her to a level III perinatal center, which is designed to react to all possible complications, it causes an increase in the probability that her child will die. Instead of an estimated crude mortality rate of 0.15 at level III, an estimated crude rate of 0.36 prevails at level II. The MTI sums up the ratios of logged mortality rates for all 32 risky conditions. Overall, the MTI can be interpreted as a measure of "risk proneness" of the respective hospital. The minimum value is 32 and is reached by all tertiary level hospitals by definition.

2.2 Fetal Risks

A second list of eight prenatally recognizable defects of the fetus (see Table 1, p. 77) was partly selected from an overview by Shulman *et al.* (1993). Three questions on skeletal deformations, chromosomal aberrations, and blood incompatibility were added on the basis of expert advice. Departmental directors were again asked to decide whether they rated these anomalies as manageable at their hospital or whether they would antenatally transfer the mother of that child to a perinatal center. In contrast to the list of maternal risks, the expected neonatal mortality rates for different levels of care are unknown for all anomalies. Thus, no scaling based on epidemiological reasoning similar to that for the maternal risks was possible. Instead, the dichotomous decisions of "accepted" (y/n) served as the basis for comparing obstetricians' risk perceptions.

The comparison method chosen is based on the representation of conceptual knowledge by line diagrams of concept lattices within the framework of formal concept analysis (FCA). It has been described in detail by Ganter and Wille (1996) and an introduction is also given by Wolff and Gabler (Chapter 7). FCA represents the relationships between objects and attributes for a given data table. Objects in our case are the obstetrics departments, and their attributes consist of accepted anomalies. Relationships can be displayed either as a table of logical implications or graphically by line diagrams of concept lattices. For this volume the latter method was chosen.

Consider the example in Figure 1. Sixteen departments (labeled with capital letters A, B, C, ...) are displayed with respect to three attributes (= decisions of acceptance): gastroschisis, myelomeningocele, and diaphragmatic hernia. The data are in columns 2 to 4 of Table 1. Departments E, J, L, and N did not accept cases with either condition, department S accepted only children with diaphragmatic hernia,

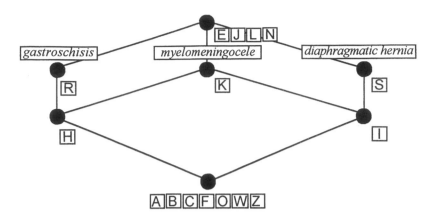

Figure 1: Reading example of a simple decision matrix analyzed by FCA.

department R only children with gastroschisis, and department K only children with myelomeningocele. Two conditions were accepted by department H (gastroschisis and myelomeningocele) and by department I (myelomeningocele and diaphragmatic hernia), whereas departments A, B, C, F, O, W, and Z accepted all three risks.

Generally, a line diagram of a formal context represents its formal concepts by nodes in the plane such that all the information of the given context is preserved. Each object name (respectively, each attribute name) is indicated in the line diagram a little bit below (respectively, above) its concept node. An object has an attribute if and only if there is an ascending path from the node of the object to the node of the attribute. The "extent" of a concept contains all objects that can be reached in the line diagram by descending paths from the concept node, and the "intent" contains all attributes that can be reached by ascending paths.

In Figure 1, for example, the concept of all departments accepting gastroschisis includes the departments R, H, A, B, C, F, O, W, and Z. All departments accepting gastroschisis and myelomeningocele (departments H, A, B, C, F, O, W, and Z) constitute a subconcept to both concepts represented by the nodes labeled "gastroschisis" and "myelomeningocele." As can be seen, the superconcept–subconcept relation is directed from top to bottom, building a conceptual hierarchy.

The aim of our analysis is to describe the interrelations between risk decisions of all obstetrics departments in Vienna. FCA simultaneously groups both objects and attributes and reveals dependences between attributes and between objects. In our application, rank ordering of perceived risk of fetal complications is visualized by the drawing direction of the graph. The lower the position of an attribute in a line diagram, the greater is the reluctance of physicians to accept this risk at their department, and the greater therefore can be regarded the perceived risk of that fetal abnormality. This is concluded from the fact that very few hospitals accept such a risk.

The possibly multidimensional nature of perceived risk can be inspected by the degree of branching out required in the line diagram. The greater the number of concepts required to represent such a data table, the smaller is the agreement on acceptable risk policies between the hospitals questioned.

3 Results

3.1 Acceptance of Maternal Risks

Figure 2 shows a plot of MTI scores, where it should be noted that a value of 32 constitutes the minimum value on the scale. This minimum level can be reached either by transferring all mothers at risk (true for department N) or by definition for all perinatal level III centers, true for departments A, B, C.

The MTI allowed a quantitative representation of all departments on a single dimension that can easily be transformed into potentially accepted additional neonatal deaths. Such a procedure also allows graphical representation on bidimensional plots as can be seen in Figure 2. Here, the number of deliveries observed in 1993 within each department is plotted against the MTI score, and one can see, for example, the high risk in admitting pregnant women for departments where many lives are at stake.

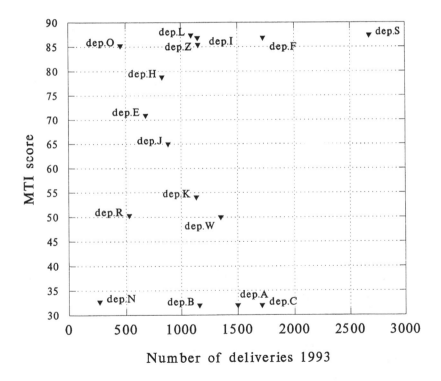

Figure 2: Departmental MTI score (risk of child's death) plotted against number of deliveries in 1993.

3.2   Acceptance of Fetal Risks

Table 1 summarizes the raw data for the analysis of fetal risks. Analyzing this matrix of 16 hospitals and 11 variables (=fetal anomalies) resulted in a complex line diagram (see Figure 5) that will be interpreted later. The complexity of this structure is not an artifact of the method: if the logical structure of the concept lattice is very simple, that is, if the decisions of all hospitals agreed to a great amount, the resulting line diagram would show an easily interpretable form. To facilitate the inspection of the hospitals' risk-taking patterns in a first step, fetal anomalies were divided into two subgroups requiring two different kinds of precautions to be taken for the delivery: "immediate" and "delayed" action, as shown in Table 1.

Figure 3 represents the decisions of Vienna's obstetrics departments concerning the six fetal anomalies that would require immediate and often serious medical action if a child with one of these anomalies is born, that is, the subgroup "immediate action."

Three departments (E, L, J) were placed at the top of the line diagram because of their refusal of all six immediate action anomalies for delivery. Five departments (the three perinatal centers A, B, C; department W offering a level II neonatal care unit; and department Z) would accept any of these anomalies and therefore were placed at the bottom of the line diagram. Three departments (K, N, R) would accept exactly one of these fetal risks. As each department accepted a different anomaly, the line diagram splits up into different branches showing this disagreement on acceptable risks. Department I accepted two risks, and departments O, H, and S rated three of six risks as manageable at their wards. Only the two most extreme risk policies (complete acceptance and complete refusal of immediate action risks) are shared by more than one department. All observed intermediate positions are held by only one single decision maker.

The second subgroup of the remaining five fetal anomalies was called the "delayed action" subgroup, for therapeutic actions either were not necessary immediately after birth or could be decided on only after a postnatal diagnosis of the intensity and quality of the anomaly. In contrast to Figure 3, the concept lattice of the delayed action anomalies resulted in a very simple line diagram (Figure 4).

If one disregards department O, all departments and risks could be placed on a simple "pearl necklace" without any branching out of the lines and nodes of this figure (omitting department O would move the risk "omphalocele" to the last node at the bottom of the line diagram). The conceptual structure of a pearl necklace is equivalent to a so-called Guttman scale (Guttman, 1944) formed by these items. Again, department L accepted no single fetal anomaly and therefore was placed at the top of the line diagram. Department E accepted only "chromosomal aberrations" on the first step of this scale. The two items "skeletal deformation" and "hydronephrosis" defined the second rank of this risk order and were accepted by departments J and K. "Esophageal atresia" on the third rank was also accepted by departments I, N, and R. The "highest" risk was associated with "omphalocele." Eight departments (A, B, C, F, H, S, W, Z) accepted omphalocele as well as the other four anomalies. Only department O deviated from this hierarchical ranking of risks by accepting an omphalocele but refusing hydronephrosis and esophageal atresia.

Table 1: Acceptance of Fetal Risks by Obstetrics Departments: Data Matrix

| | Fetal Risk (accepted: y/n) Requiring: | | | | | | | | | | |
| | Immediate action | | | | | | Delayed action | | | | |
Department care level	Blood group incompatibility	Diaphragm hernia	Myelo-meningocele	Gastro-schisis	Transposition of great vessels	Hypoplastic left heart	Chromosomal aberration	Skeletal deformation	Hydro-nephrosis	Esophageal atresia	Omphalocele
A (III)	y[a]	y	y	y	y	y	y	y	y	y	y
B (III)	y	y	y	y	y	y	y	y	y	y	y
C (III)	y	y	y	y	y	y	y	y	y	y	y
E (I)	n	n	n	n	n	n	n	n	n	n	n
F (I)	y	y	y	y	n	n	y	y	y	y	y
H (I)	y	n	y	y	n	n	y	y	y	y	y
I (I)	n	y	y	n	n	n	y	y	y	y	n
J (I)	n	n	n	n	n	n	y	y	y	y	n
K (I)	n	n	y	n	n	n	y	y	y	y	n
L (I)	n	n	n	n	n	n	n	n	n	n	n
N (I)	y	n	n	n	n	n	y	y	y	y	n
O (I)	n	y	y	y	n	n	y	y	n	n	y
R (I)	n	n	n	y	n	n	y	y	y	y	n
S (I)	y	y	n	n	y	n	y	y	y	y	y
W (II)	y	y	y	y	y	y	y	y	y	y	y
Z (I)	y	y	y	y	y	y	y	y	y	y	y

[a]y, unborn child accepted for delivery; n, decision to transfer this child's mother antenatally.

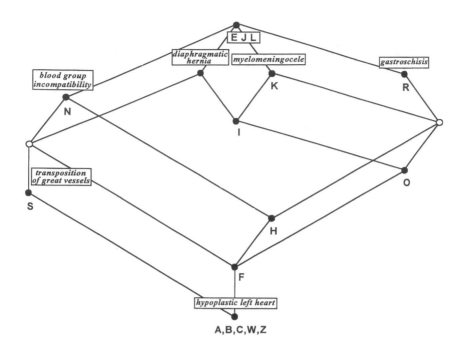

Figure 3: Acceptance of "immediate action" fetal risks at obstetrics departments in Vienna.

Figure 5 gives the joint conceptual structure of risk acceptance (11 attributes) and departments' service levels. Departments of service level I display a heterogeneous variety of all empirically observable risk policies from complete refusal (department L) to complete acceptance (department Z) of fetal risks. The four departments of the higher service levels II and III, of which one is at level II, would all accept the complete list of fetal risks and do not differ in their risk policies according to their service levels. Graphically, this can be depicted from the fact that all fetal risks occur between the node labeled "level 1" and the node "level 2" in Figure 5.

Interpreting the relationships between the attributes in Figure 5 is not easy, because of the observed heterogeneity of referral decisions. With the exception of the perinatal centers A, B, and C no two departments accept the same risks; that is, each department follows its own singular "risk concept" (node). But risk concepts can be arranged in a meaningful chain (Guttman scale) because there is a remarkably strong dependence between the attributes—there exists a very long chain of 10 attribute concepts in Figure 5 forming a strictly hierarchical line. This can be seen more easily in Figure 6, which, as a reduction of Figure 5, represents exclusively the order of all attribute concepts without the departments.

Within the central chain of Figure 6 the concepts of the delayed action risks chromosomal aberration, skeletal deformation, hydronephrosis, and esophageal atresia are

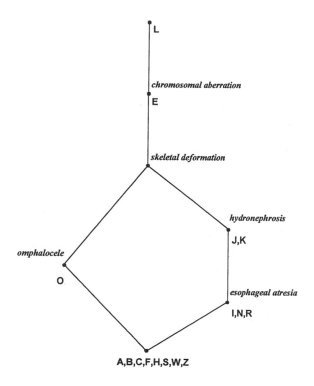

Figure 4: Acceptance of "delayed action" fetal risks at obstetrics departments in Vienna.

all superconcepts of the following immediate action risks: blood group incompatibility, transposition of great vessels, and hypoplastic left heart. The other four attributes (omphalocele from subgroup delayed action and gastroschisis, diaphragmatic hernia, and myelomeningocele from subgroup immediate action) are incompatible with one another in the sense that none of these attribute concepts is a subconcept of any other of these.

If we study the subcontext of all departments and only the attributes of the central chain, we obtain the concept lattice drawn in Figure 7. Figure 7 can be seen as the "interpretative core" of the whole context of all 11 fetal risks, as it shows only the attributes for which agreement on their logical order is unanimous among the physicians of this study.

4 Discussion

Three wards (A, B, and C) were part of perinatal centers and thus could offer the highest possible level of care (level III) for neonates. As a consequence, their MTI

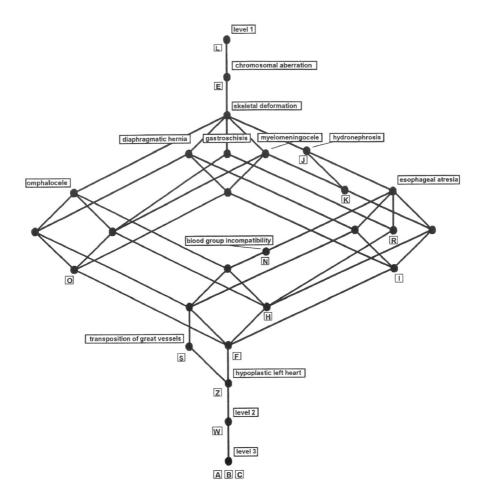

Figure 5: Complete line diagram of immediate and delayed risks' acceptance and departmental service levels.

scores could by definition not exceed the minimum value of 32. All three wards in perinatal centers would accept all mothers listed. One hospital (N) handles births strictly on an outpatient basis. In accordance with that principle, almost no avoidable risks were accepted: mothers having a symptom of perinatal risk would be transferred to a perinatal center by this hospital. Thus, its MTI score is very low (33.5).

One hospital (W) offers services of a level II neonatal unit in addition to its obstetrics ward. The MTI score for this hospital therefore reached a relatively moderate level of risk acceptance (50.0), although all listed mothers' complications would be accepted at this hospital. Six additional hospitals (F, I, L, O, S, Z) also accepted nearly all maternal complications. But because they offered no additional neonatal

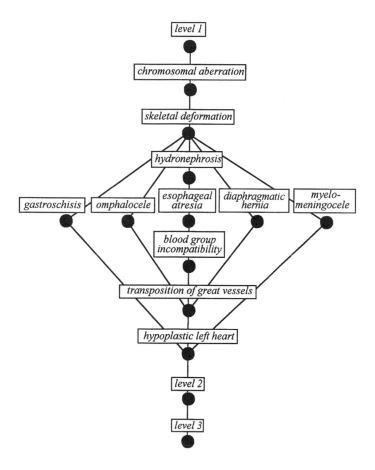

Figure 6: Acceptance of fetal risks: order of attributes.

services (not exceeding level I care), their MTI values were very high (> 85). The MTI scores of these six hospitals represent in the hypothetical case of 3200 risky mothers (100×32 symptoms) delivering in one of these hospitals about 700 additional neonatal deaths to be expected due only to this misallocation. The remaining hospitals (E, J, K, R) had MTI scores of an intermediate level. They accepted only a subset of the described maternal complications.

All in all, willingness to take risks reported by heads of obstetrics departments in Vienna showed little awareness of the problems associated with postnatal transport. Obstetricians too often rated their own department capable of managing medical problems of pregnant women, disregarding the amount and quality of their facilities for newborn infants. This tendency toward "overconfidence" based only on the condition of the mother could be shown for nearly all departments with one exception: the outpatient hospital (N). If decisions of departmental heads were similar to decisions

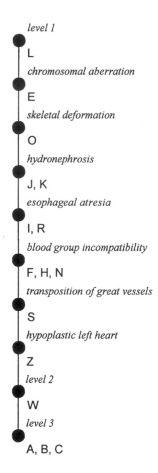

Figure 7: Reduced context of fetal risks with perfect Guttman structure.

of their subordinate physicians, a considerable amount of perinatal risk should have resulted from the described referral preferences. Indeed, Vienna's hospital discharge statistics for obstetrics departments show that the recommended cumulation of risky deliveries within perinatal centers has not been established yet: the rates of multiple birth, cesarian sections, and preterm infants do not differ significantly between obstetrics departments inside and outside perinatal centers (Frick *et al.*, 1995). Thus, self-reported risk acceptance is correlated with empirical indicators of risk dispersion over hospitals.

The quantification of the subjective variable "risk acceptance" used a risk scale in which the weight of each item was given by the "objective" epidemiological risk a priori. Problems of "homogeneity" of our risk scale thus were not relevant. No matter how a physician perceived and subjectively weighted maternal risk during a cognitive evaluation of the transferral decision, we were able to place him or her on a single, homogeneous dimension with a substantially meaningful interpretation: "number of additional deaths that are possibly accepted by the respective referral pattern." The MTI enabled governmental authorities to prioritize quality assurance measures. Figure 2 showed a visualization of the product of risk attitude and annual number of deliveries. Pressured by administrative authorities, departments situated at the upper right corner of this figure were those that had to start first with quality discussions on their referral decisions, because changing their attitudes meant a greater reduction in absolute risk for Vienna's newborns.

One should not withhold the fact that the chosen quantitative scaling of subjective risk attitudes according to epidemiological risk also has an important disadvantage: the reasons and processes underlying perinatal "overconfidence" cannot be analyzed by studying the composite MTI score. Different patterns of risky decisions could result in very similar or identical MTI scores (and thus are homogeneous in the sense of "additional deaths") but nevertheless would require different directions of attitudinal change. Thus, the decomposition and analysis of the global MTI value into its single decisions are necessary when planning and discussing quality assurance measures with medical experts.

For the case of fetal anomalies, a comparable objective interpretation of the acceptance of a single item could not be given. Therefore the scaling of fetal risks inevitably means an analysis of subjective risk perceptions by the medical experts who were questioned. Under these circumstances, a homogeneous, quantitative scale would be an unrealistic expectation (Yates and Stone, 1992). On the contrary, different obstetricians would possibly categorize fetal anomalies into qualitatively different risk groups. FCA seems to be the method of choice for dealing with that problem.

Substantively it could be shown that departments willing to accept the risk of (nearly) all maternal complications (A, B, C, F, I, L, O, S, W, Z) reacted differently in regard to fetal anomalies. The level III centers (A, B, C), department W (with a level II neonatal unit integrated), and department Z (level I) accepted *all* fetal anomalies for treatment as well. However, department Z cannot offer any specifically pediatric facility for treatment of these anomalies. Intensive care (required for a child with a hypoplastic left heart, for instance) would also exceed the capabilities of the level II unit in department W.

Departments F, I, L, O, and S also accepted all maternal risks, but showed more restrictive attitudes toward pediatric risks. F and S accepted major, although different risks, and I and O were still more conservative (but again on qualitatively different options). Department L was not willing to accept any pediatric risk but accepted all maternal risks. No medical argument can be cited to support such a policy. Outpatient clinic N, on the other hand, accepted no maternal complications but did accept some fetal anomalies; mostly delayed action risks. Among the immediate action risks, N accepted only blood incompatibility. This is a risk requiring a "standard" procedure (blood exchange) that is the least invasive procedure of the immediate actions and seems acceptable from the viewpoint of quality assurance management.

FCA showed some very helpful features for interpreting the results. When subjects agreed in their risk policies as measured by specific items, FCA revealed common patterns of perception twice in a very effective way. First, delayed action risks were perceived on a nearly perfect Guttman scale. The seriousness of subjective delayed action risks can thus be measured at least on an ordinal level for these data. Second, the distinction between immediate and delayed action risks was shared by Vienna's departmental directors with regard to four out of five delayed action risks and three out of six immediate action risks. This seems noteworthy, especially because the questionnaire was not designed to give any hints about this distinction.

Immediate action risks were perceived as more serious and disagreement on acceptability of these risks was much greater. Therefore quality assurance measures should focus on the outlier risks of Figure 6 as a first step toward clarifying the reasons for the deviating perceptions. If we were—as a result of discussions within a quality circle—able to extend the central chain of Figure 7 into a pearl necklace comprising all fetal risks (this could also mean positioning more than one risk at a single node of the current chain), a second step in quality assurance should scrutinize whether the position of each department on this "perinatal risk ladder" can be justified by its technical equipment and the medical expertise of its staff.

For the discussion of the disagreeing referral decisions (e.g., Figure 3), FCA enabled us to raise meaningful questions about the reasons for this heterogeneity: Is there a medical rationale for accepting a diaphragmatic hernia in departments I, O, S, F, A, B, C, W, and Z but not in departments E, H, J, L, N, K, and R? What are possible common elements of staff or equipment of both groups? Does it depend on the capacity to cope with other fetal complications requiring immediate action? It does not, as one can see: diaphragmatic hernia is no subconcept of any other risk in Figure 3. On the other hand, hypoplastic left heart was accepted only by departments that would also accept all other fetal anomalies (including all delayed action risks). It can be concluded that hypoplastic left heart is perceived as a very serious complication, even if we do not know enough about the expected epidemiological outcome of this anomaly. Even if different hospitals do not agree about whether they should accept this anomaly, they would perhaps agree on the seriousness of this risk. Thus, even if the objective risk of the referral decisions is unknown, FCA enables the discussion and construction of adequate measures to be taken for standardization and improvement of medical care.

Overall, the study showed serious discrepancies between risk attitudes toward maternal and pediatric complications. Some attitudinal sets were shown to be inconsistent with accepted standards of care in the field. As a result of this study, the Viennese government formed a perinatal quality assurance committee to develop guidelines for acceptance and referral of perinatal risk conditions.

Acknowledgments

This research was supported by a grant from the city of Vienna (KAV-GD-M/17.224/94/M).

Chapter 7

Comparison of Visualizations in Formal Concept Analysis and Correspondence Analysis

Karl Erich Wolff and Siegfried Gabler

1 Introduction

The development of formal concept analysis (Wille, 1982; Ganter and Wille, 1996) led to a new possibility for the visualization of data by line diagrams of conceptual hierarchies: in contrast to methods such as correspondence analysis, which represent data approximately in planar displays, line diagrams visualize data without any loss of information. For a survey of nine graphical data analysis methods the reader is referred to Wolff (1996); for an application to medical data see Frick *et al.* (Chapter 6).

The purpose of the present chapter is to compare the visualizations in formal concept analysis (FCA) and correspondence analysis in its two variants: simple correspondence analysis (CA) and multiple correspondence analysis (MCA). In the following chapter we compare these methods as to the same data sets, and discuss their respective advantages and disadvantages.

2   Examples

2.1   A Repertory Grid of an Anorexic Young Woman

The following data are from an investigation of anorexic young women by Spangenberg (1990). Using the Repertory Grid Technique, each patient was asked to name her most important persons including herself and her ideal self. The patient evaluated each of these persons according to seven bipolar constructs on a rating scale from 1 to 6; for example, one such construct is "resolute–insecure," where 1 means very resolute and 6 means very insecure. The seven constructs were as follows:

peaceful–conflicting	(pe-co)	self-confident–weak	(sc-wk)
lively–theoretical	(li-th)	dependent–independent	(dp-id)
being in want of warmth–unfamiliar	(ww-uf)	resolute–insecure	(re-is)
lonely–loved	(lo-lv)		

One of the patients generated the repertory grid in Table 1 with the persons SELF, IDEAL, FATHER, and MOTHER.

To visualize these data by FCA and by CA, we first analyze the data by selecting the extreme responses with marks 1 or 2 and 5 or 6, which leads to the incidence matrix of Table 2. For example, SELF and MOTHER are lonely (lo), the IDEAL is loved (lv), and FATHER has none of these attributes, which implies that he has one of the intermediate values 3 or 4 of the construct lonely–loved. It is possible to construct tables that represent all the information of the grid, but the chosen coding is very close to the language used in conversation between the patient and the therapist. This table will be visualized now by FCA and CA.

Table 1: The data table of an anorexic young woman

	pe-co	sc-wk	li-th	dp-id	ww-uf	re-is	lo-lv
SELF	1	6	2	1	1	6	1
IDEAL	5	1	2	6	3	1	5
FATHER	4	1	1	5	2	2	3
MOTHER	5	1	6	6	6	1	1

Table 2: The table of the extreme responses

	pe	co	sc	wk	li	th	dp	id	ww	uf	re	is	lo	lv
SELF	X			X	X		X		X			X	X	
IDEAL		X	X		X			X			X			X
FATHER			X		X			X	X		X			
MOTHER		X	X			X		X		X	X		X	

2.2 Line Diagram of the Concept Lattice

We start with the standard graphical output of FCA, namely the line diagram in Figure 1. For each person and for each attribute there is a solid circle in the line diagram and there are additional, unlabeled, solid circles. A person has an attribute in the data table if and only if there is an ascending path from the circle of the person to the circle of the attribute.

Hence SELF is peaceful, weak, dependent, insecure (and no other person has these attributes); SELF also has the attributes being in want of warmth like the FATHER (and nobody else), and lively like IDEAL and FATHER, and lonely like the MOTHER (and nobody else). The FATHER is self-confident, independent, and resolute like IDEAL and MOTHER. The IDEAL is the only person who is loved, the IDEAL is conflicting like the MOTHER, and the MOTHER is theoretical and unfamiliar and nobody else has these attributes. This is the complete information from the data in Table 2.

Further valuable information can be recognized easily from the line diagram. First, there are several meaningful partitions (called *extent partitions*) on the set of persons, meaningful in the sense that each class (cluster) of a partition is described by a subset of attributes, for example, the two-class partition consisting of the conflicting persons {MOTHER, IDEAL} and the persons {FATHER, SELF} being in want of warmth. There is another remarkable two-class partition, namely the partition SELF versus the others, with SELF described by the attributes peaceful, weak, dependent, insecure and the others by self-confident, independent, resolute. It is clear that this

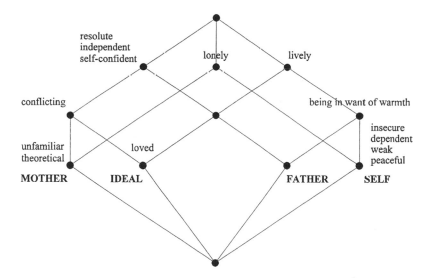

Figure 1: A line diagram representing the conceptual structure of the extreme responses of an anorexic young woman.

partition is very important for this patient (SELF), because three different bipolar constructs are used by the patient to describe this partition.

2.3 Correspondence Analysis Map

To apply CA to these data we construct from Table 2 the corresponding 0–1 matrix N; that is, we replace each cross by 1 and fill each empty cell with 0. Figure 2 shows the CA map of the table coded in this way.

In the map each attribute point lies in the barycenter of the points of the persons having this attribute. Using this *barycenter reading rule*, we see from the map very clearly that only SELF is peaceful, weak, dependent, and insecure; MOTHER is theoretical and unfamiliar; IDEAL is loved. We further see that only SELF and MOTHER are lonely, because the point lonely is the barycenter of the points of SELF and MOTHER (and of no other pair of object points); SELF and FATHER are being in want of warmth; IDEAL and MOTHER are conflicting.

The FATHER is the only person who shares each of his attributes with other persons. The point representing lively does not lie on a line between two person points; hence this attribute is associated with at least three persons. It cannot be

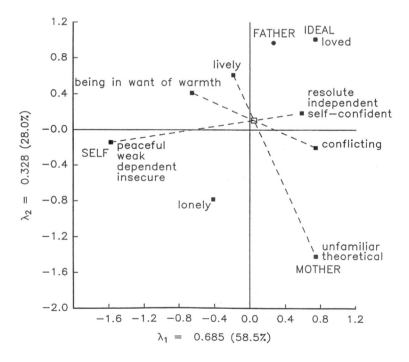

Figure 2: The asymmetric map in which attributes are at the barycenter of their objects.

associated with MOTHER, otherwise it would lie much lower on the vertical axis. Hence lively must be at the barycenter of the points of SELF, FATHER, IDEAL. The same argument shows that the common point of resolute, independent, self-confident is the barycenter of MOTHER, FATHER, IDEAL. Hence we have again been able to reconstruct the original table, but with a little more difficulty compared with the FCA diagram. In CA there are maps with more objects and attributes; however, it may happen that barycenters of different subsets of objects are very close together or even coincide, which makes it difficult or impossible to reconstruct the data from the map. Finally, we mention that the large value of 86.5% for the explained inertia indicates relatively good quality of this map in the sense that the (multidimensional) points lie close to the plane shown in Figure 2.

2.4   Comparison of Both Visualizations

We compare the FCA line diagram (Figure 1) and the CA map (Figure 2) first with respect to their intentional similarities and then with respect to their graphical differences. In Figure 2 one can see the persons and attributes in an arrangement similar to that in the line diagram of Figure 1: the unfamiliar and theoretical MOTHER, connected by the attribute conflicting with the loved IDEAL, both connected with the FATHER by the attributes self-confident, independent, resolute; IDEAL, FATHER, and SELF are the lively persons with the subgroup FATHER, SELF both being in want of warmth; the SELF, which is by itself peaceful, weak, dependent, and insecure, shares the attribute lonely with the MOTHER. This circular story starting from MOTHER over IDEAL, FATHER, and SELF back to MOTHER can be seen in both visualizations.

A CA map is a representation of the rows and columns of the table in a multidimensional metric vector space, obtained by a linear projection onto a suitable plane. It should be mentioned that the whole CA process produces not only a two dimensional display but also extensive numerical output that can be used for analyzing the data.

In contrast to this metric vector space approach, line diagrams represent conceptual hierarchies that are combinatorial ordinal structures obtained from the data table. To represent these hierarchies in the plane one needs only the usual order of real numbers in the y-direction of the plane and ascending lines connecting two points to represent the relation that a formal concept is a lower neighbor of another formal concept in the conceptual hierarchy.

2.5   Interordinal Scales and the Guttman Effect

An example that shows the difference between FCA and CA displays more dramatically is the so-called interordinal scale given in Table 3.

This cross table represents a "language" about the numbers $\{1, 2, 3, 4, 5, 6\}$ (i.e., the values in Table 1) using the attributes $\leq 1, \ldots, \leq 6, \geq 1, \ldots, \geq 6$, which

Table 3: An Interordinal scale

I(6)	≤ 1	≤ 2	≤ 3	≤ 4	≤ 5	≤ 6	≥ 1	≥ 2	≥ 3	≥ 4	≥ 5	≥ 6
1	X	X	X	X	X	X	X					
2		X	X	X	X	X	X	X				
3			X	X	X	X	X	X	X			
4				X	X	X	X	X	X	X		
5					X	X	X	X	X	X	X	
6						X	X	X	X	X	X	X

are needed to describe intervals, for example, the set of all numbers x satisfying $3 \leq x \leq 5$, which is just $\{3, 4, 5\}$, the intersection of the sets $\{x \mid x \geq 3\}$ and $\{x \mid x \leq 5\}$. The concept lattice of this interordinal scale is shown in Figure 3 and the asymmetric CA map of the indicator matrix (Table 3) is shown in Figure 4.

The CA map (Figure 4) shows the well-known Guttman effect (or "horseshoe effect"), namely the parabola-shaped configurations of the set of numbers 1 to 6 as well as the "attribute" points $\leq i$ ($i = 1, 2, 3, 4, 5, 6$) on the left side and $\geq i$ on the right side (see also Greenacre, 1984, secs. 8.3 and 8.8.2). This CA map shows how difficult it is to reconstruct the data using only the barycenter reading rule. A parabola-shaped configuration in a CA map occurs not only in ordinal and interordinal scales but also in many data with a certain "trend." The line diagram in Figure 3 allows the data to be reconstructed exactly.

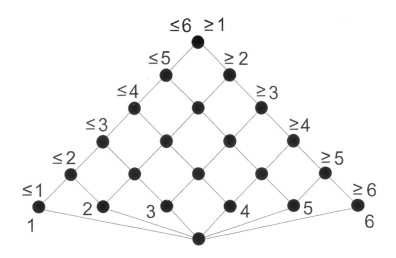

Figure 3: Line diagram of the concept lattice of the interordinal scale.

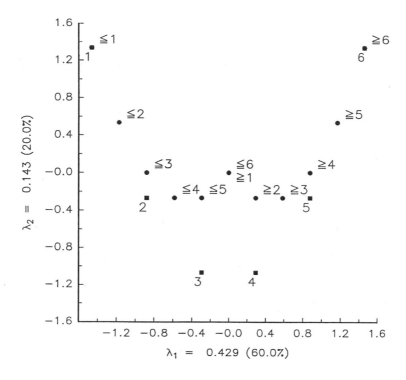

Figure 4: CA map of the interordinal scale.

3 Data Representations in Formal Concept Analysis

FCA was introduced by Wille (1982) "based on the philosophical understanding of a concept as a unit of thoughts consisting of two parts: the extension and the intension (comprehension); the extension covers all objects (or entities) belonging to the concept while the intension comprises all attributes (or properties) valid for all those objects." For detailed introductions the reader is referred to Ganter and Wille (1996) and Wolff (1994). FCA is the central theory in the field of conceptual knowledge processing with two main branches: namely, conceptual knowledge systems (see Wille, 1992) and conceptual data analysis (see Wille, 1987; Spangenberg and Wolff, 1991; Wolff *et al.*, 1994).

Here we give a short overview of the main ideas in conceptual data analysis. We start with the formal representation of data tables by *many-valued contexts*. Data are usually collected in tables with many rows and columns and possibly empty cells. Hence such a table can be described formally as a partial mapping from $G \times M$ into W, where G is called the set of objects ("Gegenstände" in German), M the set of many-valued attributes ("Merkmale"), and W the set of values ("Werte").

Attributes usually describe variables such as age, temperature, and time. Conceptual data analysis unfolds a given many-valued context in a certain conceptual frame that represents the conceptual meaning of the values, where "conceptual" is understood in the sense of "formal concepts" introduced by Wille (1982).

An example of a formal context is given in Table 2. This table can be described by three sets, namely the set G of the four persons, the set M of the 14 attributes, and the set I of all the person–attribute pairs, which are indicated by crosses in the table. In general, a *formal context* (or briefly a *context*) is defined as a triple (G, M, I) of sets where I is a subset of all possible pairs (g, m), denoted by $G \times M$. The elements of G are called *objects* and the elements of M *attributes*. If $(g, m) \in I$, we say that g has the attribute m. Each "cross-table" describes a formal context uniquely. The central definition of formal concepts and their hierarchy is formulated with respect to a given formal context. A formal concept of a context (G, M, I) arises in a natural way from a "database question" consisting of a subset Q of M by constructing first the "answer" A of all objects having all attributes of Q and second the set B of all attributes that are valid for all objects of A. Then the pair (A, B) is called a *formal concept of* (G, M, I). The set A is called the *extent* and B the *intent* of the concept (A, B).

In the context of the interordinal scale of Table 3 the pair $(A_1, B_1) = (\{3, 4, 5\}, \{\leq 5, \leq 6, \geq 1, \geq 2, \geq 3\})$ is a concept of this context. Another concept is $(A_2, B_2) = (\{1, 2, 3, 4, 5\}, \{\leq 5, \leq 6, \geq 1\})$. The first concept has a smaller extent and a larger intent than the second, which demonstrates the well-known fact that the more conditions there are, the fewer objects fulfill them. This leads to the definition that for any two concepts (A_1, B_1), (A_2, B_2) of a given context the concept (A_1, B_1) is called a *subconcept* of (A_2, B_2), briefly $(A_1, B_1) \leq (A_2, B_2)$, if $A_1 \subseteq A_2$. The ordered set of all concepts of a context is called the *concept lattice* of the context. Concept lattices can be represented in the plane by *line diagrams*.

Line diagrams are specially labeled *Hasse diagrams* of concept lattices. One can show that every finite ordered set can be represented without any loss of information by a Hasse diagram in the plane (see Davey and Priestley, 1990). There are two main steps in the construction of a Hasse diagram of a finite ordered set, denoted by $(P, \leq)$. In the first step each element p of P is represented by a planar point $h(p)$, called the *Hasse point* of p, such that smaller elements in the ordered set $(P, \leq)$ are represented by lower points in the plane. In the second step two Hasse points $h(p)$, $h(q)$ are connected by a line if and only if p is a lower neighbor of q, that is, $p \leq q$, and there is no other element of P between p and q. It is clear that an ordered set can be represented by many graphically quite different looking Hasse diagrams, and it is an art to draw "nice" Hasse diagrams.

The line diagram in Figure 1 is a Hasse diagram of the ordered set for the context represented in Table 2. In general, a line diagram is a Hasse diagram of a concept lattice labeled in the following way. The name of each object g of the given context is written a little bit below the Hasse point of the *object concept* of g and the name of each attribute m is written a little bit above the Hasse point of the *attribute concept* of m. The object concept of an object g is the smallest concept having g in its extent, and the attribute concept is the greatest concept containing m in its intent. Then the following reading rule for line diagrams holds: an object g has the attribute m if and only if there is an ascending path from the point labeled with g to the point labeled with m. Hence a context is reconstructable from any of its line diagrams.

Using the reading rule in the line diagram of Figure 1, one can see that the FATHER has the attributes self-confident, independent, resolute, being in want of warmth, and lively, but he does not have the attribute lonely, for example, because there is no ascending path from FATHER to lonely. Finally, we mention the role of unlabeled circles in a line diagram. Each of these represents a concept that is neither an object concept nor an attribute concept. The unlabeled circle in the middle of the line diagram in Figure 1 represents the concept ({IDEAL, FATHER}, {self-confident, independent, resolute, lively}). In general, a circle in a line diagram represents the concept with the extent consisting of all objects reachable from the circle by descending paths and the intent of all attributes reachable from the circle by ascending paths. Hence the circle at the top denotes the concept having all objects in its extent, and the circle at the bottom denotes the concept having all attributes in its intent. Top and bottom concepts may or may not be object or attribute concepts. To represent a many-valued context by a line diagram, we construct a "meaningful" formal context from the given many-valued context. This process is called *conceptual scaling* in FCA (see Ganter and Wille, 1989, 1996; Wolff, 1994) and is a generalization of the construction of the so-called indicator matrix in MCA (e.g., Greenacre, 1994) and the process of coding in multivariate analysis (see Gifi, 1990).

4 Data Representations in Correspondence Analysis

4.1 A Common Background for CA and FCA

To have a common background for both CA and FCA, we discuss briefly the meaning of some fundamental key words in CA. *Categorical data* are obtained in the process of describing some part of the "reality" by classifying "objects" with respect to certain "aspects" into "categories." Each aspect (e.g., age) has several categories (e.g., age groups) into which the objects are classified. The categories of an aspect are often ordered in a certain hierarchy. The two most useful, very simple, and extreme types of orders are "chains" and "antichains." In a chain (e.g., the chain of ages from 0 to 100 with the usual order) any two elements are comparable; in an antichain any two elements are incomparable in the given order relation (e.g., the antichain of the age groups $[0, 17]$, $[18, 64]$, $[65, 100]$ regarded as sets with respect to the order of set inclusion). In conceptual scaling the corresponding scale types are called one-dimensional ordinal scales and nominal scales. The classification of objects into categories of several aspects is usually described in a "data table," in which the rows are indicated by names of the objects, and the columns by names of the aspects; the cell (i, j) in row i and column j is filled with the name of the category into which object i is classified with respect to aspect j, and this cell is empty if object i is not classified into a category of j. The mapping of the objects to the categories of a given aspect is usually called a "variable" in statistics. It is obvious that categorical data can be represented by many-valued contexts. The hierarchy of categories is represented in the concept lattices of the conceptual scales.

Contingency tables are constructed from a many-valued context without missing values by selecting two attributes ("variables") a and b and by indicating the rows and the columns of the contingency table with the names of the values of attributes a and b, respectively. The entry in cell (x, y) is defined as the number of objects g in the many-valued context such that $a(g) = x$ and $b(g) = y$. Contingency tables do not represent the hierarchies of the categories of a and b, in contrast to FCA, in which both hierarchies and the contingency numbers are represented in nested line diagrams (see Wolff, 1994) of the formal context obtained from the many-valued context of the data table restricted to the categories a and b by scaling a (respectively b) with a scale for the hierarchy of a (respectively b). The same holds true for multidimensional contingency tables with more than two categories. They represent only the numbers of objects of the contingency classes of the corresponding concept lattice but not its conceptual structure.

Matrices are special many-valued contexts without missing entries. If all matrix entries are real numbers, one can use matrix algebra, but then the question arises of whether the algebraic operations have a meaning. To circumvent these problems (which are discussed from a general viewpoint in measurement theory by Krantz *et al.*, 1971), CA does not work with the given many-valued context directly, even if it is a real matrix. Instead, new matrices are generated in which all entries have the same meaning as a frequency or all entries are values of binary (dummy) variables. This leads to two main strategies in CA: simple CA, which works with contingency tables, in which all entries have the same meaning of frequencies, and MCA, where the given many-valued context is transformed into an indicator matrix Z of zeros and ones. This transformation is exactly the same as the conceptual scaling of this many-valued context with nominal scales for each many-valued attribute.

In its typical form CA generates a map in the usual Euclidean plane from a two-dimensional contingency table. The main steps in the construction of this map are described in the introduction to Part 2 later in this book.

MCA is the application of CA to the indicator matrix Z. As an example we choose the data table given in Table 1 and apply the nominal scaling to these data. This results in a table with four rows for the persons and $7 \times 6 = 42$ columns for the attributes. Table 4 shows the first six columns of this indicator matrix corresponding to the peaceful–conflicting attribute. Please note that the full table contains many empty columns, all of which have to be removed before CA can be applied. The asymmetric map of the nominally scaled Table 1, with attribute points at the barycenter of their object points, is shown in Figure 5.

The most remarkable impression in Figure 5 is that the points IDEAL and MOTHER coincide, hence seem to have exactly the same attributes, whereas they have

Table 4: The 'peaceful-conflicting' part of the nominally scaled Table 1

	pe.1	pe.2	pe.3	co.4	co.5	co.6
SELF	1	0	0	0	0	0
IDEAL	0	0	0	0	1	0
FATHER	0	0	0	1	0	0
MOTHER	0	0	0	0	1	0

the same values only at four of the seven constructs. Also the points of the attributes lively 2 and lonely 1 coincide, but they are the projections of two different points in the multidimensional space. In addition, MCA places the attributes loved 5, resolute 1, and theoretical 6 at the same projected position, even though they are different points in the full space. This can be checked without using the multidimensional space by looking at the line diagram in Figure 6, which represents the same indicator matrix without loss of information.

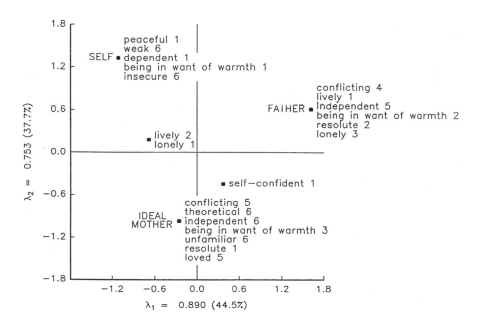

Figure 5: The asymmetric MCA map of the nominally scaled Table 1.

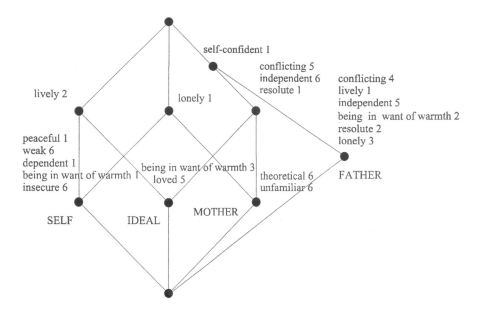

self-confident 1

conflicting 5
independent 6
resolute 1

conflicting 4
lively 1
independent 5
being in want of warmth 2
resolute 2
lonely 3

lively 2

lonely 1

peaceful 1
weak 6
dependent 1
being in want of warmth 1
insecure 6

being in want of warmth 3
loved 5

theoretical 6
unfamiliar 6

FATHER

SELF   IDEAL   MOTHER

Figure 6: Line diagram of the information-preserving nominally scaled Table 1.

5   Conclusion

In contrast to CA, the data evaluation in FCA starts with a many-valued context given by the original data and represents an ordinal structure on the values of each many-valued attribute by the scaling procedure. The graphical representation of the line diagrams of the concept lattice of the derived context contains all the information of the derived context. Hence FCA has an exact graphical data representation in contrast to the metric approximation of the data in CA. The exact representation has the disadvantage that even small many-valued contexts may have concept lattices with thousands of concepts. Then the main strategy is to use first a very rough scale and later a finer one. The main technique is the use of nested line diagrams that can be automatically generated from the small diagrams of the scales.

From our experience with both methods, we suggest using FCA for data with a small number of many-valued attributes. For data with more than 20 many-valued attributes one should first apply MCA to the suitably scaled context to find some interesting attribute clusters that may serve as the starting point for a data analysis with FCA.

Software Notes: The Programs TOSCANA and JOSICA

TOSCANA is a conceptual data management and retrieval system that uses the relational data base system MS-ACCESS. It generates nested line diagrams from a conceptual file containing the information about the scales. The conceptual file is generated with the program ANACONDA. Both programs are available from the ErnstSchröderZentrum für Begriffliche Wissensverarbeitung e.V., Schlossgartenstr. 7, D-64289 Darmstadt. E-mail: esz@mathematik.th-darmstadt.de.

JOSICA is a GAUSS program written by Siegfried Gabler that enables the user to run simple, multiple, and joint correspondence analyses. All computations and graphics in this chapter concerning correspondence analysis are generated by JOSICA. For more information one may contact the second author.

Chapter 8

The Z-Plot: A Graphical Procedure for Contingency Tables with an Ordered Response Variable

Vartan Choulakian and Jacques Allard

1 Introduction

Analysis of contingency tables with an ordered response variable has received much attention during the past two decades. Two widely known and used models are McCullagh's (1980) proportional odds model and Goodman's (1979, 1985) R or C association models. To take into account the ordinal nature of the response variable, two main approaches are used: the first one consists of modeling the empirical distribution function as done in McCullagh (1980); the second one consists of assigning scores to the categories of the ordinal response variable as done in Goodman (1979). The graphical procedure proposed in this chapter combines both approaches. The Z-plot was proposed by Choulakian *et al.* (1994) in the context of goodness-of-fit statistics. We use it as a preliminary aid to screen the data, which is a first step before applying a formal statistical analysis.

2 The Z-plot

Let $N = \{n_{ij}\}$ for $i = 1, \ldots, I$ and $j = 1, \ldots, J$ be a two-way contingency table, where the column variable is an ordinal response variable and the rows represent I different groups, I different time periods, or a combination of some explanatory variables.

Notice that the column variable j is a specified score as in Goodman's R association model. The row variable i need not be a score. Let us define $n_{i.} = \sum_{j=1}^{J} n_{ij}$, $p(j \mid i) = n_{ij}/n_{i.}$ the conditional relative frequency of the jth column category given the ith row, and $P(j \mid i) = \sum_{m=1}^{j} p(m \mid i)$, the conditional empirical distribution function given the ith row. Let us also define $n_{.j} = \sum_{i=1}^{I} n_{ij}$, $n_{..} = \sum_{j=1}^{J} n_{.j}$, $p(j) = n_{.j}/n_{..}$ the marginal relative frequency function of the ordinal response variable, and $P(j) = \sum_{m=1}^{j} p(m)$, the marginal empirical distribution function of the ordinal response variable. $P(j)$ will be used as a reference or baseline distribution. We note that the reference distribution could be any other empirical distribution function, such as the uniform $P(j) = j/J$. Finally, define $Z(j \mid i) = P(j \mid i) - P(j)$. Notice that $Z(j \mid i)$ is based on the empirical distribution function as in McCullagh's proportional odds model. The Z-plot is based on plotting for a fixed $i = 1, \ldots, I$, $Z(j \mid i)$ for $j = 1, \ldots, J - 1$, and interpreting the resulting curves. Two kinds of information can be obtained from $Z(j \mid i)$. First, for an i fixed, $Z(j \mid i)$ shows how the ith row behaves with respect to the reference distribution. In particular, if $Z(j \mid i) \geq 0$ for $j = 1, \ldots, J - 1$, it means that the reference distribution stochastically dominates the conditional distribution of the ith row, which in turn implies that the quantiles and the mean of the ith row are less than or equal to the corresponding quantiles and the mean of the reference distribution. The opposite interpretation is obtained if $Z(j \mid i) \leq 0$ for $j = 1, \ldots, J - 1$. Second, let us fix two rows i and i' and consider $D(j \mid i, i') = Z(j \mid i) - Z(j \mid i') = P(j \mid i) - P(j \mid i')$ for $j = 1, \ldots, J - 1$, the difference between two Z's. $D(j \mid i, i')$ measures the difference between two conditional empirical distribution functions. If $D(j \mid i, i') \geq 0$, then the empirical distribution function of the i'th row stochastically dominates the empirical distribution function of the ith row; and the opposite happens if $D(j \mid i, i') \leq 0$.

Goodman's R association model and McCullagh's proportional odds model imply the stochastic ordering of the rows, that is, $D(j \mid i, i') \geq 0$ [or $D(j \mid i, i') \leq 0$]. Therefore, if the Z-plot does not reflect this, then these simple models do not describe the data well and more complex models, such as Goodman's R + RC model or McCullagh's "nonlinear" model, should be fitted to the data. Now, let us present some examples.

3   Examples

3.1   Opinion of Youths on Military Service

Table 1 is a contingency table of order 7×4, taken from Gilula and Haberman (1994). It represents the opinion on military service of a sample of youths aged 14–22 in the United States who participated in the National Longitudinal Survey of Youth from 1979 to 1985. The data represent part of a panel study. One of the aims of the study was to see how the opinion of youths developed with time. Figure 1 represents the Z-plot of the data. It is evident that the curves representing the years 1979 through 1985 are almost ordered, and during this time period, on the average,

Table 1: Evolution of the opinion of a sample of youths on military service

	Response			
Year	Definitely good	Probably good	Probably not good	Definitely not good
1979	1196	4,966	1370	578
1980	930	5,441	1193	546
1981	1048	5,693	967	402
1982	1049	5,737	942	382
1983	1208	5,898	735	269
1984	1125	5,902	772	311
1985	1143	5,996	703	268
Total	7699	39,633	6682	2756

the opinion of the sampled youths changed from "not good" to "good." We note that the conditional distributions of the middle years (1981 and 1982) are close to the reference distribution, because the conditional distributions are stochastically time ordered.

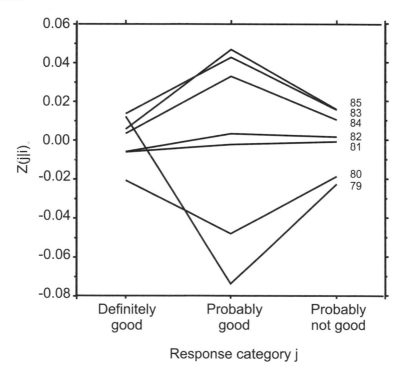

Figure 1: Z-plot of evolution of the opinion of a sample of youths on military service.

Table 2: Response frequency in a taste-testing experiment for five treatments

			Response			
Treatment	Terrible	2	3	4	Excellent	Total
1	9	5	9	13	4	40
2	7	3	10	20	4	44
3	14	13	6	7	0	40
4	11	15	3	5	8	42
5	0	2	10	30	2	44
Total	41	38	38	75	18	210

3.2 Taste-Testing Experiment

The data in Table 2 are taken from Bradley *et al.* (1962) and give the response frequency of judges in a taste-testing experiment. The five possible responses are on an ordered scale from terrible (1) to excellent (5), and the rows represent five unordered treatments. Figure 2 presents the Z-plot. We see that the curves are not ordered, which implies that McCullagh's proportional odds model does not fit the data well, as found by McCullagh (1980), who proposed a more complex "nonlinear" model to fit the data set.

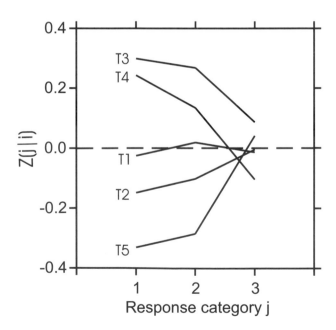

Figure 2: Z-plot representing the results of a taste-testing experiment for five treatments.

3.3 Severity of Persistent Wheeze According to Age, City, and Maternal Smoking

Table 3 is a four-way contingency table of order $3 \times 3 \times 4 \times 2$, taken from Stram *et al.* (1988). These data are from a panel study on the effects of indoor and outdoor air pollution on respiratory health. Indoor air pollution is represented by the maternal smoking (S) level, having the three categories Sl (less than 0.5 packets a day), S2 (from 0.5 to 1.5 packets a day), and S3 (more than 1.5 packets a day). Outdoor pollution is represented by the two cities (C), where the first category is Kingston-Harriman in

Table 3: Severity of persistent wheeze according to age, city, and maternal smoking status

Maternal smoking (packets a day)	Age	Severity of wheeze		
		1	2	3
Kingston-Harriman, Tennessee				
Less than $\frac{1}{2}$	9	418	70	66
	10	465	83	67
	11	458	64	64
	12	467	71	59
$\frac{1}{2}$ to $1\frac{1}{2}$	9	168	46	40
	10	177	43	35
	11	184	48	41
	12	172	45	36
More than $1\frac{1}{2}$	9	41	17	16
	10	64	9	15
	11	72	19	11
	12	35	12	16
Portage, Wisconsin				
Less than $\frac{1}{2}$	9	622	113	77
	10	788	104	91
	11	750	87	99
	12	652	56	65
$\frac{1}{2}$ to $1\frac{1}{2}$	9	225	48	22
	10	251	45	37
	11	250	35	32
	12	209	24	31
More than $1\frac{1}{2}$	9	46	14	8
	10	49	15	11
	11	36	17	16
	12	56	12	11

Tennessee, a highly polluted metropolitan area, and the second category is Portage in Wisconsin, a city with a relatively low level of pollution. The response variable is the occurrence of a persistent wheeze in the previous year, graded in severity as 1 (none), 2 (only with colds), or 3 (apart from colds). The data are gathered at examination of children of ages (A) from 9 to 12. In this example, for fixed categories s, c, and a of maternal smoking, city, and age, $Z(j \mid s, c, a) = P(j \mid s, c, a) - P(j)$ for $j = 1, 2$.

Figure 3 presents six groups of parallel Z-plots, the last three pertaining to the polluted city of Kingston-Harriman and the first three pertaining to the relatively less polluted city of Portage. For the city of Kingston-Harriman, we see that the persistent wheeze deteriorates as the maternal smoking level increases and the three maternal smoking levels are clearly separated. For the city of Portage, the effects of the first two levels of maternal smoking are less distinguishable, and the persistent wheeze deteriorates when the maternal smoking level increases to S3 (more than 1.5 packets a day).

To compare the cities, let us fix the smoking level. For low-level maternal smoking, the figure shows a slight difference in wheezing level between the cities. For the medium level of maternal smoking, the difference is greatest. Finally, for the highest level of maternal smoking, the figure does not suggest a difference in wheezing level.

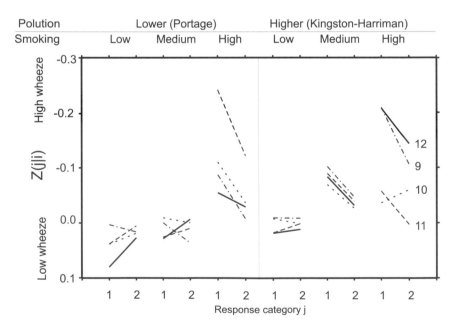

Figure 3: Parallel Z-plots for the severity of persistent wheeze according to age, city, and maternal smoking status.

We also notice that the figure does not suggest an age trend. However, we note that, during the 4 years, the variability of the persistent wheeze relative to age is greatest in both cities when the maternal smoking level is more than 1.5 packets a day (S3).

4 Conclusion

The Z-plot is a simple graphical procedure for visually displaying a contingency table having one ordered response variable. It is used as an exploratory data analytic tool to screen two-way and multiway contingency tables, but no formal inferential conclusion can be deduced from it. However, it may help to choose between formal statistical approaches, such as Goodman's R or R + RC association model, McCullagh's proportional odds model, or more complicated models for ordered categorical data that are available to analyze such data sets.

PART II

Correspondence Analysis

The following 14 chapters are concentrated on the theory and practice of correspondence analysis (CA). On the more practical side, we will see CA used as a method for interpreting political data in countries as different as Bulgaria and Luxembourg; as a method for analyzing textual data to understand American social attitudes and attitudes of French pupils to mathematics; and as a sociological tool in mapping themes used in political campaigns in Germany and comparing groups of visitors to art exhibitions in Vienna, Hamburg, and Paris. On the more theoretical side, we have a panorama of chapters that deal with aspects of interpretation, diagnostics, missing data, and three-way and nonsymmetric forms of CA. These theoretical chapters all contain substantive applications as well, again in a variety of contexts in many different countries: for example, the French Worker Survey, the German General Survey Program, the Canadian National Election Study, changes in the workforce in the Languedoc–Roussillon region of France, a survey of British premarital and extramarital relationships, and a survey of Italian parents of adopted children. Clearly, CA has come a long way since it was called a "neglected multivariate method" in the 1970s.

The variety and richness of CA can be attributed to several factors. First, it has been rediscovered and developed in various forms by researchers in different countries and has a history marked by cultural richness and diversity (see Nishisato, 1994, for a comprehensive historical overview). Second, its powerful visualization capability, based on a simple and familiar geometric paradigm, finds substantive applications in discovering the dimensions of "social space" (see Bourdieu, 1979, 1984), the scales of psychological traits, environmental gradients in ecology, and time ordinations in archeology. Third, the method is applicable to different forms of categorical data: counts, preferences, ratings, and zero/one "dummy" variables, which make it a versatile technique in many areas of research.

Although difficult to read for the uninitiated, the book by Benzécri and collaborators (1973) is a rich source of basic results and applications of CA. A simple introduction to CA is given by Greenacre (1993) in the form of structured, nontechnical explanation aimed specifically at an audience interested primarily in practical aspects. Different approaches to the theory and practice of CA can be found in Lebart *et al.* (1984), Greenacre (1984), Gifi (1990), and Nishisato (1996). The present book's antecedent, edited by Greenacre and Blasius (1994a), gives a thorough explanation of the simple and multiple forms of CA, including a step-by-step guide to all the computations and many different applications in a social science context. The following 14 chapters can be considered to be a continuation of that book and reflect developments in the subject over the past four years.

In a nutshell, CA may be described as a special type of principal component analysis of the rows and/or columns of a table, especially applicable to cross-tabulations. The most typical result of CA is a planar map on which each row and each column are depicted by a point. It is the values in each row or column relative to their respective row or column totals that are analyzed and displayed, and these vectors of relative values are called "profiles." A standardization is introduced consistent with the assumption that the elements of each profile have variances proportional to their respective means. This assumption engenders what is called the "chi-squared distance" between profiles and gives CA its particular algebraic and geometric properties, notably its symmetric treatment of rows and columns. Indeed, the results of CA are unaltered by transposing the data matrix. This property has caused a good deal of controversy, because there are only few occasions when the rows and columns are considered to be completely interchangeable. In a typical sociological cross-tabulation, for example, the rows might be family status groups used to explain the difference between different attitudes as columns, for example, attitudes to abortion, and this immediately implies an asymmetry in the two modes of the table. Nevertheless, it is true that for many cross-tabulations it makes sense to interpret both the row percentages and column percentages of the table, that is, the row and column profiles, in which case the symmetry inherent in CA is appropriate. A good example of this is a square table in which the rows and columns are the same categories applied to related groups, for example, a table of preferred leisure activities of husbands and wives, or the cross-tabulation of two multiresponse variables.

As in the case of principal component analysis, CA can be defined algebraically as the singular value decomposition of a centered and standardized form of the original data matrix. This is the matrix of standardized residuals that can be obtained by computing the difference between observed and expected frequencies, divided by the square root of the respective expected frequency. The left and right singular vectors provide the unscaled coordinates of the rows and the columns along respective principal axes, and the singular values define scaling factors for the rows and/or the columns on respective axes. The singular values are square roots of eigenvalues that decompose the total variation of the table, called the "total inertia." In the case of a cross-tabulation, the total inertia is equal to the usual Pearson chi-squared statistic divided by the total of the table.

In the so-called symmetric map, the row and column coordinates are both scaled by the singular values. The coordinates are called principal coordinates in this case, and both row and column points displayed in principal coordinates are projections of the row and column profiles onto the visualization space, usually a plane formed by the first two principal axes. In an "asymmetric map," either the row or the column coordinates are scaled by the singular values, leaving the other set of coordinates, called standard coordinates, unscaled. In a map these latter points are the projections of unit vectors, analogous to the projections of unit vectors onto principal components to obtain component loadings. Also similar to principal component analysis is the fact that the asymmetric map is a biplot where the set of points in standard coordinates is thought of as a set of directions onto which the other points (in principal coordinates) can be projected to approximate the values in the original matrix.

Apart from issues surrounding the geometric interpretation, there has been a great deal of interest in the extension of CA to situations in which more than two categorical variables are cross-tabulated. Two types of generalizations have been investigated, multiple correspondence analysis (MCA) and joint correspondence analysis (JCA) (see, for example, Greenacre, 1994). In both cases attention is usually focused on the positions of the category points and their joint two-way interactions. Although it is possible to represent a point for each individual response as well, this usually leads to so many points that it is not a practical strategy—often group mean points of individuals according to a relevant explanatory variable are interpreted instead. An interesting development, reflected by several of the following chapters, is in the interpretation of patterns of responses in MCA. This turns out to be useful in understanding the patterns of independence among the variables. Chapter 20 by Meulman and Heiser is a pioneering contribution to this development, as they show that appropriate forms of visualization of MCA results yield information about all the interactions in the multiway table, not just the two-way interactions as supposed up to now.

Let us look at the chapters in this part more closely. There are several chapters that apply CA to political data, and the first chapter in this part, Chapter 9, written by Ivailo Partchev, is about the Bulgarian political scene after the 1994 general elections and leading up to the presidential elections in October 1996. This chapter is a classic example of simple CA applied to a medium-sized cross-tabulation, just too large to be easily interpreted by scanning the values in the table itself. Here the visualization given by CA provides compact descriptions of the data in the form of maps of the candidates and their relationships to voter groups supporting the different political parties. The chapter includes the use of a clustering of the political parties in terms of their voters' profiles of support for the candidates, which neatly complements the CA results.

Chapter 10, by Bernd Martens and Jörg Kastl, looks at themes in the media leading up to and during the so-called "Superwahljahr" (super election year) in Germany in 1994. Here there are three categorical variables of substantive importance as well as four distinct periods of interest. CA shows how the thematic issues develop and dissipate during the election campaigns. This chapter also illustrates how a number of variables, four in this case, can be analyzed together by appropriate stacking of two-way cross-tabulations.

Chapters 11 and 12 deal with the analysis of textual data, by Ludovic Lebart and Mónica Bécue Bertaut, respectively. Lebart gives a comprehensive overview of ways of preparing and coding textual data for statistical analysis. He shows how effective CA is in visualizing similarities and differences between groups of texts in terms of their word or "segment" content, where groups are often defined by socioeconomic variables relevant to the study. In his example, open responses to a survey question concerning the most important things in life are compared among nine groups defined by combining three categories of age and three categories of education.

In Chapter 12, Bécue Bertaut applies lexicometric methods to a survey of French pupils' attitudes to mathematics, in which pupils have replied freely to a question about why they do or do not like mathematics. Again, the art of textual analysis seems to lie in the coding system adopted, and she proposes a more complex lexical unit called a "quasi-segment" composed of several words that are not necessarily consecutive. In this example the groups of interest are boys and girls at different levels of mathematical proficiency. Rather than directly comparing these groups, she performs a cluster analysis in the subspace defined by these groups and compares the resulting clusters.

Chapter 13, by Fernand Fehlen, is another application of simple CA to political data. Luxembourg has an unusual electoral system in which voters indicate their preferences in two different ways: first, a preferred list of candidates belonging to a particular party is chosen, giving each candidate on the list a single vote; second, each voter distributes a number of votes among the candidates of choice (the "panachage" system). CA shows visually which candidates deviate from their respective party positions and how their personal influence among the voters compares with their party's influence.

In Chapter 14, Christian Tarnai and Ulf Wuggenig use a combination of latent class analysis (LCA) and CA to analyze responses from visitors to major art exhibitions in three European cities: Vienna, Hamburg, and Paris. LCA is applied to data on basic value orientations to identify distinct classes of visitors in each city. Then the authors follow a classic strategy in interpreting the latent classes: they cross-tabulate the classes against a categorical variable that describes the visitors and use CA to visualize the tables. In addition, they add supplementary points to represent special groups in the art world, thereby enriching the interpretation of the maps.

Chapter 15, by Shizuhiko Nishisato, is the first of a set of chapters dealing with aspects of geometric interpretation. Nishisato starts by describing dual scaling of rank order data and stresses that to recover the ranks it is necessary to use the asymmetric map for the solution. When it comes to dual scaling of multiple choice data, that is, multiple correspondence analysis (MCA), he recommends coding the subject points by their response patterns, at least by the parts of their response patterns that are relevant to the interpretation.

Brigitte Le Roux and Henry Rouanet also look more closely at the interpretations of MCA in Chapter 16 and give a methodical way to interpret all the contributions to inertia by the variables themselves and by the different response categories. They give a thorough step-by-step interpretation of an MCA of four questions from the French Worker Survey, conducted in 1969, and interpret as far as the fourth dimension of the solution. They also indicate important response patterns on each dimension in order to interpret the positions of individuals with respect to the response categories.

In Chapter 17, Michael Greenacre looks at the geometric interpretation of the category points in MCA. He considers the biplot and unfolding models, based on the scalar products and distances, respectively, which underlie the CA interpretation. A measure of quality of this interpretation is proposed that measures the recovery of the scalar products and distances in a nonmetric sense and generalizes to the multiple case, both MCA and JCA. By using this type of nonmetric diagnostic, the differences between MCA and JCA results as well as between the biplot and unfolding approaches are eliminated and a more consistent measure of quality is obtained that is directly related to the interpretation.

In Chapter 18, Victor Thiessen and Jörg Blasius examine the degree to which responses in social surveys such as "no difference" or "unsure" can be regarded as substantive. They use MCA to interpret the structure of such responses to a set of questions in a survey of political parties in Canada, where the response "no difference" indicates that the respondent does not distinguish the political parties on a specific issue. They find a clear distinction between the nonsubstantive responses "don't know" and the other substantive responses, with the "no difference" responses uncorrelated with this distinction. They also display supplementary variables such as "political interest," "education," and "age" to explore their correlations with the substantive–nonsubstantive distinction, as well as the "no difference" categories.

Chapter 19, by André Carlier and Pieter M. Kroonenberg, is a thorough description of a three-way generalization of CA, with an application to a three-way contingency table of regions in southern France by occupational classes by time points, that is, a regions-by-occupations table observed at four different time points. The central idea is to decompose a measure of global dependence in the three-way table into components for the marginal (two-way) dependences and for the three-way dependence. A three-way matrix decomposition is fitted to account for these different components. Finally, visualization of different sources of dependence is possible thanks to a biplot of all three modes in a single map, where two modes (in this case, the regions and time points) have been combined to enable the biplot interpretation (between the region–time points and occupational class points).

In Chapter 20, Jacqueline J. Meulman and Willem J. Heiser come to grips with the elusive question: how can homogeneity analysis (alias MCA) shed light on interactions of order higher than two-way? As in Chapters 15 and 16, the answer lies in displaying the individual response patterns, which they call profiles, but also the centroids of the response patterns that represent various category combinations as well as the categories themselves. Using an example of four variables with two categories each, they show how structures in the odds ratios implied by different models of independence for the multiway table turn out as ratios of distances between the profiles and the centroids. Interactions can be studied by identifying additivity in the inertia contributions of the profiles and their centroids. The independence models can be diagnosed by certain patterns of parallel lines in the map.

The last two chapters of this part both deal with a variant of CA called nonsymmetrical correspondence analysis (NSCA). Whereas CA is invariant to the transposition of a data matrix, NSCA treats the sets of rows and columns differently, one being the predictor and the other the response. In Chapter 21, Simona Balbi gives an overview of NSCA and describes the geometric interpretation of NSCA maps as well as their differences from CA maps. She shows that the row and column points have different geometries in accordance with their nonsymmetric roles in the analysis. NSCA is also extended by three-way tables when two variables are considered to be predictors by combining the categories of these two variables to form a single one.

In Chapter 22, Roberta Siciliano and Francesco Mola use NSCA as a way to develop a prediction rule when there are several categorical predictors of a categorical response variable. Their approach is in the family of decision tree methods such as CART (see Aluja-Banet and Nafría, Chapter 5) and CHAID, where the prediction rule is determined by successively partitioning the observations according to the categories of a predictor. These authors propose splitting the observations into three groups at each stage, and NSCA is used to decide which predictor variable is chosen and how the splitting is performed.

Chapter 9

Using Visualization Techniques to Explore Bulgarian Politics

Ivailo Partchev

1 Introduction

Political studies provide a particularly good playground for data visualization techniques. Temptations to extract far-reaching conclusions from simple displays are less strong here than in, say, sociology or psychology, and we can use visualization techniques in the way they are intended: as efficient tools for the reduction of data.

Political life in a country is shaped by a small number of fairly invariant factors: the political parties; a not so numerous elite of leading politicians; several influential newspapers; and the electorate, classified into various social and demographic groups. Of course, we must ask many questions before we can relate these factors to one another. At which end of the political spectrum is a party situated? Who votes for it? What are the political messages of the leading newspapers? Many of these questions can be answered by running a few correspondence analyses on a survey that simply asked people about the newspapers they read, the politicians they trust, and the party they vote for. This may be the quickest way to become familiar with the political scene.

This chapter shows how visualization techniques helped to analyze Bulgarian public opinion data at different stages of the presidential race in 1996. Of the many possible data sets, we have chosen two simple cross-tabulations. In this way the political part can be kept short while still sufficiently detailed. In both tables, the intended vote for president was cross-classified against the self-reported actual vote in the most recent general election.

2   The Data

The first table comes from a public opinion poll conducted by mail in Bulgaria in February 1996. At that time, it was known only that the presidential election had to take place in autumn and that the president in office, Dr. Zheliu Zhelev, intended to run again. None of the political parties had come up with a candidate yet, and the survey tried to identify any names that were "in the air." To that end, an open-ended question asked, "Can you name a person who is most worthy to become president of Bulgaria?"

The sample and the questionnaire of the February survey were prepared by the author. A total of 2397 questionnaires were mailed to every eighth person over 18 years of age and born on a certain date; of these, 523 were returned complete. At a response rate of barely over 20%, the adequacy of the sample should be defended by practical experience rather than by the theory of random sampling. In situations in which results could be verified by a subsequent election, our practice has found similar samples to perform at least averagely compared with face-to-face interviews. Moreover, we are primarily interested in data structures, and there is no reason to believe that associations between variables could be seriously biased by nonresponse. The data have been poststratified to match self-reported voting in the latest general election in December 1994 with the actual election outcome.

The second table is based on a survey that was done in October 1996—a week before the real election—by the Laboratory of Political Behaviour, Sofia University. It used face-to-face interviews and a two-stage cluster design, sampling 160 clusters with a probability proportional to their size and then 10 persons out of each cluster. The response rate (80%) was much higher than that of the mail poll, but the question on intended voting had a flaw, reversing the percentages for the candidates who ended up second and third in the election. This may be due partly to the layout of the question but mostly to bad luck or some subtle bias in the selection of clusters. For this chapter, the table of counts has been adjusted to the real outcome of both elections by iterative proportional fitting (see Bishop *et al.*, 1975, pp. 97–102, and Friendly, Chapter 2 in this volume), a procedure that changes the marginals but preserves the association.

3   The Politics

Visualization techniques can indeed shed light on the political identity of parties and politicians, but some preliminary information will help in understanding results and evaluating the new insights gained.

3.1 Parties

The Bulgarian Socialist Party (BSP) is the direct descendant of the Bulgarian Communist Party, the undisputed ruler in an effectively one-party system that lasted from 1944 to 1989. Because the destruction of civil society after 1945 was more radical in Bulgaria than in, say, Poland or Czechoslovakia, it may be observed that the BSP stands farther apart from other parties, has undergone less change, and faces less political challenge than similar parties in other ex-communist countries. Apart from a short period in 1991–1992, it has had the greatest influence in ruling the country since 1989 and returned to power by winning 43.5% of the votes in the general election of 1994.

The Union of Democratic Forces (UDF) is a coalition of anticommunist parties that gained much popular support after 1989. It formed a short-lived government in 1991 after winning 34.4% of the votes in the general election but lost some of its backing later (24.2% in 1994). Still, it remains the second most influential political force in the country. The parties within the UDF are either new or belonged to the leftist parties that coexisted with the Communist Party after 1944 (the bourgeois parties having been immediately destroyed), until eliminated in their own turn. Although BSP propaganda tends to describe the UDF as extreme right, it is in fact supported by intellectuals, poorer people, and by those who managed—or still hope—to get back properties confiscated in the 1940s.

The Movement for Rights and Freedoms (MRF) is predominantly a party of ethnic Turks, although it claims to support human rights in general. The MRF became particularly important as a political *ballanceur* after the general election of 1991 (with 7.6% of the votes, it was the only parliamentary party except the BSP and the UDF) but fared less well in the election of 1994, winning only 5.4%.

The Popular Union (PU) is the union of the Democratic Party and one of the many Bulgarian agrarian parties. Both split from the UDF in 1991, made it back into parliament in 1994 with 6.5% of the votes, and are now attempting some political cooperation with the UDF.

3.2 Politicians

The president in office, Dr. Zheliu Zhelev, was born in 1935, joined the Communist Party in 1960, and was expelled in 1965 for "antileninism, antimarxism, antimaterialism, and antisovietism." Two years later, he wrote *Fascism*, a treatise on the totalitarian state. The book was published in 1982, then banned and confiscated, but the copies that circulated underground led many Bulgarians to reflect on the true nature of their society. Zhelev was among the founders of the UDF and became its first leader. He was elected president by parliament in 1990 and reelected by popular vote in January 1992, winning 44.7% in the first round and 52.8% in the runoff. Relations with the UDF withered in September 1992 after a press conference in which Zhelev criticized the UDF government. The government eventually failed a vote of confidence and resigned—a fact that the UDF somehow attributed to Zhelev's criticism. Apart from this episode, Zhelev persisted in his anticommunistic, free-market, pro-Western orientation and did not win any lasting sympathy among BSP supporters.

Simeon II became king of Bulgaria in 1943 with the sudden death of his father, the highly popular monarch Boris III. Exiled in the late 1940s, Simeon made his fortune as a financial and political consultant and remained a stable if not too visible influence in Bulgarian political life. His home in Madrid has been visited by politicians of all colors, and his popularity among the population became evident in May 1996, when his visit to Bulgaria made world news. Simeon did not try to run in the election and would have been prevented by the constitution from doing so. He appears here mostly because so many Bulgarians would name him their "most worthy candidate."

Georges Gantcheff is a populist figure that could be described as a Ross Perot without the money. His aggressive campaigning is not taken seriously by some but seems to appeal to many. Gantcheff appears to have easy access to television and uses it quite skilfully, attacking both the BSP and the UDF in some rather unrefined language. However, his role in politics is less than perfectly neutral. His party, the Bulgarian Business Block (BBB, 4.7% of the vote in 1994), consistently backs the BSP in parliament, ensuring a majority and safeguarding the BSP leadership against internal dissent. Gantcheff's attacks against the alleged "big deal" between the BSP and the UDF may appeal to many Bulgarians who failed to see the improvement in life conditions they had hoped for, but it is also in tune with the voices within the BSP who challenge the democratic change in principle. Gantcheff seriously hopes to become president and was third in 1992 (16.8%) and 1996 (21.9%); in either case, he carefully avoided telling his supporters how to vote in the runoff.

Against this colorful trio, the persons who actually contested the presidential election in October may pale by comparison.

The BSP initially put forward the man identified as the BSP favorite by our February poll: Georgi Pirinski, the minister of foreign affairs. His name was mentioned in about 12% of all returned questionnaires and by some 23% of BSP supporters. However, it turned out that Pirinski did not qualify as a "natural-born citizen" in the sense adopted by the constitution. He was replaced by Ivan Marazov, a professor of ancient cultures and minister of culture in the BSP government. Practically unknown to the general public until then, Marazov came in second in the October election, winning 27% in the first round and 40.3% in the runoff.

The democratic opposition opted for a strange version of the American primaries that took place in June and produced an even more unexpected result. The Popular Union, the MRF, and some smaller parties not represented in parliament supported the president in office, while the UDF put forward Petar Stoyanov, a fairly unknown lawyer who had served as vice-minister in the UDF government of 1992. Back in February, Zhelev's name had been mentioned in 14% of all questionnaires, and Stoyanov's in less than 1% (3% among UDF supporters). Capitalizing on widespread and mounting dissatisfaction with the worsening conditions in the country, Stoyanov eliminated Zhelev in the primary and was elected president in October: he won 44.1% in the first round and 59.7% in the runoff.

Finally, Alexander Tomov is a professor of economics and a former economic advisor in the last BCP government before 1989. He split from the BSP and ran in the general elections of 1991 and 1994 with various formations intended to fill up the missing political center with a social-democratic alternative, but he could never make the 4% threshold. Tomov tried his luck in the presidential election and got 3.2%.

4 February 1996

The cross-tabulation of the questions "Can you name a person who is most worthy to become President of Bulgaria?" and "For which party did you vote in the 1994 general election?" is shown in Table 1. About 6% of all responses on the most worthy candidates could not be classified into any of the seven categories in Table 1, and it was decided to drop them altogether rather than create a new group that would be sparse in numbers and scarce in meaning. This raised the percentages for other candidates—for instance, Zhelev now has 15% overall rather than 14%.

Table 1 shows column percentages because they are easier to compare across parties; correspondence analysis was actually applied to the table of counts. By choosing a display that has parties in principal coordinates and candidates in standard coordinates, we can concentrate on the same aspect of the relationship as in Table 1: the choices made by groups of a known previous electoral behavior (Figure 1).

The plot reveals a roughly triangular shape. One quadrant is empty, another is taken by the BSP and the BBB, a third one is left to the opposition parties, and the last one is clearly the province of those disaffected with politics, refusing to vote, voting for small or obscure parties, believing that nobody is worthy to be the president, or putting all hope in the king.

Table 1: Cross-tabulation of the questions "Can you name a person who is most worthy to become President of Bulgaria?" and "For which party did you vote in the 1994 general election?" Frequencies are tabulated as column percentages.

Candidate	BSP	UDF	PU	MRF	BBB	Other	Did not vote	Total
Zhelev	7	32	40	63	4	0	9	15
Simeon	3	22	28	3	4	33	13	14
UDF persons	1	17	12	3	8	3	6	6
BSP persons	56	3	4	13	21	14	20	27
Gantcheff	14	7	4	3	59	8	8	11
Nobody	8	11	4	6	0	14	17	11
Left blank	11	8	8	9	4	28	27	16
Total	**100**	**100**	**100**	**100**	**100**	**100**	**100**	**100**
$n =$	156	87	23	19	17	56	123	481

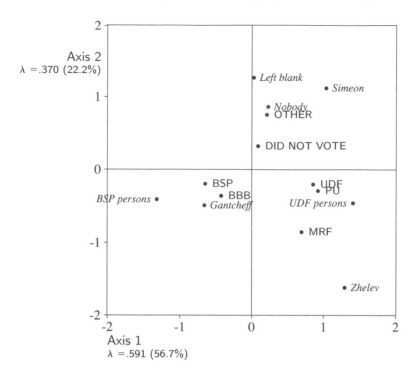

Figure 1: CA of the data in Table 1: parties in principal coordinates, candidates in standard coordinates.

The principal axes have a clear interpretation. Axis 1 is defined by the potential BSP candidate as mainly opposed to Zhelev, Simeon, and the UDF persons. Axis 2 seems to oppose the president to the king or, to put it another way, those who still believe in the existing political structure to those who have lost faith in it.

Keeping in mind that the Popular Union was the main political party that backed Zhelev in the primary, it is somewhat surprising to find PU supporters actually closer to UDF persons. In fact, MRF supporters seem to be the only group markedly in favor of Zhelev. This is explained by his personal involvement in defending the human rights of ethnic Turks, such as restoring their names after the unfortunate renaming campaign of 1984–1985, and restitution of properties lost during the mass emigration in 1989. Another explanation has to do with the prorepublican stand of MRF supporters, which places them far from the king.

The plot shows President Zhelev in an uncomfortable position. He competes with the still unknown candidate of the UDF for the anticommunist vote, but his placement along axis 2 puts him in a less favorable position to attract voters among the disaffected. This is in line with political logic and was confirmed by the primary. Widespread dissatisfaction with the conditions of life turned against the president in office, even if he had little influence in producing these conditions. The June primary was won by Stoyanov under the slogan of a fresh anticommunist start.

Zhelev, on the other hand, could enjoy the support of the MRF, which was important in the 1992 presidential election. Other opposition candidates would have to pay a higher political price for the same votes.

The quality of representation can be further raised to 93% of the total inertia by taking into consideration the third principal axis. This carries 13.9% of the total inertia and is defined primarily by one row and one column point: the Bulgarian Business Block and its leader, Georges Gantcheff. They are also the only points not sufficiently well represented on the two-dimensional plot. Because CA solutions are "nested," we may stay with the observations made so far while keeping in mind that the BBB and Gantcheff "stick out" of the plane of projection. This may be regarded as a statistical illustration of populism as a challenge to the established ways of doing and discussing politics.

Greenacre (1993, pp. 111–118) describes a way to cluster the row and the column categories in a cross-tabulation. The idea is to merge successively the two rows (or columns) that would lead to the smallest decrease in the chi-squared statistic (hence, preserve most of the association). The height of each merge in the dendrogram is determined by the difference in the chi-squared statistic before and after the merge.

Figure 2 shows the clustering of the columns of Table 1 (or rather, of the counts from which Table 1 has been calculated). The dendrogram has been enhanced by making the branches thick in proportion to the number of people represented.

The clustering translates the triangular structure in Figure 1 into a tree with three well-defined branches: the BSP and the BBB; the newly formed united opposition block, consisting of the UDF, the PU, and the MRF; and those undecided or disaffected.

Most elections are decided by the last of these three categories. Rather alarmingly for the opposition, this merges with the BSP in Figure 2. Of course, the process of clustering columns is different from the process in which people decide how to vote. The greater the distance from the most recent merge, the less likely are the persons in a branch to vote in the same way. Hence, the dendrogram does not necessarily say that the opposition parties must do badly in the presidential election: it would be safer to predict a highly contested election in which every vote will count.

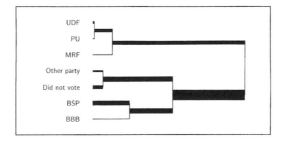

Figure 2: A clustering of the columns in Table 1.

5   October 1996

When the BSP regained executive power in 1994, it had won at least an extra 10% over its "normal" electoral support. During its first year in office, the socialist government managed to sustain modest economic growth. Evidently, this was achieved largely by pouring money into inefficient businesses, which eventually led to a collapse in the financial system and other mishaps too long to be discussed here. Our February data may still reflect relative satisfaction with the early record of the government; however, by June popular discontent was strong enough to turn the tables on Zhelev, a long-proven anticommunist who happened to be in office and was hence seen as belonging to the establishment.

In October, even the notoriously united BSP was shaken by strife and mutual suspicions as one of its top men, former prime minister Andrei Lukanov, was assassinated weeks before the presidential election. There seems to have been bad blood between Lukanov and other prominent figures in the BSP over financial interests, and the many nervous reactions to the killing exposed the magnitude of conflict within the party.

Table 2 shows the cross-tabulation of intended voting in the presidential election against self-reported voting in the general election of 1994. It is very similar to Table 1, except that it comes from a fully developed electoral situation—the data were collected a week before the election.

One row and one column have been dropped from Table 2. These correspond to the Bulgarian Communist Party, the BCP, which won about 1% of the vote in the 1994 general election, and its candidate, Vera Ilieva, who obtained 0.8% in the presidential election. The combination of low masses and perfect association would have made for a CA map with virtually two points: the BCP as an extreme outlier, and everyone else.

Table 2: Cross-tabulation of the questions "For which candidate do you intend to vote in the forthcoming presidential election?" and "For which party did you vote in the 1994 general election?" Frequencies are tabluated as column percentages.

Candidate	BSP	UDF	PU	MRF	BBB	Other	Did not vote	Total
Stoyanov	9	69	73	51	17	11	20	28
Marazov	45	2	5	2	4	2	5	17
Gantcheff	15	9	2	10	46	17	13	14
Tomov	1	1	2	1	1	8	1	2
Other	2	1	2	1	3	7	2	2
Will not vote	28	18	16	35	29	55	59	37
Total	**100**	**100**	**100**	**100**	**100**	**100**	**100**	**100**
$n =$	386	216	57	47	41	139	304	1190

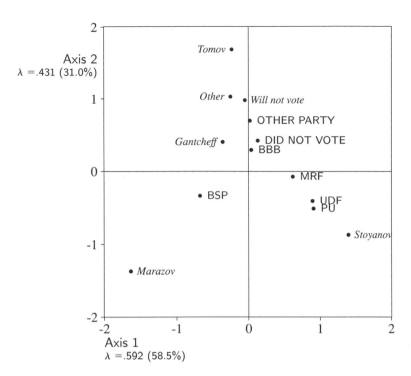

Figure 3: CA of the data in Table 2: parties in principal coordinates, candidates in standard coordinates.

The CA map of Table 2—or rather, of the counts from which it has been produced—is shown in Figure 3. The plot is broadly similar to Figure 1. The triangular shape is even more pronounced; this time, it shows Marazov (and the BSP) opposed to everyone else.

Again, axis 1 is defined by the polarity between BSP and the united opposition. This looks almost identical to axis 1 in the February plot—even the share of explained inertia is the same.

Axis 2 opposes the "protest vote" to those voting for the candidates of the two largest parties. The vertex is taken by Alexander Tomov, largely because of the relatively low mass. The important thing to note here is the general similarity between voting for Gantcheff, voting for Tomov or some other minor candidate, and not voting at all. These are all symptoms of dissatisfaction with the political establishment—the "blue–red mist," to quote from Gantcheff's favorite political imagery.

In CA of tables matching parties with politicians, MRF invariably produces a separate axis whenever its leader figures among the politicians. This is because of the ethnic factor, which is distinct from all other sources of political opposition. Traces of the same effect were evident in the CA from February, even if Zhelev is not an MRF man but just popular among MRF supporters. Now that the MRF does not have

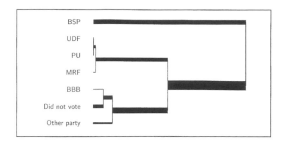

Figure 4: A clustering of the columns in Table 2.

a candidate of choice, it plots in the middle, which allows a clearer interpretation of axis 2.

Clustering of the data in Table 2 leads to Figure 4. The tree has three branches, which correspond to the triangular shape on the CA map but are different from the three branches observed in February. The three parties making up the united opposition—UDF, PU, and MRF—now merge much earlier: back in February, they formed a cluster more heterogeneous than the one produced by the merge of "Other party" with "Did not vote." The BBB, "Other party," and "Did not vote" also form a relatively homogeneous cluster with respect to their intended voting—the "vote of protest." Last but not least, the BSP now forms a cluster on its own, which is indicative of its isolation in current public opinion.

The latter could be demonstrated with regression analyses of election data (not shown here). According to these, Marazov was backed in the first round by about two thirds of those who voted for the BSP in 1994 and by virtually nobody else. Marazov's votes in the runoff could be explained statistically with his votes in the first round, about 35% of the votes cast for Gantcheff, and the votes for Vera Ilieva, the candidate put forward by the BCP.

6   Conclusion

Data visualization techniques have proved to be useful tools in following and analyzing Bulgarian surveys in 1996. The resulting displays were easier to understand and present to others than were the underlying cross-tabulations and provided many valuable insights into the general structure and some finer aspects of the data.

In the context of earlier work (Partchev, 1995), the plots presented here are quite typical and reveal some of the basic tensions in Bulgarian society. Comparing plots from early and late stages of the campaign, it is possible to identify stable and changing elements. Of the former, the most important one is the extreme polarity between the BSP and the larger opposition parties. Of the latter, it is worth mentioning the consolidation of the opposition, the crystallization of a vote of protest, and the deepening isolation of the BSP.

Chapter 10

Visualization of Agenda Building Processes by Correspondence Analysis

Bernd Martens and Jörg Kastl

1   Introduction

Public "themes" have become an important topic in the social sciences and especially in political sociology in the past decades. In the context of media research on agenda setting, important questions of research are the effects of media on political opinions, the development of themes in the public, the strategies of political actors when launching thematic issues, and the selection of different thematic topics by actors, the so-called framing of situations. By such framing processes, situations, events, and facts acquire a structure that is recognized in society. This framing also has consequences for the set of themes for which decisions of the political system are expected. Furthermore, certain themes become relevant or irrelevant and public debates can be described as struggles about the importance of issues. In recent sociological concepts of the public, the development and institutionalization of themes possess an important status. According to Lang and Lang (1981), we use the term "agenda building" for such processes of launching and stressing issues during which a hierarchy of relevant themes in the public evolves.

Prominent examples of such launching and rejection of thematic issues are election campaigns. The decision for and the maintenance of certain themes are apparently crucial points during election campaigns, at least to the way political actors see themselves. It is stressed by political managers that such themes should be "repeated continuously, but in an intelligent manner" (Bergsdorf, cited by Mathes and Freisens, 1990, p. 552; Radunski, 1980). These opinions are supported by scientific conjectures about a so-called priming effect according to which thematic priorities of political parties and candidates primarily influence the decisions during the elections. Budge and Farlie (1983, p. 84) even assert that the results of elections can be predicted by the structure of themes during election campaigns.

Public themes are extremely short-lived and certain trends of such issues can be distinguished (Luhmann, 1983). Therefore it can be supposed that the last weeks before an election are essential in order to exploit the priming effect. For example, Iyengar and Kinder (1987) emphasize that undecided voters can be influenced by short-term launching of controversial topics. The last weeks before an election should thus be the most important time period of the whole campaign. During this time significant activities of all relevant actors should be noticeable, in order to launch their specific themes.

In this context the year 1994 is an interesting case that can be used to prove the conjectures about the timing of thematic activities during election campaigns, because in 1994 an unusually high number of elections on all political levels occurred in Germany. The whole year and the sequence of elections were referred to in German newspapers as the "Superwahljahr" ("super election year"). It included nine communal elections, eight state elections, and the European parliament election and concluded with the national election in October 1994 [analyses and reports of the super election year are given by Buerklin and Roth (1994) as well as Falter (1995)].

Owing to the large number of elections, the interesting questions are: Did the intensity of thematic activities by political actors increase during the super election year in Germany? And, if that happened, when did such intensified activities start?

2   Data

This contribution is part of a research project dealing with the media coverage of the super election year. The main goal of this work was to give broad and feasible insight into the thematic development of the election campaigns. Fortunately, we were able to use a simple but presumably very meaningful indicator of such thematic trends during the time frame in question. The broadcasting station "Süddeutscher Rundfunk" in Stuttgart (Germany) allowed us access to its database "Venus." This database offers exhaustive information about different thematic issues, because all news items of important national and international news agencies are stored in it. Items can be retrieved via key words that occur in the texts. In our case, we chose all news items in which the words "Wahlkampfthem/a/en" (themes or issues in election campaigns) occurred.

We started from the assumption that the launching of themes would probably be more visible in items of news agencies than in the contents of other media (for example, articles of newspapers). This view is partially supported by the fact that most news items are related to concrete events that can be concisely described by

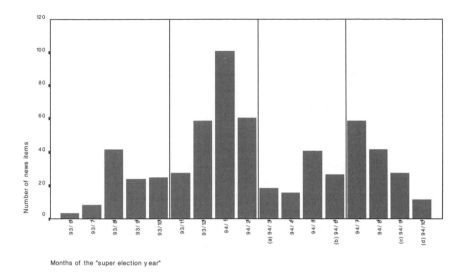

Months of the "super election year"

Figure 1: Distribution of the news items referring to issues in election campaigns stored in the database Venus, aggregated by months. The letters a–d refer to the points in time when elections took place. The numbers of elections were (a) one communal and one state election; (b) seven communal, one state, and one election to the European parliament; (c) three state elections; and (d) one communal, three state, and one national election.

key words. The texts cover the time span between June 1993 and October 1994. The retrieval yielded almost 600 news items (Figure 1).

Approximately one third of all news items are commentaries written by journalists of different newspapers. The collection of the commentaries is a special service of the news agencies. Journalists can be seen as actors on their own part in election campaigns, because they launch, take up, and stress issues by themselves. Thus, we regard the commentaries as indicators for the activities of media itself. The frequencies in these texts produced by journalists can be taken as a measure of the importance of a certain thematic issue. Because the database provides an exhaustive sample of available news agencies in Germany, it can be assumed that the collection of commentaries and the other news items are representative and valid for the issues in question and the specific time period. The remaining two thirds of the news items are related to statements on events, press conferences, interviews, or declarations.

In comparison with other sociological research about thematic developments and agenda setting processes, our empirical material is *special* in two respects:

- It is not confined to the last weeks before the national election. Instead, it includes a time span of 17 months of different election campaigns in Germany (the whole super election year and the months before it).

- The empirical material allows a "backstage glimpse" of media coverage, because we are dealing with the material that was used by newspaper and broadcast journalists. This material provides an overview of the thematic issues primarily before the selection processes of journalists take place. Therefore, we assume that it will be possible to detect thematic strategies used to launch or to reject thematic issues in the public debate by actors in the public and political sectors as well as those from the media.

In order to make the material feasible for statistical analyses, we followed classical content analytic approaches (Schrott and Lanoue, 1994), categorizing the textual data using a coding scheme. In the following, we will merely use three nominal variables of the whole scheme (thematic topics, assessments of the topics by the actor in question, and the type of actor) with altogether 20 categories (shown below). The categories comprise 12 thematic issues, 3 assessments given by the actors mentioned in the texts, and 5 types of actors. The abbreviations given here are used in Figures 2 and 3:

Thematic topics of the news items in question

   pds   Party of Democratic Socialism, its role in the political system of the Federal Republic of Germany (FRG)
   asyl   Right of asylum and foreigners in the FRG
   unem   Unemployment
   insu   Social insurance for nursing old and disabled people
   eco   Economic conditions and development concerning the FRG
   secu   Internal national security of the FRG
   abort   Legalization of abortion
   social   Social topics (for example, social welfare and conditions)
   scandals   Political scandals (for example, corruption)
   taxes   Taxes and economic situation of the state
   unific   Problems of the German unification process
   other   Other topics (for example, presidential election, armed forces)

Assessments of topics by actors

POS   The topic should become a theme during the election campaigns.
NEG   The topic should not become a theme in the election campaigns.
NEU   It will not matter if the topic becomes or does not become a theme during the campaigns.

Type of actor that occurs in the news

      PARTY   Members of a political party
      MEDIA   Journalists

> GOVERNMENT Members of government
> PARLIAMENT Members of parliament
> NGO Nongovernmental agencies

All variables refer to the main or first topic of the news item. The resulting data set has 596 cases.

With respect to agenda building processes, the dynamic developments of thematic issues are of special interest. The topics of the news items were coded by key words that occurred literally in the basic texts. This coding procedure seems to be advantageous for our purposes, because most news items can be well summarized by these concrete terms. The thematic categories that constitute the basis of the following analyses are confined to the first or the main topics of the news items. However, second or third themes that were not taken into consideration in the following analyses appeared in only 15% of the cases.

3 Results

The distribution of the news items over time is very uneven, as Figure 1 reveals. Four time periods can be distinguished:

1. In the months before the super election year (June to October 1993), 104 news items with the relevant key words occurred.
2. A first boom in media activities took place between November 1993 and February 1994. The number of news items rose to 249.
3. A period of stagnancy was indicated by rather small frequencies of issues related to election campaigns (March to June 1994; the number of cases is 103).
4. A second, only temporary peak of news items' occurrences can be detected during the summer, which was followed by a decline in launching themes that continued until the national election in October 1994 (the number of news items in the last four months is 141).

The main activities occurred at the end of the year 1993 and in the beginning of 1994. In the light of the conjectures about the timing of thematic activities during election campaigns mentioned earlier, this distribution of news items is unexpected, because it does not show an increase in thematic activities.

In contrast to other theoretical statements about agenda building and agenda setting in election campaigns, our empirical material indicates that political actors often avoid launching a theme positively but try instead to prevent a topic from becoming an issue in the campaigns. In almost 40% of all cases the assessment by the political actors mentioned in the news item is negative. The issue in question should not become an issue in election campaigns. Only in 32% of the cases is the reference to the theme positive in the sense that the actor is in favor of actively supporting the launching of a thematic issue. In 29% of the cases no explicit judgment about the theme as a topic in election campaigns is given. In these cases, the statement "x is an election theme during a campaign" is given without further assessment of it.

In light of these results, the assumption that political actors are keen on launching themes in a positive way seems to be too simple. It is obviously more realistic, at least in the case of the German super election year, that a process of selection of relevant themes took place during the run-ups to the elections. This impression is supported by the analyses described in the following, in which the thematic issues and the four periods of the super election year are simultaneously taken into consideration.

Figure 2 shows the plot of a correspondence analysis (CA) of the thematic issues by the four time periods. The first axis is determined primarily by the contrast between

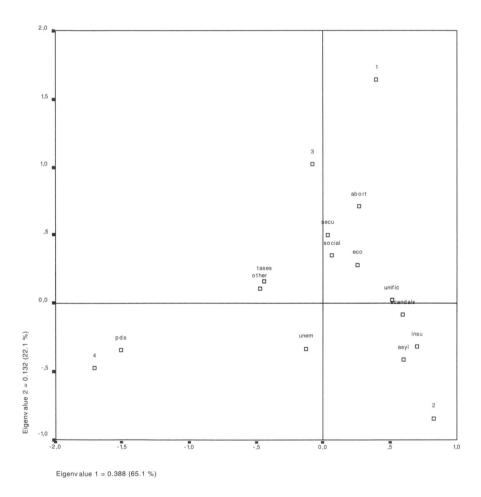

Figure 2: Correspondence analysis of the thematic issues by the four time periods (indicated by numbers), showing the asymmetric map.

periods 2 and 4. This axis represents 65% of the total inertia and two thirds of the axis' dispersion can be explained by period 4. The squared correlations of the two time categories with the first principal axis are 0.97 and 0.74, respectively. This contrast between earlier and later periods is associated with themes such as "asylum" (the squared correlation is 0.64), "social insurance" (0.81), "political scandals" (0.97), and "pds" (0.95). The theme "pds" makes a large contribution to the geometric orientation of the axis, and the last period is strongly associated with the "pds" theme.

The Party of Democratic Socialism (PDS) is the successor organization of the former communist party in the German Democratic Republic. Because of this history, the political assessment of the PDS is rather controversial in Germany and the support by voters differs extremely between East and West Germany. Also, in the beginning of the year 1994 the PDS participated in both East and West German state elections. However, the PDS emerged as a leading theme only at the end of the year—Falter (1995, p. 28) even suggests that this topic was decisive for the national election.

The second axis of the correspondence analysis represents 22% of the total inertia. It is essentially related to the first period (with a contribution of 47% and a squared correlation of 0.63) and the themes "abortion" and "internal national security" (squared correlations of 0.78 and 0.53, respectively).

The main result is that the last period before the national election was determined chiefly by the theme "pds." This was a rather formal issue, because it focused on election strategies and assumed intentions of the Social Democratic Party (SPD). It essentially dealt with the fear raised and stressed by the conservative party, CDU, that the SPD could build coalitions or would cooperate with the Party of Democratic Socialism. The news items did not treat the chances of the PDS to be successful in elections (which were indeed partially very small), but the supposed willingness of other parties to cooperate with it was thematized. It was not a controversy about different solutions of political problems. Themes that are connected with such problems could be detected in the beginning of the super election year. However, none of these themes survived during the course of the year 1994. Most of these early themes that are located at the right side of the first principal axis in Figure 2 are polarizing and politically controversial. It could be assumed that these themes would be the main topics of the following election campaigns during the course of the year—but this assumption is shown to be false by the data.

In a more detailed dynamic analysis, the connections between the thematic issues, the judgments given by the political actors, the actors themselves, and the periods of time were simultaneously taken into account. Owing to small frequencies of some categories, the data set comprises only seven thematic issues (pds, asylum, unemployment, insurance, economy, security, abortion) and 407 cases. The thematic issues, separated for each time period, form the rows of the input table, whereas assessment and actors form the columns of the table to be analyzed.

For example, the following four rows of the input matrix show the distributions of the thematic issue "unemployment" differentiated between the assessment by the actor, the type of actor, and the four periods:

Label	POS	NEG	NEU	PARTY	MEDIA	GOVERN-MENT	PARLIA-MENT	NGO
unem_1	0	5	1	1	1	4	0	0
unem_2	7	21	1	20	5	1	1	0
unem_3	2	3	0	3	2	0	0	0
unem_4	5	16	0	16	4	1	0	0

In order to give an impression about the development of themes in time, rows such as those just shown were concatenated for the seven topics. Thematic issues that did not occur during a certain period were excluded from the analysis. The resulting table with 26 rows and 8 columns was the input matrix of the next CA (Figure 3).

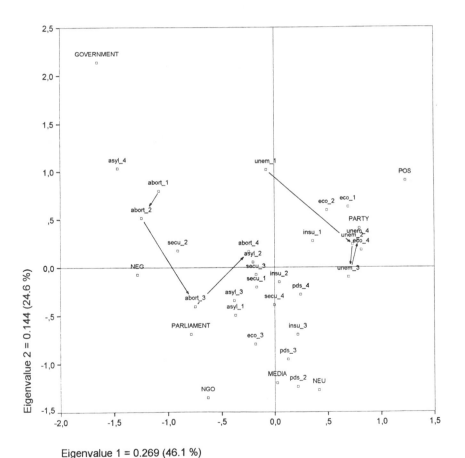

Eigenvalue 1 = 0.269 (46.1 %)

Figure 3: Correspondence analysis of the thematic developments during the time frame of the super election year. The numbers refer to the time periods. Abbreviations in capital letters refer to the categories of the column variables. Again, an asymmetric scaling is used.

The first two principal axes represent 71% of the total inertia. The first axis of the map is determined mainly by (1), the difference between governmental and political party actors who were mostly members of the SPD, the largest party in opposition to the federal government, and (2), by the difference between positive and negative assessments given by the respective actors. The inertia of the axis explained by the categories amounts to "GOVERNMENT" 0.21, "NEG" 0.34, "PARTY" 0.13, "POS" 0.27. Almost all the dispersion along the axis is therefore attributable to the dichotomy between these categories.

The vertical axis is highly correlated with the contrasting points "GOVERN-MENT" and "MEDIA" (squared correlations 0.40 and 0.73 and contributions to the inertia of 0.35 and 0.21, respectively). The category "neutral assessment" of the election themes also correlates with this axis (the squared correlation is 0.47 with an explained inertia of 18%). Other actors—especially the categories "PARLIA-MENT" and "NGO"—are represented poorly in the map (quality of 0.14 and 0.26, respectively).

The map in Figure 3 enables one to visualize the development of themes over the course of time. (A few paradigmatic developments are illustrated by arrows.)

1. Governmental actors were relatively successful in rejecting issues that were controversial topics in public debates (for example, the "abortion" and the "asylum" topics), which should not become themes during the election campaigns according to the opinion of political actors.

2. On the other hand, actors who appeared in their function as members of political parties were rather busy in launching the theme "unemployment," but this was not recognized by the media. If the association between this topic and the political parties is also taken into account, it becomes clear that the SPD tried to push this theme, which was evidently not recognized by the media. Nearly 60% of the unemployment topics refer to the SPD. The thematic restraint of the journalists seems to be a serious drawback that could not be compensated by the SPD.

3. The media itself stressed the "pds" issue, which became a prominent topic of the usual media coverage during the last weeks before the national election in October 1994. One third of the frequencies for this theme are attributable to the conservative party CDU, whereas two thirds are essentially associated with journalists. For this reason, it seems that the SPD was not able to build its own agenda but was forced to deal with topics that are not favorable to the party.

4. One general empirical pattern of the agenda building processes during the super election year is a remarkable dethematization. The "theme cycle" of the political debate about the "social insurance" is an example of this kind of pattern. It was an actual political topic during the first period. In later periods of the year it seems that it became only an issue for the media. However, in the last time period the topic totally vanished.

5. In general, polarizing and highly controversial themes (such as the questions of abortion, asylum for foreigners, and internal national security) have a strong

affinity to negative assessments given by political actors. These topics, which should obviously not become themes during election campaigns, were rather successfully avoided by the political actors.

4   Conclusions

Correspondence analysis was used to illustrate the development of thematic issues during election campaigns. The possibilities for visualizing the data provide good opportunities for displaying the evolution of themes over time. The distribution of the news items depicts a sequence of efforts to launch issues and thematic withdrawals (dethematization) over the course of time. Dethematization does not verify common models of agenda building and agenda setting processes during election campaigns. These models are based on the notion of increasing activities. Our exploratory analyses suggest that, at least according to the empirical example of the German super election year, a sequence of agenda building processes and more often an elimination of themes by political actors can be detected. Political struggles over thematic issues took place months before the national election, in the beginning of the year 1994. During this early period highly controversial themes appeared, but the political actors were very busy emphasizing that these issues should not become themes during the later election campaigns. At the end of the year, whether the SPD and the PDS would cooperate after the elections emerged as a theme. This theme is to a certain extent self-referential, because no political problems are dealt with, but a hypothetical outcome of the election is used in order to influence the voting.

One important finding is the active role of the media in these processes. The analyses reveal that the agenda building efforts of the SPD were not adopted by the media, and therefore it can be reasoned that this agenda building was not successful. On the other hand, the "pds" topic became a theme pushed by the media themselves in the last time period before the national election, and political actors were forced to deal with it.

Finally, our data show only short sequences of thematic polarization. Instead, attempts to reject certain topics were partly successful and pronounced time periods of dethematization can be seen. A thematic polarization was not the main strategy during the German super election year. On the contrary, the data reveal processes of rejecting certain themes as the essential strategy of political actors. It is an open question how these findings can be explained. One could imagine that possible explanations may be offered by the specific historical situation in Germany after unification, by a special German political culture, by structural developments of the political system, or by a more adequate sociological model of the functions that themes have during election campaigns.

Chapter 11

Visualizations of Textual Data

Ludovic Lebart

1 Textual Data and Meta-information

Our aim here is to show how correspondence analysis (CA) can help to visualize the profiles of a series of texts, whether they be literary texts, documents, or responses to open questions grouped into artificial texts (groupings based on age categories, profession, educational level, or any other relevant criterion). Which texts are most similar with respect to vocabulary and frequency of use of words? Which words are characteristic of each text, through either their presence or absence?

The reader may recognize these questions as the types of questions that may be answered with the CA of a *lexical table* (a table that cross-tabulates words and texts; see Bécue Bertaut, Chapter 12). In the case of responses to open questions in surveys, the approach we present here assumes that the responses have already been grouped according to socioeconomic variables, but we shall also briefly discuss other approaches.

Meta-information or meta-data is particularly abundant in the case of textual data. Briefly, meta-information is the information concerning a data matrix that does not appear in the matrix itself. This meta-information, which is relatively easy to formalize, is used routinely to check and clean files or to carry out consistency tests in processing survey data [see Hand (1992) and, in the context of information retrieval, Froeschl (1992)]. Attempts to formalize meta-information have been carried out by Diday (1992) in the framework of symbolic (as opposed to numeric) data analysis. The development of exploratory analyses and work done on databases have

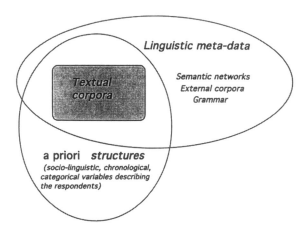

Figure 1: Textual data and external information.

accentuated interest in the concept of meta-information. In the case of textual data, every word is allocated several rows or several pages in an encyclopedic dictionary. Words belong to semantic networks that dictionaries of synonyms and partially automated morphosyntactic analyzers attempt to take into account. The rules of grammar obviously constitute basic meta-information (see Figure 1). The main issue is whether these different levels of meta-information are relevant to the problem under consideration.

On the other hand, in some information retrieval applications, for example, it is possible to work with key words that are treated as classical presence–absence qualitative variables and thus to construct tables that are wholly analogous to those encountered in other statistical applications. Then the data table obtained is no longer a text but a bundle of words, without order or syntax.

2   About Responses to Open Questions

Sociologists such as Lazarsfeld (1944) suggested the use of open questions in the preparatory phases of a study; their principal use is in developing a battery of response items for a closed question. There are three typical situations in which open questions must be used: to shorten interview time, to gather spontaneous information, and to probe and understand the response to a closed-end question (for example, the follow-up question, *why?*). The importance of the latter has been advocated by many sociologists who specialize in surveys, such as Schuman (1966), who alluded to international surveys in which problems of comparability and wording comprehension are acutely present. In international studies, it is important to know whether people interviewed in different countries understand the questions in the same way. In fact,

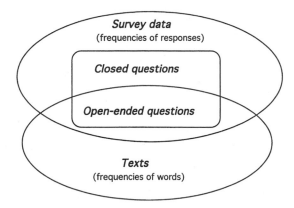

Figure 2: Status of frequency in the case of open-ended questions.

one could raise the same issue with respect to regional, generational, or sociocultural differences.

Responses to open questions, or *free responses*, are very specific elements of information. Observed lexical frequencies are artificial for the most part, because the same question is asked of hundreds or thousands of people. The juxtaposition of the responses results in a redundant text by construction where stereotypes are not uncommon. However, open questions become an essential part of questionnaires when the scope of research goes beyond a simple tally and when a complex and new topic is being explored.

A thousand responses to the question "Did you use your car yesterday?" constitute a text in which the words *yes* and *no* are predominant and the relative frequencies of these words have a simple interpretation that is familiar to survey specialists. Responses to an ancillary question "Why?" asked after a closed question have an intermediate status. Because they constitute 1000 identical stimuli, these responses can be stereotypical, or they can include original and unexpected contents and expressions. Even if expected word differences are taken into account, simple counts are notoriously inadequate.

On the other hand, when responses are grouped within categories (for example, age, gender, profession), comparisons of the mean lexical profiles of these categories can be productive. The most common approach consists of "closing" the open question a posteriori. This practice is called *postcoding*, a time-consuming but often irreplaceable technique. This technique, unfortunately, contributes to maintaining the dangerous illusion that closed questions asked during the interview are not different from questions that have been closed for coding purposes.

2.1 Open Questions Versus Closed Questions

Open and closed questions are not really comparable (see Schuman and Presser, 1981). An example of these two types leading to big differences is found in a survey on living conditions and aspirations of the French (Lebart, 1987) in which people were asked to give their comments on, "What types of people does the government spend the most money on?" The question was asked in open form in 1983, 1984, and 1987 and in closed form in 1985 and 1986 to a sample of size 2000 each year. The closed question was constructed on the basis of the main items given in the preceding years. The response item *migrant workers* obtained 4% in 1983 and 5% in 1984, when the question was open; 28% in 1985 and 30% in 1986, when the question was closed; and then 8% in 1987, when the survey reverted to an open question. The fact that the items were specified most probably had an effect on the range of responses considered "acceptable." Even though there was an increase in percentages of those choosing this item between 1983 and 1987 (4%, 5%, then 8%), the 2 years in which the question was closed gave percentages (28% and 30% in 1985 and 1986) whose order of magnitude is by no means comparable to those of the corresponding open questions. The closed form is, however, more valid if a recollection is involved in the question. It is well known that lists of items can play a positive role in such a case, according to an experiment performed by Belson and Duncan (1962), in which subjects identified newspapers read in the course of preceding days.

2.2   Grouping Responses

As free responses are entered into the computer in their original form, they can be matched with interviewees' demographic characteristics as well as their responses to closed questions. Then they can undergo data management procedures that are both useful and elementary, such as categorizations and groupings, without being altered.

For example, responses can be grouped by socioprofessional categories. Thus, responses given by farmers, housewives, workers, and executives can be examined separately. There might be categories or combinations of categories that are relevant with respect to each open question. By grouping responses within categories, "artificial speeches" are obtained that are all the more meaningful when the categories are carefully chosen. Reading and interpretation are made easier because repetitions and concentrations of certain issues appear within each category.

However, this rearrangement of raw information can be carried out in many different ways. The questions are how to group responses in a relevant manner and then how to facilitate interpreting the groupings thus generated.

First, one can use the criteria that are thought to be the most discriminating through prior knowledge of the theme being analyzed, with or without the use of cross-tabulations. If, for example, the questions are related to the evolution of the family, and if an age effect combined with a sociocultural effect is suspected, a variable combining age and educational levels can be used.

Second, a partition can be sought that is as universal as possible, within the limitations of sample size; this is the principle behind working demographics. The major characteristics that are judged to be relevant (for example, age, gender, educational level, region) are brought together through an automatic clustering technique into a single partition. This amounts to replacing several thousand individuals with about

30 or 50 groups that are as homogeneous as possible with respect to the foregoing criteria.

Third, a direct typology of responses can be carried out (without preliminary grouping) on the basis of their lexical profiles. Then categories that have the highest association with this typology can be selected before proceeding to groupings (see Lebart and Salem, 1994). To satisfy these needs, the textual material has to be prepared and segmented in such a way as to define new elements that are likely to be recognized and treated by computer software.

3 Choosing Units in a Text

Different counts have different degrees of relevance in each particular field of research. They also have different advantages as far as practical implementation is concerned. For instance, a researcher exploring a set of articles assembled from a database in the field of physics might require that the noun *particle* and its plural *particles* be grouped into the same unit, in order to be able to query all of the texts at once on the presence or absence of one or the other word. However, in the field of political text analysis, researchers have observed that singular and plural forms of a noun are often related to different, sometimes opposite, concepts (for example, the opposition in recent texts of *defending freedom and defending freedoms*, referring to distinct political currents). In the latter case, it could often be preferable to code the two types of elements separately and to include both in the analysis.

3.1 Analyses Based on Words

A particularly simple way to define textual units in a corpus of texts is to analyze words (or types). This approach can be used for various purposes depending on the objectives of the analysis: verification of data entry, inspection of vocabulary, or creation of a database for subsequent statistical comparisons.

To obtain an automatic segmentation of a text into occurrences of words, a subset of characters must be specified as *separators*. A series of characters whose bounds at both ends are separators is an *occurrence* (or a *token*). Two identical series of characters within separators constitute two occurrences (*tokens*) of the same word (*type*). The entire set of words (types) in a text is its *vocabulary*.

3.2 Lemmatized Analyses

In the lexicometric approach, the words resulting from automatic segmentation may be *lemmatized*. This means that identification rules must be established in order to group together words arising from the different inflections of one lemma (usual entry in a dictionary). The main steps in lemmatizing the vocabulary of a text written in English are to put verb forms into the infinitive, to put nouns into the singular, and to remove elisions. Although these steps are relatively easy in principle, the actual lemmatization of the vocabulary of a corpus involves some unavoidable problems that may be difficult to resolve.

3.3 Homography, Disambiguation

A systematic determination of the lemma to which each word belongs in a text often requires prior disambiguation, involving a morphosyntactic analysis and often a pragmatic analysis (see Kelly and Stone, 1975; Gale *et al.*, 1992; Charniak, 1993; Weischedel *et al.*, 1993). Some ambiguities might be the result of two words that happen to be homographs being inflections of clearly different lemmas (for example, *can* could be a verb and a noun). In other cases several derivations may exist from the same etymological source (the different meanings of the word *state*, for example). In some cases, ambiguities concerning the syntactic function of a word have to be removed, requiring a grammatical analysis. Some ambiguities of a semantic nature can be removed through simple inspection of the immediate context. Others require examining several paragraphs or even the text in its entirety. Sometimes ambiguities can exist between several meanings of a word.

4   Numeric Coding of Text

Computer-based processing of textual data is greatly simplified by giving a numeric code to each word to be used during calculations. This code is associated with each occurrence of the word. Codes are stored in a dictionary of words that is unique for each application. We illustrate our approach using the following open question, which was asked in a multinational survey conducted in seven countries (Japan, United States, United Kingdom, Germany, France, Italy, and Netherlands) in the late 1980s (for more details, see Hayashi *et al.*, 1992): "What is the single most important thing in life for you?" It was followed by the probe "What other things are very important to you?"

Our example is limited to the American sample of size 1563. Some aspects of this multinational survey concerning general social attitudes are described in Sasaki and Suzuki (1989).

Examples of answers to the first question were:

1. Family, being together as a family
2. Mother, money, peace of mind, peace in the world

4.1   Tagged Corpora

It is possible to obtain an automatic tagging of the words in such a text of responses. Some taggers provide an indication of the lemma with which each word can be associated as well as its grammatical category. In most cases the information furnished by categorizers must be subjected to careful checking before being used, because the process of automatic categorization may generate some errors.

The main grammatical codes used in this categorization are NN, noun singular; NNS, noun plural; NP, proper noun; DT, determiner; VB, verb, base form; VBD, verb, past tense; VBG, verb, gerund or present part; JJ, adjective; PRP, personal pronoun; RB, adverb; and IN, preposition or subordinating conjunction.

Examples of tagged responses, showing the grammatical category code for each word, are as follows:

1. Family/NN being/VBG together/RB as/IN a/DT family/NN
2. Mother/NN, money/NN, peace/NN of/IN mind/NN, peace/NN in/IN the/DT world/NN

Note that such lists of tagged responses provide the user with new categorical variables related to the same individuals. These variables can play alternately the roles of active and supplementary variables. They allow one to obtain a syntactic (or grammatical) point of view over the set of texts (see Salem, 1995).

4.2 Repeated Segments

Even after setting aside words with a purely grammatical role, the meaning of words is linked to how they appear in compound words or in phrases and expressions that can either inflect or completely change their meanings. For example, expressions such as *social security* and *living standard* have a meaning of their own that cannot be construed from the meaning of the words of which they are composed. It is thus useful to count larger units consisting of several words that could be analyzed in the same ways as words. These units are called *repeated segments* (Salem, 1984, 1987). Like the syntactic categories mentioned previously, these units can play the role of supplementary variables in the visualizations involving words. Such projection of segments as supplementary elements enables the reader to grasp the most frequent contexts of certain words.

4.3 Basic Lexical Tables

We saw that responses can be coded numerically in a way that is completely "transparent" to the user. The result of this numerical coding can take two different formats, coded in two tables **R** and **T**.

Table **R** has as many rows as there are respondents, say *n*. There can be missing responses, but it is convenient to reserve a row for each respondent to ensure easy merging with responses to closed questions given by the same individuals. The number of columns of **R** is equal to the length (number of tokens) of the longest response (i.e., the number of occurrences in this response). For individual *i*, row *i* of table **R** contains the addresses of the words that constitute his or her response, while respecting the order and the possible repetitions of these words. These addresses refer in the vocabulary that is inherent in the response. Table **R** thus makes it possible to reconstitute the original responses integrally. In practice, table **R** is not rectangular, because each row is of variable length. The integers of which table **R** is composed cannot be bigger than the size of the vocabulary, say V.

Table **T** has the same number of rows as table **R** but it has as many columns as the number of words used by all of the individuals. The cell defined by row i and column j of **T** contains the number of times word j is used by individual i in his or her response. This table of frequencies is called a lexical table. Table **T** can easily be derived from table **R**, but the converse is not true: information related to the order of the words in each response is lost in table **T**. Specific algorithms using **R** instead of the large sparse matrix **T** can lead to enormous computational savings (Lebart, 1982a).

4.4   Aggregated Lexical Tables

Isolated responses are often too sparse to be the object of direct statistical treatment, and then it is necessary to work on grouped responses. Let us designate by $\mathbf{Z}_q$ the indicator matrix with n rows and J_q columns that describes the responses of the n individuals to closed question q with J_q possible response categories, where the responses are mutually exclusive. In other words, each row of $\mathbf{Z}_q$ has only one 1 and $(J_q - 1)$ 0s. Table $\mathbf{C}_q$, obtained through the matrix product $\mathbf{C}_q = \mathbf{T}^\mathsf{T}\mathbf{Z}_q$, is a table with V rows and J_q columns whose general term c_{ij} is the number of times word i is used by the set of individuals having chosen response j. Each table $\mathbf{C}_q$ offers a different viewpoint, namely the viewpoint of the closed question q on the distribution of the lexical profiles of the responses to the open question being analyzed.

4.5   Frequency Threshold for Words

These comparisons of lexical profiles become meaningful from a statistical point of view only if the words appear with a certain minimum frequency. Frequency distributions of vocabularies are such that choosing a frequency threshold often drastically reduces the size of the vocabulary without reducing the size of the remaining corpus too much.

The counts for the first phase of the numeric coding were as follows. Out of $n = 1563$ responses, there were 13,999 occurrences (tokens), with 1378 distinct words (types). When the words appearing at least 16 times are selected, the vocabulary is reduced to 126 words, occuring 10,752 times in total.

Table 1 shows the alphabetical list of the 126 words that appear at least 16 times in the set of 1563 responses and their frequencies of occurrence. Note that graphical forms such as *I'm* and *don't* are considered words because the apostrophe is not designated as a separator in this example.

4.6   Grouping Responses

As an initial step it is appropriate to find groupings of responses that are pertinent to the phenomenon being analyzed. By grouping responses within categories, "artificial

Table 1: Words appearing at least 16 times (alphabetic order) in the 1563 responses to the open question

Word	Frequency	Word	Frequency	Word	Frequency
I	111	getting	29	of	289
I'm	17	God	64	on	31
a	254	good	422	other	24
able	46	grandchildren	52	others	29
about	22	happiness	228	our	44
all	40	happy	100	out	20
and	389	have	71	own	19
are	22	having	81	parents	18
as	36	health	794	peace	112
at	17	healthy	74	people	63
be	112	helping	24	personal	22
being	159	home	102	relationship	37
better	18	house	22	relationships	18
can	30	husband	69	religion	84
car	21	important	22	respect	19
care	37	in	128	safety	19
children	230	income	17	satisfaction	31
children's	18	is	48	secure	17
Christ	21	it	28	security	137
Christian	18	Jesus	24	self	28
church	77	job	209	so	18
comfortable	22	just	21	stay	21
comfortably	26	keeping	24	staying	24
country	22	kids	48	that	55
day	19	know	51	the	191
do	42	life	160	things	17
doing	25	like	24	time	50
don't	57	live	86	to	439
education	69	living	84	travel	18
enjoy	18	Lord	20	up	19
enough	59	love	53	want	28
faith	19	making	24	we	18
family	935	marriage	22	welfare	36
family's	27	me	34	well	53
financial	47	mind	50	what	36
food	23	money	199	wife	76
for	168	more	40	with	118
free	20	mother	17	work	108
freedom	63	my	1000	working	20
friends	197	myself	29	world	35
future	22	no	23	you	29
get	27	not	44	your	19

speeches" are obtained that are all the more meaningful when the categories are carefully chosen. However this rearrangement of raw information can be carried out in many different ways. In the particular case of our example, the individuals are grouped into nine subgroups that differ with respect to age in three categories [less than 30 years (denoted AGE1), between 30 years and 55 years (AGE2), over 55 years (AGE3)] and education at three levels [no degree or low (denoted E1), medium (E2), and high level (E3)].

The two combined criteria had the advantages of being common to the seven surveyed countries and having a straightforward interpretation (it is much more difficult, for example, to compare socioprofessional categories from one country to another). However, it must be kept in mind that this particular partition provides a specific (and not unique) viewpoint on the set of responses.

To read the information contained in this table effectively, the row profile and column profile tables are calculated, and the distances between words, on the one hand, and between age–education categories, on the other, are displayed. It is precisely the purpose of correspondence analysis to provide the user with such a dual visualization.

5   Correspondence Analysis of the Lexical Table

Figure 3 shows the plane of the first two principal axes of the correspondence analysis of the aggregated table C with 126 rows and 9 columns. The first two eigenvalues are 0.054 and 0.028, respectively, and account for 32.6 and 16.9% of the total inertia. In Figure 3, categories belonging to the same level of education are connected by a bold line, and categories belonging to the same age category are connected by a dashed line. The arrangement of the column points is remarkably regular: on the basis of purely lexical information (elements of column profiles), the composite character of the partitioning of the individuals into nine categories is recreated. Individuals with a higher educational level are situated toward the upper part of the graph; whatever their educational level, the older respondents appear along the right side of the horizontal axis.

Thus, these vectors that describe the frequency of 126 words (chosen according to a simple frequency criterion) for each category can reconstitute approximately the gradations of ages (within each educational level category) and the gradations of educational levels (within each age category). It is more difficult to obtain a clear-cut distinction between the first two age categories. However, within each age group, the level of education increases from the bottom to the top of the graph.

This visualization can be enhanced through further modulations of the original display: adjectives, verbs, and pronouns could be identified. The graphical display can also be enriched by identifying the words according to general semantic categories. For example, it appears that all words related to general concerns (country, others, world, religion, welfare) characterize older well-educated respondents located in the upper left part of the display, whereas words such as Lord, church, Christian are more frequently encountered in the responses of older less educated persons.

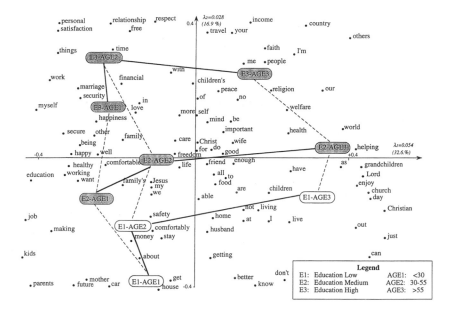

Figure 3: Correspondence analysis of Table **C** (126 × 9).

Words describing a secure and happy family life (marriage, secure, security, happy, happiness, comfortable, comfortably, safety, family) concentrate in the left-hand side of the map, with the younger age groups.

The automatic indexing of words and frequency computations purposely ignore much information of a semantic or syntactic nature that is available to any reader. Neither synonyms nor homonyms are accounted for. Applications of this type of analysis to large samples, such as the present study involving 1563 responses, show that these objections can easily be waived in the case of artificial texts constructed by juxtaposition of responses, where the main purpose is to find repeated elements.

In this statistical context, analyzing words often gives more interesting results than analyzing lemmas or groups of words established on a semantic basis. The words *happy* and *happiness* occupy similar positions in Figure 3 (upper left), which shows that it would have been possible to combine them beforehand for substantive reasons.

One may think that including function words such as *for, in, of,* and *as* burdens the analysis. In fact, these words appear to be significant in a CA only if their distribution is not uniform in the texts, in other words, if they are characteristic of some grouping of responses; if this is the case, it is interesting to place them among the other words. For instance, the word *as* is located in the right part of the display, close to the horizontal axis. It thus appears frequently in responses of older interviewed persons. Examples of such responses are "that we have as good of health as we have" and "to continue to work as long as I can."

Again, with analyses based on frequencies, finding repeated segments makes it possible to take into account occurrences of units that are richer at the semantic level than isolated words. The selection of *modal responses* (Lebart, 1982b; Lebart and Salem, 1994), which is discussed later, also answers several of the preceding objections by highlighting the most frequent contexts of some of these words.

An internal lemmatization procedure and elimination of the function words applied to the subcorpus make it possible to evaluate the stability of the structures obtained. The question is whether the observed pattern (i.e., in Figure 3, the relative positioning of the nine category points) depends on the presence of distinct inflections of the same lemma and of particular grammatical words. If that were the case, the categories would be distinguished primarily through their use of certain parts of speech and not solely through the content of their responses.

We thus eliminated the following function words: *a, and, at, for, in, of, on, the, to*. In addition, we concatenated into single units the following words: *be, are, is, being* into *be*; *live, living* into *live*; and so forth. The vocabulary is then reduced from 126 words to 101 "pseudolemmas." In this particular example, a description of the new reduced table through correspondence analysis produces a pattern of category points similar to that of Figure 3. The point of view selected for this experiment is to test the internal stability of the results. We may consider this transformation of the data set as a perturbation allowing the user to assess the patterns obtained.

We have seen that the category points are positioned in a way that respects the order of the age and educational level categories. This adds to the conjecture (but does not prove) that there is a connection between the categories and the content of the responses.

6   Characteristic Words and Modal Responses

6.1   Characteristic Words

It is useful to complement the visualizations provided by correspondence analysis with a few parameters of a more probabilistic nature: the *characteristic words*. These are words that are abnormally frequent (or abnormally rare) in the responses of a group of individuals (see, for example, Lafon, 1981).

A *test value* measures the deviation between the relative frequency of a word within a group and its global frequency calculated on the entire set of responses or individuals. This deviation is normalized so that it can be considered as an (asymptotically) standardized normal variable under the hypothesis of random distribution of the word in the groups. Under such a hypothesis, the test value lies between -1.96 and $+1.96$ with a probability of .95. However, since this calculation depends on a normal approximation of the hypergeometric distribution, it is used only when the counts are not too small.

For example, the most characteristic words of the category E1–AGE1 (lowest level of education, age less than 30) are (with the corresponding test values in parentheses) car (3.3), mother (2.9), house (2.6), job (2.4), money (2.3), parents

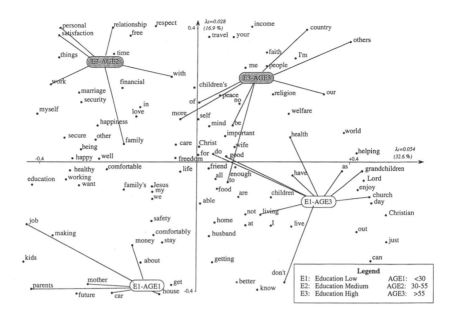

Figure 4: Some characteristic words from the margin Figure 3.

(2.1). For the opposite category E3–AGE3 (highest level of education, age over 55), the sequence of characteristic words is country (3.6), our (3.1), of (2.9), be (2.5), more (2.4), to (2.3), others (1.96). For illustrating the characteristic words we concentrate on four peripheral categories of Figure 3 (E1–AGE1, E3–AGE2, E1–AGE3, E3–AGE3). The words are connected to their respective categories in Figure 4.

On the one hand, there is a large compatibility between the proximities observed on the map and the links induced by the computation of the characteristic words (obviously, these characteristic words are not distributed at random on the display); we note that these characteristic words are situated around their corresponding categories, which is also consistent with the usual simultaneous representation in CA. On the other hand, many proximities on the map have no counterpart in terms of characteristic words; for instance, me, faith, people, religion do not specifically characterize the group E3–AGE3. In this sense, indicating characteristic elements complements the usual display of CA.

6.2 Modal Responses

A simple technique simultaneously addresses a number of objections that could be raised concerning the fragmentation inherent in any analysis that is limited to isolated words without placing them in their immediate context. This technique consists of the automatic selection of *modal responses*. There are two ways in which modal responses can be chosen: (1) on the basis of calculations that make use of characteristic elements and (2) on the basis of distance computations according to simple geometric criteria, for example, the chi-squared distance.

Modal Responses Based on Characteristic Elements   A modal response of a grouping is a response that contains, as much as possible, the most characteristic words of this grouping. For each grouping, these words are ranked by degree of significance: the greater the test value of a word in this ranking, the more significant it is. A simple empirical formula consists of associating with each response the mean test value of the words it contains: if this mean test value is large, it means that the response contains only words that are very characteristic of the grouping.

This calculation mode (criterion of characteristic words) has the property of favoring short responses, whereas the chi-square criterion described next tends to favor lengthy responses.

Modal Responses Based on Chi-squared Distances   These distances express the deviation between the profile of a response and the mean profile of the group to which the response belongs. The preferred distance is the chi-squared distance, because of its distributional properties. For each grouping, these distances can be sorted in increasing order. Thus the most representative responses with respect to the lexical profile, that is, those whose distances are the smallest, can be identified. Whatever the mode of calculation, several characteristic responses are printed out for each grouping. It is indeed highly improbable that there should exist among all of the original responses a single response that summarizes by itself all of the properties of a category.

Examples of modal responses for the group (E1–AGE1) are:

- my children, the car, the house, my family (my parents, grandparents, whole family)
- live well. Having house and car, job with a future where I can hope to earn more money
- my daughter, a decent home, money, decent neighbors, my husband, my mother, my cat, my nephew

Examples of modal responses for the group (E3–AGE3) are:

- my family, the welfare of our country, the welfare of the individual, that's all
- inner peace, family, health, moral stability of our country
- health, to know more, to be more compassionate, more understanding, more tolerant, more forgiving

We can find in these responses most of the characteristic words mentioned previously (car, house, money, job, parents, mother for the group E1–AGE1; country, our, of, to, more, be for the group E3–AGE3). Such listings of modal responses summarize each of the main themes for each category.

In summary, by combining the three approaches (visualization of proximities between words and categories through correspondence analysis, selection of characteristic words, selection of modal responses) we can obtain, without any need for preprocessing and without precoding, the main features of the differences between responses or texts.

Acknowledgments

The author wishes to express his grateful thanks to Professor C. Hayashi for the opportunity of working on the free responses of the Cross-Cultural Survey and to E. Berry for her help in the translation of this chapter.

Chapter 12

Visualization of Open Questions: French Study of Pupils' Attitudes to Mathematics

Mónica Bécue Bertaut

1 Introduction

Lexicometric methods, including multidimensional descriptive statistical methods such as correspondence analysis and cluster analysis, constitute a tool for analyzing textual corpora. Lexicometry concerns the count of lexical units and the distribution of these units in the various parts (or texts) of the corpus (see Lebart and Salem, 1994). The usual unit for dividing the textual chain is the word, defined as a succession of characters delimited by blanks or punctuation marks. Other units, such as a lemma or complex units (consisting of several words or lemmas), can be selected but must be invariant and identifiable without either ambiguity or the intervention of the researcher. In effect, the aim is to produce a formal treatment, without subjective interpretation prior to the analysis, enabling the text produced—not the text received by the reader—to be studied and to relegate subjectivity to the later stage of interpretation. This objective makes it impossible, at the initial stage, to take into account the "meaning" of the words, that is, the content of the texts. The procedure is systematic and requires counts to be exhaustive. Everything must be counted, as there is no way to know a priori which words are significant.

Descriptive statistical methods such as cluster analysis and correspondence analysis allow us to synthesize these counts from a large corpus of textual data in an exhaustive and systematic way (Benzécri *et al.*, 1981; Lebart and Salem, 1994). They offer visualizations of similarities or dissimilarities between texts and/or words and also of associations between vocabulary and authors' characteristics.

The application of the same methods to corpora of responses to open questions in surveys raises specific problems, because the texts analyzed in this way are usually very short (a few lines at the most for each individual). To obtain larger texts, individual responses are usually amalgamated according to characteristics of the individuals, which are known through their responses to closed questions. It is then possible to study the relationship between individuals' characteristics and their "language." However, there is still interest in studying the full spectrum of individual responses, and constructing typologies based on similarities and differences between individuals constitutes a powerful tool.

Several authors have attempted to apply cluster analysis to open questions (Haeusler, 1984; Establet and Felouzis, 1993). To reduce the dimensionality of the table, a correspondence analysis (CA) is first performed on the table of counts of words used by each individual. Then the individuals are clustered according to their coordinates on the first principal axes, using a hierarchical clustering algorithm. Ward clustering (Ward 1963; Lebart *et al.*, 1995, pp. 191–195) is a commonly used procedure in which cluster analysis is performed in a reduced space. The partition thus obtained is improved by means of several iterations of the K-means algorithm, which reassigns the individuals to their closest centroids in an attempt to increase the variance between clusters. This strategy is not entirely satisfactory: in particular, it often leads to clusters that are determined solely by the use of one very frequent word. Furthermore, the role of single words in isolation is more important the greater the number of axes preserved in the original CA solution (Haeusler, 1984). The difficulty lies in the nature of the data analyzed; in effect, comparing individual lexical profiles—profiles of word frequencies—when the responses are short is quite different from comparing lexical profiles of long texts or comparing average profiles of groups of individuals. Responses are distinguished by the presence or absence of a word rather than differences between frequency profiles. Indeed, responses with similar meanings may have no unit in common. Conversely, two responses may differ only in the negation included in one of them, thus altering the meaning of the utterance completely. The problem lies in the inherently sparse structure of the lexical table (Lebart and Salem, 1994, pp. 152–154).

Our approach to this problem is new in two respects. First, because many words can have a very different meaning depending on the context, we will define a complex lexical unit, called a "quasi-segment," composed of several words, not necessarily consecutive, in order to force the words into a context. Second, we will find subspaces that are directly related to specific individual characteristics judged to be most relevant with regard to the objective of the study. In these subspaces, the better differentiated the use of a lexical unit in different groups, the greater the importance of that unit in defining distances between individuals for eventual clustering.

2 Data

To illustrate the method, we will use a corpus of responses to an open question, collected by Baudelot (1990) for a research project on differences in attitudes observed in boys and girls when choosing what to study at university. Baudelot based his work on the fact that in French secondary education selection according to performance in mathematics leads to sexual and social selection. He was interested in ascertaining why so few girls, whose results in mathematics are as good as those of boys in the third-last year of secondary education, choose degrees demanding mathematics.

Among other studies, a survey was conducted by means of a questionnaire in the city of Nantes; 974 secondary school pupils were interviewed 2 years before the academic year in which they write the final examination for this level of education. The questionnaire included a closed question asking them how they felt about mathematics, followed by an open question to elucidate their choice from among the possibilities provided as responses to the closed question. The two questions are reproduced in Table 1.

Some of the responses are reproduced in their entirety in Table 2. The 974 open responses—some of them empty—make up a corpus with 16,851 occurrences and 1496 different words. Some of the most frequent words are listed in Table 3.

In his research, Baudelot compared boys and girls taking into consideration their real level in mathematics, classified as bad, medium, and good. By studying the responses to the open question, Baudelot singled out the most characteristic lexical

Table 1: Questions regarding feelings about mathematics

Quel est, parmi les sentiments suivants, celui dont tu te sens le plus proche?

1. *Je déteste les maths*
2. *J'aime peu les maths*
3. *J'aime bien les maths*
4. *J'adore les maths*

Pourquoi? Peux-tu indiquer dans les lignes qui suivent les principales raisons pour lesquelles tu aimes ou tu n'aimes pas les maths? Essaye en particulier de préciser les aspects de cette discipline qui te plaisent ou te déplaisent le plus.

(in English:
Among the following expressions, which is the one which you feel closest to?

1. *I hate math*
2. *I like math a little*
3. *I like math*
4. *I adore math*

Why? Can you describe in the following lines the main reasons for liking or disliking mathematics? Try to specify the aspects of this discipline that please you or that displease you.)

Table 2: Some examples of responses

C'est intéressant.

J'aime les maths parce que c'est méthodique, c'est amusant; il faut beaucoup creuser pour trouver.

J'ai toujours été terrorisée par mes profs de maths. L'enseignement lui-même est assez rébarbatif.

Le langage employé par les professeurs de maths ne me convient pas. Je ne trouve pas les raisons de ma motivation et de ma passion. Pour apprécier une matière, j'ai besoin de la comprendre et de faire des liens logiques et je n'en trouve aucun en maths. On ne peut me fournir des explications sensées dans ce cours. Je n'y trouve aucune utilité. Elles ne m'apportent rien en connaissances et en logique.

Les maths sont: parfois intéressants, souvent ennuyeux, ils vous font coucher tard.

Table 3: Examples of words and, in parentheses, number of repetitions in the corpus

---

abstrait (21)	*comprendre (37)*	*intéressant (57)*	*me (152)*
adore (17)	*déteste (34)*	*je (570)*	*moi (28)*
aim (488)	*difficile (22)*	*jeu (13)*	*problème (16)*
algèbre (99)	*esprit (36)*	*logique (205)*	*que (439)*
apprendre (16)	*faut (73)*	*logiques (13)*	*réfléchir (32)*
avoir (33)	*géométrie (146)*	*matière (162)*	*vois (19)*

---

features of boys and girls and the attitudes these features reflected, according to their level in mathematics.

In this work we shall attempt to detect whether groupings exist beyond groups preconstituted according to sex and mathematics level. In particular, in order to answer a question that is implicit in the hypotheses formulated by Baudelot, we will study the existence of a subgroup of girls who are good at mathematics and display a language similar to that of boys at the same level.

3   Methodology and Results

3.1   Repeated Quasi-segment Definition

A quasi-segment is defined as a repeated ordered succession of words, not necessarily consecutive, but within the same sentence and not separated by any punctuation mark. Table 4 shows all the sequences of the corpus that contain the quasi-segment *il. . . faut. . . logique* and also the frequency of each sequence.

An algorithm has been elaborated to identify and list all the repeated quasi-segments of any length (Bécue and Peiró, 1993). Each interval length between words is limited by a maximum value; for example, *il faut beaucoup de logique* has an

Table 4: Corpus sequences containing the quasi-segment "il... faut... logique"

Sequence	Frequency
il faut avoir de la logique	1
il faut avoir beaucoup de logique	1
il faut avoir l'esprit de synthèse, de logique	1
il faut avoir un esprit logique	1
il faut avoir un esprit droit et logique	1
il faut avoir une logique	1
il faut constamment démontrer ce qui est logique	1
il faut beaucoup de logique	3
il faut de la logique	3
il faut énormément de logique	1
il faut être logique	4
il faut procéder avec méthodes, avoir de la logique	1
il faut toujours avoir la même logique	1

interval length of two. Specifying a maximum of two limits the sequences to those underlined in Table 4, in which the quasi-segment *il...faut...logique* is repeated $1 + 3 + 3 + 1 + 4 = 12$ times. A second parameter, which is important for applying statistical methods, in particular cluster analysis, is the frequency threshold, the minimum number of times a quasi-segment is repeated. Finally, the total length, that is, the number of words in the quasi-segment, can be fixed. The values of these parameters are chosen depending on objectives and previous knowledge of the corpus. The stability of the results can be investigated for different parameter values.

In the results that are presented in the following, the maximum interval length is two words long, the frequency threshold is fixed to be seven, and, finally, the quasi-segments are three words long. With these parameter values, a total of 717 repeated quasi-segments are identified. Table 5 shows some of the more frequent repeated quasi-segments.

Table 5: Some of the more frequent quasi-segments (in parentheses, number of repetitions)

aime bien algèbre (12)	*il faut être* (7)	*je aime pas* (80)	*ne comprends pas* (11)
aime la logique (14)	*il faut logique* (12)	*je ne pas* (99)	*ne m'intéresse* (13)
aime les car (122)	*j'ai toujours* (9)	*je ne suis* (29)	*ne vois pas* (17)
aime les maths (247)	*je ai jamais* (9)	*les maths sont* (42)	*peu les maths* (51)
c'est une (69)	*j'aime bien* (167)	*logique des maths* (10)	*pour plus tard* (7)
déteste les maths (11)	*j'aime les* (247)	*m'intéresse pas* (12)	*que je suis* (7)
il faut avoir (12)	*j'aime maths* (213)	*mais j'aime* (14)	*suis pas bonne* (7)

To apply cluster analysis, the corpus is represented by a table with 717 columns whose ith row contains the relative frequencies of these 717 quasi-segments in the ith response, that is, the lexical profile of the ith response over the quasi-segments. This Individuals × Quasi-segments table is very large and sparse. The total inertia of the row points, when using chi-squared distance and weighting each row by the relative frequencies of quasi-segments in the response, is equal to 49.2.

3.2 Projection onto a Reference Subspace

A subspace is now sought such that the projection of the individual responses on the subspace assists the interpretation of the responses. One way to reduce the dimensionality is to group the individuals according to some characteristics related to the objective of the study. For example, the pupils can be placed in six groups corresponding to the six combined categories of the variable Sex × Mathematics. Table 6 shows the composition of these groups, together with the number of responses given in each group.

The centroids of each group can be represented by the average lexical profile of the individuals belonging to it, that is, by a vector that contains the relative frequency of the 717 quasi-segments in the corresponding amalgamated responses. The subspace generated by the six centroid vectors—which has dimensionality five—constitutes a reference subspace. The position of an individual response profile, as projected onto this subspace, depends on its similarity to each group average profile.

The inertia of individual responses projected onto the reference subspace is equal to 1.6. It is well known that the subspace with dimensionality five accounting for the greatest part of the inertia would be generated by the first five principal axes obtained from the CA of the Individuals × Quasi-segments table. In fact, the inertia accounted for in the five-dimensional CA solution is equal to 3.1. As shown in Lebart and Salem (1994, p. 91), the percentage of explained inertia contitutes a pessimistic measure of explained variance. Many examples show that low values can still lead to a satisfactory representation and interpretation of the information with respect to these principal axes.

Table 6: Groups according to sex and level at mathematics

Sex × Mathematics	Number of indivs.	Responses given
Boys good at mathematics	52	40
Girls good at mathematics	83	62
Boys medium at mathematics	144	101
Girls medium at mathematics	221	166
Boys bad at mathematics	195	136
Girls bad at mathematics	279	222
Total	**974**	**727**

3.3 Cluster Analysis of Projected Table

The distance between individuals in the reference subspace is the chi-squared distance between projected profiles. A hierarchical clustering is performed using Ward's criterion. The hierarchical tree is cut, so as to obtain a relatively fine partition. Given the great number of lexical units, it is very rare to find big homogeneous clusters. The partitions consisting of between 13 and 16 clusters were studied and we decided to retain the partition consisting of 15 clusters.

The intercluster inertia is equal to 65% of the projected inertia of 1.6. The partitions can be improved by means of several K-means clustering iterations that reassign individuals to their closest centroids. This increased the intercluster inertia percentage to 68%. The number of individual responses per cluster varies from 33 to 73.

3.4 Characterization of the Clusters

The partition can be interpreted by cross-tabulating the individuals according to cluster and characteristics (Table 7). For example, of the 33 individuals assigned to cluster 14, 14 are boys good at mathematics. From other question responses not repeated here, we find that 17 want to choose a mathematics speciality in the secondary school final examination, and 13 estimate they have a good or very good level at mathematics. It is possible to summarize the vocabulary of each cluster in terms of the quasi-segments (and/or words) over—or under—used in it, using the marginal relative frequencies of the quasi-segments as a reference value.

But the most important result is the list of the "modal responses," complete responses selected according to a criterion that measures their power to characterize the cluster (Lebart and Salem, 1994; Lebart, Chapter 11). For example, the chi-squared distance can be calculated between each response profile and the average profile of the cluster; and the responses can be ordered by increasing magnitude of distance. It is useful to reorder all the responses within each cluster according to this criterion, thereby returning to the data when interpreting the clusters.

Table 7: Results for each sex × mathematics category in each of the 15 clusters

	Clusters														
Groups	1	2	3	4	5	6	7	8	9	10	11	12	13	14	15
Boys, good	0	0	0	3	0	0	0	1	4	1	1	0	1	14	14
Girls, good	2	0	2	2	1	1	2	21	4	0	6	0	10	9	1
Boys, medium	0	3	1	0	2	18	4	1	15	5	9	18	14	7	3
Girls, medium	22	5	2	7	4	4	29	10	19	16	15	8	8	9	8
Boys, bad	15	22	23	3	6	3	0	7	14	8	7	10	5	10	2
Girls, bad	17	7	16	22	42	23	24	12	17	5	5	3	9	11	5
Total	**56**	**37**	**44**	**37**	**55**	**49**	**59**	**52**	**73**	**35**	**43**	**39**	**47**	**60**	**33**

Table 8: Modal responses for groups of boys and girls good at mathematics

Boys good at mathematics in cluster 15	Girls good at mathematics in cluster 8
• C'est un peu un jeu, et cela peut être très intéressant si l'on comprend	• Il faut beaucoup de logique, de rigueur, de précision
• Exercice de la logique, du raisonnement, développement des facultés intellectuelles. Matière intéressante	• Il faut de la logique pour la géométrie et j'aime les chiffres
• On ne peut pas s'exprimer par soi-même, on est obligé de suivre des règles strictes et ennuyeuses (...)	• Les maths nécessitent un esprit de logique (...)
• Les maths sont représentatives de la logique, donc bonnes	• Positif: je préfère la géométrie et j'adore trouver les exercices. Négatif (...)
• J'aime les maths parce que c'est une gymnastique de l'esprit	• Les aspects qui me plaisent le plus sont ceux de la logique, l'analyse, l'intuition (...)
• J'adore les maths à cause de la logique (...)	• J'adore l'algèbre car il faut être logique. J'aime aussi la géométrie (...)
• Je trouve cette discipline intéressante, car variée. Elle fait travailler la logique plus que la mémoire (...)	• J'aime les maths car je trouve cela intéressant et pour l'instant ce n'est pas trop difficile
• On peut arriver à de bons résultats en ne retenant pas grand chose	• J'aime l'algèbre car je trouve (...)
	• J'aime car c'est logique
	• J'aime l'algèbre car je trouve cela facile
	• J'aime bien cette matière (...)

We can also group the ordered list of responses according to the categories used to build up the reference subspace. For example, Table 8 lists the modal responses of clusters 15 and 8, respectively, for boys and girls good at mathematics.

4   Interpretation of Results

To interpret the results, we compare modal responses between clusters or between subgroups of clusters of interest (e.g., Table 8). We also visualize proximities between clusters and position these clusters relative to all quasi-segments. To achieve the latter, the contingency cross-tabulating quasi-segments with Sex × Mathematics groups can be submitted to CA and then the centroids of the 15 clusters projected onto the resulting principal as supplementary points (Figure 1).

In accordance with the objective of studying the differences between boys and girls who are good at mathematics, we must study the clusters with a high proportion of pupils who are good at mathematics. Boys and girls who are good at mathematics are located mainly in clusters 8, 13, 14, and 15 (see Table 7). Cluster 14 is the only one that contains a significant number of both boys (14) and girls (9) good at

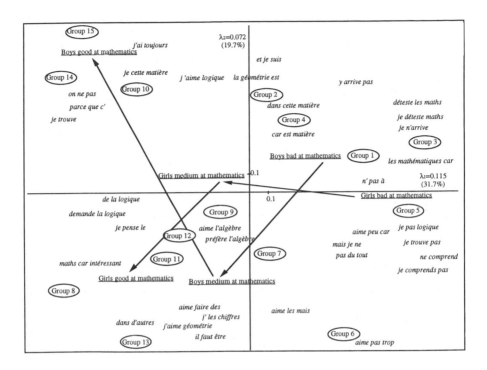

Figure 1: Principal plane resulting from the correspondence analysis of the contingency table Quasi-segments (Sex × Mathematics). The centroids of the clusters are projected as illustrative elements.

mathematics. We can, therefore, draw a preliminary conclusion: the great majority of boys and girls who are good at mathematics do not use the same language.

Furthermore, bearing in mind their position in relation to the centroids of these clusters (Figure 1), boys good at mathematics in cluster 15 and girls good at mathematics in cluster 8 can be regarded as the most typical subgroups of these two groups. From Table 8, which lists the most characteristic responses of these two subgroups, we can deduce the following. Girls use "Je" ("I") and "J'aime" ("I like"), although less so the rather weaker "J'aime bien" ("I quite like"), they qualify their responses with "Je trouve" ("I find") or "Je pense" ("I think"), and they consider that "Il faut de la logique" ("You need logic") or "Les maths nécessitent un esprit de logique" ("Math needs a feel for logic"). In other words, the ego, one's own qualities, abilities, and tastes, have a high profile. Boys are more neutral: "C'est" ("It's"), "Les maths sont" ("Math is"), "Les maths sont représentatives de la logique" ("Maths is representative of logic"), "Cette discipline est intéressante" ("This discipline is interesting"), "On n'a pas" ("You don't have to"). The ego is not totally absent, but it is expressed through the form "On" ("One," or the impersonal "you"). Any qualities mentioned pertain to mathematics rather than to themselves.

Among girls belonging to clusters 14 and 15 (nine in cluster 14, but only one in cluster 15), we find a language more typical of boys (the model responses of these subgroups are not reported here). They say "C'est stimulant, c'est intéressant" ("It's stimulating, it's interesting"), "C'est logique" ("It's logical"), "J'aime les maths parce que c'est (...)" ("I like math because it's (...)"), "J'aime bien les maths car c'est intéressant" ("I quite like math because it's interesting"), "J'aime bien parce que c'est logique, on n'a pas besoin de réfléchir" ("I quite like math because it's logical, you don't need to think"), "On n'a pas à retenir les dates" ("You don't have to remember dates"), "Parce que c'est vivant" ("Because it's alive").

Although also not reported here, cluster 10 contains the responses with the greatest breadth of vocabulary and expression. The use of nouns is more frequent, which tends to indicate verbal richness. Rather than "C'est intéressant" ("It's interesting"), we find "C'est une matière intéressante, logique, qui fait appel... " ("It's an interesting, logical subject that appeals"), "C'est une matière qui apprend à avoir un raisonnement rigoureux" ("It's a subject that teaches you to use rigorous reasoning"), "C'est une matière, une discipline" ("It's a subject, a discipline"). The responses are longer and better thought out. It is a cluster that contains many girls with a medium mathematics level (16 of the 35 belonging to this category), often with favorable opinions of mathematics, but not always. This cluster contains no girls who are good at mathematics and only one boy at that level. In Figure 1, this cluster occupies a position close to clusters 14 and 15, reflecting certain similarities of language. Cluster analysis, however, distinguishes this group of responses, which are particularly rich in the quality of their expression and precision, and differentiates them from the responses in clusters 14 and 15.

5   Conclusion

The clustering method presented here highlights the existence of groups of individuals with a relatively similar language, thus facilitating the identification of significant features of the discourse. In the example used, it can be seen that the results obtained complete Baudelot's study, giving it a finer edge. "Logique" ("logical") is a word that is used by pupils with a good level, but whereas girls find that "Il faut de la logique" ("One needs logic"), boys say that "C'est logique" ("It's logical"). For girls it is a quality of the individual; for boys it is a feature of mathematics. Pupils who use richer forms of expression involving nouns, for example, "C'est une matière, une discipline" ("It's a subject, a discipline"), are often females with a medium mathematics level.

An essential aspect of our approach is the projection of the individuals onto the subspace generated by the centroids of the groups defined by combining the variables Sex and Level of Mathematics. Even though a relatively small part of the inertia of the individual profiles across lexical units is contained in this subspace, the computation of distances between individuals and the subsequent clusters will be directly related to the differentiation of the Sex $\times$ Mathematics groups.

Chapter 13

The Cloud of Candidates. Exploring the Political Field

Fernand Fehlen

1   Introduction

In this chapter correspondence analysis is used as a tool of exploratory data analysis to investigate voting results, allowing identification of the mechanism ruling the political field of Luxembourg and revealing capital specific to this field (Bourdieu, 1981). We shall analyze the results of the 1989 parliamentary elections for the city of Luxembourg.

The structure of the voting system is a determining factor in the functioning of the political field because it defines the rules that allow admittance to power, which ultimately is the justification for the existence of the political field. One could think, for instance, of the various majority systems that lead to bipolarization. Luxembourg introduced a proportional system of universal suffrage for men and women in 1919. This replaced a system of census suffrage (in which a minimal level of tax assessment was required in order to vote), which had hitherto guaranteed political power to a small group of leading citizens. Faced with a challenge to the very existence of the Luxembourg state, the liberal party accepted universal suffrage only on the condition that an essential characteristic of the previous system, namely the direct bond between the voter and "his" representative, would be maintained in the new system (Fehlen, 1993). This feature was seen as a desirable weakening of the role of political parties.

As it is applied today in Luxembourg, the panachage system remains unusual. Elsewhere it can be found only in Switzerland and in Germany at some local levels. Every voter has as many votes to distribute as there are seats to represent his or her constituency. Each of the four constituencies in Luxembourg directly elects its own representatives. The number of representatives of each constituency is related to the number of inhabitants of the constituency. In all, there are 60 seats to be distributed in the Luxembourg parliament. Every voter can spread his or her votes over all parties, giving a candidate up to two votes. Therefore the same ballot can carry votes for candidates belonging to different parties.

The data we use here are the results for *Luxembourg City* which belongs to the constituency center, where each voter has 21 votes to distribute. The panachage system allows three different types of votes:

- List votes: each party presents a list that generally includes as many candidates as there are seats to be distributed in the constituency. By choosing a party list, the voter gives one vote to each candidate on that list.

- Intraparty panachage: the voter can also distribute his or her votes unevenly over the candidates of one list by giving one or two votes to some individual candidates.

- Interparty panachage: the voter can also distribute his or her votes over the lists of different parties.

Of course, the voter cannot distribute more votes than of right, otherwise the ballot is not valid. But by not using all of his or her votes, it is possible for the voter to practice a very subtle form of gradual abstention, which is widespread, especially because voting is compulsory in Luxembourg.

Since the introduction of the panachage system, the political life has been dominated by the Christian democratic party (Chrëschtlech Vollekspartei), which, with only some minor exceptions, has participated in the government ever since. Since 1984 the socialist worker party (Lëtzebuerger Sozialistesch Arbechteer Partei) has formed a coalition government with the Christian democratic party. The most important opposition party is a liberal party (Demokratesch Partei), which in the past formed several coalitions with the Christian democratic party. The communist party, which can look back on a great history, has lost its influence and has not been represented in the parliament since 1994. Since the 1984 election, some new parties have made their appearance on the political scene, such as the "Aktiounskomitee 5/6 Pensioun," a rural and "poujadiste" protest party, as well as two ecologist parties, "Greng Alternativ Partei" and "Greng Lëscht Ekologesch Initiativ," which have unified in the meantime, and a nationalist party, (National Bewegong), which dissolved itself after the 1994 elections.

2   Correspondence Analysis and Voting Behavior

Correspondence analysis (CA) provides a natural framework in which to analyze voting behavior as it is associated with the sociological approach of Pierre Bourdieu. CA features a twofold congruence with Bourdieu's understanding of society as a social space of relative positions and as a series of partly autonomous but homologous fields. First, there is the relational aspect, which is reflected in Bourdieu's own words:

CA "is essentially a relational procedure, whose philosophy corresponds completely to what in my opinion constitutes social reality. It is a procedure, that 'thinks' in relations" (Bourdieu, 1991; English translation by Wuggenig and Mnich, 1994, p. 304). Second, there is the correspondence between the analysis of the row profiles and the column profiles that has given CA its name: the understanding of the structure of the space of the rows leads to the comprehension of the space of the columns and vice versa. Or, to take the example in "La Distinction" (Bourdieu 1979), the space of the social positions reveals the space of the lifestyles. This approach transcends the deterministic view of society that is often incorporated in statistical models with their limited number of variables classified as either dependent or independent. See also The BMS (1994) and their attempt to reconcile what they call the "French data analysis" and statistical modeling approaches. Another development of the "French approach" is represented by Rouanet and Le Roux (1993), who try to reorganize the whole range of the multidimensional data analysis in the light of Benzécri's geometrical formalism using the work of Bourdieu as epistemological background.

The results of an election, presented in contingency tables with the parties as columns and the different districts of the constituency as rows, seem well suited to CA. This approach will result in two spaces that will mutually explain each other. The column points will materialize the space of the parties and the row points will materialize the socioadministrative space. For Luxembourg, at the national level we find four voting wards: the industrialized south, the central region surrounding Luxembourg City, and two rural districts.

The main result for all elections we have analyzed, either for communal or for national elections, is that the first axis always represents the left–right opposition, that is, the socialist and communist parties on the one hand and the Christian democratic and liberal parties on the other. This is also true for the study presented here, which analyzes the 31 polling wards of the city of Luxembourg. As a rule, these wards have very distinct but subtle social characteristics.

3 The List Votes

The data we analyze are the number of list votes obtained by eight parties in the 31 polling wards of Luxembourg City in the 1989 parliamentary election. The numerical results of the CA of this 31 × 8 contingency table can be found in Table 1. Figure 1 shows the first axis, containing 47.2% of the total inertia, as a vertical axis. The 31 ward points are labeled on the left side and the eight parties on the right. Both are positioned by their coordinates on the axis (the horizontal shift of the labels is done only to improve readability of those that have similar coordinates). Concentrating on the wards, along this axis we can identify the opposition of working class areas with wealthier neighborhoods. This outcome confirms the existence of a political duality, which today is often denied, especially for an economically booming city such as Luxembourg, whose inhabitants at first sight seem to be all wealthy middle class. As far as the parties are concerned, the first axis is defined by the opposition between

Table 1: Numerical output from CA of the list votes (8 parties and 31 wards)

ROWS Name	$K = 1$	COR	CTR	$K = 2$	COR	CTR	$K = 3$	COR	CTR
beg	107	207	13	−193	668	145	63	70	16
bel	−181	683	99	23	11	5	83	143	72
bis	296	340	24	379	557	132	−61	14	3
bon	230	965	84	−12	2	1	−17	5	2
bos	136	800	97	−49	105	43	2	0	0
cer	−161	200	6	7	0	0	13	1	0
ces	−111	214	16	16	5	1	108	203	54
cls	152	155	8	291	569	106	−127	109	21
con	−113	227	3	−125	276	14	−12	2	0
dom	142	517	12	−84	182	14	−59	90	7
eic	31	25	1	78	155	14	39	38	3
epa	−126	362	12	73	123	14	−119	326	38
fet	−147	437	36	−94	179	50	111	249	71
gas	−50	164	5	−19	23	2	−92	553	57
ham	393	633	103	−103	44	24	70	20	11
hol	−46	176	5	21	38	3	32	86	8
kay	−177	239	18	190	277	70	204	318	81
kir	−173	234	15	112	98	21	−248	481	108
kie	111	178	5	87	108	11	−57	47	5
lic	−296	679	102	−68	36	18	−40	12	6
lih	−232	902	107	−27	12	5	62	64	26
mer	−142	497	30	−23	13	3	−114	318	65
muh	−43	15	1	−75	47	9	−261	563	109
neu	221	827	66	−37	23	6	1	0	0
paf	56	24	2	276	571	170	115	99	30
pes	−372	335	53	13	0	0	−341	282	155
rol	5	1	0	−57	139	11	−119	604	51
str	106	444	30	69	187	43	3	0	0
wal	124	791	14	21	23	1	8	3	0
wei	193	765	32	16	5	1	−4	0	0
yol	59	41	2	201	474	62	−5	0	0

COLUMNS Name	$K = 1$	COR	CTR	$K = 2$	COR	CTR	$K = 3$	COR	CTR
CO	153	159	42	231	362	326	−165	184	169
LP	−69	187	41	101	400	297	88	305	231
CD	−129	576	173	−23	18	18	−87	264	274
SO	268	904	502	−69	60	114	8	1	2
GL	−173	314	68	−69	50	37	57	34	26
NB	293	657	150	66	33	26	20	3	2
GA	−51	35	4	−115	181	78	−124	211	92
AD	−75	93	19	−97	151	104	134	291	204

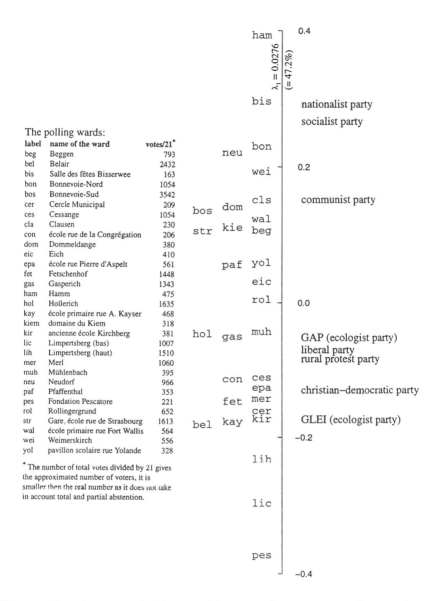

The polling wards:

label	name of the ward	votes/21[*]
beg	Beggen	793
bel	Belair	2432
bis	Salle des fêtes Bisserwee	163
bon	Bonnevoie-Nord	1054
bos	Bonnevoie-Sud	3542
cer	Cercle Municipal	209
ces	Cessange	1054
cla	Clausen	230
con	école rue de la Congrégation	206
dom	Dommeldange	380
eic	Eich	410
epa	école rue Pierre d'Aspelt	561
fet	Fetschenhof	1448
gas	Gasperich	1343
ham	Hamm	475
hol	Hollerich	1635
kay	école primaire rue A. Kayser	468
kiem	domaine du Kiem	318
kir	ancienne école Kirchberg	381
lic	Limpertsberg (bas)	1007
lih	Limpertsberg (haut)	1510
mer	Merl	1060
muh	Mühlenbach	395
neu	Neudorf	966
paf	Pfaffenthal	353
pes	Fondation Pescatore	221
rol	Rollingergrund	652
str	Gare, école rue de Strasbourg	1613
wal	école primaire rue Fort Wallis	564
wei	Weimerskirch	556
yol	pavillon scolaire rue Yolande	328

[*] The number of total votes divided by 21 gives the approximated number of voters, it is smaller then the real number as it does not take in account total and partial abstention.

Figure 1: One-dimensional CA map of list votes for 8 parties and 31 wards. CD, Chrëschtlech Volkspartei (Christian-democratic party); LP, Demokratesch Partei (liberal party); GP, Greng Alternativ Partei (left-ecologist party); GL, Greng Lëscht Ekologesch (Initiative ecologist party); CO, Kommunistesch Partei Lëtzebuerg (communist party); SO, Lëtzebuerger Sozialistesch Arbechteer Partei (socialist party); NB, National Bewegong (nationalist party); AD, Aktiounskomitee 5/6 Pensioun (rural and 'poujadiste' protest party).

the socialist party at the top (*SO*: CTR = 502) and the Christian democratic party and the ecologist party (GLEI) at the bottom (*CD*: CTR = 173 and *GL*: CTR = 68). On the same side of the axis as the socialist party, we find south and north Bonnevoie (*bos*: CTR = 97 and *bon*: CTR = 84), a neighborhood behind the railway station where railroad workers used to live and which today is generally a cheap but decent area. At the same extreme we find Hamm (*ham*: CTR = 103), a formerly rural area with an industrial past rooted in the last century and a large workers' colony built before World War II. At the other extreme of the axis, we find the two polling wards of the Limpertsberg (*lih*: CTR = 107 and *lic*: CTR = 102), as well as Belair (*bel*: CTR = 99), some very pleasant neighborhoods with many small family houses near the center, an agreeable but expensive area. Thus this first axis represents the rich–poor hierarchy of the neighborhoods, which can be confirmed by the positions of the other polling districts.

While this axis identifies the local roots of the different parties, which are also their socioeconomic roots, it also appears to reproduce the classical left–right spectrum, at least for the parties that are well represented on this axis. One exception is the "National Bewegong" (*NB*: COR = 657), a right-wing, nationalist party that can be identified as getting its votes from the same working class, "leftist" neighborhoods as the socialists. One of the extremes on the bottom of the first axis is taken by the voting district "Fondation Pescatore" (*pes*). This is a very small bureau situated in an old people's home, where the overwhelming majority of the voters are the inhabitants of this residence. The extreme position of this point reflects the exceptional success of the Christian democrats in this bureau, where they won 55% of the votes.

Axes two and three (Figure 2) have almost equal contributions to the total inertia (13.9% and 13.6%). We have plotted all parties, but only the wards that are well represented in this display (COR2 + COR3 > 500).

Along axis one, both the communist and liberal parties were not well represented (*CO*: COR = 159 and *LP*: COR = 187). However, both parties have a high contribution to axis two (*LP*: CTR = 297 and *CO*: CTR = 326) and are situated on the same side, which suggests that they share some common strongholds. At first sight this seems paradoxical. But the districts Clausen, Bisserwee, and Pfaffenthal, which are well represented on this axis (*cls*: COR = 569, *bis*: COR = 557, *paf*: COR = 571), do indeed have a common characteristic: they are situated in the Alzette valley, which is populated by modest families, often considered more subproletarian than workers. Their adherence to the liberal party reflects the small jobs they hold at the lowest levels of the municipal services and may also express their protest against the larger established parties. But this is only one element shared by the electorate of the communist and liberal parties. Axis three differentiates these two parties, as we find the communist party at one side (*CO*: CTR = 169) and the liberal party (*LP*: CTR = 230) and the rural protest party (*AD*: CTR = 204) at the other.

It is interesting to note that the two ecologist parties seem quite distinct in the factorial space: although they are both on the political right on the first axis, the GAP is closer to the center than the GLEI. On axis two they have roughly the same position, but on the third axis the GLEI is on the same side as the liberal party, while

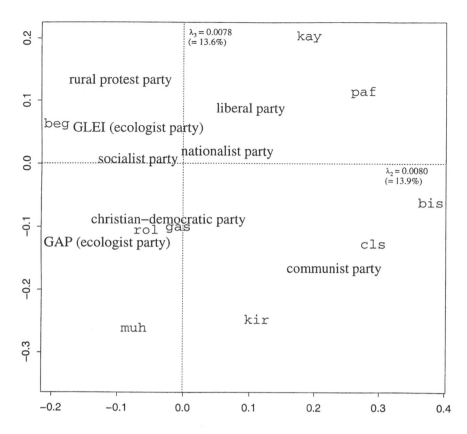

Figure 2: List votes for 8 parties and 31 wards. Two-dimensional display (axes 2 and 3).

the GAP is on the same side as the communist party. So we can assume that they have somewhat different electorates.

4 The Personal Votes

So far, we have considered only party list votes. But as we said at the beginning, list voting is only one possible way of voting in Luxembourg. In Luxembourg City, of the 28, 412 valid ballots counted, 41% were completed using some sort of panachage. In all, there were 212 candidates on 12 lists. However, we will limit ourselves to the eight largest parties, who put forward 168 candidates and who received 98.5% of all votes.

To investigate the phenomenon of panachage, we can analyze two different 168 × 31 matrices. Each cell corresponds to the votes one of the 168 candidates gained in one of the 31 polling districts. The first data set considers only the personal votes, and the second also includes the list votes.

There was a total of $227,075$ personal votes. The sum of each row represents the total number of personal votes obtained by a candidate. For instance, the candidate who received the most personal votes is Jacques Santer, the former prime minister of Luxembourg, with $10,497$ votes. At the other end of the distribution we find an unknown candidate of the right-wing party with 38 personal votes; the median for the 168 candidates was 400 votes.

Turning to the list votes, there were $15,513$ list votes given to the eight parties, summing to $15,513 \times 21$ individual votes that can be distributed equally to all the candidates of the relevant list. We call the sum of the personal votes and the distributed list votes the total votes for a candidate. For example, as the Christian democratic party, to which Jacques Santer belongs, had 4378 ($\times 21$) list votes to distribute, Jacques Santer had $10,497 + 4378 = 14,875$ total votes.

The panachage system is especially criticized by politicians bound to the party organization, who like to disparage panachage as an immature habit consisting of haphazardly spreading votes on disparate candidates. If this were true, there would be no systematic pattern to the voting results. But our CA of the personal votes established the contrary: the first axis, with its contribution to the total inertia of 26.3%, can be identified as the same opposition between working class areas and wealthier neighborhoods found in the analysis of the list votes. In fact, the overall structure of the 31 wards was practically identical to the structure revealed in Figure 1. This shows, therefore, that voting behavior according to the panachage system follows the same logic as list votes. In fact, a direct investigation of a sample of polling cards (CRISP, 1989) revealed that the normal behavior of the *panachage* voter is to concentrate his or her votes on one or two parties. Often, one party gets the majority of the votes, while a few remaining votes are cast elsewhere. Although we have only aggregate data in the 168×31 matrix, we can confirm the results of the CRISP research. For instance, candidates that often gained panachage votes on the same ballot according to the study cited are close in the factorial space. As a rule, the row points of the candidates are attracted by the column points of the districts where their party has its strongholds. Exceptions to this rule can often be explained, as can be shown by an almost caricatural example. Théid Stendebach of the liberal party has an atypical position in the socioadministrative space: he is clearly situated in the direction of working class neighborhoods and finds himself together with candidates of the socialist party and the national movement. The results show that, unlike the rest of his party, he obtains his personal votes in this area. In fact, he is well known in this neighborhood, where he owns a garage and has won the esteem of the locals as a former football player.

We shall not concentrate on this CA, as the next consists of its superposition with the CA of the parties presented in Section 2. By the way the matrix of the total votes is defined, the candidates of a party are represented by points that are weighted averages of the points representing the personal and list votes gained by the party. The data set of total votes is clearly structured, because the 21 candidates of each party have gained the same list votes in each case. Therefore it is not surprising that the inertia for the first axis is 44.5%, a value near that of the CA for the list votes. We will present the CA of the total votes by two charts: Figure 3 representing axis one and Figure 4 representing axes two and three. The cumulated percentage of the inertia of these three axes is 65%.

Figure 3 shows the first axis vertically with the labels in a rather unusual format. The 31 polling wards appear on the left side of the plot and the 168 candidates on the right. The points are symbolized by triangles pointing to their coordinates on the first axis. The labels on the left have been shifted horizontally to improve readability, as before. On the right-hand side the candidates of each party have been aligned under the denomination of their party. Two sorts of supplementary row points have also been added: the total votes gained by the 21 candidates of each party are represented by empty squares and the sum of the list votes of each party is represented by a solid square.

Note that the first axis reveals the same left–right opposition that we found for the first CA, with the 31 polling wards in about the same order. The socialist party candidates are the most stretched out on this first axis. In fact, this shows that the candidates of this party belong either to a "worker" faction or to a circle of social-liberal lawyers and intellectuals, partly from the leading citizens of the town. As the latter group draws their personal votes from the bourgeois neighborhoods, these points are attracted by the clouds of the Christian democratic and liberal candidates.

Comparing the two sorts of supplementary points (the sum of all votes won by a party and the sum of the list votes won by that party) shows the different local origins of the parties. The greater the distance between these two squares, the more candidates of the given party gain their personal votes out of the party strongholds. This is especially true for the socialist party. The general shift toward the wealthier neighborhoods seems to indicate that panachage is more frequent in these areas.

Figure 4 presents all the row points (i.e., the 168 candidates) each symbolized by a point. The 21 candidates of each party are surrounded by a convex hull. The acronyms of the eight parties have been added to identify the eight separate candidate clouds. The column points are displayed with the height of the labels proportional to the quality of representation on the two axes shown. The advantage of this pseudoperspective plot is illustrated by the position of Beggen (*beg*) on the far left: even though it is very close to axis two, it is at some distance to the other axes that are not represented.

Axis two opposes the candidates of the communist party on the right to the candidates of the ecologist GLEI and the poujadist ADR on the left, while the third axis opposes some Christian democrats at the bottom to some candidates of the liberal party at the top, relativizing the overlapping of these two clouds of points on axes one and two. This is due to the exceptional performances of some Christian democrats at the bottom of the graph in Rollingergrund (*rol*) and Muhlenbach (*mul*) and to a lesser extent in the "Fondation Pescatore" (*pes*).

Axes two and three confirm in some sense what we have seen for the parties in Figure 2: once again the voting districts of the Alzette valley (*cls* and *bis* on axis two and *paf* on axis three) have a high contribution to the definition of these axes and

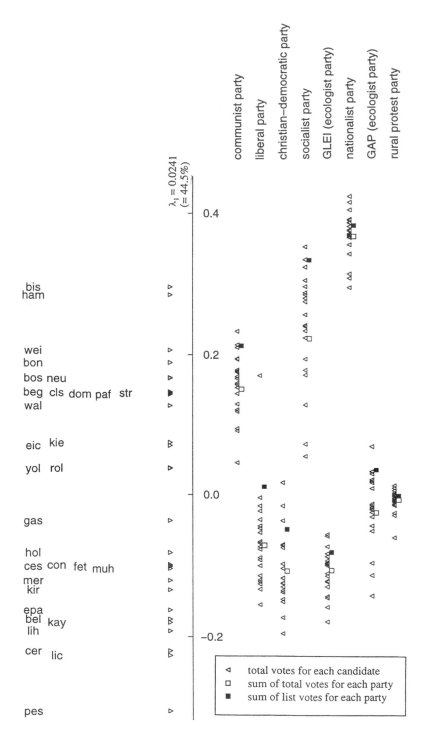

Figure 3: Total votes for 168 candidates and 31 wards. Axis 1: parties at the right, wards at the left.

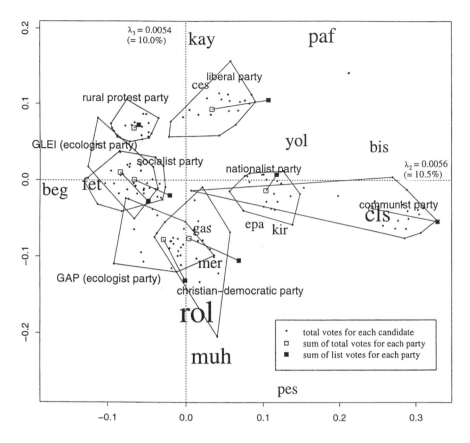

Figure 4: Total votes for 168 candidates and 31 wards. Axes 2 and 3: candidates represented as points and wards by their label (height proportional to QLT23).

many candidates of the communist party and some of the liberal party are attracted in their direction. On axis two we have an overall shift to the left of the supplementary points, that is, toward the more "bourgeois" neighborhoods, which confirms the result for axis one.

5 The Panachage Capital

Our mapping approach can uncover many examples in which the personal reputation of the candidate or even the presence of his or her family in a certain neighborhood—whether in the present or in the past—has a great influence on the voters. Apart from regional or local affinities, there are other strategies used by candidates to obtain panachage votes. These range from traditional political activities to attending specific, clientele-like groups or to participation in sports and the like. Sometimes, such activities, although they are not on a local level, are reflected in the socioadministrative space: for instance, if a candidate is famous for a certain popular sport, like Stendebach, his panachage votes will probably come from more working class wards.

Certainly, the CA map of the polling results is not a photograph of the political field, but it represents the starting point for the construction of the political field. Appealing to the general laws of fields (Bourdieu, 1980) and the specific aspects of the political field (Bourdieu, 1981), we could combine our CA with other quantitative and qualitative work to construct the political field, that is, to identify the agents, the factors of differentiation, the distribution of the various forms of capital, and, most important, the specific capital of this particular field.

CA emphasized the central role played by panachage in the political field in Luxembourg. The divergence of individual candidates from the barycenter of their party cloud required an explanation. This led us to consider the different strategies that candidates adopt to gain personal votes and helped us to understand how the political field is influenced by these strategies. We identified panachage capital as the specific capital of this particular field. The capacity of a politician to attract personal votes can even decide the composition of the government, as this traditionally consists of the candidates who won the most personal votes. As our analyzes have demonstrated, influential statesmen have a large proportion of interparty *panachage* votes. This confirms the existence of a large, state-supporting electorate and partly explains the success of a policy aimed at consensus politics. In fact, since World War II, the three major parties (the Christian democratic party, the liberal party, and the socialist party) have alternated in two-party coalitions. The smallness of a country that counts only 400,000 inhabitants is a fundamental characteristic of Luxembourg society and the *panachage* capital is a logical transposition of this smallness to the political field.

Software Notes

CA and the graphics of this contribution were produced with CORA-library, a set of S-Plus functions written to perform computer-aided interpretation of CA. Further information is available from the author.

Chapter 14

Normative Integration of the Avant-garde? Traditionalism in the Art Worlds of Vienna, Hamburg, and Paris

Christian Tarnai and Ulf Wuggenig

1 Introduction

Contemporary fine arts is a field largely unexplored by sociology. One of the factors that impeded research was a distrust of sociology on the part of many people in the art world, who see the social sciences as disciplines "bent on depriving art of its sacred status" (Moulin, 1987, p. 3).

In the nineties, however, there emerged a neoconceptualist movement that shows a critical attitude in the manner of sociology. This movement around avant-garde artists based in New York such as Andrea Fraser, Clegg & Guttman, Cristian Philipp Müller, and Renée Green emerged in leading institutions of the international art world in 1993, for example, in the 45th Biennale of Venice and the Whitney Biennial in New York, and was soon labeled "contextual art" (Weibel, 1994). Contextual art, which was a kind of fad in the art world up to 1995, explores the artistic field and its institutions. Some of these "artist-researchers" even directly borrow methods from the social sciences, for example, interviews, questionnaires, and photo elicitation method (see von Bismarck *et al.*, 1996).

This artistic movement created a climate more favorable for doing sociological field research in the world of visual art than ever before. It is the background for our empirical study, which explores habits, preferences, and value orientations of people who belong to this social world.

The empirical investigations took place in three well-known cities of the West. One is Vienna, the former center of the Habsburg monarchy, which has a long tradition in the arts. It is famous for its turn-of-the-century *art nouveau* modernism and notorious for its *actionist art* in the sixties, which—from a sociological point of view—was perhaps the most deviant and radical art movement the world had seen in this century. The second is Hamburg, the second biggest city of Germany, which is nearly the same size as Vienna. Hamburg is widely known as a center of trade and commerce. In the more recent past Hamburg postmodernized in a rather quick way, extending its media, entertainment, and service sectors and also building up an institutional structure for showing contemporary visual art, the "Kunstmeile," which is now one of the most important in Europe. The third is Paris, the city where the "dealer–critic" system emerged at the end of the last century (White and White, 1993). It was the center of aesthetic modernism up to the fifties. At this time New York "stole the idea of modern art"(Guilbaut, 1983) and Paris began to lose its position as a center of artistic production. Paris, however, remained one of the world's most important places for the distribution and consumption of art (see Moulin, 1992).

The study is based on random samples of visitors at important exhibitions with international contemporary art in these three cities in 1993 (Vienna), 1993–94 (Hamburg), and 1995 (Paris). These exhibitions, such as the show "The Broken Mirror" curated by Kasper König and Hans Ulrich Obrist and presented in Vienna as well as in Hamburg, attracted a public of specialized insiders (artists, critics, curators, and dealers), as well as a general art public, which to a high proportion is an intellectual and academic population. Comparability between exhibitions is, of course, always a problem. We tried to solve it by concentrating on the same kind of art and on institutions that stand in a relation of homology. Comparability was improved by the fact that one of the exhibitions was shown in Vienna as well as in Hamburg. Thus, about half of the sample in Hamburg and about a third of the sample in Vienna were taken at the same exhibition. In Paris, the curator Obrist, who cocurated "The Broken Mirror," was also responsible for the show "$1 - 1 = 2$" of Fabrice Hybert in the ARC, the department of contemporary art of the *Musée d'Art Moderne de la Ville de Paris*, where our investigation took place.

It has to be emphasized that our study is not representative of the whole field of contemporary art. Verger (1991) demonstrated empirically that one of the field's main oppositions is the contrast between the national and the international market. Our study refers to the international market only. The economists Rouget *et al.* (1991) subdivide the field from a somewhat different perspective, which is more in line with Bourdieu's (1996, p. 141 ff.) distinction between autonomous and heteronomous production and consecrated and (as yet) nonconsecrated art. They suggest four submarkets, each with its own laws. With regard to this theory of art market segmentation, the study is aimed at the group of people who produce, broker, buy, or sell the art of the submarket termed the "market of the mediated avant-garde" or who at least are attracted by this kind of art to the extent that they go to exhibitions in galleries or public institutions. It is the market about which the bourgeois press and the specialized art journals write the most, the part of the dealer–critic system where artists of "high visibility" (Moulin, 1996, p. 160) are struggling for status after death, that is, a place in international art history.

Samples of the visitors at the exhibitions in the three cities were approached and asked to participate in the research. The questionnaires were completed at home. Compared with past experience, the response rates for the largely standardized instruments were relatively high. They amounted to 42% in Vienna, 55% in Hamburg, and 36% in Paris. The samples were restricted to Austrians, Germans, and French in the three respective cities: $n = 616$ in Vienna, $n = 583$ in Hamburg, and $n = 358$ in Paris.

We use latent class analysis (LCA) to reduce the complexity of the manifest attitude space, to test the ordinality of the response format of the attitude items, and to differentiate between subgroups, for whom the items have different meanings. Correspondence analysis (CA) is also used to explore in a visual way whether some assumptions about associations between social positions in the art field and value orientations are valid. Both LCA and CA lead to visualizations of results, which can be communicated to members of the social worlds we are investigating.

DiMaggio (1996) drew on the General Social Survey (GSS) of 1993 of the NORC to test the hypothesis that visitors to art museums have the same social and political attitudes as everyone else. Even on the basis of a highly inclusive definition of the boundaries of the art world (historical, modern and contemporary art, avant-garde and commercial art, self-report of art participation), it turned out that the art public generally takes more politically liberal positions and is significantly more secular, more tolerant of nonconformists, and more open to other cultures and lifestyles.

In our research we are interested in a similar question, restricted to the much smaller world of contemporary art and to the internal differentiation of that field. With regard to internal social differences, a distinction between "center" and "periphery" is often drawn in art criticism as well as in economics and sociology of art (Frey and Pommerehne, 1989; Anheier *et al.*, 1995). If an avant-garde subculture still exists, the values and preferences that are thought to be constitutive for the social system of avant-garde art, for example, individualism, antitraditionalism, moral agnosticism, antieconomism, should be clearly more widespread in the "center" than in the loosely involved "periphery" of the art field. Bourdieu (1993) refers to the center of the field and not to the general public when he underlines the ascetism, the moral agnosticism, and the negation of bourgeois and petit bourgeois values and tastes. The same is true of Bell's (1976) descriptions of the radical individualism, the hedonism, and the hostility toward bourgeois values of postmodernist art and culture. On the other hand, if it is true that producers and mediators of avant-garde art are also socially integrated and conformist to the extent that writers such as Gablik (1985) or Crane (1987) assume, differences in value orientations between center and periphery should more or less have vanished.

In this context we will concentrate on one of our scales developed to represent value dimensions, in order to judge these controversial assumptions on an empirical basis. This scale, labeled "traditionalism," consists of six items referring to traditional bourgeois and petit bourgeois social values and symbols of status. Four of the items represent such communitarian or altruistic values as partnership, children, religion, and nation in the sense of Durkheim (1961), and the other two items refer to mainstream status symbols, the importance of owning an apartment or house and of owning a car. Persons high on individualism and with antibourgeois tendencies should identify less with these values and objects.

The specific question referring to basic value orientations posed to the art publics of Vienna, Hamburg, and Paris was: "What makes life worth living? What do you find especially important and what less important?" The six items of the traditionalism scale read as follows: "Having children," "Having a strong religious conviction," "Having a motherland," "Involvement in a partnership," "Owning an apartment or house," and "Owning a car." For each of the six items the three art world samples were asked to rate importance on a four-point Likert-type scale with four categories: (1) very important, (2) fairly important, (3) fairly unimportant, and (4) totally unimportant.

2 Latent Class Analysis

Latent class analysis is used to identify subgroups of persons, called classes, who are homogeneous in their value structures (see, for example, McCutcheon, 1987b, and Chapter 32). LCA for ordinal data, an extension of the pioneering work of Lazarsfeld (1950) and an integration of latent trait and latent class models by Rost (1988a, 1988b), opens the possibility of checking to what extent the gradations of the four response categories of the six items of the traditionalism scale are interpreted in the same way.

The basic concept of LCA for ordinal data is the concept of thresholds. A threshold is the point at which the probability of two adjoining response categories is equal. The category probabilities are parameterized by the thresholds. A high response probability corresponds to a large difference between threshold values. Different models are distinguished by their restrictions on the thresholds. We use the program LACORD (Rost, 1990) to estimate the threshold parameters and also to search for the appropriate number of classes.

Ordinality of the manifest response categories is given empirically if the estimated thresholds are ordered. Similarly, the latent classes can be ordered if a rank order is observed for all items between the classes. LCA was applied to the individuals with no missing values on the six items of traditionalism, leading to reduced sample sizes of Vienna, $n = 501$; Hamburg, $n = 524$; and Paris, $n = 306$.

On the whole, the results show a remarkable similarity for the art worlds of Vienna and Hamburg. For both samples the method identifies three latent classes, whereas for Paris only two classes are identified. Figure 1 gives the expected scores in each class for the six items as well as the item means in each of the three samples.

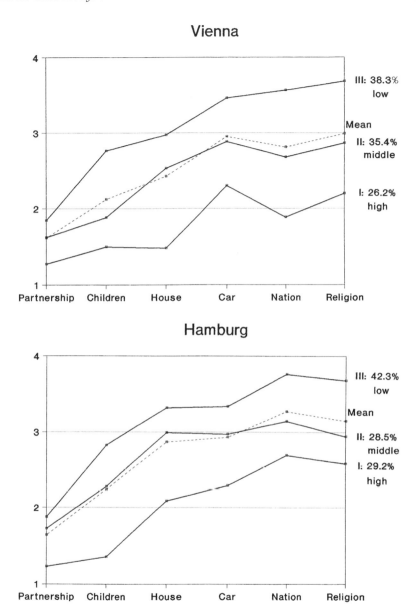

Figure 1: Traditionalism in the art worlds of Vienna, Hamburg, and Paris. Profiles of expected values. Latent class analysis ($n = 1311$).

Figure 1 shows that the expected item scores for all classes in each of the three samples have the same order across all six items of the scale. This means that the classes themselves are ordinal. Thus for the art worlds of Vienna and Hamburg it is possible to speak of these classes as being characterized by "high" (I), "middle" (II),

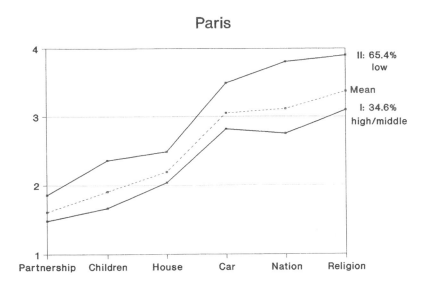

1= very important; 2= fairly important;
3= fairly unimportant; 4= totally unimp.

Figure 1: (*continued*)

and "low" (III) traditionalism. The sizes of the latent classes, which are estimates of the proportions in the populations, are given as percentages in Figure 1. The class sizes are rather similar in both art worlds. The largest groups are those with low traditionalism. The Vienna art world is characterized by a larger "middle class" of traditionalism than the art world of Hamburg (35.4% vs. 28.5%). Both latent middle classes display a profile that is similar to the profile of the manifest item means. The class with low traditionalism, which represents best the values ascribed to the avant-garde by Bourdieu or Bell, is a bit larger in Hamburg than in Vienna (42.3% vs. 38.3%). The rank order of the expected mean scores for the six items is approximately the same in the three cities, the only differences lying in the ordering of the two least-valued items in all three art worlds, nation and religion. In Vienna and in Paris nation is valued higher than religion; in Germany the reverse holds true. With regard to Austria and Germany, these differences may reflect the divergent political reactions of the two countries to the experience of National Socialism, which also had far-reaching implications for the reconstruction of the art systems in these countries. Germany was decentralized and national identification deemphasized, whereas in Austria the political parties in power tried to construct a new national identity especially in contrast to the former widespread German one. The biggest difference between the two art worlds of Vienna and Hamburg is the degree of national identification. In Vienna, the importance of nation in the middle class, II, is as high as in the high-traditionalism class I of the Hamburg sample. In a similar but less pronounced manner, this is also true for holding a strong religious conviction and for owning an apartment or house.

In Paris the group with low traditionalism (II) amounts to nearly two thirds of the art world (65.4%). In comparison with the low-traditionalism groups of Vienna and Hamburg, there is no difference with regard to partnership or car. Children and ownership of a house or apartment, however, are clearly more important for the low-traditionalism group in Paris. On the other hand, religion is even less important. It nearly reaches the extreme value of 4 ("totally unimportant"). Identification with the "motherland" is extremely unpopular in this group as well. In this respect, the difference from Vienna is stronger than that from Hamburg. Since this group is much larger in Paris than in Vienna and Hamburg, one of the main differences between the French- and the two German-speaking art worlds is the greater number of persons who do not identify at all with religion, nation, the church, and the state. Group I, with high traditionalism, is also different from the high-traditionalism groups in the other art worlds in being less traditional with regard to partnership, children, car, nation, and religion. The group in Paris, characterized by a higher degree of traditionalism, is less conservative than group I in Hamburg and group I in Vienna. LCA shows that the Paris art world is much more homogeneous with respect to social values and symbols of status than those of Hamburg and Vienna.

3 Results of Correspondence Analysis

Statistical models imply a certain philosophy of the social, of action, and of causality. CA is a method that, as Bourdieu (1991, p. 277) put it, "thinks" in relations. CA is especially attractive for our purposes, because it allows us to represent and explore the relations between social positions and value orientations in a graphical way at a low level of abstraction. Apart from the study of the basic relations between center and periphery with regard to high, middle, and low traditionalism, the inclusion of secondary factors that might differentiate between the value orientations (e.g., center and periphery in a geographical sense) is possible. The rows of the tables analyzed by CA are the latent classes identified by LCA and our additional group reincorporating respondents who gave no answer to one or more of the attitude items. Thus there are four rows in the case of Vienna and Hamburg and three rows for Paris.

The column variable in our analyses is based on the four categories combining two dichotomous measures. One refers to the social position in the art world. We differentiated between social center and periphery on the basis of a question that asked how intensively one is occupied with contemporary fine arts. Those who responded "almost every day" were classified as center, the rest as periphery. The validity of this measure was tested with the help of many indicators, such as being an artist, having studied fine arts, having many artists as friends, or attending many openings. All these indicators turned out to be highly correlated with that measure. The other variable differentiates those living in the city where the exhibition was shown from those coming from outside. Exhibitions of international art attract not only local residents but also visitors from all over the country. The proportions of those coming from outside (but from the same country) amount to about a third (34.6%) of the audience in Vienna, to 39.5% in Hamburg, and to 38.8% in Paris.

The proportion specializing in contemporary art and thus classified as center is lowest among the local residents of Hamburg (22%) and the highest among the local population of Paris (46%), with the Viennese in between (30%). The proportions of the center persons among the external visitors are 34% for Hamburg, 24% for Vienna, and 39% for Paris. Thus, in Paris and Vienna there are more professionals among the locals, whereas for Hamburg, which has no comparable position of structural, strategic, and cognitive dominance in the country (see DiMaggio, 1993, p. 195 ff.), this relation is reversed.

The CA visualization can be enriched by the inclusion of supplementary points (e.g., Greenacre, 1993a, p. 96 ff.). Thus, in addition to the four general social positions, we consider four special art world groups as supplementary points: artists, art critics, curators, and collectors. Most of the members of these groups belong to the center of the art world. Because they represent the spheres of production, distribution, and consumption of art, or, in a different theoretical frame, the fractions of cultural capital (artists, critics, curators) and of economic capital in the field of art, some differences between them on the level of value orientations are to be expected. Bourdieu's theory of the homologies of social position, habitus, attitudes, signs, and practices (Bourdieu, 1984, p. 128 ff.) implies that the cultural capital groups should identify less with traditional values than the collectors. In social space most collectors represent the art-consuming part of the class fractions with high economic capital.

The biggest of these groups is the artists. The cumulative sizes of the four art world groups selected reveal the high degree of self-referentiality of contemporary art in a social sense. Consumers are at the same time producers and mediators of art to a high extent. Seen from a comparative perspective, the Paris audience with these four groups constituting 52.2% of the total sample is clearly the most specialized and the Hamburg audience with 28.5% the least (Vienna, 36.2%). The position of Paris corresponds to Fleck's (1996, p. 25) description of the Paris art field as a "self-referential system" and a scene "nearly exclusively turning around itself." The percentages for the single groups are: (1) artists: Hamburg 16.9%, Vienna 20.2%, Paris 24.6%; (2) collectors: 2.9%, 5%, 10.5%; (3) art critics: 3.8%, 6.7%, 8.7%; (4) curators: 4.9%, 8.8%, 8.4%. Collectors we term the small self-defined part of the buyers of art in the audience who indicate that they are buying art objects not spontaneously or temporarily only but in a systematic way in order to build up a "collection."

3.1 The Art World of Vienna

Figure 2 refers to the Vienna art world. It is a symmetric display based on a simple CA. The four social positions in the art world—Vienna and center in the art world (labeled VIENNA CENTER), Vienna and periphery in the art world (VIENNA PERIPHERY), other Austrian cities and center in the art world (ELSE CENTER), and other Austrian

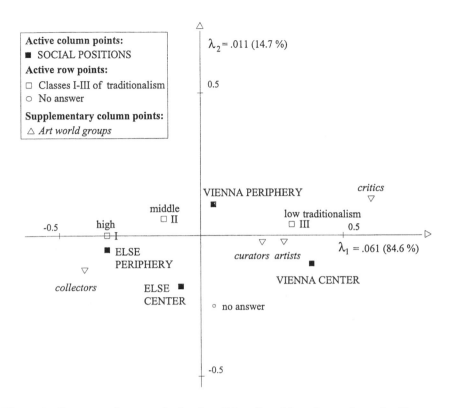

Figure 2: Correspondence analysis of traditionalism (three latent classes) with center and periphery of the art world in Austria (Vienna vs. else). First and second axes, 99.3% of total inertia represented (Vienna 1993, $n = 616$).

cities and periphery in the art world (ELSE PERIPHERY)—are represented as black squares. High, middle and low traditionalism are represented as white squares with numbers corresponding to the labeling in Figure 1, and "no answer" is indicated by an empty circle. The four art world groups projected in as supplementary column points are represented by empty triangles.

Nearly all the total inertia is represented in the plot. The first axis is much more important than the second, explaining about 85% of the total inertia. The first axis is determined on the left side by high traditionalism (CTR = 0.36) and on the right side by low traditionalism (CTR = 0.54), thus showing the opposition between high and low traditionalism in which we are mainly interested. Correspondingly, both groups of visitors living in the Austrian provinces are situated on the left side, opposing the Viennese visitors on the right side. That means that center and periphery matter in a social as well as in a geographical sense; the effect of center versus province is stronger than the effect of center versus periphery of the art world. Within the two geographical groups, the center art world subgroups are both farther to the right in the direction of low traditionalism, especially VIENNA CENTER.

The projection of the four special art world groups shows that groups with high specific cultural capital are situated on the right side of the first axis, tending toward low traditionalism, whereas collectors are on the left side, where traditionalism is high. In their value orientations collectors are similar to nonspecialized visitors from the province. Curators, artists, and critics represent the low degree of traditionalism characteristic of the center of the Viennese art world. Among them, art critics are the least traditionally orientated group. Curators and artists in the Austrian art world are "neighbors," indicating a high degree of similarity with regard to traditionalism.

3.2   The Art World of Hamburg

In the Hamburg sample (Figure 3), again almost all of the inertia is represented by the first and second axes. It also shows the contrast of high and low traditionalism. On the left side it is strongly determined by high traditionalism (CTR = 0.73) and on the right side by low (CTR = 0.18). An important difference from the Vienna

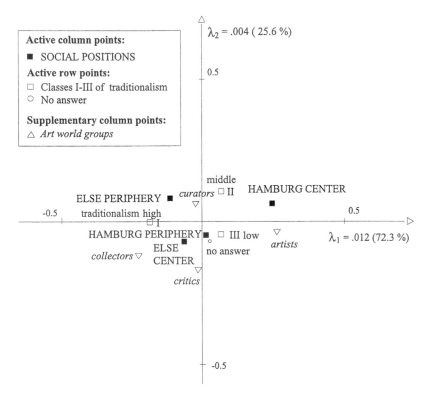

Figure 3: Correspondence analysis of traditionalism (three latent classes) with center and periphery of the art world in Germany (Hamburg vs. else). First and second axes, 97.9% of total intertia represented (Hamburg 1993–94, $n = 583$).

sample concerns the amount of variation in the data. In the Austrian sample the inertia explained by the first and the second axes amounts to 0.072, in the German sample only to 0.016. This indicates that the associations in Hamburg are much weaker than in Vienna.

Otherwise, the basic results are quite similar, with the center of the Hamburg art world on the side of low traditionalism, a greater contrast being between the geographical groups, and, as in Austria, the differences between center and periphery in the art world more pronounced among those living in the city.

Considering the supplementary art world groups, there are signs of inhomogeneity of the center again, in this case explaining the low associations between center versus periphery and value orientations. Collectors are clearly the most traditional group. In contrast to Vienna, critics and curators do not differ much from the average of the members of the art world. Only among artists is traditionalism low.

3.3 The Art World of Paris

For Paris, represented in Figure 4, all the inertia is explained, because the data are two-dimensional. The first axis explains 95.8% of the inertia, but its value (0.014) is nearly as small as in Hamburg. This axis, however, is determined not by one of the latent classes of traditionalism but by the group "no answer" (CTR = 0.83). The tendency not to respond to all six attitude items is highest among the specialized individuals living in Paris. There is nearly no differentiation of value orientations with respect to social position in the art world, apart from the tendency to express these clearly in a questionnaire. The "no answer" proportions are higher among artists and collectors than among curators and critics.

The second axis shows a very small difference between low and high traditionalism. In contrast to Vienna and to Hamburg, in Paris the collectors are not the group that shows the highest degree of traditionalism. It comes perhaps as a surprise as well that artists in Paris are not less conservative than the average of the members belonging to the "community of taste" (Becker, 1986, p. 76). Fleck (1996, p. 25), a well-informed art critic, in his report on the Parisian subfield that we investigated, hints at processes of "self-provincialization" in the Paris art world since the end of the eighties due to state intervention, which might partly explain these results.

4 Conclusion

The assumptions, mainly based on the American experience, regarding the assimilation of the artistic and intellectual milieu of avant-garde art to the values of the "middle class" Crane (1987) or to the "predominant values" Gablik (1985) were reformulated and applied to three European art worlds. Whereas the assumption of a dissolution of the boundaries between art and society is clearly refuted by DiMaggio's (1996) analyses of the GSS data in the American case, we found mixed evidence concerning the internal differentiation of the European art worlds. Our findings show that Vienna

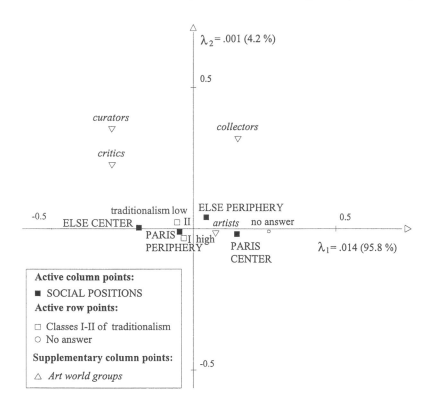

Figure 4: Correspondence analysis of traditionalism (two latent classes) with center and periphery of the art world in France (Paris vs. else). First and second axes, 100% of total inertia represented (Paris 1995, $n = 358$).

is still characterized in its center by the individualistic and antibourgeois tendencies described by Bell and Bourdieu, for example. On the other hand, social position does not differentiate the Paris art world in this respect. In view of the small weight of the vertical distances between center and periphery in the Paris sample, the hypothesis that the center of the art world forms a subculture, in which Durkheimian egoism is much higher and negation of communitarian bourgeois values and mainstream status symbols is much lower than in the groups much less involved in art could not be confirmed in any convincing way.

The Hamburg art field is neither as strongly differentiated as the Vienna field nor as homogeneous as the field of Paris. The basic associations about differences between center and periphery are supported by systematic distinctions in value orientations among special art world groups. The opposition between cultural and economic capital in the field of visual art, emphasized by Bourdieu, is still characteristic for the Hamburg art world, at least when producers and economic appropriators of art are considered. Art mediators (critics and curators) in that social world, however, do not differ from the average.

What can be learned from our findings is that specifications of the "death of the avant-garde" proclamations according to social and cultural context are of great importance. One reason for the differences found might be that the German and the French art worlds show more features of postmodernism, with traditional "symbolic boundaries" (see Lamont and Fournier, 1992) being eroded to a higher extent than in Austria, which culturally is a rather conservative country. Another reason might be connected to the histories and the structural frames of the art worlds themselves. Whereas in Austria the central state intervenes heavily in the art system, in Germany— because of the experiences with National Socialist cultural centralism—state funding as well as control of the visual arts is still rather negligible. Because especially in Vienna much public money goes into the arts, Austrian artists and their constituencies are public persons much more under social and mass media control than those in Germany. The history of Austrian avant-garde art after World War II is also a history of artistic scandals and revolts against bourgeois society—from Viennese activism in the sixties, to criticism of the actions of former President Kurt Waldheim and the restrictive immigration laws of the nineties (e.g., "Art and Politics," an exhibition in 1994 in the hall of the Austrian parliament in Vienna, partly censored by the state), and neoactivist sexual transgressions in exhibition spaces funded by public money (e.g., "Jetztzeit," Kunsthalle Wien, 1994). Nothing comparable could be observed in the past two decades in Hamburg, an autonomous town, where the art institutions represent only the local community, not the central state. Thus confidence in our findings in this respect is enhanced by external, historical data.

One might argue that the field of Paris is characterized by a high degree of state intervention, too. In Paris, however, modern art and its symbolic transgressions have a long history. It is much more absorbed and tolerated than in Austria, where nothing like the "reeducation" of Germany, including taste (e.g., the "documenta" exhibitions at Kassel), took place. That there are no differences between local and external visitors in Paris seems to be more difficult to explain. Fleck (1996) emphasizes that due to heavy cultural political interventions since the early eighties (the era of Mitterand and Jack Lang), France had constructed "the best decentralized exhibition landscape in Europe." This diffusion of contemporary art and its institutions all over the country (e.g,. the 22 FRAC—"Fonds Régionaux d'Art Contemporain"—and the network of "Centres d'Art"), which Chirac and the conservatives have begun to stop recently, might partly explain that the old distinction between capital and province, at least in the field of visual art, was not as important in the last decade as before. We conclude that pure social-structural explanations are relevant and fruitful but that cultural and historical factors have to be considered as well.

Acknowledgments

The study "Art worlds—a comparative study," directed by the second author, was supported by grants of the Ministry of Science, Art and Traffic (Vienna), by grants of the University of Lüneburg, Department of Cultural Studies, and by grants of the

Cultural Administration of Hamburg. We thank Vera Kockot for producing the figures and Beatrice Prent, Raphale Jeune, and Hans Ulrich Lë Lobrist, who at the time of our survey were curators at the Musée d'Art Moderne de la Ville de Paris, for their kind support and cooperation.

Software Notes

The models of LCA for ordinal data are programmed in the computer program "LACORD" (Rost, 1990), available from I. P. N. Kiel, Olshausenstr. 40, D-24098 Kiel, Germany.

Chapter 15

Graphing Is Believing: Interpretable Graphs for Dual Scaling

Shizuhiko Nishisato

1   Introduction

Visual display of quantified rows and columns in a joint space has been an almost routine procedure for data analysis. As mentioned several times in this book, there are three widely accepted choices of coordinates:

1. Asymmetric mapping: standard (normed) coordinates for rows and principal (projected) coordinates for columns
2. Asymmetric mapping: principal coordinates for rows and standard coordinates for columns
3. Symmetric mapping: principal coordinates for both rows and columns

The first two are true joint maps in that they are visualizations of projections of row and column points in the same space and that column points in 1 (respectively, row points in 2) are at average positions of row points (respectively, column points). The third choice is often used for the reason that row points and column points have the same norm for each axis. The fourth possible choice, which uses standard coordinates for both rows and columns, must be rejected because joint display can neither reproduce input data nor reflect the relative importance of axes.

No matter which one of the three acceptable choices one may adopt, the ultimate goal of visual display must be to facilitate an interpretation of quantified outcome. Furthermore, the graph should be such that it enhances the meaning of the old adage "seeing is believing." To this end, we will consider the joint graphical display for dual scaling of two types of categorical data, dominance data (e.g., rank order, paired comparison, successive category data) and incidence data (e.g., multiple-choice data, sorting data, contingency tables) (Nishisato, 1993). As typical examples of the respective data types, we will look at rank order and multiple-choice data.

2   An Interpretable Graph for Rank Order Data

Coombs (1950) proposed a model for analyzing rank order data, called the unfolding model, in which he postulated a unidimensional continuum, called a J scale, along which both subjects and objects are located. The decision rule is that a subject ranks first the stimulus that is located closest to him and ranks the rest of the objects according to the order of their distances from him. In this model, a given set of objects ranked by a subject, called an I scale, can be regarded as the ranking of the objects on the continuum folded at the subject's position, called his ideal point. Depending on the location of an ideal point, it is easy to see that a folded continuum results in a different ranking of the objects. Coombs extended the underlying continuum to multidimensional space, leading to the problem of multidimensional unfolding, in which the main task is to determine subject's positions and positions of objects in such a way that the rank order of distances from each subject to the objects is the same as the ranking of the objects by that subject.

Historically, the problem of multidimensional unfolding has been investigated by a number of researchers (e.g., Coombs, 1964; Coombs and Kao, 1960; Schönemann, 1970; Gold, 1973; Heiser, 1981). The same problem has also been handled as a quantification problem of rank order data by such investigators as Guttman (1946), Slater (1960), and Nishisato (1978, 1994, 1996). In particular, it was noted (Nishisato, 1994, 1996) that dual scaling of subject-by-object rank order data always recovers the data perfectly in the full space solution *provided* that asymmetric mapping is used. The unfolding framework can therefore be used as one in which the joint graph is to be interpreted.

Following Nishisato (1978), the ranking of objects j and k by subject i is coded as follows:

$$f_{i,jk} = \begin{bmatrix} 1 & \text{if subject } i \text{ judges} & j > k \\ 0 & \text{if the judgment is} & j = k \\ -1 & \text{if the judgment is} & j < k \end{bmatrix}$$

The basic unit of analysis is called the dominance number of object j for subject i, e_{ij}, which is defined as the number of times object j is ranked earlier than the other $(n - 1)$ objects minus the number of times it is ranked after them by subject i. For N subjects and n objects, the $N \times n$ dominance matrix is denoted by $\mathbf{E}$, with (i, j)th element

$$e_{ij} = \sum_{k=1}^{n} f_{i,jk}$$

In the case of rank order data e_{ij} can be simplified to the following form (de Leeuw, 1973; Nishisato, 1978):

$$e_{ij} = n + 1 - 2K_{ij}$$

where K_{ij} is the rank of object j given by subject i.

Assuming that each element e_{ij} is the outcome of $(n - 1)$ comparisons, the optimal score vector (in standard coordinates) for subjects and the corresponding score vector for objects on dimension k are given by

$$\mathbf{x}_k = \frac{c}{\rho_k} \mathbf{E} \mathbf{y}_k, \qquad \mathbf{y}_k = \frac{c}{\rho_k} \mathbf{E}^T \mathbf{x}_k$$

where

$$c = \sqrt{\frac{1}{Nn(n-1)^2}}$$

and ρ_k is the square root of the eigenvalue for dimension k.

It is known (Nishisato, 1994, 1996) that only the asymmetric mapping of $(x_k, \rho_k y_k)$ for all dimensions k provides a perfect solution to the problem of multidimensional unfolding and further that the existence of a perfect solution does not depend on the relative sizes of the number of subjects, N, and the number of objects, n. The last statement may be difficult to accept, considering that a number of papers on the problem of multidimensional unfolding have discussed the conditions under which a perfect solution might be obtained. Let us look at an example to see what a "perfect solution" means.

Ten subjects ranked the following six plans for a Christmas party according to the order of their preference:

1. Potluck in the group room during the day
2. Pub/restaurant crawl after work
3. Reasonably priced lunch in an area restaurant
4. Evening banquet at a hotel
5. Potluck at someone's home after work
6. Ritzy lunch at a good restaurant

Table 1 contains the 10×6 input data, the dominance matrix $\mathbf{E}$, and the first two solutions, each consisting of two columns: standard coordinates of subjects and the corresponding positions of the six Christmas party plans in the principal coordinates. The first two solutions account for 72% (45% and 27%, respectively) of the total information. Figure 1 shows the plot of those subjects and the party plans using the

Table 1: Rank order data, dominance table, scores

		Subjects (rows)		Objects (columns)	
Data	Dominance table	Sol. 1	Sol. 2	Sol. 1	Sol. 2
6 1 5 4 3 2	−5 5 −3 −1 1 3	1.06	0.84	−0.63	−0.11
2 6 3 5 4 1	3 −5 1 −3 −1 5	−0.16	−1.74	0.38	0.57
6 1 5 4 2 3	−5 5 −3 3 −1 1	1.17	1.16	−0.15	−0.37
3 5 2 4 1 6	1 −3 3 −1 5 −5	−1.28	0.04	0.24	0.17
3 4 2 6 1 5	1 −1 3 −5 5 −3	−1.11	−0.08	−0.48	0.20
5 3 1 4 6 2	−3 1 5 −1 −5 3	0.89	−0.93	0.63	−0.46
1 2 4 5 3 6	5 3 −1 −3 1 −5	−0.99	0.93		
4 3 2 6 5 1	−1 1 3 −5 −3 5	0.63	−1.11		
2 1 4 5 3 6	3 5 −1 −3 1 −5	−0.67	1.29		
6 1 4 3 5 2	−5 5 −1 1 −3 3	1.40	0.53		

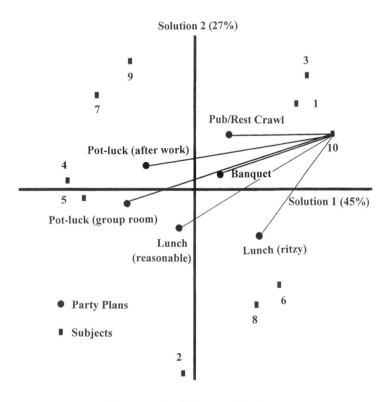

Figure 1: Rank-2 approximation.

coordinates listed in Table 1. Solution (axis) 1 divides the parties into those that are expensive (ritzy lunch, banquet, pub/restaurant crawl) and those that are inexpensive (potluck in group room, potluck in someone's home, reasonably priced lunch), and may therefore be called the cost factor. Solution 2 categorizes the plans into daytime parties (ritzy lunch, reasonably priced lunch, potluck in the group room) and evening parties (potluck after work, banquet, pub/restaurant crawl), indicating the time factor. Considering that 72% of the total information is accounted for by these solutions, one may conclude that subject's rankings largely reflect those two underlying factors, cost and time.

Table 2 contains Euclidean distances, calculated from Figure 1, between subjects and party plans and the rankings of these distances within each subject, which is referred to as the rank-2 approximation to the input ranking. Without further investigation, the rank-2 approximation looks good, corroborating that the first two solutions account for a substantial amount of information.

We can explore other possible approximations, starting with the rank-1 approximation, using only the first solution, all the way up to the rank-5 approximation to the input data. Because the rank of the dominance matrix is five, the rank-5 approximation is the highest degree we can consider. To indicate goodness of approximations, let us calculate the sum of squared discrepancies between the input data and the approximated ranks for each subject and each approximation. Table 3 shows the summary. It is interesting to note that goodness of the approximation shows individual differences. For example, subject 3 needs only the first two solutions to recover the ranking, and subjects 8, 9, and 10 can fit perfectly in the three-dimensional space. Notice also that the rank-5 approximation perfectly reproduces the input ranks. In other words, it is a perfect solution to the problem of multidimensional unfolding.

In conclusion, when we have an $N \times n$ rank order matrix, we can always map subjects and objects in $(n - 1)$-dimensional space or N-dimensional space, whichever is the smaller, in such a way that the rankings of the distances between subjects and objects are exactly the same as those in the input data, provided that asymmetric

Table 2: Euclidean distances and ranking

Distances						Ranking of distances					
1.94	0.73	1.72	1.06	1.67	1.37	6	1	5	2	4	3
1.72	1.58	1.18	1.27	1.77	0.54	5	4	2	3	6	1
2.20	0.98	2.02	1.36	1.91	1.71	6	1	5	4	2	3
0.67	1.75	1.21	1.53	0.81	1.98	1	5	3	4	2	6
0.49	1.63	1.01	1.38	0.69	1.79	1	5	3	4	2	6
1.70	2.38	1.37	1.95	1.96	1.51	3	6	1	4	5	2
1.10	1.42	1.55	1.45	0.89	2.14	2	3	5	4	1	6
1.61	1.70	1.07	1.33	1.71	0.65	4	5	2	3	6	1
1.40	1.28	1.74	1.45	1.11	2.18	3	2	5	4	1	6
2.13	1.02	1.79	1.21	1.91	1.25	6	1	4	2	5	3

Table 3: Sums of squares of discrepancies in ranks between input ranks and rank-K approximation

Subject	$K = 1$	2	3	4	5
1	8	6	6	6	0
2	42	22	6	4	0
3	8	0	0	0	0
4	6	6	6	0	0
5	12	12	8	0	0
6	14	14	4	4	0
7	12	8	6	0	0
8	18	14	0	0	0
9	20	8	0	0	0
10	2	2	0	0	0

mapping is used. In contrast, symmetric mapping does not provide a perfect solution even when the inertia is comparatively high.

When the researcher has biographical information about subjects (e.g., gender, age group, socioeconomic class), the mapping of subjects provides an opportunity to identify any clusters of them in terms of such information. Although traditionally many researchers may be interested only in the configuration of objects, there exists a definite opportunity for finding additional clues for interpreting the outcome by looking at joint graphs of both subjects and objects.

As for other examples of dominance data such as paired comparisons and successive categories data, the same idea of the joint graph can be extended to them because dual scaling of those data can be formulated by handling ranking information or, more specifically, ordinal information contained in pairwise and between-set rankings (Nishisato and Sheu, 1984).

3   An Interpretable Graph for Multiple-Choice Data

Within the framework of Coombs' multidimensional unfolding analysis the asymmetric graph for rank order data, discussed earlier, can be characterized as the only legitimate graph. When we consider multiple-choice data, however, there seem to be more possibilities for choosing a graph than in the case of dominance data, and the problem of interpretation therefore becomes even more relevant to and important than that of dominance data.

Suppose that we adopt symmetric mapping with principal coordinates, because this is one of the most widely used methods. With this assumption, Nishisato (1988) presented some 20 measures of badness of joint graphical display. Greenacre (Chapter 17) shows that the graph can be evaluated on how many items of input data are recovered by identifying which option in standard coordinates from each question lies closest to the subject point in principal coordinates. These ideas are helpful for evaluating a given choice of mapping, that is, the symmetric mapping, but what if some of those measures indicate that symmetric graphs are problematic most of the time? What if asymmetric mapping, too, suffers from similar problems? What characteristics should a graph for multiple-choice data possess?

In the current chapter, we will look at a different approach to an interpretable graph, which meets some desiderata for graphical display of multiple-choice data, as summarized by Nishisato and Nishisato (1994, p. 124):

1. From the graph, we should be able to see the information contained in the data matrix—for example, plot subjects and response options of a given data set, and ask if the graph can tell you which options a particular subject has chosen.
2. The position of each point in the graph should not be unduly influenced by the frequency of the data point.
3. The space for the graph should be a well-defined one such as Euclidean space; overlaying two different spaces onto one is out of the question.

What appears to be a plausible method satisfying these points is the one used in Nishisato (1990). Since the method is not well known, let us use a numerical example to introduce it, discuss its possible criticisms as well as justifications, and reassess its potential as an interpretable graph for multiple-choice data.

The method considers information contained in subject's response patterns for graphical display. Bahadur (1961) presented a model for binary response patterns, in which he expressed the probability of each response pattern over n binary items as the sum of the item means, two-item interactions, three-item interactions, and all the way up to the n-item interaction. His model shows an example in which response patterns contain all conceivable multiway associations of the variables. To see how informative response patterns are, try to rearrange the rows (subjects) of the input data matrix in order of their coordinates on axis 1 and see a systematic change of response patterns as a function of the coordinates (see Nishisato, 1994, p. 159). Noting this, one can consider an informative graph as plotting subjects only and labeling them by their response patterns.

This simple method was used by Nishisato (1990). Notice that subjects are plotted in multidimensional Euclidean space using principal coordinates and that the label of each subject tells us which options the person has chosen and which subjects have chosen a specific option.

Example: Before evaluating this simple idea critically, let us look at an application of the method to the data in Table 4. Because there are four items with three options per item, one can expect $12 - 4 = 8$ solutions. Of these solutions, three show a correlation ratio greater than the expected value, that is, $1/n$ (Nishisato, 1980, 1994). The correlation ratios are 0.65, 0.45, and 0.29, respectively. For illustrative purposes, we will examine only the first two solutions, which account for 79% of the total of the three values. Figure 2 shows the plots of subjects in principal coordinates in the two-dimensional solution. Boundaries of clusters are determined by common sets

Table 4: Adults' views of children (Singapore data)

Questionnaire	Subject	Data Item 1	2	3	4
[Item 1]:					
How old are you?					
(1) 20–29	1	3	1	2	1
(2) 30–39	2	2	1	3	2
(3) 40 or older	3	2	1	2	2
[Item 2]:	4	1	2	2	3
Children today are not	5	3	1	2	2
as disciplined as when I	6	1	3	1	2
was a child.	7	2	1	2	2
(1) agree	8	2	1	1	2
(2) disagree	9	1	2	3	1
(3) I cannot tell	10	3	1	2	1
[Item 3]:	11	1	2	2	3
Children today are not	12	2	1	1	1
as fortunate as when I	13	2	1	3	3
was a child.	14	3	1	2	1
(1) agree	15	1	1	2	3
(2) disagree	16	3	1	2	1
(3) I cannot tell	17	3	1	1	1
[Item 4]:	18	2	3	2	2
Religions should be	19	3	1	2	1
taught at school.	20	2	1	2	2
(1) agree	21	1	3	3	3
(2) disagree	22	2	1	2	2
(3) indifferent	23	1	3	3	3

of response patterns. In other words, the subjects in a cluster share the same subset of identical responses, a source of information used to maintain interpretability. As illustrated in Nishisato (1994), subjects who have chosen particular options of an item are distinctly and tightly clustered if the item is highly correlated with the two graphed solutions. This is clearly shown in [A] and [B] in Figure 2. Age groups 20–29 (1***), 30–39 (2***), and 40+ (3***) are distinctly separated in [A], where an asterisk indicates a choice of any option of the corresponding item. Similarly, patterns (***1), (***2), and (***3) clearly partition the subjects in [B], indicating that those are subjects who agree with teaching of religions, disagree, and are indifferent, respectively. By the same token, if the item has low correlations with

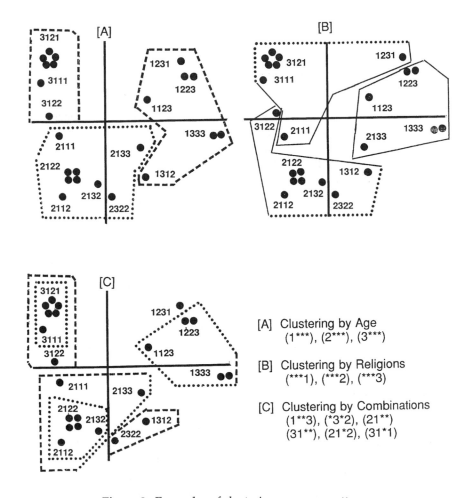

Figure 2: Examples of clustering response patterns.

the two solutions, subjects cannot be cleanly clustered in terms of the options of the item.

From this finding, we can go one step further and find a set of items that cluster subjects into comparatively tight subgroups in terms of combinations of their options, as shown in [C]:

- Age 20–29 and indifferent to teaching religions (1**3)
- Unsure if children are less disciplined but against teaching of religions (*3*2)
- 30–39 years of age and children are not disciplined (21**)
- Age 40+ and children are not disciplined (31**)
- Age 30–39, children are not disciplined, but are against teaching religions (21*2)
- Age 40+, children are not disciplined, and religions should be taught (31*1)

These are only several of many possible examples, which include overlapping clusters such as (21**), (1**3), and (**33), the last one overlapping with the first two clusters.

Criticisms of this method are not so much from the theoretical point of view but more in terms of the implementation of the method in practice. The following are some of the conceivable criticisms and possible remedies for them:

1. When the number of items is large, the use of response patterns as labels is too cumbersome, if not impossible. This criticism is right, and it can be mitigated by introducing a key consisting of a set of codes to replace long strings of chosen options, such as A = (12*3*11***5) and B = (**33**111**).

2. There are too many ways to cluster subjects in terms of their responses to a particular set of items. This is also right. An alternative to the subjective clustering is to use a method of cluster analysis that provides an objective way to partition subjects into groups.

3. The method allows one to look at two or three solutions at a time, and for higher dimensional solutions there are too many combinations for a single graph (e.g., solution 1 versus solution 2, or 1 versus 3). This criticism is right, too. But, what alternatives do we have for a single graphical display of multidimensional solutions? To mention a few, there are Andrews curves (Andrews, 1972), the alternating monotone graph, the parallel graph, and the semicircular incremental radial graph (see examples of these in Figure 3). These graphs indeed show multiple solutions in the two-dimensional plane, but they are difficult to interpret and are typically used to depict only one set of variables, rows or columns. In other words, these are not for joint graphical display. It is nearly impossible to infer the relation between row variables and column variables, not to mention the communicability between the graphs and the input data. Therefore these multidimensional graphs are currently at best only of theoretical interest, and their interpretability aspect needs to be investigated further.

4. What if there are some missing responses? As long as the number of missing responses is very small, say a few percent, missing responses will not affect the respondents' positions in any serious way. With more missing data, one would have to employ a method of imputation for missing responses or include special categories for missing responses (see, for example, van Buuren and van Rijckevorsel, 1992; Nishisato and Ahn, 1994; and Chapters 16 and 18 in this volume).

4   Discussion and Conclusion

We have looked at two kinds of interpretable graphs. The asymmetric mapping of rank order data can be extended without difficulty to paired comparison data and successive categories data. We often collect only ranking of objects by subjects, but it would be particularly useful to obtain some background information about the subjects and supplementary information about objects. For instance, consider collecting not only

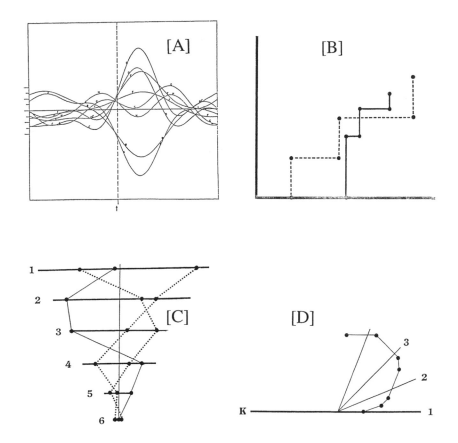

Figure 3: Multidimensional graphs: [A] the Andrews curves, [B] alternating monotone graph, [C] parallel graph, [D] semicircular incremental radial graph.

ranking of political issues from subjects but also information about subjects' political affiliations. In the asymmetric map of the political issues and subjects, we can now introduce a coding system that identifies the subjects' political affiliations. Similarly, ranked objects can also be coded to enrich the interpretations.

We have also looked at a graph that allows us to examine the relations between subjects and chosen options without overlaying two spaces (i.e., one for subjects and the other for options). The study shows that the graph usually provides interpretable clusters in terms of common options that characterize them; that is, the interpretation comes directly from those options. One can envisage computer software that can plot subjects who choose individual options of a given item as well as plot subjects who share specified response patterns. Such a program would be useful when the graph is intended for exploratory investigation of the data. Our experience suggests that the method presented here can easily be implemented for practical use and should be preferred to the popular method of joint mapping in principal coordinates.

In concluding, let us pose a final question. Can we replace "seeing is believing" with "graphing is believing"? No matter what answers we may hear, let us hope that it should be a goal for quantification research.

Acknowledgment

This study was supported by a research grant from the Natural Science and Engineering Research Council of Canada to the author.

Chapter 16

Interpreting Axes in Multiple Correspondence Analysis: Method of the Contributions of Points and Deviations

Brigitte Le Roux and Henry Rouanet

1 Introduction

In geometric data analysis, once a "cloud" of points has been constructed, as the outcome of correspondence analysis (CA), for example, or principal component analysis, the phase of interpretation follows. This phase is always a delicate one; at this point, the need to fill the gap between theory and practice appears essential—a need well reflected in the book edited by Greenacre and Blasius (1994a). In the French tradition of data analysis, *aids to interpretation* have been devised, such as the familiar table of contributions and supplementary elements. The method we will present in this chapter, namely the method of the *contributions of points and deviations*, directly extends the existing aids to interpretation. It stems from the following remark: in analysis of variance (ANOVA) terms, contributions of points to an axis are simply *parts of variance* accounted for by points. This leads to considering other parts of variance that are also used in ANOVA; for example, those that express *contrasts* among groups of observations. That is, it leads us to study the contributions of *deviations* between points. Indeed, all those who practice geometric data analysis are accustomed to think intuitively in such terms ("axis 1 opposes rich vs. poor, axis 2 old vs. young, etc."). From a theoretical viewpoint, the statistical interpretation of CA in ANOVA terms is well known; see Fisher (1940) and Tenenhaus and Young (1985). But we feel that the idea deserves to be fully elaborated.

This chapter will be devoted mainly to the first and basic phase of interpretation, namely that of the principal axes, in the case of multiple correspondence analysis (MCA). Henceforth we assume the data structure of a *questionnaire* in standard form; that is, there is a set of *questions*, together with, for each question, a set of *response modalities* (also called response categories)—including nonresponse whenever relevant—and each individual chooses one (and only one) modality of each question. Then consider the following two ideas taken from nested designs in ANOVA:

1. With each modality is associated one and only one question; in ANOVA terms, this means that the set of all modalities is *nested* in the set of questions. This prompts us to investigate—in addition to contributions of modalities—the *contributions of questions to axes* and also the *contributions of modalities to questions*.

2. For each question, each individual chooses one and only one modality, which means that for each question the set of individuals is *nested* in the set of the observed modalities of the question. In other words, each question generates a partition of the individuals indexed by the modalities of the question. This suggests that we investigate the *cloud of individuals* and its *subclouds* associated with modalities of interest.

The method of the contributions of points and deviations will be illustrated with data taken from the French Worker Survey.

2   The French Worker Survey

2.1   The Survey

The French Worker Survey (Adam *et al.*, 1970) was conducted in July 1969 on a representative sample of French workers—unskilled, specialized, and technicians—using a thorough battery of 70 questions, with the overall objective of "analyzing the political and social behavior of the working class."

At the time of the survey, presidential elections had just taken place, opposing the candidates of the four main political families, along the traditional range from left to right: Communist (Duclos), Socialist (Defferre), Center (Poher), Gaullist (Pompidou, who won the election). One objective of the survey was to inquire about this traditional dimension in the specific population of workers; other objectives were to identify and interpret other important dimensions, possibly specific to this population. For instance, the communist dominance among workers was beyond doubt (although already on the decline), but the influences and roles of the noncommunist left and of center were not so well delineated. Also, what needed to be clarified were the relations and interplay between political attitudes and attitudes toward trade unions. The leading trade unions were the CGT (with notorious links with Communist party), then—far behind—CFDT, FO (both loosely linked with the noncommunist left), and "autonomous" (inclined toward right wing).

Table 1: The four basic questions and their relative frequencies

Professional elections ($q1$). In professional elections in your firm, would you rather vote for a list supported by:

1. CGT	.3298
2. CFDT	.0877
3. FO	.0782
4. CFTC	.0248
5. Autonomous	.1077
6. Abstention	.1525
7. Nonaffiliated list	.1049
8. NR	.1444

Union affiliation ($q2$). At the present time, are you affiliated to a Union, and in the affirmative, which one:

1. CGT	.2107
2. CFDT	.0524
3. FO	.0229
4. CFTC	.0048
5. Autonomous	.0210
6. CGC	.0114
7. Not affiliated	.6663
8. NR	.0105

Presidential election ($q3$). On the last presidential election [1969], can you tell me the candidate for whom you have voted?

1. Jacques Duclos (Comm)	.2221
2. Gaston Defferre (Soc.)	.0467
3. Alain Krivine	.0095
4. Michel Rocard	.0286
5. Alain Poher (Center)	.1420
6. Louis Ducatel	.0067
7. Georges Pompidou (Gaullist)	.2336
8. NRAbst	.3108

Political sympathy ($q4$). Which political party do you feel closest to, as a rule?

1. Communist [PCF]	.1935
2. Socialist [SFIO+PSU+FGDS]	.1697
3. "Left" ("Party of workers",...)	.0429
4. Center [+MRP+RAD.]	.1192
5. RI	.0086
6. Right [+INDEP.+CNI]	.0381
7. Gaullist [UNR]	.1335
8. NR	.2946

Note. Within union questions $q1$ and $q2$, there are correspondences between modalities (except 6), reflected by label numberings. Similarly for modalities 1, 2, 7, and 8 of political questions $q3$ and $q4$, the other label numbers being arbitrary. There are no such correspondences between union and political parties, except for the well-known affinities between CGT and Communist (modalities 1).

The analysis to be presented in this chapter is based on 1049 respondents and concentrates mainly on two questions about trade unions and two questions about political preferences. The four basic questions, each with eight modalities of response, are presented in Table 1, together with the associated relative frequencies. In Table 2 the 319 observed *response patterns* are given, along with their frequency counts.

(Let us briefly comment on the one-way tables.) From the union questions ($q1$ and $q2$), we see that 63% of workers vote for a list sponsored by some union ($q1$, modalities 1–5 and 7), more than half of them for CGT; 67% of workers, however, are not affiliated with any union ($q2$, modality 7). From the two political questions ($q3$ and $q4$), we see the high percentages of nonresponses, NR (31% and 29%). Among expressed sympathies, the communist party indeed comes first (19%) but is exceeded by noncommunist left-wing sympathies pooled together (21%, $q4$, modalities 2 and 3); also, the Gaullist pooled with other right-wing parties ($q4$, modalities 5 and 6) come up to 18%. Duclos' (22%) score is exceeded by Pompidou's (23%), and so on.

Table 2: 319 response patterns with frequency counts

1111	81	1712	4	2234	1	3356	1	4722	1	5751	1	6776	4	7787	4
1112	9	1717	1	2242	2	3357	1	4732	1	5752	3	6777	19	7788	16
1113	7	1718	7	2251	1	3358	1	4753	1	5754	10	6778	5	8111	1
1114	2	1721	1	2252	6	3374	1	4756	1	5756	3	6781	1	8113	1
1118	7	1722	5	2254	8	3377	2	4766	1	5757	3	6782	5	8152	1
1122	5	1728	1	2258	2	3378	1	4773	1	5758	7	6783	4	8154	1
1126	1	1738	1	2261	1	3384	1	4774	2	5772	1	6784	8	8181	2
1128	2	1742	2	2274	3	3388	2	4777	7	5774	5	6786	4	8182	1
1132	1	1748	1	2276	2	3554	1	4778	3	5775	1	6787	4	8188	2
1142	4	1751	3	2282	3	3614	1	4782	1	5776	2	6788	50	8288	1
1146	1	1752	5	2284	1	3662	1	5113	1	5774	14	7111	2	8322	1
1148	2	1754	3	2285	1	3711	2	5132	1	5778	4	7112	1	8588	1
1151	3	1757	1	2286	1	3712	1	5142	1	5781	2	7154	1	8677	2
1152	3	1758	4	2287	1	3713	1	5161	1	5782	2	7177	1	8678	1
1153	2	1771	1	2288	1	3714	2	5174	1	5784	1	7181	1	8711	3
1154	2	1772	3	2711	3	3722	3	5184	2	5787	3	7522	1	8712	4
1158	3	1774	3	2728	3	3724	1	5187	1	5788	9	7582	1	8713	1
1161	1	1775	1	2737	1	3732	1	5354	1	5876	1	7588	1	8718	4
1162	1	1776	2	2738	1	3751	1	5382	1	6116	1	7711	9	8741	1
1171	1	1777	7	2742	3	3752	2	5512	1	6172	1	7712	2	8742	1
1172	3	1778	5	2744	1	3754	4	5513	1	6178	1	7713	1	8751	1
1177	5	1781	8	2752	1	3755	1	5518	1	6181	1	7716	1	8752	1
1178	3	1782	9	2754	3	3756	1	5522	2	6182	1	7718	1	8753	1
1181	10	1783	3	2756	2	3758	5	5548	1	6188	2	7722	2	8754	1
1182	7	1784	4	2772	1	3774	4	5574	2	6528	1	7742	2	8757	1
1183	5	1786	1	2774	3	3775	2	5575	1	6676	1	7752	2	8758	4
1184	1	1787	2	2777	7	3776	1	5577	4	6711	8	7754	6	8765	1
1188	13	1788	26	2778	5	3777	7	5584	1	6712	1	7756	1	8774	2
1218	1	1858	1	2782	1	3778	4	5588	1	6714	1	7758	5	8776	2
1272	1	1881	2	2784	1	3782	3	5672	1	6718	5	7772	1	8777	12
1288	1	2111	1	2787	2	3783	1	5674	1	6722	3	7774	1	8778	9
1311	1	2132	1	2788	3	3784	3	5677	1	6742	2	7775	1	8781	2
1381	1	2154	1	3122	1	3787	1	5711	1	6752	3	7776	2	8782	2
1418	1	2178	1	3182	1	3788	6	5712	5	6753	3	7777	22	8783	5
1481	1	2211	2	3277	1	4241	1	5713	2	6754	6	7778	11	8784	3
1552	1	2214	1	3311	2	4254	1	5722	1	6756	2	7781	2	8788	37
1611	1	2218	1	3312	1	4274	1	5728	1	6758	4	7782	3	8822	1
1673	1	2222	7	3322	3	4441	1	5732	1	6771	1	7783	2	8878	1
1677	1	2223	1	3342	2	4477	2	5742	1	6772	3	7784	3	8888	5
1711	33	2224	1	3354	1	4712	1	5744	1	6774	5	7786	1		

We might continue by commenting on two-way and higher way tables. Looking at the four-way table amounts to considering response patterns (Table 2). The most frequent pattern (81 individuals) is 1111, describing the CGT–communist "hard core": CGT vote and affiliation, Duclos vote and communist sympathy. Next comes the pattern 6788 (50 individuals), that is, abstention and nonaffiliation for union questions and nonresponse for the political ones. The 14 most frequent patterns together represent about one third of the total number of respondents.

In the book by Adam *et al.* (1970), the reader will find one-way and two-way tables for the most important questions, with extensive sociological comments, organized by topics—for example, attitudes toward unions, electoral behavior—based on careful examination of tables. The alternative approach that we will follow in this chapter, along the line of geometric data analysis, is to construct a relevant "social space" (as Bourdieu would call it), a "union–political space" for the French workers in 1969, applying MCA to the responses to the four questions. The study of maps yielded by MCA amounts to a synthesis of analyses of the conventional kind.

2.2 Multiple Correspondence Analysis

From the responses of the individuals, we construct the *disjunctive table* (Benzécri, 1992, p. 392; Lebart *et al.*, 1995, p. 108), also called an indicator matrix, crossing the 1049 individuals and the $8 \times 4 = 32$ modalities. The principle of construction of this table is recalled by Table 3.

Correspondence analysis of the disjunctive table, that is, multiple correspondence analysis, yields two clouds of points, namely the cloud of 32 modalities, and the cloud of 1049 individuals—or equivalently of 319 weighted response patterns. In numerical terms, each cloud is defined by a table of principal coordinates, where for each axis the weighted average of the squares of principal coordinates is equal to the eigenvalue associated with the axis.

Here we will interpret the first four axes; the corresponding eigenvalues are given in the first row of Table 4. For the cloud of modalities in the plane 1–2 (Figure 1):

- On the left, a compact group of four modalities emerges: vote and affiliation CGT, Duclos, Communist.
- On the lower right, there are the various NR and abstention modalities, together with the two nonaffiliated modalities, Pompidou and Gaullist. Moving up, we find Center and Poher, Socialist and Defferre, and then CFDT vote and affiliation.

Table 3: Disjunctive table

Patterns		$q1$	$q2$	$q3$	$q4$
1111 ⎫ · · · ⎬ 81 1111 ⎭		10000000 · · · 10000000	10000000 · · · 10000000	10000000 · · · 10000000	10000000 · · · 10000000
	Disjunctive encoding $\Rightarrow$				
8888 ⎫ · · · ⎬ 5 8888 ⎭		000000001 · · · 00000001	000000001 · · · 00000001	000000001 · · · 00000001	000000001 · · · 00000001

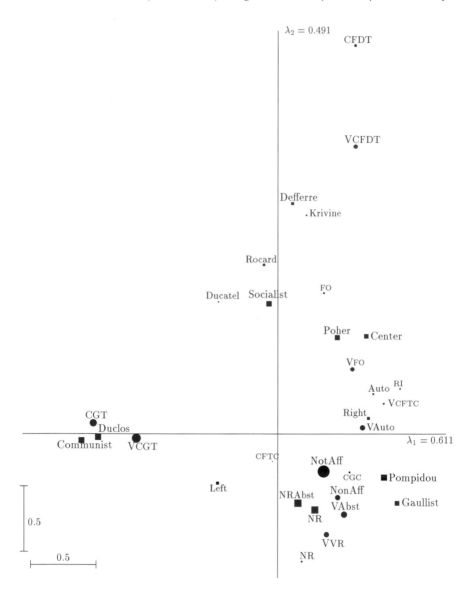

Figure 1: Cloud of 32 modalities in plane 1–2. Modalities that contribute most to axes 1 and 2 are in large characters; modalities of the two union questions are represented by circles and those of the two political questions by squares, whose areas are proportional to frequencies. CGT voting is denoted VCGT, as distinct from CGT affiliation, denoted CGT, and so on.

3 Contributions of Points and Deviations

3.1 Basic Formulas

A cloud of weighted points being given, the *variance* (also called inertia) of the cloud is the weighted mean of the squares of the distances between the points and the mean point of the cloud (Benzécri, 1992, p. 36). The *absolute contribution* of a point to the cloud is defined as the product of the weight of the point by the square of its distance from the mean point (Benzécri, 1973a, p. 38; 1992, p. 61). In this chapter, we will be mainly interested in contributions to an axis; accordingly, distances will be measured along the axis under consideration.

1. *Contribution of a point* (Cta). Let us consider a point of weight, or mass, p and coordinate y along the axis. The *absolute contribution* of the point to the axis will be denoted Cta; it is given by the formula (Benzécri, 1992, p. 340; Greenacre, 1984, p. 67):

$$\text{Cta} = py^2 \quad \text{(point)}$$

2. *Contribution of the deviation between two points* (Cti). Let us now consider two points. Let p and p' denote the weights of the points and y and y' their coordinates along the axis. The absolute contribution of the deviation, also called the *intra (within) contribution*, will be denoted Cti and is given by the following formula (Rouanet and Le Roux, 1993, p. 268):

$$\text{Cti} = \frac{pp'}{p + p'}(y - y')^2 \quad \text{(deviation)}$$

These notions of contribution readily extend to a subset of points, or subcloud. With a subcloud are associated its *weight* (sum of the weights of its points), its *weighted mean point* (barycenter), and its *variance*, and the following three types of contribution:

- Its (global) contribution (Cta), which is the sum of the contributions of its points
- The absolute contribution (Cta) of its mean point, which is the product of its weight by the square of the principal coordinate of its mean point
- Its intra-contribution (Cti), which is the weighted sum of squares of distances from the points to their mean point

By the classical Huyghens property, the Cta of a subcloud is the sum of its Cti and the Cta of its mean point, which shows that Cta and Cti are equal if and only if the mean point of the subcloud coincides with the mean point of the cloud (Rouanet and Le Roux, 1993, p. 118).

3.2 Application to the Cloud of Modalities

In MCA the weight of a modality is the relative frequency of the modality divided by the number of questions. Hereafter we illustrate the calculations for axis 1.

- *Contribution of modality* (Cta). Taking CGT vote (denoted VCGT) as an example: the relative frequency is 0.3298 (Table 1), hence the weight $p = 0.3298/4 = 0.0825$. The coordinate along axis 1 is $y = -1.090$ (see Table 5). Hence the absolute contribution of VCGT: $py^2 = 0.0825 \times (1.090)^2 = 0.0980$.

- *Contribution of deviation between modalities* (Cti). Take VCGT (coordinate $y = -1.090$, weight $p = 0.0825$) on the one hand and VAuto and VAbst on the other hand; the barycenter of VAuto and VAbst has a weight equal to $p' = 0.0269 + 0.0381 = 0.0650$ (weights add up), and its coordinate is $y' = (0.0269 \times 0.659 + 0.0381 \times 0.513)/0.0650 = 0.573$ (coordinates average up). One has $pp'/(p + p') = (0.0825 \times 0.0650)/(0.0825 + 0.0650) = 0.0364$. Hence the absolute contribution of the deviation $(-1.090 - 0.573)^2 \times 0.0364 = 0.1014$.

- *Contribution of modality to axis* (Ctr). Let us divide the contribution of VCGT, namely 0.0980, by the *sum of the contributions of all modalities*, that is, $\lambda_1 = 0.6113$; we get $0.0980/0.6113 = 0.160$, which means that VCGT contributes to 16% of axis 1. This ratio is often denoted by Ctr.

We further define two other ratios that will be directly useful in the interpretation process; for clarity, we will always express them as percentages.

- *Contribution of question to axis.* By definition, the Cta of a question is the sum of the Ctas of its modalities. For example, the Cta of $q1$ (Professional Elections) for axis 1 is the sum of the eight Ctas: $0.0980 + \cdots + 0.0041 = 0.1482$ (see Table 5). If we now divide the contribution of $q1$ by the sum of the contributions of questions, that is, $\lambda_1 = 0.6113$, we get $0.1482/0.6113 = 0.24$; accordingly, we state that question 1 *accounts for* 24% of axis 1.

- *Contribution of modality (and of deviation) to question.* If we divide the contribution of VCGT by the *contribution of the question* it belongs to, namely Professional Elections ($q1$), we get $0.0980/0.1482 = 0.66$; therefore we state that VCGT contributes to 66% of question $q1$ (for axis 1). Similarly for the contributions of deviations. The deviation VCGT versus VAuto and VAbst contributes to $0.1014/0.1482 = 68\%$ of question $q1$ (for axis 1).

3.3   Cloud of Individuals and Cloud of Modality Mean Points

In the cloud of individuals, with each observed modality is associated the subcloud of the individuals who have chosen that modality. The mean point of this subcloud will be called the *modality mean point*. For each axis, the coordinate of the modality mean point is the mean of the principal coordinates of the individuals who have chosen this modality, and this can be shown to be equal to $\sqrt{\lambda}\, y$, where y is the principal coordinate of the corresponding modality (Benzécri, 1992, p. 410).

The cloud of all modality mean points can be obtained from the CA of the *Burt table*, which has as eigenvalues the squares λ^2. As a consequence, if one divides the contribution of a modality mean point—or of a deviation between modality mean points—by λ^2, one again finds the relative contribution (Ctr) of modality, or of deviation, and consequently, the relative contribution (Ctr) of a question to an axis.

Each question q induces a partition of individuals into as many subclouds as there are observed modalities for that question. Consider the derived cloud of the modality mean points for question q. For each axis, the inertia of this cloud, or *interclass (between-class) inertia*, is equal to λ times the absolute contribution (Cta) of question q in the cloud of modalities. As a consequence, if one divides the contribution of a modality mean point—or of a deviation between modality mean points—by the interclass inertia, one again finds the relative contribution of the modality—or of the deviation—to the question.

As a conclusion, it will be equivalent to interpret axes in the cloud of modalities or in the cloud of modality mean points.

4   Interpreting Axes

Benzécri (1992, p. 405) gives the following guideline: "Interpreting an axis amounts to finding out what is similar, on the one hand, between all the elements figuring on the right of the origin and, on the other hand between all that is written on the left; and expressing with conciseness and precision, the contrast (or opposition) between the two extremes." The method of contributions of points and deviations has been devised as a guide along this line.

4.1   The Method of Contributions of Points and Deviations

As far as MCA is concerned, the method consists of the following four steps.

Step 1. *Important questions.* In the cloud of modalities, look for the questions whose contributions to the axis are important. This leads to a first overall interpretation of the axis.

Step 2. *Important modalities.* Select modalities—or groups of modalities of the same question that are close on the axis—whose contributions to the axis exceed some threshold (average contribution is a rule of thumb, but when the cumulated amount is not sufficient, a less severe threshold may be in order).

Step 3. *Contributions of modalities to questions.* For each question retained at step 1, calculate the relative contribution to the question (on the axis) accounted for by the modalities retained in step 2. When for the question under study, those modalities separate into several groups—often, for the first axes, into two groups on the two sides of the origin—determine the barycenters of groups, then their intra-contribution, and express this contribution as a percentage of the contribution of the question. For each question, the content of groups is a concise summary of the interpretation of the axis, whereas the relative intra-contribution to the question is a quantitative appraisal of the precision of that summary.

Table 4: Contributions (Cta) of the four questions

	Axis 1	Axis 2	Axis 3	Axis 4
Eigenvalue λ	**0.611**	**0.491**	**0.416**	**0.373**
$q1$ Professional elections	0.148	0.149	0.078	0.162
$q2$ Union affiliation	0.137	0.141	0.049	0.162
$q3$ Presidential election	0.157	0.105	0.148	0.024
$q4$ Political sympathy	0.169	0.096	0.141	0.026

Step 4. *Composite modalities or patterns.* The interpretation will be usefully complemented by the examination, in the cloud of individuals, of the composite modalities or patterns brought out at step 3. When interpreting a specific response pattern, be aware that its frequency count can be quite low.

4.2 First Overview

In the cloud of modalities, the contributions of the four questions to the first four axes are given in Table 4. The relative contributions of $q1$ through $q4$ to axis 1 lie between 22% and 28%; for axis 2, they lie between 19% and 31%. Therefore the interpretation of axes 1 and 2 will be based on the four questions. For axis 3, questions $q3$ and $q4$ contribute to 70% of the axis; therefore the interpretation of axis 3 will be based predominantly on the two political questions. For axis 4, $q1$ and $q2$ contribute to 87% of the axis; therefore the interpretation will be essentially based on the two trade union questions.

4.3 Interpretation of Axis 1

The interpretation of axis 1, in the cloud of modalities, is based on the results shown in Table 5, which may be used for checking the numerical values, with Figure 1 serving as an intuitive guide.

Step 1. *Important questions.* All four questions are important for axis 1; axis 1 is a general axis, that is, its interpretation involves all four questions.

Step 2. *Important modalities.* There are four very important modalities, namely Communist (Cta $= 0.1111$, i.e., 18% of axis), CGT (17%), Duclos (17%), VCGT (16%). Those four modalities together account for 69% of axis 1. They are all on the left side of axis 1. Three other modalities have contributions exceeding average $(0.6113/32 = 0.0191)$, namely Pompidou (7%), Gaullist (5%), and NotAff (3%), all three on the right side of the axis. The previous seven modalities together contribute to 84% of axis 1. Let us add to them the two modalities VAuto and VAbst, which are close to each other on the axis and together contribute 3%; with the nine modalities we come up to 87%.

Table 5: Axis 1: weights, coordinates, and absolute contributions (Cta) of modalities ($\lambda_1 = 0.61132$)

	Professional elections				Union affiliation		
q1	Weight	Coord.	Cta	q2	Weight	Coord.	Cta
1. VCGT	.0825	−1.090	.0980⋆	1. CGT	.0527	−1.425	.1069⋆
2. VFDT	.0219	0.605	.0080	2. CFDT	.0131	0.602	.0047
3. VFO	.0195	0.578	.0065	3. FO	.0057	0.356	.0007
4. VCFTC	.0062	0.824	.0042	4. CFTC	.0012	−0.040	.0000
5. VAuto	.0269	0.659	.0117+	5. Auto	.0053	0.741	.0029
6. VAbst	.0381	0.513	.0100+	6. CGC	.0029	0.557	.0008
7. VNonAff	.0262	0.463	.0056	7. NotAff	.1666	0.355	.0210⋆
8. VNR	.0286	0.377	.0041	8. NR	.0026	0.186	.0001
	.2500		.1482		.2500		.1373

	Presidential election				Political sympathy		
q3	Weight	Coord.	Cta	q4	Weight	Coord.	Cta
1. Duclos	.0555	−1.387	.1069⋆	1. Comm.	.0484	−1.516	.1111⋆
2. Defferre	.0117	0.114	.0001	2. Soc.	.0424	−0.069	.0002
3. Krivine	.0024	0.221	.0001	3. "Left"	.0107	−0.460	.0027
4. Rocard	.0072	−0.108	.0001	4. Center	.0298	0.687	.0140
5. Poher	.0355	0.461	.0075	5. RI	.0022	0.950	.0019
6. Ducatel	.0017	−0.452	.0003	6. Right	.0095	0.705	.0047
7. Pompidou	.0584	0.826	.0398⋆	7. Gaull.	.0334	0.926	.0286⋆
8. NRAbst	.0777	0.156	.0019	8. NR	.0737	0.286	.0060
	.2500		.1568		.2500		.1690

Stars (⋆) refer to modalities whose contributions exceed the average of the axis ($.61132/32 =$ .0191). Plus (+) refers either to modalities close (on the axis) to a starred modality or to clustered modalities whose grouped contribution exceeds average.

Step 3. *Contributions of modalities to questions.*

Professional elections (q1). The sum of the Cta of VCGT (on the left side), VAuto, and VAbst (on the right side) is $0.0980 + 0.0117 + 0.0100 = 0.1197$; that is, those three modalities contribute to $0.1197/0.1482 = 81\%$ of the question on the axis. The intra-contribution (Cti) of the deviation VCGT vs. VAuto with VAbst is found to be 0.1006; that is, it accounts for 68% of the question on the axis.

Union affiliation (q2). CGT (left) and nonaffiliated (right) together contribute 93% to the question on axis 1. The opposition between these two modalities accounts for 92% of the question.

> *Presidential election (q3).* Duclos (left) and Pompidou (right) contribute to 94% of the question. The opposition Duclos vs. Pompidou accounts for 89%.
>
> *Political sympathy (q4).* Communist (left) and Gaullist (right) contribute to 83% of the question. The opposition Communist vs. Gaullist accounts for 70%.

Step 4. *Relevant patterns.* The foregoing results suggest considering the composite modalities that emerge for axis 1. Since all (four) questions are involved in the interpretation of the axis, the relevant composite modalities are patterns obtained by combining the cells of Table 6.

Hence the three relevant patterns (with frequency counts, out of a total of 1049): 1111 (81); 5777 (14); 6777 (19). Figure 2 gives the simultaneous representation of relevant modalities and patterns for axis 1. It provides a graphical summary of the interpretation of axis 1, and the summary in words may read as follows. Axis 1 opposes the left profile VCGT–CGT–Duclos–Communist (1111) vs. the right profile [VAuto or VAbst]–nonaffiliated–Pompidou–Gaullist (5777, 6777).

4.4   Interpretation of Axis 2

Applying our four-step interpretation to axis 2 leads to the following results.

Step 1. Axis 2 is also a general axis (involving all four questions).

Step 2. Important modalities are CFDT, VCFDT, Socialist, and Defferre (upper side of axis), then NR to $q4$, NRAbst, Poher, VNR (i.e., NR to $q1$), Center. Adding VAbst and NotAff (which are nearly average) and Gaullist (near NR to $q4$ on axis), one arrives at 90% of axis 2.

Step 3. For $q1$, $q2$, and $q4$, there are well-marked oppositions: VCFDT (upper side) vs. VNR and VAbst (lower side) (92% of $q1$); CFDT (upper) vs. NotAff (lower) (92% of $q2$); Socialist and Center (upper) vs. NR and Gaullist (lower) (97% of $q4$).

> Question $q3$ (presidential election) calls for a more detailed interpreta- tion. Defferre, Poher, and NRAbst together contribute 74% of $q3$. However, Poher lies halfway between Defferre and NRAbst, which means that those three modalities do not lend themselves easily to a grouping into two opposed

Table 6: Relevant modalities for axis 1

	Prof. vote	Union aff.	Pres. vote	Polit. symp.
Left	1. VCGT	1. CGT	1. Duclos	1. Comm.
Right	5. VAuto 6. VAbst	7. NotAff	7. Pompidou	7. Gaull.

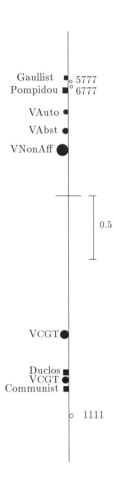

Figure 2: Axis 1: simultaneous representation of relevant modalities and patterns.

classes. This difficulty is confirmed by the weakness of the contribution of the opposition Defferre and Poher vs. NRAbst (only 65%). To get a more substantial contribution to the question, one must resort to the "ternary" comparison Defferre vs. Poher vs. NRAbst, which accounts for 74% of question $q3$.

Step 4. The composite modalities that emerge from the analysis of axis 2 are obtained by combining the cells of Table 7.

Hence there are eight patterns (with frequency counts): 2222 (7); 2224 (1); 2252 (6); 2254 (8); 6787 (4); 6788 (50); 8787 (0) (a nonobserved pattern!); 8788 (37). Figure 3 gives the simultaneous representation of relevant modalities and patterns for axis 2. On the whole, axis 2 reflects the opposition between noncommunist left workers, with CFDT vote and affiliation, and nonrespondent nonaffiliated workers.

Table 7: Relevant modalities for axis 2

	Prof. vote	**Union aff.**	**Pres. vote**	**Polit. symp.**
Above	2. VCFDT	2. CFDT	2. Defferre 5. Poher	2. Socialist 4. Center
Below	6. VAbst 8. VNR	7. NotAff	8. NRAbst	7. Gaull. 8. NR

Figure 3: Axis 2: simultaneous representation of relevant modalities and patterns.

4.5 Interpretation of Axis 3

We summarize the results.

Step 1. Axis 3 is predominantly a political axis.

Step 2. The important modalities are Gaullist, Pompidou (on one side of the axis), NRAbst, and NR (on the other side), all four belonging to $q3$ and $q4$; then come three modalities of $q1$: VCFTC and VAuto (on the Gaullist side) and VNR (on NR side). Those seven modalities together account for 76% of axis 3.

Step 3. The opposition Pompidou vs. NRAbst contributes to 90% of $q3$, the opposition Gaullist vs. NR to 89% of $q4$.

Step 4. In the cloud of individuals, the important modalities of questions $q3$ and $q4$ induce a subcloud of 114 Pompidou-Gaullists (patterns xx77), and a subcloud of 177 "political nonrespondents" (patterns xx88). Figure 4 shows the simultaneous representation of important modalities and of those two subclouds with their mean points. As may be seen, the separation between the two subclouds is perfect. Notice the "union-committed" patterns 4x77 and 5x77 (among Pompidou–Gaullists), and the noncommitted patterns 8x88 (among political nonrespondents).

Axis 3 is predominantly political and opposes politically committed Pompidou–Gaullist workers to political nonrespondents.

4.6 Interpretation of Axis 4

Step 1. Axis 4 is predominantly a union axis.

Step 2. The important modalities are FO, VFO (on one side of the axis), VCFDT and CFDT (opposite side), Socialist, and Defferre: together 87% of the axis.

Step 3. The opposition VFO vs. VCFDT contributes to 86% of $q1$, FO vs. CFDT to 91% of $q2$.

Step 4. The important modalities of $q1$ and $q2$ induce a subcloud of 19 FO-affiliated voters (patterns 33xx) and a subcloud of 47 CFDT-affiliated voters (22xx). Figure 5 shows the simultaneous representation. Again, the separation between the two subclouds is perfect.

Axis 4 is union dominated and opposes CFDT-affiliated voters to FO ones.

4.7 Synopsis

The synopsis is shown in Table 8.

4.8 Plane 1–2

The interpretation of axes 1 and 2 leads to allocating the relevant modalities for those axes to three classes corresponding to three polar areas: A (communist left, CGT

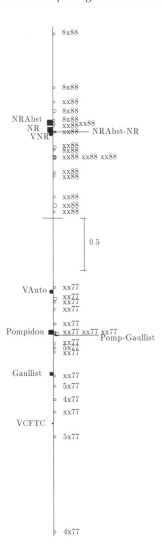

Figure 4: Simultaneous representation on axis 3 with patterns Pompidou–Gaullist (xx77) and NRAbst–NR (xx88).

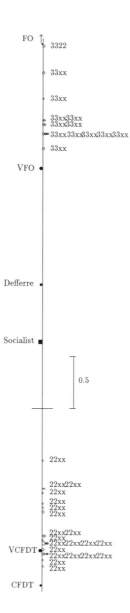

Figure 5: Simultaneous representation on axis 4 with patterns FO–VFO (33xx) and CFDT and VCFDT (22xx).

Table 8: Synopsis

Axis 1: $\lambda_1 = 0.611$	Axis 2: $\lambda_2 = 0.491$	Axis 3: $\lambda_3 = 0.416$	Axis 4: $\lambda_4 = 0.373$
q1 24% of axis VCGT vs. VAuto-Vabst: 68% of question	30% of axis VCFDT vs. VAbst-VNR: 92% of question	[19% of axis]	43% of axis VCFDT vs. VFO: 86% of question
q2 22% of axis CGT vs. 7 NotAff: 92% of question	29% of axis CFDT vs. NotAff: 92% of question	[12% of axis]	43% of axis CFDT vs. FO: 91% of question
q3 26% of axis VDuclos vs. Pompidou: 89% of question	21% of axis Defferre vs. Poher vs. NRAbst 74% of question	36% of axis Pompidou vs. NRAbst: 90% of question	[6% of axis]
q4 28% of axis Commun. vs. Gaullist: 70% of question	20% of axis Socialist-Center vs. Gaullist-NR: 97% of question	34% of axis Gaullist vs. NR: 89% of question	[7% of axis]

All comparisons are oppositions (1 d.l.) except the ternary comparison (2 d.l.) for question q3 and axis 2.

affiliated), B (Gaullist together with nonaffiliated and NR), and C (noncommunist left, CFDT affiliated).

The modalities in Table 9 lead to defining the $1 + 12 + 4 = 17$ following response patterns (with their frequency counts, total 269): 1111 (81); 5777 (14); 5778 (4); 5787 (3); 5788 (9); 6777 (19); 6778 (5); 6787 (4); 6788 (50); 8777 (12); 8778 (9); 8787 (0); 8788 (37); 2222 (7); 2224 (1); 2252 (6); 2254 (8).

Figure 6 shows the simultaneous representation on plane 1–2, with the relevant modalities and patterns used as landmarks. This figure shows all 1049 individuals in their 319 unique positions.

Table 9: Relevant modalities for plane 1–2

	Prof. vote	Union aff.	Presid. elect.	Polit. sympathy
A	VCGT	CGT	Duclos	Communist
B	Auto Abst NR	NotAff	Pompidou NRAbs	Gaullist NR
C	VCFDT	CFDT	Defferre Poher	Socialist Center

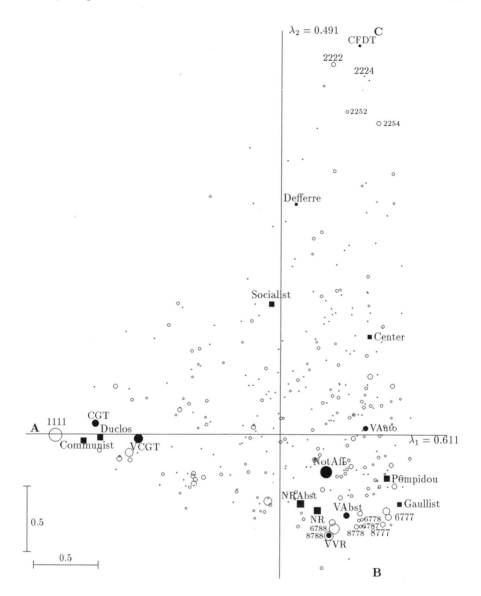

Figure 6: Simultaneous representation in plane 1–2: cloud of 319 weighted patterns (1049 individuals) and 16 relevant modalities.

5   From Interpretation to Exploration

Henceforth we place ourselves in the cloud of individuals. Considering this cloud opens new opportunities for interpretation—to begin with, the possibility of representing any patterns of interest, for instance, those that contribute most to an axis or typical patterns chosen by the specialist as landmarks to enhance interpretation. Further, the interpretation of axes may be prolonged by the exploration of the cloud and enlarged to planes or higher order spaces, always making use of the structures of the questionnaire. In this section, we will suggest—without trying to be systematic— some lines for cloud exploration. Exploration will often be motivated by specific interrogations (i.e., pertaining to parts or to groupings of data), which may be raised either before gathering data or when examining results.

5.1   Composite Modalities

The cloud of individuals enables one to go farther than the cloud of modalities, because individuals carry all the information of the data (see Chapters 15 and 20). In particular, the concept of the subcloud associated with a modality also applies to a *composite modality* (also called an "interactively coded category"). That is, with each observed pair of modalities (k, k') (with k belonging to question q and k' to question q') is associated the subcloud of the individuals who have chosen both k and k'. The derived cloud of mean points now corresponds to the composite modalities of questions q and q'.

  For example, the two political questions $q3$ and $q4$ induce 51 subclouds (among $8^2 = 64$ possible subclouds). The derived cloud of 51 mean points contributes to 81% of axis 3. Now consider the deviation between the mean points of the two composite modalities: Pompidou–Gaullist vs. political nonrespondents (see Figure 4). It is found that this deviation contributes 73% to the inertia of this derived cloud (Rouanet and Le Roux, 1993, p. 295). This result reinforces and refines the interpretation of axis 3.

5.2   Correlation Ratios and Supplementary Questions

Every question of a questionnaire generates a partition of individuals, with a cloud of modality mean points, whose variance defines the interclass (between class) variance of the question. For each axis, dividing the interclass variance by the total variance yields a ratio denoted by η^2, which expresses the correlation between the question and the numerical variable of principal coordinates of individuals on the axis. The η^2 ratios can be calculated for active questions, as well as for *supplementary questions*.

  For example, for the supplementary question "personal political situation" with five modalities after recoding—left communist (175), left noncommunist (237), center (254), right (154), and NR (229)—the graphs of Figure 7 show, in plane 1–2, the derived cloud of the five mean points (Figure 7a) and the five subclouds (Figures 7b through 7f).

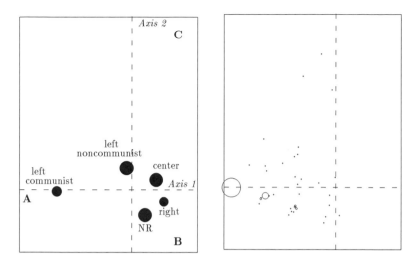

Figure 7: Personal political situation (plane 1–2). (a) five mean points; (b) left communist (175).

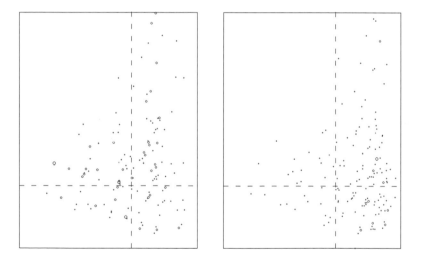

Figure 7: Personal political situation (plane 1–2). (c) left noncommunist (237); (d) center (254).

Along axis 1, the interclass variance of this supplementary question is found to be 0.347. Dividing by $\lambda_1 = 0.6113$ yields $\eta^2 = 0.57$. Then calculating the contribution of the deviation between the mean points left communist vs. right and center yields the value 0.332; that is, this opposition accounts for $0.332/0.347 = 96\%$ of the correlation ratio η^2 between axis 1 and this question.

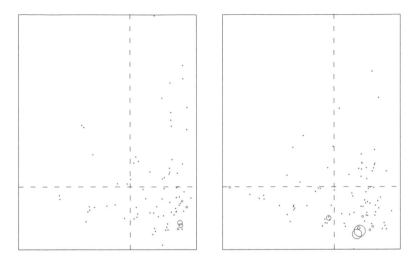

Figure 7: Personal political situation (plane 1–2). (e) right (154); (f) NR (229).

The exploratory process may extend beyond mean points. For instance, a look at the five subclouds reveals striking disparities among dispersions in plane 1–2. The most concentrated subcloud is left communist, whose variance (in plane 1–2) is equal to 0.395; the most scattered subcloud is left noncommunist, whose variance is equal to 1.018.

5.3   Crossing Relationship and Interaction

When all pairs of modalities of two questions (whether active or supplementary) are observed, it may be said (adopting ANOVA language) that there is a "crossing relationship" between the questions. Then the concept of *interaction between questions* may be formally defined as in ANOVA with unbalanced designs (Bernard *et al.*, 1989; Le Roux, 1991; Le Roux and Rouanet, 1984).

For example, let F denote the question "personal political situation" and L denote the question "trust toward unions," with three modalities high, moderate, and low or none or NR. For any axis, a diagram akin to the interaction diagrams familiar in experimental data analysis can be constructed. Figure 8 shows the interaction diagram for axis 1. Abscissas correspond to the five modalities of question F. Ordinates are the coordinates along axis 1 of the $3 \times 5 = 15$ mean points corresponding to the crossing of questions F and L. For each modality of L, the points of the five modalities of F have been joined. The three lines appear to be nearly parallel, which means that there is virtually *no interaction* between the two questions F and L with respect to axis 1. The η^2 ratio associated with the crossing $F \times L$ for axis 1 is equal to 0.64; calculation shows that the interaction accounts for only 1% of η^2.

In plane 1–2, the visualization of interaction—or the weakness of interaction, for that matter—can be performed similarly by constructing the modality mean points

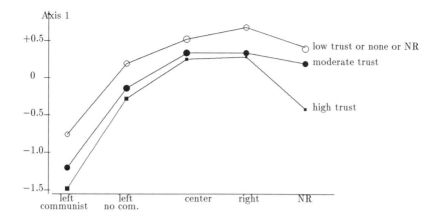

Figure 8: Interaction for axis 1.

corresponding to the crossing and joining the points corresponding to one of the questions. In Figure 9, the points of the five modalities of question F have again been joined. The quasi-parallelism of the three lines now means that there is virtually no interaction between the two questions with respect to the plane. The η^2 ratio associated with the crossing for plane 1–2 is equal to 0.45, and calculation shows that the interaction accounts for only 1% of this η^2.

6 Concluding Comments

After presenting this guide for interpretation of axes in MCA, several points are worth stressing, all directly bearing on the topic of the visualization of data.

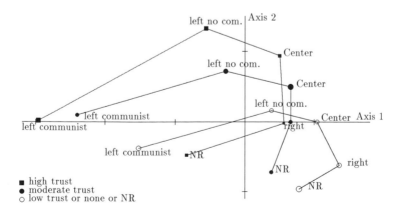

Figure 9: Interaction in plane 1–2.

1. The method of the contributions of points and deviations, developed in this chapter for MCA, readily applies, with appropriate modifications, to the interpretation of principal axes of all kinds of structured multidimensional data (Le Roux and Rouanet, 1984).

2. The interpretation of axes of higher order may reveal important findings.

3. Simultaneous representation in CA has been recognized as a most powerful visualization tool to sustain interpretations; see Benzécri (1969, 1973a, especially pp. 330–331 and pp. 468–469). This is all the more important in the case of MCA, where simultaneous representation brings together two radically different entities, namely individuals and modalities—or in other terms, objects and descriptors of objects.

4. In MCA, investigating the cloud of individuals, together with its subclouds and derived clouds (modality mean points), leads to detailed interpretations, in the first place by the examination of composite modalities.

5. A general claim underlying this chapter is that the use of specific comparisons, a tool borrowed from ANOVA, should considerably enrich the usual aids to interpretation in geometric data analysis. The method of the contributions of points and deviations provides a first step in this direction. Another step would be the investigation of the interactions between questions, a topic we have just touched upon in this chapter.

Software Notes

The strategy of data analysis that we have presented can easily be performed by combining any standard software for CA with the EyeLID software developed in our research group.

For the data of this chapter, we performed MCA using ADDAD (Association pour le Développement De l'Analyse des Données, 22 Rue Charcot, Paris 75013), then all subsequent analyses, such as derived graphs and computations of contributions and interaction effects, through EyeLID (Bernard, Rouanet & Baldy, Université René Descartes, 45 rue des Saints–Pères, Paris 75270 Cedex 06. E–mail: eyelid@math–info.univ–paris5.fr).

The EyeLID software combines the following two features: a *Language for Interrogating Data* (LID), which designates relevant data sets in terms of *structuring factors*, formally analogous to factors in an experimental design, and the *visualization* ("Eye") of the derived clouds designated by a LID request. For detailed applications of the LID language to sociological examples, see Bernard *et al.* (1989), and Bonnet *et al.* (1996). A demonstration version of the software EyeLID applied to the data of the present chapter is available by `ftp` at the address:

```
ftp.math-info.univ-paris5.fr
```

under the directory `/pub/MathPsy/EyeLID`.

Chapter 17

Diagnostics for Joint Displays in Correspondence Analysis

Michael Greenacre

1   Introduction

One of the main advantages of correspondence analysis (CA) is its simultaneous visualization of the row and column categories of a table or, in the multiple case, the simultaneous visualization of all the categories of a set of discrete variables. The visualization is achieved by projecting points that represent the categories in multidimensional space onto a subspace, usually a plane, resulting in an approximate map of the categories. We cannot tell from the projections of the points in the map whether the points are displayed accurately or not—some points might be far from the planar map while others might be close to the plane and thus more accurately represented. Supporting the interpretation of such maps are sets of diagnostics that measure the overall quality of display of the points as well as the quality of individual points. These diagnostics, usually called "contributions," are based on the decomposition of inertia in CA. They measure how much each point contributes to the map and how much the map contributes to each point. We call these contributions "within-variable" diagnostics because they involve categories of a variable by themselves, without direct reference to categories of other variables.

The intepretation of the map, however, goes beyond studying individual points. Although usually not stated explicitly, the interpretation of a joint CA map relies implicitly on one of two geometric concepts: either scalar products between category points for different variables, which we refer to as an underlying *biplot* model, or distances between such points, which is a type of *unfolding* model. In this chapter we look at a different way to measure the quality of display that is specifically aimed at the scalar-product or distance-based "between-variable" way we interpret the joint display. This is a "nonmetric" measure in that it is based on the rank ordering of the scalar products or distances, rather than their actual values. The properties of this measure make it suitable for simple CA as well as the variants of CA for the multiple case: CA of "stacked" tables, multiple correspondence analysis (MCA), and joint correspondence analysis (JCA).

2   Data Set on Cultural Competences

We shall illustrate our between-variable diagnostics using a set of categorical data from the ALLBUS '86 survey (ALLBUS stands for "Allgemeine Bevölkerungsumfrage der Sozialwissenschaften," part of the German General Social Survey Program). Our interest here centers on 25 variables measuring what we call "cultural competences":

a	dance waltz	b	put on bandage	c	fill in tax form
d	play chess	e	adjust quartz watch	f	play musical instrument
g	fix lamp	h	use PC	i	take photographs
j	hang wallpaper	k	swim	l	change spark plugs
m	read city map	n	read timetable book	o	use typewriter
p	knit	q	ride bicycle	r	cook
s	fix tire	t	use calculator	u	use video recorder
v	use tape recorder	w	sew on button	x	shorten trousers
y	dance to pop music				

Each of 3092 respondents was classified into one of four categories for each of these cultural competences: (1) yes, (2) somewhat, (3) no, or (4) don't know/missing. Thus $n = 3092$, $Q = 25$, $J_q = 4$ for all q, and $J = 100$. Notice that the fourth category of missing or "don't know" responses was generally of quite low frequency. Other variables in the survey available as explanatory variables were, for example, gender, religious affiliation, age group, income group, and type of housing.

This data structure is frequently found in social surveys. First, we have a battery of questions, each with responses on the same scale, measuring some phenomenon—in this case cultural competence—and second, we have a number of biographical and demographical variables that we would like to relate to the battery of questions as a whole. To visualize these data we can form various cross-tabulations, or groups of cross-tabulations, to be analyzed by different forms of CA.

3 Diagnostics in Simple CA

For the theoretical explanation we shall use the following standard notation:

- N is an $I \times J$ contingency table with grand total n.
- $P = (1/n)N$ is the correspondence matrix, or discrete bivariate frequency density.
- r and c are the row and column margins of P, respectively.
- D_r and D_c are the diagonal matrices with r and c on the diagonal.
- F and G are the principal coordinates of the rows and columns, respectively, with normalization: $F^T D_r F = G^T D_c G = D_\lambda$, where D_λ is the diagonal matrix of principal inertias $\lambda_1, \lambda_2, \ldots$.
- X and Y are the standard coordinates of the rows and columns, respectively, with normalization: $X^T D_r X = Y^T D_c Y = I$.

The relationship between principal and standard coordinates of the same variable is $F = XD_\lambda^{1/2}$, $G = YD_\lambda^{1/2}$, while the "between-variables" relationship, or transition formula, between principal and standard coordinates of different variables is $F = D_r^{-1} P Y$, $G = D_c^{-1} P^T X$.

Using the cultural competences data, we can choose one of the explanatory variables, say age with five categories, and cross-tabulate it with all 25 variables, giving a 100×5 matrix consisting of 25 contingency tables, each of size 4×5, stacked one on top of the other. The most basic diagnostic is the set of principal inertias in descending order. In this case the four principal inertias and their percentages of inertia are calculated as 0.05401 (80.3% of the total inertia), 0.01068 (15.9%), 0.00175 (2.6%), and 0.00085 (1.3%). These values summarize the overall quality of the display of the profile points along the principal dimensions, which can be used to form a map of the category points. For example, if we choose a two-dimensional map, then the quality is 80.3% + 15.9% = 96.2%, which means that 96.2% of the inertia of the profile points is explained in the map (see Figure 1).

The contributions to inertia provide a similar decomposition of inertia for individual profile points; see, for example, Greenacre, 1984, p. 91; 1993, chap. 12; Blasius, 1994. Le Roux and Rouanet (Chapter 16) discuss how the contributions apply to MCA. When using these diagnostics we tend to think of the row points or the column points separately. An overall quality of 96.2% means that on average the row profiles (equivalently, the column profiles) have a quality of representation sometimes greater than this percentage, sometimes lower. Even if a profile has a display quality of 100%, however, this does not mean that all other points are correctly interpreted with respect to this profile. Interpretation of a CA map, for example, the one in Figure 1, involves seeing how the row points lie relative to the column points. There are two customary ways of thinking of this between-variable (row-to-column) relationship: the biplot model, based on scalar products between points, and the unfolding model, based on interpoint distances. We shall look at each of these in turn and develop a diagnostic of the quality of the interpretation that is applicable to both of them.

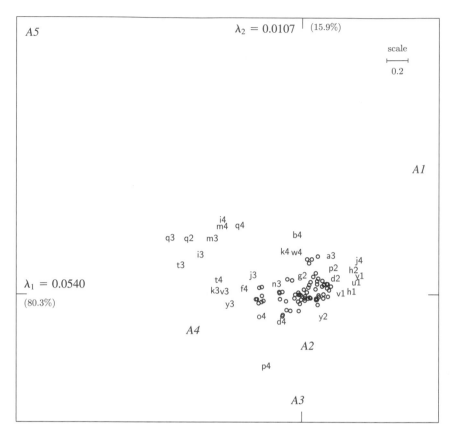

Figure 1: Correspondence analysis of cultural competence categories by age groups, asymmetric map with age points (*A1* to *A5*) in standard coordinates (many overlapping points indicated by ∘).

4   CA as a Biplot Model

A map is a biplot when the values x_{ij} in the data matrix, usually standardized in some way, are approximated by scalar products $\mathbf{f}_i^T \mathbf{g}_j$ between the corresponding row and column points in the map (see Gabriel *et al.*, Chapter 27). A scalar product between two point vectors is equal to the product of their lengths times the cosine of the angle subtended by the vectors. This is equivalent to projecting one of the vectors, say $\mathbf{f}$, onto the other one, $\mathbf{g}$, and calculating the scalar product as the projected length, multiplied by the length of $\mathbf{g}$, with sign depending on the angle between the vectors. To interpret a column of the data matrix, we would interpret the projections of all the row points (points $\mathbf{f}$ representing the rows) onto the vector $\mathbf{g}$ through the column point, called a "biplot axis." Since each projection is multiplied by the same scaling factor, the length of $\mathbf{g}$, the scalar products are proportional to their projections on the biplot axis, so the axis can be calibrated in order to read off the values of the scalar products and hence the approximations to the data values.

In CA the biplot relationship can be deduced from the so-called reconstitution formula for recovering the elements p_{ij} of the correspondence matrix from the row principal coordinates **F** and column standard coordinates **Y**:

$$p_{ij} = r_i c_j \left(1 + \sum_{k=1}^{K} f_{ik} y_{jk} \right)$$

[a similar formula in terms of the g_{jk} and x_{ik} is possible; see, for example, Blasius and Greenacre (1994, p. 76)]. When the first K^* dimensions are retained (often $K^* = 2$), then we replace the equality above by $\approx$, which denotes "is approximated by weighted least squares as":

$$p_{ij} \approx r_i c_j \left(1 + \sum_{k=1}^{K^*} f_{ik} y_{jk} \right) \tag{1}$$

In this case the weighting factor for the (i, j)th squared term is $1/(r_i c_j)$.

Dividing (1) throughout by r_i and rearranging terms, we obtain:

$$\left(\frac{p_{ij}}{r_i} - c_j \right) \bigg/ c_j \approx \sum_{k=1}^{K^*} f_{ik} y_{jk} = \mathbf{f}_i^\mathsf{T} \mathbf{y}_j \tag{2}$$

where $\mathbf{f}_i^\mathsf{T} \equiv [f_{i1} \cdots f_{iK^*}]$ and $\mathbf{y}_j^\mathsf{T} \equiv [y_{j1} \cdots y_{jK^*}]$. Formula (2) shows that the difference between the row profile element p_{ij}/r_i and its average c_j, relative to the average c_j, is approximated by the scalar product between the row point in principal coordinates and the column point in standard coordinates. In other words, the joint display of the $\mathbf{f}_i$'s and the $\mathbf{y}_j$'s, called the *asymmetric map* (Greenacre, 1993b), constitutes a biplot for the matrix $(\mathbf{D}_r^{-1}\mathbf{P} \quad \mathbf{1}\mathbf{c}^\mathsf{T})\mathbf{D}_c^{-1}$. Further details of the biplot for simple CA are given by Greenacre (1992), who also shows how the directions defined by the $\mathbf{y}_j$'s may be considered biplot axes that can be calibrated in profile units, that is, on a zero-to-one scale. This reduces the calculation of scalar products between points $\mathbf{f}_i$ and $\mathbf{y}_j$ to simply projecting the point $\mathbf{f}_i$ onto the biplot axis and reading off the profile value on the scale.

In the simple CA of the 25 cultural competences cross-classified with the five age groups, shown in Figure 1, the age groups *A1* to *A5* are displayed in standard coordinates and the 100 cultural competence category points in principal coordinates, labeled a1, a2, a3, a4, b1, b2, . . . , and so on (where there are many overlapping points in the center, the positions of the points are indicated by a ∘, without a label). The roughly parabolic curve traced by the five age groups is a phenomenon often observed in a CA map, called the "horseshoe effect." In this case it is due to the gradual change in cultural competences as age increases. The competence category points fall roughly into an arch as well, with categories such as "use video recorder" (u1) and "use tape recorder" (v1) at the right-hand end, associated with the youngest age group, and categories such as "cannot ride bicycle" (q3) and "cannot use calculator" (t3) at the left, associated with the oldest age group. Between these ends the cultural competence categories form a continuum that gradually moves from the youngest to the oldest end of the spectrum.

To interpret Figure 1 as a biplot, scalar products would be computed between the 100 cultural competence categories and the five age points, giving estimates $(\hat{p}_{ij}/r_i - c_j)$ of the deviations of the profile values from their respective averages. This can be performed equivalently by drawing an axis through each of the five age groups and calibrating it in profile units, so that the estimated profile values are obtained by projecting the cultural competence categories on the biplot axes. The total inertia is equal to the weighted sum of squares of the exact deviations, in the chi-squared metric (dividing each squared deviation by c_j), and this can be decomposed into two parts:

$$\sum_i \sum_j r_i \frac{(p_{ij}/r_i - c_j)^2}{c_j} = \sum_i \sum_j r_i \frac{(\hat{p}_{ij}/r_i - c_j)^2}{c_j} + \sum_i \sum_j r_i \frac{(p_{ij}/r_i - \hat{p}_{ij}/r_i)^2}{c_j}$$

(3)

The first part is the weighted sum of squares of the estimated deviations, which is equal to the sum $\sum_{k=1}^{K^*} \lambda_k$ of the first K^* principal inertias, which is the inertia accounted for in the map. The second part is the total error, the weighted sum of squared errors (also in the chi-squared metric), equal to the sum of the remaining principal inertias. We can study the individual errors by expressing each term in the second summation on the right-hand side of (3) as a percentage of the total error. Such an analysis of individual errors can help us to locate outliers in the data. For diagnostics more directly related to the way we interpret the map, we can consider the accuracy of recovering the ordering in the profile elements, rather than their actual values. We shall introduce such a criterion after discussing the unfolding interpretation of the joint map.

5   CA as an Unfolding Model

An alternative way to interpret the joint map is in terms of distances between points. When distances between one set of points approximate the data (or some transformation thereof), the mapping technique is called "multidimensional scaling." When the distances in the map are between points from two different sets, this is a special case of multidimensional scaling called "unfolding." In an asymmetric CA map there is some justification in looking at row-to-column distances, since we can write the criterion in simple CA as

$$\text{minimize} \qquad \sum_i \sum_j p_{ij}(\mathbf{f}_i - \mathbf{y}_j)^{\mathsf{T}}(\mathbf{f}_i - \mathbf{y}_j)$$

(4)

subject to the identification condition $\mathbf{Y}^{\mathsf{T}}\mathbf{D}_c\mathbf{Y} = \mathbf{I}$ on the standard coordinate vectors. In other words, we want to minimize the weighted squared distances between row and column points, where the weights are the observed frequencies of co-occurrence and where the normalization of the column points has been fixed. For given $\mathbf{y}_j$ the minimum of (4) with respect to $\mathbf{f}_i$ is

$$\mathbf{f}_i = \frac{\sum_j p_{ij}\mathbf{y}_j}{\sum_j p_{ij}} = \sum_j \frac{p_{ij}}{r_i}\mathbf{y}_j \tag{5}$$

which is the usual barycentric property between rows and columns or *transition formula*. This formula can be used to substitute for $\mathbf{f}_i$ in (4) so that the minimization is with respect to the $\mathbf{y}_j$ only.

To see that (4) is equivalent to the usual formulation of CA, we expand the unfolding criterion (4) and use (5):

$$\sum_i \sum_j p_{ij}(\mathbf{f}_i - \mathbf{y}_j)^\mathsf{T}(\mathbf{f}_i - \mathbf{y}_j) = \sum_i \sum_j p_{ij}\mathbf{f}_i^\mathsf{T}\mathbf{f}_i \tag{6}$$

$$+ \sum_i \sum_j p_{ij}\mathbf{y}_j^\mathsf{T}\mathbf{y}_j - 2\sum_i \sum_j p_{ij}\mathbf{f}_i^\mathsf{T}\mathbf{y}_j$$

$$= \sum_i r_i\mathbf{f}_i^\mathsf{T}\mathbf{f}_i + \sum_j c_j\mathbf{y}_j^\mathsf{T}\mathbf{y}_j - 2\sum_i r_i\mathbf{f}_i^\mathsf{T}\mathbf{f}_i$$

$$= K^* - \sum_i r_i\mathbf{f}_i^\mathsf{T}\mathbf{f}_i$$

For example, when $K^* = 2$, this criterion has a minimum value of $2 - \lambda_1 - \lambda_2$ corresponding to the maximum of $\sum_i r_i\mathbf{f}_i^\mathsf{T}\mathbf{f}_i$ being $\lambda_1 + \lambda_2$. Thus the unfolding objective of minimizing the row-to-column squared distances in the asymmetric map is equivalent to maximizing the row (or column) inertia displayed, and the minimum and maximum of the respective objective functions sum to a constant, the dimensionality of the solution.

Notice in the preceding development the use of the identity:

$$\sum_i \sum_j p_{ij}\mathbf{f}_i^\mathsf{T}\mathbf{y}_j = \sum_i r_i\mathbf{f}_i^\mathsf{T}\mathbf{f}_i \tag{7}$$

The left-hand side of (7) is the biplot criterion, which is required to be maximized. From (6) and (7) we have

$$\sum_i \sum_j p_{ij}(\mathbf{f}_i - \mathbf{y}_j)^\mathsf{T}(\mathbf{f}_i - \mathbf{y}_j) = K^* - \sum_i \sum_j p_{ij}\mathbf{f}_i^\mathsf{T}\mathbf{y}_j \tag{8}$$

which illustrates the complementary objectives of the biplot and unfolding models. In words, we can say that the biplot criterion is "display rows by $\mathbf{f}_i$ and columns by $\mathbf{y}_j$ so that their scalar products $\mathbf{f}_i^\mathsf{T}\mathbf{y}_j$ are, for high p_{ij}, as large positive as possible and for low p_{ij}, as large negative as possible" and the unfolding criterion is "display rows by $\mathbf{f}_i$ and columns by $\mathbf{y}_j$ so that their interpoint (squared) distances $(\mathbf{f}_i - \mathbf{y}_j)^\mathsf{T}(\mathbf{f}_i - \mathbf{y}_j)$ are, for high p_{ij}, as small as possible and for low p_{ij}, as large as possible."

6 A Nonmetric Graphical Diagnostic

We first consider the full space geometry of simple CA where, in the case of the biplot model, the geometry is well known. The biplot axes, defined by the vertex points, are the original coordinate axes of the space, so that the projections of the profile points onto the biplot axes give the exact profile values. In the reduced space, the profile values are approximated by the projections onto the calibrated biplot axes.

Let us consider the unfolding interpretation in a similar way by looking at the row–column distances in the asymmetric map in the full space. Let $\mathbf{a} = [a_1 \cdots a_J]^T$ be any profile point, where $\mathbf{1}^T\mathbf{a} = 1$, and $\mathbf{e}_j = [0 \cdots 0\ 1\ 0 \cdots 0]^T$ any unit point, where the 1 is in the jth position. The squared chi-squared distance between the profile point and vertex point is

$$\|\mathbf{a} - \mathbf{e}_j\|_\mathbf{c}^2 = \sum_{i \neq j} a_i^2/c_i + (a_j - 1)^2/c_j = \|\mathbf{a}\|_\mathbf{c}^2 + (1 - 2a_j)/c_j \qquad (9)$$

where $\| \cdots \|_\mathbf{c}$ denotes the chi-squared metric with respect to the average profile point $\mathbf{c}$, for example, $\|\mathbf{a}\|_\mathbf{c}^2 = \sum_{i=1}^{J} a_i^2/c_i$. We now compare the distance from two different profile points, $\mathbf{a}$ and $\mathbf{b}$, to the same vertex point. Specifically, we want to know what we can infer about the profile values if the profile point $\mathbf{a}$ lies closer than profile point $\mathbf{b}$ to the vertex point. Using (9) we thus obtain the following equivalent inequalities:

$$\|\mathbf{a} - \mathbf{e}_j\|_\mathbf{c}^2 < \|\mathbf{b} - \mathbf{e}_j\|_\mathbf{c}^2 \Leftrightarrow \frac{(2a_j - 1)}{c_j} - \|\mathbf{a}\|_\mathbf{c}^2 > \frac{(2b_j - 1)}{c_j} - \|\mathbf{b}\|_\mathbf{c}^2$$

The terms $\|\mathbf{a}\|_\mathbf{c}^2$ and $\|\mathbf{b}\|_\mathbf{c}^2$ are the squared distances of $\mathbf{a}$ and $\mathbf{b}$ from the origin $\mathbf{0}$ of the multidimensional space. Especially if the inertias are low, these distances are practically the same, so that the following equivalence holds approximately:

$$\|\mathbf{a} - \mathbf{e}_j\|_\mathbf{c}^2 < \|\mathbf{b} - \mathbf{e}_j\|_\mathbf{c}^2 \Leftrightarrow \frac{(2a_j - 1)}{c_j} > \frac{(2b_j - 1)}{c_j} \Leftrightarrow a_j > b_j$$

In words, if $\mathbf{a}$ is closer than $\mathbf{b}$ to the vertex point $\mathbf{e}_j$, then the jth profile element of $\mathbf{a}$ is greater than that of $\mathbf{b}$.

These results give some theoretical justification for the following general rule in the full profile space, where the rows are depicted as profiles and the columns as vertices:

> *Use each column vertex point one at a time and refer the set of row profile points to the column point. As the profile points come closer to the vertex point, the profile element with respect to that column category is increasing.*

This rule is very similar to the one that we had in the biplot case, because there is an almost monotonic relationship between the distances from a profile point to a vertex point and the position of the profile point projected onto the corresponding biplot axis.

The difference is that in the biplot situation the recovery of the profile elements is exact in the full space, whereas in the unfolding situation the recovery is approximate. It is apparent that the only difference between the biplot and unfolding criteria is that distance in the biplot model is measured along the biplot axis, whereas distance in the unfolding model is absolute distance between the profile and vertex points. Therefore, differences between the two criteria occur for points along perpendiculars to the biplot axis (Figure 2), for which the projections onto the biplot axis are identical, whereas the distances to the category point are different.

In practice, when confronted with a joint representation, we are seldom interested in trying to recover the data exactly. Rather, we would be satisfied in knowing that the *rank ordering* of the positions of the profile projections on a biplot axis (in the biplot interpretation), or the distances from the profile points to the vertex points (in the unfolding interpretation), agrees with the rank ordering of the profile elements. This leads us to propose a simple nonmetric measure of rank correlation as a diagnostic for the joint display.

Consider a vertex point j and I profile vectors in a joint reduced-space map (Figure 3). Suppose that the profile elements of the I vectors with respect to the vertex category point are $a_1, a_2, \ldots, a_I$ (we omit a subscript j referring to the vertex category because this is fixed throughout the ensuing discussion). In the display we can consider either (a) the projections of the profile points onto the biplot axis, leading to values $s_1, s_2, \ldots, s_I$ on any scale that increases toward the vertex, or (b) the distances $d_1, d_2, \ldots, d_I$ between the profile points and the vertex point. In the case of the biplot interpretation (a), a correct display would require that as the a_i's increase, so do the s_i's. Hence the display is exact in a nonmetric sense when for all i and i':

$$\text{(biplot)} \qquad (s_i - s_{i'})(a_i - a_{i'}) \geq 0 \qquad (10)$$

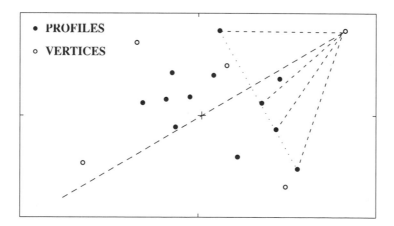

Figure 2: Biplot axis through a vertex point, showing that the projections of several points perpendicular to the axis can give the same approximate profile estimates, whereas the distances to the vertex vary.

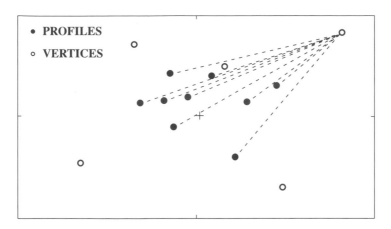

Figure 3: Measuring the quality of interpretation of the joint map by comparing distances from the profile points to a specific vertex point.

For the unfolding interpretation (b), the reverse is needed, that is, as the a_i's increase, so the d_i's decrease. Hence the display is exact in a nonmetric sense when for all i and i':

$$(\text{unfolding}) \qquad (d_i - d_{i'})(a_i - a_{i'}) \leq 0 \qquad (11)$$

In practice, of course, the scalar products s_i and the distances d_i are not in a perfect monotonic relationship with the profile values a_i, but we can measure the quality of the display by counting the number of unique pairs (i, i') (say where $i < i'$) for which either condition (10) or (11) is satisfied and expressing this count as a proportion of the total number $\frac{1}{2}I(I - 1)$ of pairs. In the language of nonparametric statistics, the pairs satisfying the particular inequality condition are called *concordant* pairs, and those not satisfying the condition are called *discordant* pairs. Let C denote the number of concordant pairs and D the number of discordant pairs, where $C + D = \frac{1}{2}I(I - 1)$, so that what we have proposed is to calculate $C/\frac{1}{2}I(I - 1)$. If the projections of the profiles onto a biplot axis or the profile-to-vertex distances were random, then we would expect as many concordant pairs as discordant pairs in the respective cases, that is, $C/\frac{1}{2}I(I - 1) = 0.5$. To measure success we should therefore subtract 0.5 from $C/\frac{1}{2}I(I - 1)$, and then to allow the measure to have an upper bound of 1, we should scale this difference up by a factor of 2, leading to the following final form:

$$2 \times \left(\frac{\text{number of concordant pairs}}{\text{total number of pairs}} - 0.5 \right) = 2 \left(\frac{C}{\frac{1}{2}I(I - 1)} - 0.5 \right) \qquad (12)$$

Using $C + D = \frac{1}{2}I(I - 1)$, this index can be shown to be identical to Kendall's tau coefficient $\tau = (C - D)/(C + D)$, a rank correlation coefficient defined by Kendall

Table 1: τ indices of quality ($\times 1000$) for each age group vertex point in Figure 3, for scalar products (biplot interpretation) and distances (unfolding interpretation), and the usual qualities (QLT) of display of the age group profiles as well as the age group vertices as supplementary points

	τ indices		Qualities	
	Scalar prods	**Distances**	**Profiles**	**Vertices**
A1	905	908	982	747
A2	649	662	856	186
A3	691	631	899	383
A4	758	772	954	291
A5	901	902	990	815

(1948). The tau coefficient can also be written in the following convenient form:

$$\tau = 1 - 2 \left(\frac{D}{C + D} \right) \tag{13}$$

In other words, the τ index of success, or quality, lying between -1 and 1, is 1 minus twice the ratio of discordant pairs to the total number of pairs.

Considering again the asymmetric CA map in Figure 1, with the five age groups as vertex points in standard coordinates, we refer all 100 row profile points to one vertex point at a time and compute the τ index to measure the quality of the biplot and unfolding interpretations respectively for each age group point. The results are summarized in Table 1. The extreme, most outlying, age group points *A1* and *A5* show tau coefficients of over 0.9. These are also the points that have the highest quality of display in the two-dimensional map, according to the usual contributions, which are given in the third column of the same table for purposes of comparison. The profile qualities are all higher than the τ indices.

7   Extensions to MCA and JCA

Here we consider the extension of the biplot and unfolding definitions of CA to the multiple case. We have seen two separate but equivalent versions of the biplot definition, namely the matrix approximation (2) on the one hand, based on the SVD, and, on the other hand, the function (7) to be maximized. We extend the latter version of the biplot and the unfolding criterion (4) to the multiple case.

7.1   MCA of Burt Matrix

The Burt matrix $\mathbf{B}$ is:

$$\mathbf{B} = \mathbf{Z}^{\mathsf{T}}\mathbf{Z} = n \begin{bmatrix} \mathbf{D}_{(1)} & \mathbf{P}_{(12)} & \cdots \\ \mathbf{P}_{(21)} & \mathbf{D}_{(2)} & \cdots \\ \vdots & \vdots & \ddots \end{bmatrix}$$

where $\mathbf{P}_{(qs)}$ is the cross-tabulation of the qth and sth variables, and $\mathbf{D}_{(q)} = \mathbf{P}_{(qq)}$ is the diagonal matrix of marginal relative frequencies of the qth variable. MCA can also be defined as the CA of the Burt matrix, and it is known that the standard coordinates of the rows of $\mathbf{B}$ (or of its columns, $\mathbf{B}$ is symmetric) are identical to the standard coordinates of the columns of $\mathbf{Z}$. It is also well known that the principal inertias of $\mathbf{B}$ are the squares of those of $\mathbf{Z}$; see, for example, Greenacre, 1984, p. 140.

Extending the biplot and unfolding definitions, (7) and (4), respectively, to $\mathbf{B}$, we obtain similar criteria except that there is an extra double summation over the block matrices that constitute the Burt matrix:

$$\text{maximize} \qquad \frac{1}{Q^2} \sum_q \sum_s \sum_i \sum_j p_{(qs)ij} \mathbf{f}_{(q)i}^{\mathsf{T}} \mathbf{y}_{(s)j} \tag{14}$$

$$\text{minimize} \qquad \frac{1}{Q^2} \sum_q \sum_s \sum_i \sum_j p_{(qs)ij} (\mathbf{f}_{(q)i} - \mathbf{y}_{(s)j})^{\mathsf{T}} (\mathbf{f}_{(q)i} - \mathbf{y}_{(s)j}) \tag{15}$$

where $p_{(qs)ij}$ denotes the ijth element of $\mathbf{P}_{(qs)}$, and $\mathbf{f}_{(q)i}$ and $\mathbf{y}_{(s)j}$ denote the principal and standard coordinate vectors, respectively, of the ith category of variable q and the jth category of variable s.

Objectives (14) and (15) are similar in aspect to those of (7) and (4) for a single table, apart from the following three crucial differences. First, the single table has two sets of points for the categories of the row and column variable, respectively, and these points are vectors of "free parameters" whose estimates will capture the association between the row and column variables. In the case of the Burt matrix we have Q^2 cross-tables; in fact, each categorical variable q is cross-tabulated with each of the other $Q - 1$ ones, as well as with itself. Separate CAs of these $Q - 1$ tables are possible but would proliferate the number of graphical displays so much as to defeat the object of the data reduction that the visualization hopes to achieve. Such analyses would lead to $Q - 1$ different solutions for the J_q categories of variable q. In analyzing the Burt matrix, however, we restrict the solution to just one set of category points for each variable, as can be seen in (14) and (15), where the coordinate vectors $\mathbf{f}_{(q)i}$ and $\mathbf{y}_{(s)j}$ each contain only one index per variable. This means that we have a much simpler map with only one point per category, which can be considered as an average display of the individual association patterns.

Second, the free parameters are not as numerous as they seem. Since $\mathbf{B}$ is a symmetric matrix, the row and column solutions are identical, so that there is a very simple relationship between the principal and standard coordinates for both rows and columns, for example:

$$\mathbf{f}_{(q)j} = \mathbf{D}_\lambda^{1/2} \mathbf{y}_{(q)j} \qquad (16)$$

where $\mathbf{D}_\lambda$ is the diagonal matrix of K^* principal inertias of $\mathbf{B}$ for the K^*-dimensional solution.

Third, (14) and (15) include terms for each variable cross-tabulated with itself, when $q = s$, for example, for the unfolding criterion (15):

$$\frac{1}{Q^2} \sum_q \sum_i \sum_j p_{(qq)ij}(\mathbf{f}_{(q)i} - \mathbf{y}_{(q)j})^\mathsf{T}(\mathbf{f}_{(q)i} - \mathbf{y}_{(q)j})$$

$$= \frac{1}{Q^2} \sum_q \sum_j c_j(\mathbf{f}_{(q)j} - \mathbf{y}_{(q)j})^\mathsf{T}(\mathbf{f}_{(q)j} - \mathbf{y}_{(q)j}) \qquad (17)$$

Here we have further simplified the summation by noting that $\mathbf{P}_{(qq)}$ is a diagonal matrix with the marginal relative frequencies c_j, $j = 1, \ldots, J_q$, down the diagonal. This shows that in addition to trying to display all the associations between different variables, the category points $\mathbf{f}_{(q)j}$ and $\mathbf{y}_{(q)j}$ in principal and standard coordinates corresponding to the same category are being coerced to lie as close to each other as possible. From (16) this can be equated to coercing the principal inertias to be as high as possible, to display the perfect association between a variable and itself that is embodied in the diagonal tables $\mathbf{P}_{(qq)}$. The terms (17) clearly have a very strong influence on the solution in MCA and explain the anomalies regularly noticed in MCA solutions—low percentages of inertia and low relative contributions (Greenacre, 1990, 1991).

Figure 4 shows the two-dimensional CA map of the 100×100 Burt matrix formed of all the cross-tabulations of the 25 cultural competence variables, where all points are in principal coordinates. On the left-hand side of the map the inabilities to "read city map" (m3), "ride bicycle" (q3), "use calculator" (t3), and "take photographs" (i3) separate out, and on the right-hand side we have the ability to "play chess" (d1), "change spark plugs" (l1), "use calculator" (t1), and a nonresponse to "knitting" (p4). The dimensions of the map are somewhat difficult to interpret if there are no external variables such as age or sex to refer them to. The utility of such an analysis would rather be to reduce the dimensionality of the data set and to notice such aspects as the association of many of the nonresponse categories at the top of the map, indicating that nonresponses are confined mostly to a particular subgroup of respondents. The usual overall quality measure of the display would be the sum of the percentages of inertia on the two axes, namely $33.4\% + 8.5\% = 41.9\%$.

To apply the tau coefficient to measure the success of the joint display, we do not have a variable such as age (see Figure 1) to which we refer the 25 cultural competence variables. Rather, each of the variables serves as a reference to the other 24 variables, because the information in the data matrix is the association among all 25 variables. Thus we successively place the four categories of each of the 25 variables in their vertex positions and then verify scalar products and distances of the other 96 points with respect to these four vertices. For example, the four categories of the variable "use PC" (h) are starred and italicized in Figure 4. These should be placed in their vertex positions (not shown here) and the same procedure followed as before. The τ indices for the four categories are found to be 0.660, 0.523, 0.734, and 0.467, respectively, for the scalar product interpretation and 0.653, 0.525, 0.744, and 0.456, respectively, for the distance interpretation. These and some other values can be inspected in Table 2, where we have also included the τ's for one- and three-dimensional solutions. Notice that the qualities increase as dimensionality increases, although small exceptions can occur. Also notice, as in the case of category f4, that a τ index can attain a negative value—recall that the range of τ is from -1 to 1, where 0 represents a completely random relationship between the profile elements and the scalar products (or distances).

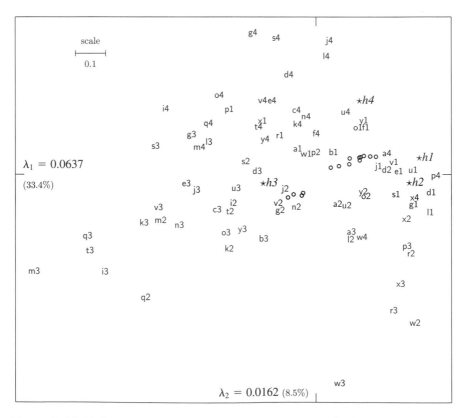

Figure 4: Multiple correspondence analysis of Burt matrix of cultural competences, with all category points in prinicipal coordinates; categories of "use PC" are in italics.

Table 2: τ indices of success ($\times 1000$) for each category point, for MCA solutions dimensionality one, two, and three, and the usual quality measures from MCA (MCA analysis of the Burt matrix)

	Scal. prods			Distances			Qualities		
	1-d	**2-d**	**3-d**	**1-d**	**2-d**	**3-d**	**1-d**	**2-d**	**3-d**
⋮	⋮	⋮	⋮	⋮	⋮	⋮	⋮	⋮	⋮
e1	848	850	864	848	844	822	818	819	838
e2	33	107	543	216	232	551	4	12	284
e3	736	752	804	736	751	796	817	821	837
e4	284	480	527	300	476	521	16	62	114
f1	405	586	661	405	571	653	154	306	324
f2	374	384	561	374	390	569	31	38	97
f3	439	692	693	244	589	588	196	256	259
f4	−67	262	277	50	259	281	0	4	34
g1	798	820	821	798	820	810	828	895	918
g2	146	272	446	151	299	436	22	44	223
g3	775	800	807	775	800	809	808	888	888
g4	264	455	599	264	446	598	12	73	173
h1	678	660	696	678	653	685	474	486	501
h2	519	523	529	519	525	545	220	221	251
h3	706	734	758	695	744	722	632	648	655
h4	223	467	590	223	456	582	18	60	160
⋮	⋮	⋮	⋮	⋮	⋮	⋮	⋮	⋮	⋮

7.2 JCA of Burt Matrix

JCA avoids the complication of the diagonal terms by excluding them from the objective criteria, biplot or unfolding. Because the tables above the diagonal of the Burt matrix are just transposes of those below the diagonal, we can express the objective criteria in terms of the $\frac{1}{2}Q(Q-1)$ tables in the upper or lower half-triangle of the $Q \times Q$ block matrix $\mathbf{B}$. Thus, the JCA criteria for the biplot and unfolding models are, respectively,

$$\text{maximize} \quad \frac{2}{Q(Q-1)} \sum_q \sum_{s>q} \sum_i \sum_j p_{(qs)ij} \mathbf{f}_{(q)i}^{\mathsf{T}} \mathbf{y}_{(s)j} \tag{18}$$

$$\text{minimize} \quad \frac{2}{Q(Q-1)} \sum_q \sum_{s>q} \sum_i \sum_j p_{(qs)ij} (\mathbf{f}_{(q)i} - \mathbf{y}_{(s)j})^{\mathsf{T}} (\mathbf{f}_{(q)i} - \mathbf{y}_{(s)j}) \tag{19}$$

One of the advantages of this definition is that JCA has as an exact special case the simple CA problem (where $Q = 2$), because the summation in (18) and (19) involves only one term corresponding to the single cross-tabulation $\mathbf{P}_{12}$ in simple CA, denoted previously as $\mathbf{P}$. This is not so for the MCA definition, where the special case for $Q = 2$ does not reduce to the simple CA definition exactly.

As in Section 6.1, we can apply the tau coefficient to measure the quality of each category's success in the joint map. Figure 5 shows the JCA of the same 100×100 Burt matrix as before. Notice first how much the percentages of inertia explained have increased compared with the previous MCA—from 33.4% and 8.5% for the two axes in Figure 4 to 68.2% and 13.7% in Figure 5. The scale of the two displays is the same, and the cloud of points in Figure 5 is very similar to that of Figure 4, with a slight reduction in scale and a flattening of the cloud in the vertical direction. The computation of τ indices, shown in Table 3, is done exactly as before, and because of the similarity of the two configurations, it is to be expected that these diagnostics will be quite similar to those computed for MCA. Comparing Table 3 and Table 2 confirms that there is hardly any difference in the quality of the joint display when it is measured in this way. This demonstrates that evaluating the quality of the display

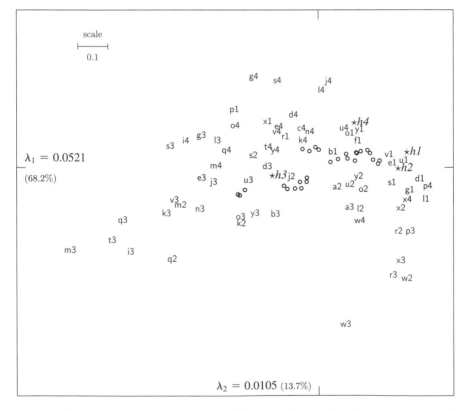

Figure 5: Joint correspondence analysis of Burt matrix of cultrual competences, with all category points in principal coordinates; categories of "use PC" are in italics.

Table 3: τ indices of success ($\times 1000$) for each category point, for JCA solutions dimensionality one, two, and three, and the usual quality measures from the JCA solution (including estimated diagonal blocks, perfectly fitted by JCA)

	Scal. prods			Distances			Qualities		
	1-d	2-d	3-d	1-d	2-d	3-d	1-d	2-d	3-d
⋮	⋮	⋮	⋮	⋮	⋮	⋮	⋮	⋮	⋮
e1	844	853	858	844	859	875	969	973	983
e2	207	249	559	33	100	550	20	38	284
e3	730	749	802	730	744	802	969	976	985
e4	291	479	492	286	474	493	178	270	381
f1	397	601	670	397	620	681	487	862	893
f2	369	395	584	369	388	586	298	359	680
f3	251	599	629	431	676	696	538	879	883
f4	51	213	258	−69	207	239	0	6	85
g1	810	836	856	810	839	857	906	962	979
g2	148	286	440	148	264	447	115	170	666
g3	787	827	829	787	829	825	906	972	972
g4	271	454	605	271	471	614	67	196	390
h1	677	667	685	677	672	695	863	889	898
h2	518	516	547	518	518	533	768	768	251
h3	694	738	761	700	728	755	893	918	927
h4	217	403	594	217	407	589	77	168	373
⋮	⋮	⋮	⋮	⋮	⋮	⋮	⋮	⋮	⋮

in a nonmetric way avoids the issue surrounding the usual way of quantifying the map quality.

8 Conclusion

The usual diagnostics in CA are based on the decomposition of the total inertia into components. These diagnostics suit the dimensional interpretation of CA; when the principal axes are interpreted one at a time, each point's contribution to an axis is evaluated as well as each axis' contribution to the points.

Our distance interpretation of a map, however, is based on comparing relative positions of categories, and the tau index is proposed as a summary measure of the validity of this interpretation. If a profile point representing a specific category is well represented, then it follows that recovery of data values relative to this category will tend to be of higher quality than that of a category that is poorly represented. Thus, there will be a close monotonic relationship between the tau indices and the usual quality indices based on inertia contributions of points.

An advantage of the tau index is that it can be used across all types of CA. Since it is not concerned with recovery of data values, it quantifies the interpretational validity of MCA and JCA, which is independent of different scaling factors. Nishisato (1988) also proposes a measure for evaluating the quality of the joint display, but uses a metric criterion rather than a nonmetric one. Using a nonmetric quality measure is one way of showing that MCA and JCA can give similar results, even though the explained inertia in MCA is low.

Finally, the same ideas can be applied to measuring quality of fit in an MCA where the rows and the columns of an indicator matrix are displayed. The data values are just zeros and ones, and the biplot and unfolding objectives would be, respectively, to ensure the largest scalar products and smallest distances between individuals and their response categories. The same tau index can be used to measure success of representation with respect to each vertex point j. Notice that we are not trying to approximate the values 0 or 1 in the map. As de Leeuw says in Section 1.4, we are not interested in approximating these values strictly but rather in the "qualitative relations" in the data. Total success in the map is therefore not measured in terms of percentage of inertia displayed, which has to do with approximating the zeros and ones, but rather by a criterion such as our tau coefficient, which does measure the success of recovering the qualitative relationships between rows and columns.

Chapter 18

Using Multiple Correspondence Analysis to Distinguish between Substantive and Nonsubstantive Responses

Victor Thiessen and Jörg Blasius

1   Introduction

More than a quarter-century ago, Bogart, in his presidential address to the American Association of Public Opinion Research, astutely remarked that "our opinions invariably transcend our knowledge" (Bogart, 1967, p. 337). This led him to conclude that the typical emphasis on substantive answers elicited from the public was fundamentally misguided: "We measure public opinion for and against various causes with the "undecided" as the residue. Often what we should be doing instead is measuring the degrees of apathy, indecision, or conflict on the part of the great majority, with the opinionated as the residual left over" (Bogart, 1967, p. 337). Our chapter aims to distinguish substantive from nonsubstantive responses (NSRs).

Two traditions characterize research on opinion holding. The one attempts to distinguish respondents who actually have an opinion from those who do not. Research in this tradition follows on the heels of the pioneering work of Converse (1964); it takes advantage of patterns of consistent and inconsistent responses in panel studies, for example, to distinguish opinion holders from respondents who do not hold an opinion on the issue at hand. Models such as Converse's (1970) black and white (BW), which divides respondents into two groups—those holding opinions and those not having opinions—are assessed in this tradition. With synchronic data, constraints between items are introduced to distinguish acceptable from unacceptable response patterns, along the lines of Guttman scales.

Closer scrutiny of respondents classified as non–opinion holders reveals that their response patterns are far less random than implied by the theory. One reaction was to introduce some "gray" into the BW model (Brody, 1986). A second was to distinguish between several forms and/or sources of NSRs. Coombs and Coombs (1976) distinguished "don't know" (DK) responses they felt reflected item ambiguity from those indicative of item-specific equivocation (due to response uncertainty). Likewise, Duncan *et al.* (1988) documented what they call a "transitory response set" that operated to produce consistent responses within a given wave of a panel study but inconsistent responses between waves. In all of these models, the focus is on the latent distribution of opinion holding in a population.

Researchers in the second tradition concern themselves more with the "problem" of NSRs. Here the focus is usually with the predictors of NSRs. Researchers in this tradition often regress a dependent variable (either the number or the proportion of DK responses) against respondent attributes such as sex, age, and education. In these approaches, only the number or the proportion, but not the structure of the DK responses, is considered. The implicit assumption is that tendency to use NSRs is a respondent attribute, essentially independent of substantive content of the issues being examined. Research such as Rapoport's (1985) finding of a correlation of 0.38 between parent–child dyads in the frequency of using a DK response supports this assumption.

2 Correlates of "Don't Know" Responses

The literature is quite consistent with respect to the correlates of DK responses. The strongest predictor is political interest: the lower the interest, the greater the tendency to give an NSR (Faulkenberry and Mason, 1978; Rapoport, 1982). Education is the next strongest factor, having an inverse relationship with the tendency to respond with DK (Francis and Busch, 1975; Converse, 1976/77; Faulkenberry and Mason, 1978). The third factor is sex: women more than men respond with DK (Ferber, 1966; Francis and Busch, 1975; Ferligoj *et al.*, 1991). Part of this relationship may be due to the greater exclusion of women from the political process. Although the relationship was indeed reduced when education and political interest were controlled, the gender effect remained statistically significant (Francis and Busch, 1975). Ferber's (1966) study dealt with opinions about durable goods, which are likely to interest women more than men. Hence it is unlikely that the gender effect is totally a function of interest differences.

Finally, age also seems to be a predictor of DK response, with older respondents more likely to utilize the DK response (Ferber, 1966; Francis and Busch, 1975;

Rapoport, 1982). But a careful review of the empirical findings suggests that age is only weakly related to DK at best. Ferligoj *et al.* (1991, p. 13), for example, find no age trends in DK responses up to the age of 60 in any of their four Slovenian national surveys. Likewise, Rapoport's (1982) data show that the only difference is essentially between those over and under 60 years of age. Also, in two of the three waves in the Francis and Busch (1975) analyses, age does not reach statistical significance, despite relatively large samples. Ferligoj *et al.* (1991, p. 16) provide evidence for a sex–age interaction effect: older women are consistently and significantly more likely than younger women to respond with DK. Among men, there is no comparable discernible age effect.

In this chapter we start with the topic of distinguishing substantive responses from NSRs. This includes an examination of whether responses such as "no difference" (ND) or the middle category of attitude scales (such as "undecided" or "unsure") should be treated as substantive. In line with studies such as those of Converse (1970), Coombs and Coombs (1976), Brody (1986), and Duncan *et al.* (1988), we describe the latent structure that characterizes the response categories. This means we will assess whether the NSRs, such as don't know or not sure, are located close to each other in a multidimensional space. For example, if we want to describe how the categories of different items (including DK and ND responses) are related to each other on any topic, we have to create a common latent space in which each response category for each item in the analysis is located. Furthermore, this space should provide the possibility of interpreting distances between response categories as (dis)similarities of meaning. The difference between substantive responses and NSRs should manifest itself in the latent space by large distances between the respective categories. This means the substantive responses should be clustered at one point of the latent space and the DK responses in a quite different part.

In the next stage we would like to incorporate the findings on the correlates of DK into a common model, which includes the assumptions of both research traditions we have mentioned. Thus we will describe how respondent attributes, such as age, education, and sex, fit into the space determined by the attitude and opinion items of the domain of inquiry. In this connection, we expect the probability of giving an NSR to decrease with interest in the topic at hand. Thus, the greater the interest in a topic, such as abortion, the lower the likelihood of responding with NSRs. In this extended model, the reported strength of association of the correlates of NSR should be replicated. This means that political interest should have the highest association with NSR in the latent space, followed by education and sex. We will examine these expectations using data from the 1984 Canadian National Election Study (CNES).

3   Data

The 1984 CNES is based on a large ($N=3377$) multistage weighted probability sample; provinces with low populations are oversampled (these data are available through the Inter-University Consortium for Political and Social Research at the University of Michigan, Ann Arbor). Face-to-face interviews were conducted following the 1984 federal election exploring a number of social and political issues. This study contains information appropriate to exploring distinctions between substantive and nonsubstantive responses. In this chapter we focus on questions concerning the perceived (in)competence of the federal political parties to deal with a number of issues: "Now I'm going to ask you about a number of tasks that the federal government has to deal with. Forget for a moment the likelihood of each party getting elected to government. I'd like you to tell me which of the three major federal parties would probably do the best job and which would probably do the worst job on each task if it were the government."

The tasks (together with the alphabetic characters used to identify them in subsequent graphical displays) were:

controlling inflation (I)

dealing with the USA (A)

running the government competently (C)

providing social welfare measures (S)

limiting the size of government (L)

working for world peace (P)

dealing with the provincial governments (G)

handling relations with Quebec (Q)

dealing with unemployment (U)

protecting the environment (E)

dealing with women's issues (W)

handling the deficit (D)

No response categories were provided for the respondents. Rather, as the lead-in to the question indicated, they were expected to name one of the three federal political parties of Canada, which are the Progressive Conservatives, the Liberals, and the New Democratic Party. Most respondents did indeed name exactly one party as the best/worst for a given task. However, sizable numbers stated they did not know; another substantial group indicated they felt there was no difference between the three parties; and finally, a relatively small number of respondents felt that two of the three parties would be equally best/worst for the given task. From these responses we constructed the following categories:

Differentiated (D): one of the three parties was named as best/worst.

Semidifferentiated (SD; in the figures denoted by *S* only): two of the three parties were considered equally best/worst.

No Difference (ND or *N):* the three parties were not considered to be distinguishable on this task.

Don't know (DK or *K):* it was not known which party would do best/worst on that task.

It is clear that respondents who named a given party provided a substantive response. Likewise, respondents who considered precisely two parties to be equally best/worst in dealing with a specific issue are providing substantive responses. On the other side, DK is by definition a nonsubstantive response.

The ND response is potentially problematic. On the one hand, it may well be that knowledgeable observers of the political scene arrive at the conclusion that the political parties are not differentiated with respect to their ability to handle an issue such as working for world peace. That would indicate that an ND response is indeed a substantive response. On the other hand, a respondent could have insufficient information to discriminate between the political parties' abilities to deal with a given issue. In such instances, a response of ND would have a nonsubstantive meaning similar to that of a DK; only in the latter case could the two categories be combined. In the analysis that follows, the main question will be which of the two interpretations is more defensible. Practically speaking, is it permissible to combine the two categories or should they be kept separate?

In addition to this question, we will extend previous findings on the correlates of DK responses. Variables used for this purpose are age, sex, educational level, political interest, and political knowledge. A measure of political knowledge was constructed by counting how many of the 10 Canadian provincial premiers a respondent could name. The individual mean of a battery of items on political participation (reading newspapers, watching programs, discussing politics, etc.) was used to measure political interest (response alternatives ranged from "never" [1] to "often" [4]).

4   Results

Table 1 shows the levels of differentiation and opinionation on the relative (in)ability of the Canadian federal political parties to deal with the foregoing tasks. Overall, this table shows that the number of DK responses is higher than the number of ND responses, and both are higher than the number of semidifferentiated responses. When differentiating which party would handle a given task best, the highest incidence of DK response belongs to the item "protecting the environment," followed by "limiting the size of government," "dealing with women's issues," and "handling the deficit"; focusing on which party would be worst, the decreasing order starts with "working for world peace," "protecting the environment," "dealing with women's issues," "providing social welfare measures," and "limiting the size of government." That is, the inner order of the two versions (best and worst) is similar. This suggests that the DK response is to some extent item specific: items on which there is a high proportion of DK responses as to which party would do best tend to be the issues on which there is also a high proportion of DK responses concerning which party would do worst. Although the inner order of the four response categories is quite similar, the levels of usage are quite different. In the extreme, for the item "working for world peace" 17.2% of the respondents chose the DK response for which party would do best, whereas 38.5% gave this response for which party would do worst.

Table 1: Level of differentiation and opinion among Canadian federal parties on their ability to deal with various issues[a]

	Best				Worst			
	D^b	SD	ND	DK	D	SD	ND	DK
Inflation	66.2	4.9	12.1	16.3	63.4	2.8	11.3	22.1
Provincial governments	71.6	2.9	7.7	17.4	63.9	3.2	8.0	24.5
U.S.	71.1	3.0	6.7	18.9	61.2	2.5	6.8	29.1
Quebec	74.2	2.5	6.8	16.0	62.6	2.8	7.1	27.1
Competent	68.8	3.1	9.1	17.7	60.2	3.6	9.7	26.1
Unemployment	69.8	2.6	10.5	16.7	59.9	3.9	10.1	25.6
Social welfare	70.4	2.9	8.0	18.2	55.4	3.5	8.6	32.0
Environment	52.5	2.2	17.1	27.8	42.0	3.2	16.3	38.1
Limit government	60.8	1.5	10.5	26.8	55.3	2.5	10.3	31.5
Women's issues	62.5	2.3	11.4	23.4	48.5	4.0	11.7	35.4
World peace	65.8	2.5	14.1	17.2	42.3	3.7	15.1	38.5
Deficit	66.4	1.9	10.3	21.0	59.0	2.8	9.9	27.9

[a]$N=3377$. However, there are either 14 or 15 cases of "no opinion" for each of the issues. These have been excluded from the table.
[b]D, Differentiated; SD, Semidifferentiated; ND, no difference; DK, don't know.

Turning to the ND responses, between 6.7% and 17.1% claimed the parties did not differ in their ability to deal with a given issue. Overall, the ND responses are consistently less likely to be used than the DK responses. Also, there is substantially less variation in the use of the ND response than of the DK response.

We turn now to the correlates of DK and ND responses. For this purpose we counted the number of DK and ND responses separately for best, worst, and their total. The 1984 CNES data replicate almost perfectly the pattern of DK correlates found in the literature (see Table 2). Political interest has the strongest relationship with the number of DK responses and age has the weakest, with political knowledge,

Table 2: Pearson correlations of "don't know" and "no difference" responses with selected respondent attributes

	Don't know			No difference		
Attribute	Best	Worst	Total	Best	Worst	Total
Political interest	−.32	−.30	−.32	−.07	−.05	−.06
Political knowledge	−.24	−.25	−.25	−.01	.00	−.01
Education	−.23	−.24	−.24	−.01	.00	.00
Sex (0 = male, 1 = female)	.15	.18	.17	.05	.03	.04
Age	.08	.10	.10	.01	−.01	.00

education, and sex in between. In contrast to these patterns, the number of ND responses shows no substantial correlation with any of the variables typically connected with NSRs. That is, the ND responses do not share any of the patterns of bivariate correlations typically found for DK responses.

Although correlations are useful for finding associations, they cannot detect latent structures between a set of variables, which is the concern of our chapter. One of our main questions is whether a response tendency exists to answer the 12 tasks in one of the four ways: D, SD, DK, or ND? If so, a graphical display should mirror four clusters reflecting the four categories within both the "handle best" and the "handle worst" sections.

The more individuals fluctuate between DK and ND responses within the item battery, the closer the clusters should be to each other. If ND and DK are interchangeable responses of being an NSR, there should be a common cluster of DK and ND responses. If there is no tendency to answer the questions in any one of the four ways (D, SD, ND, DK), there will be no cluster constituted by one of the four categories. In addition, by simultaneously analyzing "best" and "worst" items we can determine whether there is a common structure between all 24 items or 96 categories, respectively. Finally, we will describe the association of political interest, education, political knowledge, sex, and age with the structure defined by the $2 \times 12 \times 4 = 96$ categories.

An appropriate method for visualizing the 12 (respectively 24) items is either multiple correspondence analysis (MCA) or the "Netherlands version" of this method called homogeneity analysis, known as HOMALS in SPSS (see Gifi, 1990; Meulman and Heiser, Chapter 20), which provides the same solution. Input data for MCA are either the indicator matrix containing all items or the Burt matrix, the cross-tabulations between all pairs of items to be included in the analysis concatenated in a square block matrix. In MCA, no distinction between independent and dependent variables needs to be made (see Greenacre and Blasius, 1994b). In our case, for example, all questions are on the theme of the perceived (in)competence of the three main Canadian political parties to handle various tasks. This situation is reminiscent of principal components analysis (PCA), which explores the structure of associations among a set of variables by identifying underlying dimensions.

When using respondent-level data (using the indicator matrix as input data), rows describe respondents and columns provide the response categories in the form of dummy variables. A 1 in a given column means that the response category associated with that column was used, and a 0 means that it was not used. For each item there are four columns, one for each of the four possible response categories of D, SD, DK, and ND. If there are no missing data, each issue has to have a 1 in exactly one of its four columns, and a 0 in the remaining three. For our problem only the spatial positions of the variable categories are of interest; the spatial positions of the individuals will be ignored. The locations of all categories can be compared with each other, where short distances mirror high similarities and long distances high dissimilarities.

Once the locations of the item categories have been computed, we can place axes into the space. These axes are chosen under a least-squares criterion. As in PCA, the first axis is located so as to explain maximum variation in the data; the second axis is orthogonal to the first axis and is chosen so as to explain a maximum of the remaining variation, and so on. Again as in PCA, it is possible to interpret the variable categories in relation to the axes, which can be treated as latent variables. The closer the categories are to a given axis, the more they are determined by it (the cosines between the vector endpoints of the variable categories and the axes can be interpreted as factor loadings).

Starting with the tasks handled best, we have a total of 48 columns (12 tasks with 4 response categories each) and 3362 rows (the number of respondents having ascertained responses on any of the tasks). This data matrix forms the input for the first MCA, which permits a maximum of 36 dimensions (number of categories minus number of variables). The eigenvalue for the first axis is 0.51, which explains 17.0% of the total variation; the corresponding values for the second axis are 0.40 and 13.2%. Without discussing the statistical details, it can be shown that this underestimates the amount of explained variances attributable to the first few axes. We will forgo recomputing the explained variances (for adjusting the explained variances see Greenacre, 1993) because the decreasing order of the axes including the variable categories belonging to them is retained; differences in the maps using several possible adjustments are so small they can be neglected. Figure 1 provides a visualization of the locations of the 48 item categories for this MCA solution.

Figure 1 shows a clear distinction between four clusters of variable categories, which indicates four patterns of response behavior. Projecting the categories onto the axes, on the left, or negative, side of the first axis are the DK responses. Negatively correlated with these categories are the items of the cluster "differentiated" and, to a lesser degree, the items of the cluster "semidifferentiated." Projecting the centroids or the range (not shown in Figure 1) of the four clusters onto the main axis, one gets a line from D through SD and ND to DK. This permits an interpretation of the first axis as the degree of differentiation, with D at the one extreme and DK at the other. This means that the higher the value on the first latent dimension, the higher the score along the scale ranging from NSR to substantive response.

The second axis is populated mainly by the ND responses, which are located on the upper part of this axis. In addition to the reported differences along the first axis, the long distances between DK and ND responses imply that individuals frequently employing the ND responses are distinct from individuals answering relatively often with DK. Furthermore, individuals who differentiate between the competences of the three parties seem to be distinct from those who respond with ND. Between these two clusters are individuals who tend to differentiate partially between the federal parties; that is, who gave responses classified as SD.

Focusing on the clusters themselves, Figure 1 shows that the responses characterized as D are relatively close to each other. This indicates that there is an overlapping latent structure between all items: the tendency to differentiate parties is not item specific—either one does or does not distinguish between the competences of the parties. The cluster of DK responses is also relatively homogeneous, with only the three items protecting the environment, dealing with women's issues, and limiting

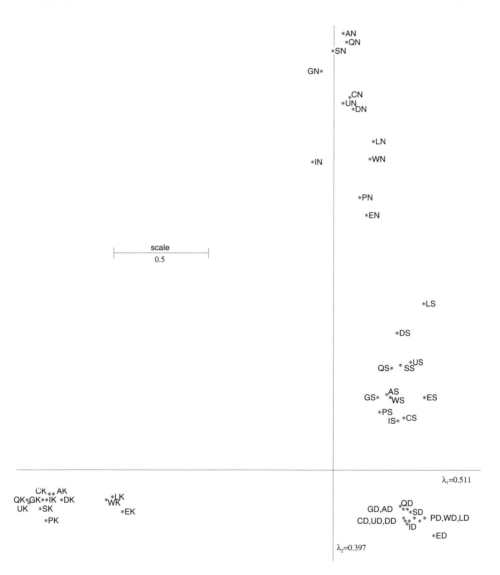

Figure 1: MCA of government tasks handled best.

the government size positioned somewhat outside the cluster centroid. In any event, some respondents answered DK relatively consistently and an additional group failed to differentiate the parties only on these three issues (see also the higher proportion of DK responses for these three items in Table 1, column 4).

In the two-dimensional solution, the variation within the ND cluster is much higher than that within the DK cluster. Therefore the likelihood of giving an ND response to one item, given that an ND response was given to any other item(s), is lower than that for the DK responses. Furthermore, the order of the ND responses along the second axis reflects almost perfectly the frequency with which resondents indicated that there was no difference between the three parties (see Table 1, column 3). The clustering in conjunction with the variation along the second axis indicates that there is an increasing "difficulty" in differentiating the three parties on the issues that are closer to the centroid. Because the axes are orthogonal, it follows that the ND response tendency is independent of the DK response tendency. The categories belonging to the SD cluster spread out; referring to the first two dimensions (Figure 1), the variation within this cluster seems to be higher than the variation within the D and DK clusters but less than that of the ND cluster. The high variation within the SD cluster becomes clearer when considering the third and fourth axes (not shown here), which explain 8.2% and 3.3% of total inertia, respectively: far away from the centroid the SD responses are clearly separated from one another.

Parallel to which party would handle various tasks best, we performed an MCA for which tasks would be handled worst (see Figure 2). As in the previous MCA, the first axis (explained variation: 18.3%) is determined by the opposition of DK and D responses, the second axis (explained variation: 14.0%) can be described especially by ND responses, and the cluster SD is located between the clusters D and ND. In general, the solution for the response behavior of party incompetence is a near clone of that found for party competence: the overall structures of the two MCAs reveals only small differences.

The purpose of the final analysis is to describe the responses to party competence (which party would handle the issues best) with those for party incompetence (which party would handle the issues worst) simultaneously, thus searching for a common latent structure between all 96 categories of the 24 items. In addition, we are interested in the associations of this structure with the respondent attributes discussed earlier. This allows us to describe which attributes, and in which forms, are related to the response categories ND, DK, SD, and D. The additional variables included in the model are sex (abbreviated by [F], [M]), age (subdivided into six categories, 18 to 24 [A1], 25 to 34 [A2], 35 to 44 [A3], 45 to 54 [A4], 55 to 64 [A5], and 65 and older [A6]), education (five categories, from grade school or less [E1] to university graduates [E5]), political knowledge (four categories, from none or one premier [K1] to five or more premiers [K4]), and political interest (four categories, from least interested [I1] to most interested [I4]). The latter variables are to be included without influencing the geometric structure of the space formed by the 24 government tasks (12 best and 12 worst). This is accomplished in correspondence analysis with the projection of supplementary information into the space of a prior solution. In general, correspondence analysis permits a distinction between active variables (or variable categories), which determine the geometric orientation of the axes, and variables (or variable categories) used for supplementary information only. These supplementary variables have no effect on the geometric orientation of the axes (see Greenacre, 1984, p. 73). The whole set of 24 government tasks supplemented by five respondent attribute variables (with a total of 21 categories) produces an indicator matrix of 3356

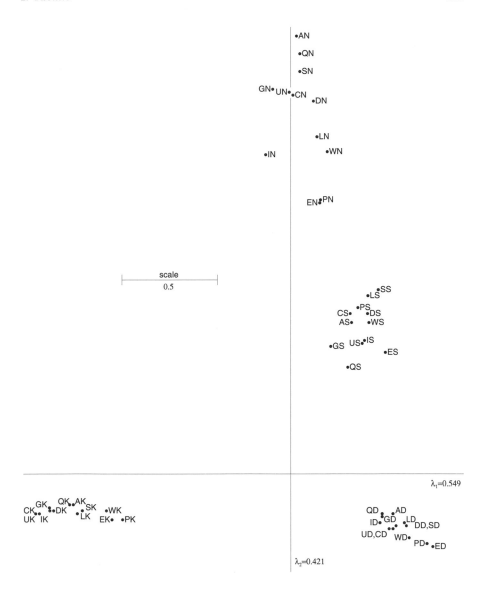

Figure 2: MCA of government tasks handled worst.

rows (the number of cases without any missing values) and 117 columns. This matrix will be used as input data for the final MCA. The solution is given in Figure 3.

Although the solution of the final MCA has 72 possible dimensions (96 categories − 24 items), the first axis explains 16.2% (unadjusted) of the total variation, the second one an additional 13.1%. Because these values are almost as high as the

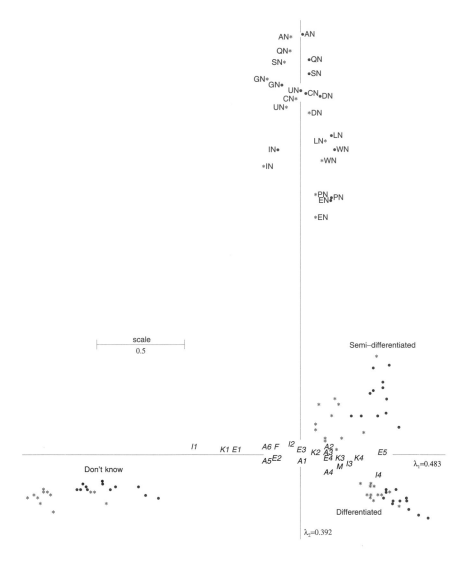

Figure 3: MCA of government task with supplementary variables.

previous ones, one can assume a high similarity in the response structure of the "handle best" and the "handle worst" items. This conclusion is corroborated by the graphical display—as in the previous solutions, there are four distinct clusters that include the respective categories. In the DK, D, and SD clusters there are only small differences between the competent (indicated by a star) and the incompetent (indicated by a circle) items. But these differences reflect mainly the different percentages of response: the DK responses are more pervasive for the incompetence question, whereas the SD and the D responses have higher proportions in the competence questions (see Table 1). In the case of ND the order of the categories is item specific—if one felt that there was no difference among the three parties' abilities to handle a task best, then it is a foregone conclusion that none should be found for their inability to handle that task.

Turning to the locations of the supplementary variables, Figure 3 shows that all 21 categories are either strongly connected with the first axis or located close to the centroid of the map. This means that the supplementary variables are not correlated with the second dimension. Therefore, sex, age group, education, political knowledge, and political interest are not recommendable predictors of ND responses; the supplementary variables can be used fruitfully only to distinguish individuals answering DK from those classified as either D or, to a lesser degree, SD.

The most important variable for describing the group with above-average DK responses seems to be least political interest (I1), which has the shortest distance to this cluster (Figure 3). Furthermore, little political knowledge (knowing none or one premier only [K1]) and a low formal education (grade school or less [E1]) are relatively good predictors for these variable categories. Two age categories are somewhat poorer indicators: persons 55 years and older (A5, A6) answered relatively often that they did not know which parties would handle the selected tasks best or worst, respectively. On the other side, especially persons with high political interest (I4) differentiated between the parties. Furthermore, high formal education (university graduates [E5]) and high political knowledge (naming five or more premiers [K4]) are relatively good predictors for both D and SD responses.

The orders of the categories of the supplementary variables along the first axis confirm the ordinality of education, political knowledge, and political interest with regard to the 24 items used in the analysis—projecting them onto the first axis, all categories belonging to these supplementary variables are in a line from high to low. Therefore, it can be concluded that the lower the political interest, the lower the political knowledge, or the lower the education, respectively, the higher the probability of a DK response.

Also in line with the literature is the response behavior of males and females: men more often differentiated or partially differentiated; women answered relatively often with DK. The age groups are not ordered in one line along the first dimension: the youngest and the oldest age groups (A1, A5, A6) are relatively close to the DK cluster, and the remaining groups are close to each other on the positive part of the first axis. This indicates that the relationship between age and response behavior is not linear, which also explains the low correlation coefficients of age with DK in Table 2.

5   Conclusion

Our analyses of DK and ND responses—which is only one example for distinguishing NSR from substantive responses—have several theoretical ramifications. First, the MCA results show a clear distinction between NSRs and substantive responses.

Furthermore, the ND responses are primarily substantive ones; on the main axis, they are much closer to the D and SD clusters than to the DK one. In the given analysis, the ND and the DK categories form relatively homogeneous clusters that are located on different orthogonal axes. This independence suggests that if respondents are given the choice between admitting they have no opinion (by responding with a DK) and masking their nonopinionation in a relatively easy, socially acceptable way (by responding with a ND), few choose the masking route. Apparently, respondents did not feel pressured into giving substantive responses. Of course, it remains a hypothetical question whether respondents would choose the ND response if they had not been given the DK option.

Our approach can also be used when the items tapping opinions in a given domain have a Likert response format. With such a format, the direction of opinion has to be removed from the intensity with which the opinion is held. That is, the response categories of "strongly agree" would be collapsed with those of "strongly disagree," as well as "agree" with "disagree." This is analogous to our ignoring which particular political party the respondent named. If the middle categories can be treated as substantive responses, in an MCA solution these substantive responses should be located in a different position from the NSRs. In general, in any case in which substantive responses with "less meaning" exist, these categories will be located in a different position from the DK responses in a common latent space. This implies that any substantive ND responses should be kept separate from DK responses.

In our final analysis we included political interest, political knowledge, education, sex, and age as supplementary variables. By doing this we could show that political interest has the strongest association with the first latent dimension that distinguishes NSRs from substantive responses. This was followed by political knowledge, education, and sex, respectively. Overall, for our topic, individuals most marginal to the political process are also least likely to express opinions in that issue domain. In general, interest in and knowledge of the topic domain should most clearly distinguish respondents who hold substantive responses from those who do not. Sociodemographic attributes can also be used as predictors, but their association with the main dimension is lower.

In our analysis we found a nonlinear relationship of age with the first axis, whereby the youngest and the oldest gave the most NSRs. This nonlinear correlation is shown by the nonordinal positions of age groups along the first dimension. In general, if we expect a linear relationship of age with the "degree of differentiation" scale of a topic, then the MCA should confirm this by an ordinal order of the age groups along the axis that distinguishes NSRs from substantive responses.

Chapter 19

The Case of the French Cantons: An Application of Three-Way Correspondence Analysis

André Carlier and Pieter M. Kroonenberg

1 Introduction

Getting insight into the structure of large contingency tables with multicategory variables has always been a difficult problem. Simple rejection of the hypothesis of independence of the variables does not bring the desired insight but only indicates that there is something to be investigated further. It is not sufficient to know that the distributions over occupations of a workforce changed over time and that these changes are different in different regions. The nature of the change in the regions is what should be the focus of an investigation. Graphical displays that depict both the general patterns and the details are desired to provide both an overview and a microscopic view of the regional changes. By applying three-way correspondence analysis (CA) to a 42 (regions) by 9 (occupational classes) by 4 (time points) contingency table about the workforce in the Languedoc–Roussillon area (southern France), we will demonstrate how such an investigation may proceed and what types of conclusions about the structure in the table can be made.

Thus the major aim of this chapter is to provide an overview of the capabilities of CA of three-way tables to investigate the structure of large three-way contingency tables. Biplots will be used rather than the usual simultaneous displays (Benzécri, 1973a) because of their more attractive properties in revealing the relationships between rows and columns. To evaluate the dependence in three-way contingency tables, special biplots have to be used, because the information of all three modes has to be displayed simultaneously.

2 Overview of Two-Way Correspondence Analysis

CA is a technique that is applied to contingency tables in order to depict the dependence between the row and column categories. The matrix of deviations from independence is decomposed into two sets of components, one for the rows and one for the columns, in such a way that the dependence can be portrayed as well as possible in low-dimensional space. To this end, the best low-rank approximation to the matrix with dependences is sought, and this approximation is displayed in a biplot.

2.1 Measuring Dependence

Let us consider an $I \times J$ table of relative frequencies p_{ij}, with marginal row and column sums $p_{i.}$ and $p_{.j}$, respectively. A global measure of dependence is given by $\Phi^2 = \chi^2/n$ where χ^2 is Pearson's chi-squared statistic and n the total number of observations. Φ^2 is *Pearson's mean-square contingency coefficient*, called the *total inertia* in CA:

$$\Phi^2 = \sum_{i,j} \frac{(p_{ij} - p_{i.}p_{.j})^2}{p_{i.}p_{.j}} = \sum_{i,j} p_{i.}p_{.j} \left(\frac{p_{ij} - p_{i.}p_{.j}}{p_{i.}p_{.j}} \right)^2 = \sum_{i,j} p_{i.}p_{.j}\Pi_{ij}^2 \quad (1)$$

The measure Π_{ij} may also be written

$$\Pi_{ij} = \frac{p_{ij} - p_{i.}p_{.j}}{p_{i.}p_{.j}} = \frac{p_{ij}}{p_{i.}p_{.j}} - 1 = \frac{Pr[i \mid j]}{Pr[i]} - 1 = \frac{Pr[j \mid i]}{Pr[j]} - 1 \quad (2)$$

where, for example, $Pr[i \mid j] = p_{ij}/p_{.j}$ is the conditional probability of row category i given column category j and $Pr[i] = p_{i.}$ is the unconditional probability of i. Equation (2) shows that $1 + \Pi_{ij}$ is equal to both the ratio of $Pr[i \mid j]$ to $Pr[i]$ and the ratio of $Pr[j \mid i]$ to $Pr[j]$. Therefore, Π_{ij} measures the *attraction* between categories i and j if $\Pi_{ij} > 0$ and the *repulsion* between the two categories if $\Pi_{ij} < 0$. It is the matrix $\Pi = (\Pi_{ij})$ that is to be decomposed into its components and displayed in a biplot.

2.2 Modeling Dependence

In order to inspect the dependence in a contingency table graphically, we have to find an optimal representation in low-dimensional space. The appropriate tool for this is based on a generalization of the singular value decomposition (see, e.g., Greenacre, 1984, p. 39).

If we indicate the rank of the matrix $\mathbf{\Pi}$ by S_0, then the *generalized singular value decomposition* (GSVD) of the matrix $\mathbf{\Pi}$ is defined as

$$\Pi_{ij} = \sum_{s=1}^{S_0} \lambda_s a_{is} b_{js} \tag{3}$$

where the scalars $\{\lambda_s\}$ are the singular values arranged in decreasing order of magnitude, and the a_{is} and b_{js} are the elements of the singular vectors, or *components* $\mathbf{a}_s$ and $\mathbf{b}_s$, respectively. The set of components $\{\mathbf{a}_s\}$ are pairwise orthonormal with respect to the inner product weighted by $(p_{i.})$, and a similar property holds for $\{\mathbf{b}_s\}$ with respect to the weights $(p_{.j})$.

If we want a small number, S, of components that explain most of the dependence and thus approximate the full solution as well as possible, we should choose them such that we get the best low-dimensional approximation to $\mathbf{\Pi}$. This means that we have to use $\widehat{\mathbf{\Pi}}^{(S)}$, which is the sum of the first S terms of equation (3) (Eckart and Young, 1936), that is,

$$\Pi_{ij}^{(S)} = \sum_{s=1}^{S} \lambda_s a_{is} b_{js} \tag{4}$$

The result of this choice is that the total Φ^2 can be split into a fitted part and a residual part. This can be used to assess the overall quality of a solution via the proportion explained Φ^2.

2.3   Plotting Dependence: Biplots

Given that we have decomposed the dependence, we want to use this decomposition to graph the dependence. The appropriate graph is the biplot (Tucker, 1960; Gabriel, 1971). In CA the biplot displays in S dimensions *markers* for the rows and columns of the matrix $\widehat{\mathbf{\Pi}}$. Using the positions of the markers of the ith row and the jth column, it is possible to approximate the value of the element $\widehat{\Pi}_{ij}$ of $\widehat{\mathbf{\Pi}}$ and to interpret geometrically the directions of the markers in the plot.

In the simultaneous representation of CA, the markers are often presented in a symmetric way with (two-dimensional) coordinates $(\lambda_1 a_{i1}, \lambda_2 a_{i2})$ for the row markers i and coordinates $(\lambda_1 b_{j1}, \lambda_2 b_{j2})$ for the column markers j. However, the biplot technique generally uses one of two asymmetric mappings of the markers. For example, in a "row-metric preserving" two-dimensional biplot (Gabriel and Odoroff, 1990) the row markers have so-called *principal coordinates* $(\lambda_1 a_{i1}, \lambda_2 a_{i2})$ and the column markers have *standard coordinates* (b_{j1}, b_{j2}) (see Greenacre, 1993, chap. 4). The advantage of the asymmetric over the symmetric representation is that the relationships between the row and column markers can be more precisely interpreted.

3   Three-Way Correspondence Analysis

Important properties of three-way CA are that the dependence between the variables in a three-way table can be measured and displayed. Previous work on extending CA to three-way tables mostly reduced such tables to two-way tables, using so-called "interactive coding" to define "composite variables"; see Van der Heijden (1987) and Van der Heijden, *et al.* (1989) as well as Le Roux and Rouanet (Chapter 16) for overviews of this approach. Papers in which three-way tables were analyzed without reducing them to two-way tables are by Choulakian (1988), Kroonenberg (1989), and Carlier and Kroonenberg (1996).

3.1   Measuring Dependence

Whereas in two-way tables there is only one type of dependence, in three-way tables one can distinguish (1) global dependence, which is the deviation from the three-way independence model; (2) marginal dependence, which is the dependence due to the two–way interactions; and (3) three-way dependence, which is due to the three–way interaction.

Measuring Global Dependence   Three-way contingency tables have orders I, J, and K with relative frequencies p_{ijk}. Dependence in the table is again measured by Φ^2, which is defined in the three-way case as

$$
\Phi^2 = \sum_i \sum_j \sum_k \frac{(p_{ijk} - p_{i..}p_{.j.}p_{..k})^2}{p_{i..}p_{.j.}p_{..k}}
$$

$$
= \sum_i \sum_j \sum_k p_{i..}p_{.j.}p_{..k} \left[\frac{p_{ijk} - p_{i..}p_{.j.}p_{..k}}{p_{i..}p_{.j.}p_{..k}} \right]^2
$$

$$
= \sum_i \sum_j \sum_k p_{i..}p_{.j.}p_{..k}(\Pi_{ijk})^2 \tag{5}
$$

Φ^2 is based on the deviations from the three-way independence model, and it contains all two-way interactions and the three-way interaction. The measure for the dependence of cell (i, j, k), Π_{ijk}, may be rewritten as

$$
\Pi_{ijk} = \frac{Pr[ij \mid k]}{Pr[ij]} \cdot \frac{Pr[ij]}{Pr[i]Pr[j]} - 1 \tag{6}
$$

The quantity $1 + \Pi_{ijk}$ is the product of, first, the ratio $Pr[ij \mid k]/Pr[ij]$, which measures the relative increase or decrease in the joint probability of the categories i and j given category k, and, second, the ratio $Pr[ij]/Pr[i]Pr[j]$, which measures the relative increase or decrease in the deviation from the marginal independence.

If the conditional probability for all k is equal, $Pr[ij \mid k] = Pr[ij]$ and the first ratio is 1. Then $\Pi_{ijk} = \Pi_{ij.}$, and the three-way table could be analyzed with ordinary two-way CA. The symmetric statement after permutation of the indices holds as well. Therefore the Π_{ijk} measure the global dependence of the cell (i, j, k).

The elements of the two-way marginal totals are defined as weighted sums over the third index. Thus for the $I \times J$ margins these elements are

$$\Pi_{ij.} = \sum_k p_{..k} \Pi_{ijk} = \sum_k p_{..k} \frac{p_{ijk} - p_{i..}p_{.j.}p_{..k}}{p_{i..}p_{.j.}p_{..k}} = \frac{p_{ij.} - p_{i..}p_{.j.}}{p_{i..}p_{.j.}} \tag{7}$$

The elements of the other two-way margins, $\Pi_{i.k}$ and $\Pi_{.jk}$, are similarly defined. One-way marginal totals are summed over two indices and they are zero due to the definition of Π_{ijk}; hence the overall total is zero as well. For instance, in the case of the one-way row margin i:

$$\Pi_{i..} = \sum_j \sum_k p_{.j.}p_{..k} \Pi_{ijk} = 0 \tag{8}$$

Measuring Marginal and Three-Way Dependence   The global dependence of the cell, Π_{ijk}, can be split into separate contributions of the two-way interactions and the three-way interaction,

$$\Pi_{ijk} = \frac{p_{ij.} - p_{i..}p_{.j.}}{p_{i..}p_{.j.}} + \frac{p_{i.k} - p_{i..}p_{..k}}{p_{i..}p_{..k}} + \frac{p_{.jk} - p_{.j.}p_{..k}}{p_{.j.}p_{..k}} + \frac{p_{ijk} - p^*_{ijk}}{p_{i..}p_{.j.}p_{..k}} \tag{9}$$

where $p^*_{ijk} = p_{ij.}p_{..k} + p_{i.k}p_{.j.} + p_{.jk}p_{i..} - 2p_{i..}p_{.j.}p_{..k}$. The terms referring to the two-way margins are equivalent to those defined by expression (2). The quantity $p_{ijk} - p^*_{ijk}$ measures the size of the three-way interaction for cell (i, j, k). Darroch (1974) provides a comparative discussion of this *additive* definition of interaction and the *multiplicative* definition as used in log-linear analysis.

Due to the additive splitting of the dependence of individual cells, Φ^2, the measure for global dependence of the table can be partitioned (see Lancaster, 1951) as

$$\Phi^2 = \sum_i \sum_j p_{i..}p_{.j.} \left(\frac{p_{ij.} - p_{i..}p_{.j.}}{p_{i..}p_{.j.}} \right)^2 + \sum_i \sum_k p_{i..}p_{..k} \left(\frac{p_{i.k} - p_{i..}p_{..k}}{p_{i..}p_{..k}} \right)^2$$

$$+ \sum_j \sum_k p_{.j.}p_{..k} \left(\frac{p_{.jk} - p_{.j.}p_{..k}}{p_{.j.}p_{..k}} \right)^2 + \sum_i \sum_j \sum_k p_{i..}p_{.j.}p_{..k} \left(\frac{p_{ijk} - p^*_{ijk}}{p_{i..}p_{.j.}p_{..k}} \right)^2$$

$$= \Phi^2_{IJ} + \Phi^2_{IK} + \Phi^2_{JK} + \Phi^2_{IJK}. \tag{10}$$

The importance of decomposition (10) is that it provides measures of fit for each of the interactions and thus their contribution to the global dependence.

3.2 Modeling Dependence

Given measures for global dependence, marginal dependence, and three-way dependence, a model for these measures has to be found with which it will be possible to construct graphs depicting the dependence. For the three-way case, a three-way analogue of the GSVD is desired. There are, however, several candidates, of which we will consider only the so-called Tucker3 model (see Carlier and Kroonenberg, 1996, for other possibilities). This model is also referred to as the "three-mode factor analysis model" (Tucker, 1966; see also Kroonenberg, 1983).

Modeling Global Dependence   A three-way version of the GSVD will contain at least an additional term for the third variable. In the Tucker3 model each of the modes has its own components and the generalized singular values are different in that they are indexed by the components of all three modes.

$$\Pi_{ijk} = \sum_{p=1}^{P}\sum_{q=1}^{Q}\sum_{r=1}^{R} g_{pqr} a_{ip} b_{jq} c_{kr} + e_{ijk}. \tag{11}$$

In this decomposition, the a_{ip} are the elements of the components $\{\mathbf{a}_p\}$, which are pairwise orthonormal with respect to the weight $(p_{i..})$. Similarly, the b_{jq} are the elements of the components $\{\mathbf{b}_q\}$, which are pairwise orthonormal with respect to $(p_{.j.})$, and the c_{kr} are the elements of the components $\{\mathbf{c}_r\}$, which are orthonormal with respect to $(p_{..k})$. The g_{pqr} are the three-way analogues of the singular values, and they are often referred to as elements of the *core matrix*. The e_{ijk} represent the errors of approximation. In three-way CA, a weighted least-squares criterion is used: the parameters g_{pqr}, a_{ip}, b_{jq}, and c_{kr} are those that minimize $\sum_i \sum_j \sum_k p_{i..} p_{.j.} p_{..k} e_{ijk}^2$.

As in two-way CA, Φ^2 can be split into a part fitted with the three-way SVD or three-way model and a residual part.

Modeling Marginal Dependence   One of the attractive features of the additive partitioning of the dependence in Section 3.1 is that the single decomposition of the global dependence can be used to model the marginal dependence as well.

The marginal dependence of the rows i and columns j is contained in a matrix $\mathbf{\Pi}_{IJ}$ with elements $\Pi_{ij}^{IJ} = (p_{ij.} - p_{i..} p_{.j.})/(p_{i..} p_{.j.}) = \sum_k p_{..k} \Pi_{ijk}$ [see (7)] with similar expressions for the other two matrices $\mathbf{\Pi}_{IK}$ and $\mathbf{\Pi}_{JK}$.

By performing the weighted summation over k for Π_{ijk} in the Tucker3 model (11), we obtain the model for marginal dependence:

$$\Pi_{ij}^{IJ} = \sum_{p=1}^{P}\sum_{q=1}^{Q}\sum_{r=1}^{R} g_{pqr} a_{ip} b_{jq} c_{.r} + e_{ij.}. \tag{12}$$

where $c_{.r} = \sum_k p_{..k} c_{kr}$ and $e_{ij.} = \sum_k p_{..k} e_{ijk}$. Inspecting this formula leads to the conclusion that the marginal model is derived from the overall model by weighted averaging of the appropriate components, in this case c_r. This will turn out to be extremely effective in the displays we intend to make.

Modeling Partial Dependence   In our application the mode k is a time mode, and we are interested in investigating the part of the dependence that explicitly depends on time. Thus the dependence not associated with time, that is, the dependence due to the $I \times J$ margin, has to be removed from the global dependence. In our particular case, this *partial dependence* has the form

$$\Pi_{ijk} - \Pi_{ij}^{IJ} = \sum_{p,q,r} g_{pqr} a_{ip} b_{jq} (c_{kr} - c_{.r}) + (e_{ijk} - e_{ij.}) \tag{13}$$

As this equation shows, the modeling of the partial dependence is achieved by centering the components of one of the modes, here the time mode c_r. In Carlier and Kroonenberg (1996) we discuss what happens when one wants to remove more than one marginal dependence.

3.3   Plotting Dependence: Interaction Biplots

With respect to dependence and its modeling, the three ways of the contingency table behave in an entirely symmetric fashion. This symmetry can, however, not be maintained when graphing the dependence, because no spatial representations exist to portray all three ways simultaneously in one graph. To display the dependence or its approximation in three-way CA, two kinds of biplots may be considered: the joint biplot, discussed by Carlier and Kroonenberg (1996), and the interaction biplot, discussed here.

The *interaction biplot* aims to portray all three modes in a single biplot. As a biplot has only two types of markers, two modes have to be combined into one. In our application each pair of indices of the canton and time modes, (i, k), will be represented by a single marker. We refer to this as interactive coding. The remaining occupation mode supplies the other set of markers j and will be called the *reference mode*. The choice of reference mode depends on the research objective. Given that an ordered mode (in this case, time) will always be coded interactively, the choice between the remaining two depends on which of two modes produces the clearest patterns in their changes over time.

The construction of the biplot for the global dependence $\widehat{\Pi}_{ijk}$ follows directly from the three-way SVD of the global dependence:

$$\widehat{\Pi}_{ijk} = \sum_{q=1}^{Q} \left[\sum_{p=1}^{P} \sum_{r=1}^{R} g_{pqr} a_{ip} c_{kr} \right] b_{jq}$$

$$= \sum_{q=1}^{Q} d_{(ik)q} b_{jq} \tag{14}$$

If we replace the composite index (ik) with a new index l we see that the coordinates of the lth row markers are the d_{lq} and those of the jth column markers are the b_{jq}. Note that the g_{pqr} are absorbed in the coordinates of the row markers l. Therefore, it can be shown that the interaction biplot is a row-metric preserving one with respect to the weights $p_{.j.}$. The number of two-dimensional biplots depends not on P or R but only on Q, the number of components of the reference mode. The choice between three or four components in the reference mode could be guided by whether it is easier to inspect a three-dimensional plot or two independent two-dimensional plots.

The interaction biplot is especially useful when the number of elements in $I \times K$ is not too large or when one of the two modes is ordered. Assuming k is an ordered mode, for example time, trajectories can be drawn in the biplot by connecting, for each i, the points (i, k) in their given order.

4 Changes over Time in the Languedoc Workforce

During the censuses of 1954, 1962, 1968, and 1975, the people of the cantons in Languedoc–Roussillon (southern France) were asked to state their occupations. The occupations could be grouped into nine major occupational classes: farmers (AF), agricultural laborers (AL), owners of small and medium-sized businesses (SB), professionals and senior managers (PS), middle managers (MM), employees (white-collar workers, WC), laborers (blue-collar workers, BC), employees in the service sector (SE), and other occupations (OO). The present data consists of 42 rural cantons or rural parts of cantons in the Languedoc–Roussillon region. Cities (or communities) of more than 5000 inhabitants (in 1954) have been excluded from the cantons in which they are located. For example, for cantons such as Montpellier, Nîmes, Narbonne, Perpignan, and several others, only the suburban communities and the more rural communities of these cantons are included. Full details as well as the data themselves can be found in Bernard and Lavit (1985), and another detailed analysis of these data can be found in Carlier and Ewing (1992). It is evident that in the period in question major changes took place in the workforce, which we aim to describe using three-way CA. The three factors or variables of interest are cantons (42 categories), occupations (9 categories), and time (4 categories).

4.1 Measuring Dependence

Table 1 shows the partitioning of χ^2 according to the different interactions. Of the total variability in the table, the largest amount is in the canton-by-occupation interaction (57%), followed by the occupation-by-time interaction (22%). If the degrees of free-dom (df) are taken into consideration as well, the occupation-by-time interaction has by far the largest contribution per df, which indicates that the occupational distribu-tions have undergone considerable changes over time. Also the canton-by-occupation interaction has a sizable contribution per df, showing that there is considerable diver-sity among the cantons. The smaller contribution of the cantons-by-time interaction

Table 1: Decomposition of χ^2

Source	df	χ^2_{total}	% of total χ^2	χ^2_{error}	% of total χ^2	% of total χ^2_{error}	$\chi^2_{fit}/\chi^2_{total}$
		χ^2 components of 42 × 9 × 4 table		χ^2 components for 5 × 4 × 2 Tucker3 model			
Main effects		0	0.0	31	0.0	0.2	—
Two-way interactions							
Canton × Occupation	328	114307	57.4	9471	4.8	47.8	92%
Canton × Time	123	16278	8.2	1061	0.5	5.4	94%
Occupation × Time	24	43469	21.8	2306	1.2	11.6	95%
Three-way interaction	984	25249	12.7	6954	3.5	35.1	73%
Total	1459	199303	100.0	19823	9.9	100.0	90%

suggests that, even though for the cantons there are changes in the overall size of their workforce over time, this is not the main feature of the data. This interaction contains the differential increase and decrease of the workforce in the cantons. The three-way interaction is not large, and it has by far the smallest contribution per *df*.

To describe the patterns in the data set, we have fitted a Tucker3 model (11) with $P = 5$ components for the cantons, $Q = 4$ components for the occupations, and $R = 2$ components for the time mode. The number of components to retain was guided by the search for a large amount of explained variability coupled with a reasonable parsimony. As the occupations were chosen to be the reference mode, retaining four occupation components meant that two interaction biplots could be constructed, one for the first two dimensions and one for the last two dimensions. The numbers of components for the time mode and the cantons mode do not have an influence on the dimensionality of the biplots but only on the amount of smoothing of the results or equivalently the amount of structural information included in the solution [see (14)].

The 5 × 4 × 2 model fits very well, leaving only 10% unexplained, and Table 1 shows that canton-by-occupation and occupation-by-time interactions are very well explained (92% and 95%, respectively). The relatively unimportant three-way interaction has the smallest fit (73%). This points to the fact that in this example it could be sufficient to examine the two-way margins. Using a two-way approach, two separate two-way CAs could be performed on the canton-by-occupation and occupation-by-time margins.

4.2 Plotting Global Dependence

As already explained, before the results of a three-way CA can be plotted, a reference mode has to be selected. As we intended to study the changes in the distribution of the cantons over time, the occupations have to be chosen as the reference mode, so that the changes in workforce of the cantons over time can be plotted. Figures 1 and 2 show the results for the first two dimensions, and Figures 3 and 4 show the results with respect to the third and fourth dimensions. In Figures 1 and 3, the four time points of each canton are connected by a line ending in an arrowhead for 1975, called a *trajectory*. Only the 16 most characteristic trajectories are displayed in the graphs. Trajectories can be interpreted in terms of the distributions of occupations in the cantons at each occasion. If we take Chateauneuf de Randon (M3) as an example, we see that the trajectory begins in 1954 with a high proportion of independent farmers (AF) but that over the years the canton moves away from the AF point and ends up in 1975 much closer to the origin or mean point. Nevertheless, M3 and M4 continue to be the cantons that have the largest proportions of independent farmers. Furthermore, the category agricultural laborers (AL) also has a sizable but diminishing projection on the trajectory.

The first interaction plot showing the first two occupation axes (Figure 1), explaining 71% of the inertia, is based on the first two terms of the decomposition of $\widehat{\Pi}$

$$\widehat{\Pi}_{ijk} = \sum_{q=1}^{2} d_{(ik)q} b_{jq} + \sum_{q=3}^{4} d_{(ik)q} b_{jq} \tag{15}$$

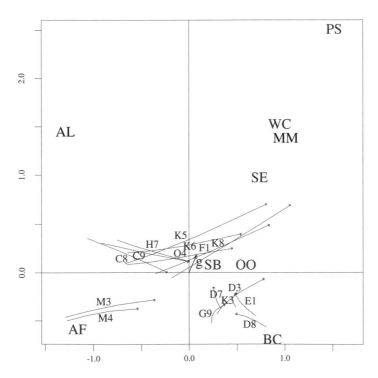

Figure 1: Global dependence of cantons, occupations, and time (axis 1 versus axis 2).

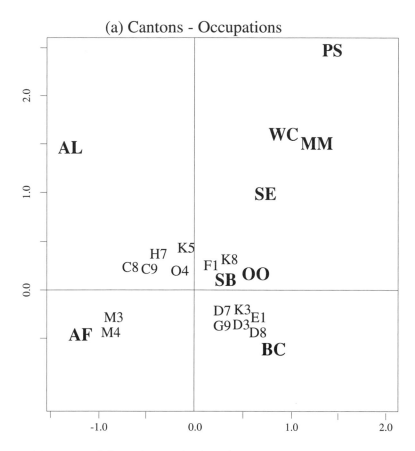

Figure 2(a): Marginal dependences displayed in the graph of the global dependence (axis 1 versus axis 2). Cantons × occupations.

with $d_{(ik)q} = \sum_{p=1}^{5} \sum_{r=1}^{2} g_{pqr} a_{ip} c_{kr}$ [see (14)]. The column markers (j) in the interaction biplot have coordinates (b_{j1}, b_{j2}), and the row markers (i, k) have coordinates ($d_{(ik)1}, d_{(ik)2}$). The second interaction biplot (Figure 3) explains an additional 19% of the inertia and is based on the third and fourth terms of the decomposition (15). Due to the additivity, the information in the second biplot can be considered as a correction or refinement of the major part of the information contained in the first biplot, because the terms in the second sum of equation (15) are on the average four times smaller than those in the first sum. The information of both plots is necessary to reproduce the estimated dependence $\hat{\Pi}_{ijk}$, as is evident from equation (15).

In principle, the figures display the global dependence, but they can also be used to study the three two-way interactions, because these interactions can be derived from the global dependence by weighted averaging and then displayed and interpreted in the same display (Figures 2 and 4). In the present data set the three-way

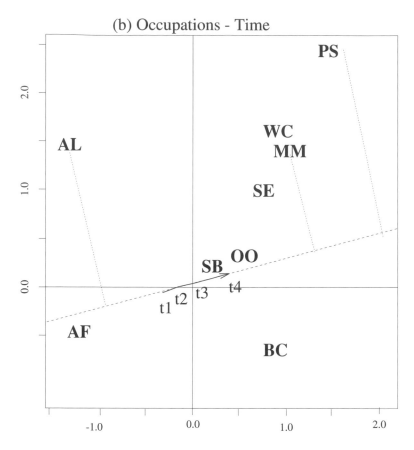

Figure 2(b): Marginal dependences displayed in the graph of the global dependence (axis 1 versus axis 2). Occupations × time.

interaction is rather small, and therefore the main patterns of the biplot can be interpreted via the three two-way interactions. In our approach we will analyze the global dependence within the perspective of the two-way dependence, in order to visualize what information the three-way display adds.

To inspect the patterns of dependence, we present two plots displaying the global dependence, namely Figure 1 for the first and second axes ($q = 1, 2$) and Figure 3 for the third and fourth axes ($q = 3, 4$). To facilitate the inspection of the two-way marginal dependences we have reproduced each of these figures three times, once for each marginal dependence (Figure 2a,b,c and Figure 4a,b,c, respectively). To avoid clutter on each of these plots, we have deleted the features not relevant for inspecting the two-way dependences, and have added markers for the centroids and/or lines and vectors to facilitate the interpretation. Please note, however, that the plots in Figures 2 and 4 display the same space as the ones in Figures 1 and 3 respectively.

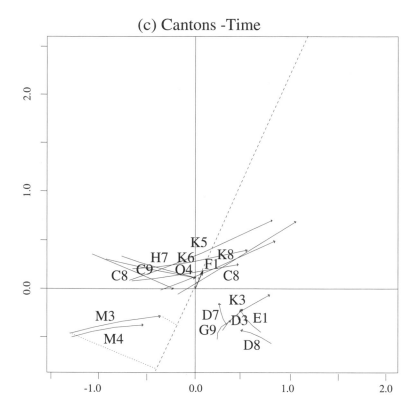

Figure 2(c): Marginal dependences displayed in the graph of the global dependence (axis 1 versus axis 2). Cantons × time.

4.3   Plotting Marginal Dependence

Canton-by-Occupation Interaction   The canton-by-occupation interaction ($I \times J$) in the interaction biplot is based on the weighted mean of equation (15) with respect to k:

$$\widehat{\Pi}_{ij.} = \sum_{q=1}^{4} d_{(i.)q} b_{jq} \qquad \text{with } d_{(i.)q} = \sum_{k=1}^{4} p_{..k} d_{(ik)q}$$

The term on the left-hand side is an approximation of the Π_{ij} used in two-way CA of the $I \times J$ margin. The occupations have the same markers as in the global dependence biplot, but the cantons have as coordinates those of the centroids of the trajectories.

One of the interesting aspects of the description of the two-way interactions within the framework of the global dependence is that the centroids of the cantons can be displayed in the same graph as the global dependence itself. In Figure 1 this is done by marking these centroids with the abbreviations of the cantons on the trajectories. Thus the label M3 is at the centroid of the trajectory for M3. The biplot

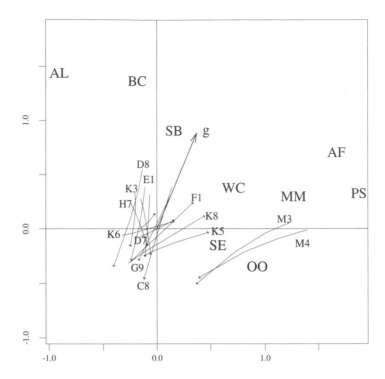

Figure 3: Global dependence of cantons, occupations, and time (axis 3 versus axis 4).

with only the canton centroids and the occupations (i.e., Figure 2a and Figure 4a) can be interpreted exactly as the comparable biplot from two-way CA. To assist in the interpretation of the patterns of the canton-by-occupation interaction, the canton-by-occupation margin with the occupations expressed as percentages of the workforce in the cantons is presented in Table 2.

This table has been arranged to highlight the patterns emerging from the interaction—high percentages are indicated in boldface. Some of the more extreme features of the interaction as evident from Figure 2a are the following.

1. The strong rural nature of the cantons Fournels (M4) and Chateauneuf de Randon (M3) with, respectively, on the average 81% and 78% of the people employed in agriculture (overall 35%) with a heavy emphasis on independent farmers (AF).

2. The similarly agricultural nature of Lézignan-Corbières (C8), Narbonne (C9), and Capestang (H7) with, respectively, on the average 66%, 58%, and 53% of the workforce employed in agriculture, but with a larger number of agricultural laborers (possibly due to the viticulture in those areas) (AL).

3. The strong industrial nature of La Grand'Combe (D8), St Ambroix (E1), Ganges (K3), Sumène (G9), and Alès (D3), which is indicated by high percentages of

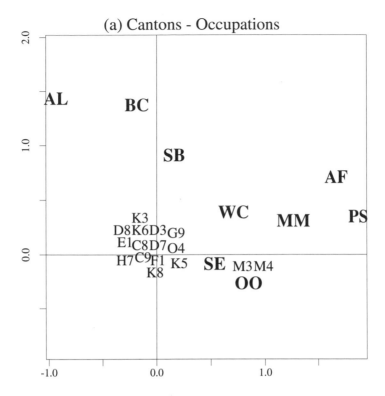

Figure 4(a): Marginal dependences displayed in the graph of the global dependence (axis 3 versus axis 4). Cantons × occupations.

blue-collar workers (71%, 58%, 56%, 54%, and 54% respectively); the overall percentage of blue-collar workers is 30%.

4. The strong tertiary sector in Montpellier (K8), Nîmes (F1), and Les Matelles (K5) (a suburb of Montpellier) with, respectively, 38%, 30%, and 29% employed in the tertiary sector (i.e., services, SE; middle management, MM; white collar workers, WC; professional and senior management, PS) compared with 21% overall.

In Figure 4a all but two cantons are located close to the origin, indicating that the higher dimensions do not contribute to their interactions. Primarily the distributions from Chateauneuf de Randon (M3) and Fournels (M4) need adjustment, with positive corrections for the categories professional and senior management (PS), middle management (MM), and white collar (WC) that correct too large negative values read on the first biplot. Furthermore, the corrections emphasize the differences between the two agricultural categories and indicate that in these cantons the proportion of independent farmers (AF) is even larger and the proportion of agricultural laborers (AL) slightly smaller than one would have deduced from Figure 2a.

Table 2: Row percentages of cantons by occupations[a] averaged over time (high percentages are indicated in boldface)

| | Agriculture | | | | | | Tertiary sector | | | | | |
| | Blue collar | Farmer | Agric. labor. | Total agric. | Small business | White collar | Middle manag. | Services | Profes. sen. m. | Total tertiary | Other occup. | Cantons |
	BC	AF	AL	AL	SB	WC	MM	SE	PS		OO	
D8	**70.7**	6.5	2.1	8.6	5.4	5.8	5.2	2.3	1.1	14.4	1.0	La Gr'Combe
E1	**58.2**	6.5	1.6	8.1	12.8	7.7	6.6	2.8	1.7	18.8	1.9	St. Ambroix
K3	**55.7**	6.3	5.1	13.4	13.3	7.8	5.1	2.8	2.1	17.8	1.8	Ganges
G9	**53.8**	20.4	3.1	23.5	6.8	6.4	4.3	1.5	2.0	14.2	1.7	Sumène
D3	**53.6**	9.4	2.7	12.1	9.8	10.2	7.2	3.1	2.4	22.9	1.7	Alès
C3	**47.8**	13.2	3.3	16.5	14.1	7.7	6.5	3.1	2.6	19.9	1.7	Quillan
D7	**45.8**	17.7	5.3	23.0	10.2	6.8	7.4	3.4	1.5	19.1	1.9	Génolhac
H3	**45.0**	10.9	4.6	15.5	13.3	8.5	7.8	3.8	3.4	23.5	2.7	Le Vigan
M4	4.5	**76.0**	5.2	**81.2**	6.6	1.5	3.7	0.8	0.5	6.5	1.2	Fournels
M3	5.9	**70.5**	7.0	**77.5**	5.7	1.7	4.0	2.3	0.4	8.4	2.4	Ch.n.Randon
M9	10.7	**53.4**	11.0	**64.4**	5.9	5.3	5.6	2.8	2.1	15.8	3.2	Mende
J6	14.1	**51.0**	17.1	**68.1**	6.1	3.8	4.7	1.1	1.0	10.6	1.1	Lodève
K1	10.5	**42.8**	**27.1**	**69.9**	6.2	3.9	4.3	1.5	2.2	11.9	1.4	Claret
F9	23.6	**40.8**	12.4	**53.2**	8.7	4.1	5.2	2.2	1.0	12.5	2.0	Uzès
C8	12.1	31.8	**34.4**	**66.2**	9.9	4.2	3.3	2.2	1.0	10.7	1.1	Lézignan-C.
C9	15.8	27.6	**30.5**	**58.1**	10.2	5.9	4.9	2.4	1.5	14.7	1.1	Narbonne
H7	18.6	22.7	**30.5**	**53.2**	10.8	7.2	5.0	2.4	1.6	16.2	1.1	Capestang
H9	15.7	31.9	**29.0**	**60.9**	8.8	5.4	4.4	2.0	1.2	13.0	1.6	Montagnac
K4	27.5	21.1	**27.0**	48.1	7.6	6.8	4.8	2.9	1.2	15.7	1.1	Lunel
H4	25.3	17.4	**26.5**	43.9	14.3	6.2	4.7	3.3	1.6	15.8	0.7	Agde

Code												Place
H6	23.2	22.1	22.6	44.7	12.0	8.4	5.4	3.4	1.7	18.9	1.1	Béziers
A5	22.6	36.6	17.2	**53.8**	10.5	4.1	4.3	2.1	1.0	11.5	1.6	Castelnaudary
A7	17.1	31.0	18.7	49.7	12.4	5.8	5.6	2.4	1.4	15.2	**5.6**[b]	Fanjeaux
G4	28.0	34.6	11.0	45.6	9.0	4.9	5.1	2.4	2.7	15.1	2.3	Lassalle
G6	27.8	31.8	12.9	44.7	11.2	4.7	4.6	3.0	1.4	13.7	2.7	St. André d.V.
K8	29.8	9.0	10.5	19.5	10.9	**14.3**	**11.1**	**5.9**	**6.3**	**37.6**	2.2	Montpellier
N8	35.6	9.6	3.7	13.3	18.4	9.6	7.8	**9.8**	2.8	**30.0**	2.8	Arles-s/Tech
K5	18.7	20.2	19.0	39.2	8.4	9.2	9.4	3.4	**7.5**	**29.5**	**4.2**	Les Matelles
F1	29.9	13.9	16.1	30.0	8.8	**12.3**	**10.3**	1.7	**4.4**	**28.7**	2.6	Nîmes
N7	30.7	13.6	11.0	24.6	**15.1**	**10.5**	6.7	**6.0**	3.2	**26.4**	3.3	Argeles s/Mer
J9	26.0	19.3	18.9	38.2	8.3	**10.2**	8.8	3.4	3.5	**25.9**	1.7	Castries
K2	40.8	8.1	12.9	21.0	11.0	**11.1**	7.7	4.7	2.2	**25.7**	1.6	Frontignan
L2	15.9	34.6	8.0	42.6	**15.3**	7.1	**8.5**	4.0	3.4	23.0	3.3	Florac
K6	23.5	17.0	25.0	42.0	9.3	8.8	7.0	4.0	3.9	22.7	1.4	Mauguio
O4	22.5	21.8	20.4	42.2	12.3	9.0	6.6	3.0	2.6	21.2	1.9	Perpignan
G7	40.4	14.4	8.9	23.3	13.6	6.4	6.5	3.0	2.0	17.9	**4.8**	St. Hyppolyte
N4	34.3	28.3	2.9	31.2	11.9	7.7	8.0	3.0	2.2	20.9	1.7	St. Chély d'A
G1	32.3	16.3	21.5	37.8	9.4	9.2	5.6	2.4	1.9	19.1	1.6	Vauvert
I8	36.8	19.2	8.9	28.1	13.0	6.2	8.3	3.0	2.7	20.2	2.0	St Pons
F2	32.3	24.3	8.5	32.8	11.9	7.6	6.6	3.5	2.4	20.1	2.9	Pont-St-Esprit
M6	21.9	30.8	4.6	35.4	**18.2**	7.9	7.9	3.4	2.4	21.6	2.9	Langogne
K7	28.4	17.7	15.2	32.9	**21.5**	6.1	5.0	2.8	1.6	15.5	1.7	Méze
TOTAL	30.4	20.2	14.8	35.0	11.7	8.3	6.6	3.4	2.6	20.9	2.0	

[a]For a further explanation of the occupational categories, see Section 19.5. The equivalent *French* abbreviations in Bernard and Lavit (1985) are:
AF = *EA*; AL = *OA*; SB = *AC*; PS = *PL*; MM = *CM*; WC = *EM*; BC = *OU*; SE = *SE*; OO = *CP*.
[b]A religious community settled in Fanjeaux at the beginning of the period.

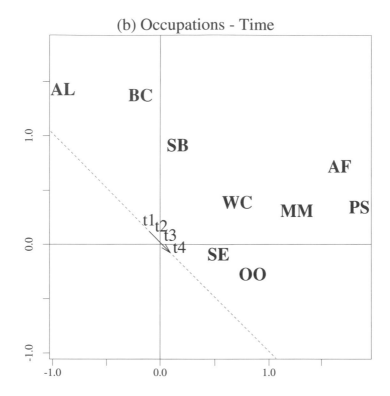

Figure 4(b): Marginal dependences displayed in the graph of the global dependence (axis 3 versus axis 4). Occupations × time.

Occupation-by-Time Interaction   The occupation-by-time interaction $(J \times K)$ can be visualized in a similar manner by displaying centroids of the row markers (i, k) for each fixed value of k, that is, weighted averaging performed on the canton points. The four time centroids constitute the *average trajectory,* which is displayed as an arrow in Figure 2b, rather than the original trajectories. A dotted line has been added to emphasize the direction of the centroid trajectory. To interpret this interaction, one has to project the occupations on this average trajectory. It shows the contrast between the two agricultural classes (farmers, AF, and agricultural laborers, AL), which are decreasing, and the tertiary occupations (especially professionals and senior managers, PS), which are increasing. Occupations such as owning a small business (SB) and blue-collar workers (BC) have only a small increase, as their projections on the average trajectory are close to the origin.

As before, Figure 2b and Figure 4b contain different visual information and one has to "add" their contributions. Figure 2b shows that both the class of independent farmers (AF) and that of agricultural laborers (AL) are decreasing, but Figure 4b shows that the decline is even more serious for the agricultural laborers than for the

(c) Cantons - Time

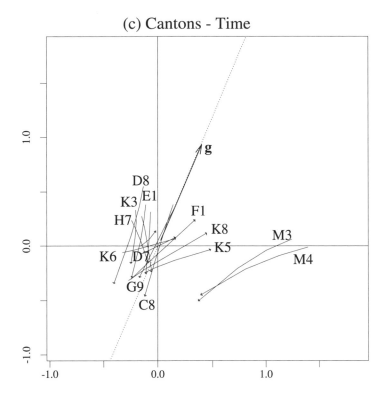

Figure 4(c): Marginal dependences displayed in the graph of the global dependence (axis 3 versus axis 4). Cantons × time.

independent farmers because of its large projection on the negative side of the time axis in Figure 4b. Furthermore, the class of blue-collar workers (BC) shows a small increase over time in Figure 2b. Figure 4b shows a small decrease for this category, but it does not reverse the trend observed in Figure 2b, because the average trajectory of the second biplot is much shorter than that of the first biplot. Finally, Figure 4b shows that the proportion of people working in the service sector is even stronger than one would have derived from Figure 2b alone.

Canton-by-Time Interaction   The canton-by-time interaction ($I \times K$) reflects the overall changes in the workforce of the cantons. It shows which cantons decrease and which cantons increase in working population. As the occupations constitute the reference mode, the centroid consists of the vector $(b_{.1}, b_{.2}, b_{.3}, b_{.4})$ as can be seen by averaging expression (15) over occupations j,

$$\widehat{\Pi}_{i.k} = \sum_{q=1}^{4} d_{(ik)q} b_{.q} \tag{16}$$

In Figure 1 and Figure 2c the vector $(b_{.1}, b_{.2})$ is indicated by an arrow with the letter **g**, and in Figure 2c it is extended in two directions by a dashed line to indicate the axis defined by this vector. In Figure 3 and Figure 4c the vector $(b_{.3}, b_{.4})$ is similarly indicated. The small size of the mean vector $(b_{.1}, b_{.2})$ in Figure 2c, with respect to that of $(b_{.3}, b_{.4})$ in Figure 4c, indicates that the contribution to the change, described by the first two dimensions displayed in Figure 2c, is much smaller than that of the third and fourth dimensions shown in Figure 4c. Therefore, for most cantons, the larger part of the changes over time is better represented in the latter graph, and the former one will be used as a correction. The complete canton-by-time interaction of a canton is derived from the projection of its trajectory on the axes **g** [see equation (15)]. As an example of such a projection, that of Chateauneuf de Randon (M3) is shown in Figure 2c.

Figure 4c separates the cantons in two classes. The first is a small class of more urban cantons with a large overall growth, for example, Montpellier (K8), Les Matelles (K5), Nîmes (F1). The second class is larger and contains both industrial cantons such as La Grand'Combe (D8), St. Ambroix (E1), and Ganges (K3), and Sumène (G9) and rural cantons with an overall decrease of population over time, such as Chateauneuf de Randon (M3) and Fournels (M4).

To make more precise our conclusions about Chateauneuf de Randon, for example, we see in Figure 2c that it has a small increase in its working population, but in Figure 4c we see that the canton has a much larger projection on **g** pointing in the other direction. Thus the sum of the projections indicates that there was an overall decline in population in this canton. Numerically, applying expression (16) and measuring graphically these projections, we obtain the contribution in Figure 2c as the product of 0.2 (length of **g**) and 0.5 (length of the projection of the trajectory of Chateauneuf de Randon). Similarly, the contribution in Figure 4c is approximately 1×-0.9. Thus the overall change for this canton is $0.1 - 0.9 = -0.8$, estimating an overall decline in the working population.

Three-Way Interaction   As mentioned before, the three-way interaction is relatively small, and it will not be analyzed on its own but in conjunction with the interactions involving time, which are investigated in the next section.

4.4   Plotting Partial Dependence: Analysis of Change

The analysis of the complete Figure 1 leads to the study of the global dependence as expressed in the Π_{ijk}. By using the difference between Π_{ijk} and $\Pi_{ij.}$, we may remove the part of the dependence that is not influenced by time, that is, the interaction between the cantons and occupations $(I \times J)$. This allows the study of only the part of the global dependence that depends on time. This difference has the form [see equation (6)]

$$\Pi_{ijk} - \Pi_{ij.} = \frac{\Pr[ij]}{\Pr[i]\Pr[j]} \left(\frac{\Pr[ij \mid k]}{\Pr[ij]} - 1 \right) \tag{17}$$

The difference is the product of $(1 + \Pi_{ij})$ with the factor in parentheses, which can be considered as a growth index for the canton–occupation category (i, j). If there is no dependence on time, that is, the distribution of occupations among the cantons remains the same over time, then the conditional probability of (i, j) given time k, $Pr[ij \mid k]$, is equal to $Pr[ij]$, and the difference in equation (17) is zero for all canton–occupation combinations i and j. In such a case each trajectory will consist only of its centroid. Instead of removing the interaction $(I \times J)$, that is, centering the trajectories, we analyze the variations of a trajectory around its centroid. These variations take into account all interactions involving change. The advantage is that on the same graph the average situation of a canton can be interpreted along with the changes around these average positions. The importance of the change of the workforce in a canton, as defined by equation (17), is indicated by the length of its trajectory; the nature of the change is indicated by the direction of its trajectory. In the graph, the interpretation of the changes of a canton–occupation category (i, j) over time can be obtained by projecting the associated trajectory onto the axes defined by the occupation markers j. The increase is proportional to the product of the length of the vector of the jth marker and the component of the projected trajectory onto the biplot axis defined by this marker. If the vector and the trajectory point in the same direction, there is an increase over time; if they point in opposite directions, there is a decrease over time.

The major pattern in Figure 1 is a general shift of all cantons toward the urban categories and away from the independent farmer (AF) category. The tertiary sector is rapidly increasing, especially in cantons such as Montpellier (K8), Les Matelles (K5), and Nîmes (F1). Many other cantons have a similar but smaller growth in the tertiary categories (Alès, D3; Ganges, K3). The position of the tertiary categories PS, WC, MM, and SE on a single line through the origin indicates that these categories are more or less proportional but that the ones farther away, such as PS, are much more important in size than the others (SE). The trajectories of the industrial cantons La Grand'Combe (D8) and St Ambroix (E1) move toward the origin, which indicates that their populations become more uniform over time. They lose their specific industrial characteristic; that is, their blue-collar (BC) population is decreasing.

Figure 3 corrects the previous interpretations in two ways. Some of the corrections could be called "technical." As an example, the move of some agricultural cantons, in Figure 1, away from the independent farmer category is also a move in the direction of the tertiary categories. The strict interpretation of Figure 1 would imply an increase in these categories for such cantons. In Figure 3, the move of the two rural cantons M3 and M4 away from the tertiary categories corrects this interpretation. Another example is related to cantons such as La Grand'Combe (D8). Its move toward the origin on Figure 1 is also a move toward the agricultural laborers (AL), which would imply that this category becomes more numerous. This interpretation is clearly negated in Figure 3. Other corrections, with respect to the interpretation of the first biplot, are the greater decrease of the cantons La Grand'Combe (D8) and St Ambroix (E1) in blue-collar workers, the greater increase of the urban cantons (Montpellier, K8; Les Matelles, K5; and Nîmes, F1) in the tertiary categories, and the stronger decrease in the two rural canons M3 and M4 of the population of independent farmers (but not of the agricultural laborers, as the latter vector is more or less perpendicular to the trajectory of M3 and M4 in that figure).

5 Conclusion

In this chapter we have made a case for analyzing large three-way contingency tables with three-way CA. Two major aspects of the technique stick out. First, as with log-linear modeling, it is possible to assess the relative sizes of the marginal dependences and that of the three-way interaction, and it is also possible to assess how well the model fits those two-way and three-way interactions. Second, in contrast to log-linear analysis, the dependence can be analyzed as a whole, and the marginal dependence can be directly assessed from the global one without further models or decompositions. In log-linear modeling there are no facilities in the model for inspecting or evaluating the interaction parameters, which would have to be done separately for each interaction. Third, biplots provide a visual analysis of the nature of the dependence in the data, and both the global dependence and the marginal dependence can be portrayed and evaluated in the same graph in a completely natural way. Finally, special problems can be handled within this framework, such as analyzing change after partialing out those parts of dependence not related to change.

In our analysis of the changes in the workforce of the Languedoc–Roussillon we have shown that the large-scale patterns can be presented in an insightful way by using various biplots. In addition, several smaller patterns can be discerned. Obviously, for a full-fledged analysis of these data, we should look at the relationships of the patterns found with external information about the actual events in several cantons.

Appendix: Computational Aspects and Software

Two-Way Correspondence Analysis

As explained, for instance, by Greenacre (1984, p. 40), the estimated parameters in two-way CA can be computed via the regular SVD by pre- and postmultiplying Π_{ij} with $p_{i.}^{1/2}$ and $p_{.j}^{1/2}$, respectively. Thus the SVD is applied to

$$p_{i.}^{1/2} p_{.j}^{1/2} \Pi_{ij} = \frac{p_{ij} - p_{i.}p_{.j}}{p_{i.}^{1/2} p_{.j}^{1/2}} = X_{ij}$$

where X_{ij} is the standardized residual from the model of two-way independence for cell (i, j). If we write the regular SVD for X_{ij} as

$$X_{ij} = \sum_s \lambda_s \tilde{a}_{is} \tilde{b}_{js}$$

then the coefficients for the generalized SVD are $a_{is} = \tilde{a}_{is}/\sqrt{p_{i.}}$ and $b_{js} = \tilde{b}_{js}/\sqrt{p_{.j}}$ with the same λ_s.

Three-Way Correspondence Analysis

In three-way CA the procedure is completely analogous. Regular three-way models may be used to calculate the estimates for the parameters of the generalized three-way SVD. In particular,

$$p_{i..}^{1/2}p_{.j.}^{1/2}p_{..k}^{1/2}\Pi_{ijk} = \frac{p_{ijk} - p_{i..}p_{.j.}p_{..k}}{p_{i..}^{1/2}p_{.j.}^{1/2}p_{..k}^{1/2}} = X_{ijk}$$

where X_{ijk} is the standardized residual from the model of three-way independence for cell (i, j, k). If we use the Tucker3 model for X_{ijk},

$$X_{ijk} = \sum_p \sum_q \sum_r g_{pqr}\tilde{a}_{ip}\tilde{b}_{jq}\tilde{c}_{kr}$$

then the coefficients for this version of the generalized three-way SVD are $a_{ip} = \tilde{a}_{ip}/\sqrt{p_{i..}}$ and $b_{jq} = \tilde{b}_{jq}/\sqrt{p_{.j.}}$, $c_{kr} = \tilde{c}_{kr}/\sqrt{p_{..k}}$ with the same g_{pqr}.

In the two-way case, the way λ_s are associated with the row or column markers determines whether biplots are row-metric preserving or column-metric preserving graphs. The situation is more complicated in the three-way case, especially for the Tucker3 model, because several kinds of biplots can be made.

Software Note

The methods and graphical procedures described above have been programmed by the first author in S-Plus (see, for example, Becker *et al.*, 1988) and can be supplied upon request. A FORTRAN implementation is being developed and will be included in the next release of the three-way data analysis package 3WAYPACK available from the second author (see Kroonenberg, 1996). The technical basis for the algorithms can be found in Kroonenberg (1983).

Chapter 20

Visual Display of Interaction in Multiway Contingency Tables by Use of Homogeneity Analysis: the $2 \times 2 \times 2 \times 2$ Case

Jacqueline J. Meulman and Willem J. Heiser

1 Introduction

Multiway contingency tables express the relationships between the categories of several categorical variables $A, B, C, \ldots$ at several levels of complexity. In the body of the multiway table, all these relationships are confounded. By adding over all variables except A and B, we obtain a bivariate marginal table, showing the bivariate relationship between A and B; by adding over all variables except A, B, and C, we obtain a trivariate marginal table, showing the trivariate relationship between A, B, and C, and so on. Of course, it is of interest to know whether these relationships, once separated, perhaps are still simpler than they look; in particular, to ask whether or not the higher order ones are simple combinations of the lower order ones. From this question, several natural forms of dependence and independence arise.

Suppose for the moment that we restrict ourselves to the case of three categorical variables, with n_A, n_B, and n_C categories. Lack of independence implies the presence of interaction between the categories. Let π_{ijk} denote the probability that an individual unit of observation (in the sequel denoted by the neutral term object) falls in category i of variable A, in category j of variable B, and in category k of variable C. We consider π_{ijk} as a probability defined over all cells of the three-way contingency table, which implies that the total $\sum_i \sum_j \sum_k \pi_{ijk} = 1$. As will be demonstrated in the next section, we can discuss several concepts of independence structure without referring, for example, to log-linear modeling, which is perhaps the most common approach to analyzing data of the present type, or to any other form of modeling. The reason is simple: models involve assumptions to relate concepts of structure to observed counts, but the concepts exist regardless of the additional assumptions. Modeling is a way to *smooth* empirical frequencies, that is, to replace them by frequencies satisfying certain regularities. Log-linear modeling just uses various types of independence as a set of possible structures for the expected values μ_{ijk} of a multinomial sampling process with $n_A \times n_B \times n_C$ categories. Then, by a famous result of Birch (1963), the maximum likelihood fitted values $\hat{\mu}_{ijk}$ are smoothed versions of the counts in the observed multiway contingency table that match them in specified marginal distributions but have higher order interactions that satisfy the chosen independence patterns. For choosing between submodels with a different independence structure, the likelihood ratio statistic G^2 or Pearson's chi-squared χ^2 are used.

Once we know the most likely (in)dependence structure among the variables, how do we interpret the interactions? In log-linear modeling, interactions correspond to groups of model parameters. To interpret the model parameters of a log-linear model, we have to express them in terms of odds and odds ratios (also called cross-product ratios; see Fienberg, 1980), which are ratios of (smoothed) frequencies or probabilities. This reformulation is not easy; it involves taking the exponential of a model parameter and describing the corresponding odds verbally. A verbal description of a three-way interaction can become incomprehensible rather quickly, because it consists of a nesting of conditional statements. The main thesis of this chapter is that the (in)dependence structure can *also* be represented in a spatial model, in which categories are mapped as points and variables as groups of points. It will be shown that in this spatial representation odds are ratios of distances, a property that offers the possibility of visual display of interaction.

The spatial representation will be obtained through the use of homogeneity analysis (Gifi, 1990, chap. 3 and sect. 8.6), also called multiple correspondence analysis (Benzécri, 1973a; Greenacre, 1984), or dual scaling (Nishisato, 1994). In the present context, the technique will be regarded as a method that maps the rows of a profile frequency table into points in a low-dimensional space (often, but not necessarily, a two-dimensional space). The profile frequency table is the multiway contingency table turned inside out: it codes the cells by listing, in some predetermined order, which category of each variable is involved (forming the profile) and attaches to each profile the cell frequency. Points representing the profiles are called profile points, and category points are obtained as centers of gravity of certain subsets of profile points.

Homogeneity analysis was developed with a focus on bivariate marginal tables. If all variables are mutually independent, all eigenvalues (the usual summary statistics) will be equal to $1/m$, and an analysis result in which the first p eigenvalues deviate substantially from $1/m$ implies that the first p dimensions account for all two-way interactions. It is asserted in Gifi (1990) that homogeneity analysis relies on the assumption that in most cases the total structure in the data can be sufficiently captured in the joint bivariate (first-order) interactions. This assumption can be considered equivalent to assuming a log-linear model that includes the pairwise associations only. If, however, these are not sufficient to produce a decent fit, the conclusion would be that homogeneity analysis should be discarded in favor of a log-linear analysis that includes the higher order interactions. The major purpose of the present chapter is to show that—in contrast to this widespread idea—homogeneity analysis includes the representation of the higher order interactions as well; an important idea is the balanced use of cross-classification variables.

To conclude the introduction, we briefly describe the data that will be used throughout for empirical illustration. The data pertain to a sample of men and women who had petitioned for divorce; a similar number of married people were asked the following questions:

1. "Before you married your (former) husband/wife, had you ever made love with anyone else?"

2. "During your (former) marriage, (did you have) have you had any affairs or brief sexual encounters with another man/woman?"

The variables in the $2 \times 2 \times 2 \times 2$ cross-tabulation (with total sample size $N = 1036$) are gender (G), premarital sex (P), extramarital sex (E), and marital status (M). The associated profile frequency matrix will be given in the following. The multiway contingency table is analyzed in Agresti (1990, sect. 7.2.4); the original British study was reported by Thornes and Collard (1979) and described by Gilbert (1981).

2   Independence Structures and Odds Structures

The following cases of simplification of a three-way contingency table are commonly distinguished (e.g., see Agresti, 1990). Three variables A, B, and C are called mutually independent if

$$\pi_{ijk} = \pi_{i++}\pi_{+j+}\pi_{++k} \tag{1}$$

for all categories i of A, j of B, and k of C, where, as usual, π_{i++} indicates that we have summed over j and k, giving the univariate marginals for variable A. Under mutual independence, there is no association whatsoever, and all cells of the three-way table can be constructed by the simple product of the univariate marginals. Variable A is called jointly independent of B and C when, again for all categories,

$$\pi_{ijk} = \pi_{i++}\pi_{+jk}. \tag{2}$$

This decomposition corresponds to ordinary two-way independence for A and a new variable, called the cross-classification variable BC, which is composed of the $n_B \times n_C$

combinations of the categories of B and C. Under joint independence, the association between two variables is distributed proportionally over the levels of the third variable to obtain the three-way probability.

Next, consider the relationship between A and B, controlling for the contribution of C. Here the concept of control implies that we study the conditional probability that two categories, say i of A and j of B, are present in the same object, given the fact that we know the object is in category k of C. The usual notation for this event is $\pi_{ij|k}$, defined as $\pi_{ij|k} = \pi_{ijk}/\pi_{++k}$, where the division by the univariate marginal π_{++k} ensures that the $\pi_{ij|k}$ form a proper set of probabilities summing to one within the subtable indexed by k. From this definition it follows that the cell probabilities can be expressed in terms of conditional probabilities as

$$\pi_{ijk} = \pi_{++k}\,\pi_{ij|k}.$$

Now if, for all k, the conditional probabilities $\pi_{ij|k}$ are independent, we must have the marginal decomposition

$$\pi_{ij|k} = \left(\pi_{i+k}/\pi_{++k}\right)\left(\pi_{+jk}/\pi_{++k}\right)$$

and therefore we obtain, combining the last two equations, conditional independence of A and B given C when

$$\pi_{ijk} = \left(\pi_{i+k}\pi_{+jk}\right)/\pi_{++k}. \tag{3}$$

Under conditional independence of A and B, each of this pair of variables is associated with C; these associations, together with the univariate marginal, completely account for the apparent association between A and B in the original table and in the bivariate marginal π_{ij+}.

Note that cases (3), (2), and (1) are fundamentally different only in terms of the number of two-way cross-classification variables that are needed to account for the cell probabilities. For conditional independence we need two cross-classification variables, for joint independence we need one such variable, and for mutual independence we need none. Conversely, it is also useful to think of the situation in terms of conditionally dependent variables. The strongest case is (1), in which there are no conditionally dependent variables. In case (2), B and C are conditionally dependent given A, implying that the conditional probability $\pi_{jk|i} = \pi_{+jk}$ does not simplify, while $\pi_{ij|k} = \pi_{i++}\pi_{j|k}$ and $\pi_{ik|j} = \pi_{i++}\pi_{k|j}$ are independent. In case (3), only the conditional probability $\pi_{ij|k} = \pi_{i|k}\pi_{j|k}$ can be decomposed into the product of its marginals, while $\pi_{jk|i}$ and $\pi_{ik|j}$ depend on π_{+jk} and π_{i+k}, respectively; so both B and C and A and C are conditionally dependent.

It is natural also to consider the case where π_{ijk} depends on three double subscripted quantities,

$$\pi_{ijk} = \alpha_{ij}\beta_{ik}\gamma_{jk}, \tag{4}$$

a case for which no closed-form expression in terms of marginal probabilities exists. Here, none of the pairs of variables is conditionally independent, yet there is

Table 1: Odds structures in a three-way table under different forms of independence

		$\theta_i = \dfrac{\pi_{i11}\pi_{i22}}{\pi_{i12}\pi_{i21}}$	$\theta_j = \dfrac{\pi_{1j1}\pi_{2j2}}{\pi_{ij2}\pi_{2j1}}$	$\theta_k = \dfrac{\pi_{11k}\pi_{22k}}{\pi_{12k}\pi_{21k}}$
I	Mutual independence	1	1	1
II	Joint independence	$\dfrac{\pi_{+11}\pi_{+22}}{\pi_{+12}\pi_{+21}}$	1	1
III	Conditional independence	$\dfrac{\pi_{+11}\pi_{+22}}{\pi_{+12}\pi_{+21}}$	$\dfrac{\pi_{1+1}\pi_{2+2}}{\pi_{1+2}\pi_{2+1}}$	1
IV	No three-way interaction	$\dfrac{\gamma_{11}\gamma_{22}}{\gamma_{12}\gamma_{21}}$	$\dfrac{\beta_{11}\beta_{22}}{\beta_{12}\beta_{21}}$	$\dfrac{\alpha_{11}\alpha_{22}}{\alpha_{12}\alpha_{21}}$

still a typical form of simplification: odds ratios between two variables are identical for each (given) category of the third variable. The odds ratio is a classic way of measuring association (Yule, 1912) that compares two ratios of probabilities (odds) by forming a ratio again. Thus, the odds of being in category 1 of A rather than in category 2 of A are compared for those who are in category 1 of B against those who are in category 2 of B. In a 2×2 table, the odds ratio θ is defined as $\theta = (\pi_{11}\pi_{22})/(\pi_{12}\pi_{21})$. For three variables, with π_{ijk} satisfying the stated condition, we find $\theta_k = (\pi_{11k}\pi_{22k})/(\pi_{12k}\pi_{21k}) = (\alpha_{11}\alpha_{22})/(\alpha_{12}\alpha_{21})$, showing that θ_k is independent of the chosen category k (by symmetry, the effect is the same if the categories of the other variables are kept fixed). After Bartlett (1935), this case is usually called "no three-way interaction."

In summary, all cases of independence have a typical odds structure, which is shown in Table 1, displaying the result of inserting (1)–(4) into the definition of the odds ratio. Under mutual independence, all odds ratios are equal to one. Under joint independence (of A with respect to B and C), there is one set of odds ratios that does not become equal to one: all θ_i become equal to the marginal odds ratio, that is, the two-way tables conditioned on category i are equal to the marginal table. Under conditional independence (of A and B upon C), both the θ_i's and the θ_j's become equal to the marginal odds ratio, while $\theta_k = 1$ for all k. Under lack of three-way interaction, all odds ratios for different categories of the same variable are equal, but unequal to one. Finally, three-way interaction implies that all odds ratios are different, both within and across variables. As we shall see shortly, these various odds structures each have a distinctive spatial pattern.

3 Odds as Distance Ratios

In this section it will be shown that odds are distance ratios between category points and how this leads to additivity of category quantifications. Homogeneity analysis finds the spatial representation of the profile frequency table by *projection*. Projection is a linear transformation that intuitively involves dropping points onto a line (or plane), along a direction perpendicular to the line (or plane). We will first describe what is projected (a high-dimensional representation of the table), then show some of the properties that hold in this high-dimensional space, and next indicate which properties remain (approximately) preserved under projection. For more technical details on projection, the reader is referred to Gifi (1990) or van de Geer (1993).

In the high-dimensional representation of the profile frequency table, all objects with the same profile coincide in one point, called the profile point $z_{ijk...}$. We associate with each profile point a mass (also called weight), equal to the cell frequency $\pi_{ijk...}$ of that profile. Note that our starting point is the cells of the multiway contingency table itself, not any of its marginal tables. We also assume that the number of variables is much smaller than the number of objects, $N \gg m$, a condition similar to what is required for a log-linear analysis, and that all frequencies are strictly greater than zero (although this is not necessary for the spatial method). If $n_A, n_B, \ldots, n_m$ are the number of categories of the m variables, this construction generates $n_A \times n_B \times \cdots \times n_m$ profile points. In the following discussion, we limit ourselves to the $2 \times 2 \times 2$ case.

In the binary case, the 2^m profile points are the vertices of an m-dimensional (hyper)cube associated with some probability mass. Thus, three variables are represented as eight profile points on a cube in three dimensions. Focusing on the edges between the two faces of the cube that correspond to the categories 1 and 2 of variable A, we may locate on each edge between the vertices z_{1jk} and z_{2jk} the point z_{*jk}, defined as

$$z_{*jk} = \frac{\pi_{1jk}}{\pi_{1jk} + \pi_{2jk}} z_{1jk} + \frac{\pi_{2jk}}{\pi_{1jk} + \pi_{2jk}} z_{2jk} \tag{5}$$

which is the center of gravity (or centroid) of all objects in category j of B and k of C, of which there are π_{1jk} in 1 of A and π_{2jk} in 2 of A. Because we know that the points z_{1jk}, z_{*jk}, and z_{2jk} are located on a line, in that order [because (5) is a convex combination], we may write $d(z_{1jk}, z_{2jk}) = d(z_{1jk}, z_{*jk}) + d(z_{2jk}, z_{*jk})$, where the notation $d(x, y)$ is used for the ordinary Euclidean distance between two points x and y. So the edge between two profile points is divided by the center of gravity into two parts. Using (5), the lengths of these two segments are

$$d(z_{1jk}, z_{*jk}) = \frac{\pi_{2jk}}{\pi_{1jk} + \pi_{2jk}} d(z_{1jk}, z_{2jk}) \tag{6}$$

$$d(z_{2jk}, z_{*jk}) = \frac{\pi_{1jk}}{\pi_{1jk} + \pi_{2jk}} d(z_{1jk}, z_{2jk}). \tag{7}$$

From (6) and (7) it follows that the odds of being in category 2 of A against being in category 1, given the fact that the object is in j of B and k of C, are equal to:

$$\frac{\pi_{2jk}}{\pi_{1jk}} = \frac{d(z_{1jk}, z_{*jk})}{d(z_{2jk}, z_{*jk})} \tag{8}$$

that is, the odds are displayed in the spatial representation of the profile frequency table as a reverse distance ratio (larger probabilities corresponding to smaller distances between $\mathbf{z}_{*jk}$ and the profile point). From (8) we can now derive novel expressions for the odds ratios in subtables of a three-way table; for example, for the association between A and C given category j of B we obtain, by putting $k = 1$ and $k = 2$ and dividing the odds,

$$\theta_j = \frac{\pi_{1j1}\pi_{2j2}}{\pi_{1j2}\pi_{2j1}} = \frac{d(\mathbf{z}_{1j2}, \mathbf{z}_{*j2})d(\mathbf{z}_{2j1}, \mathbf{z}_{*j1})}{d(\mathbf{z}_{1j1}, \mathbf{z}_{*j1})d(\mathbf{z}_{2j2}, \mathbf{z}_{*j2})}. \tag{9}$$

Thus, the odds ratio θ_j is a multiplicative combination of four distances, defined between four profile points and two centroids. It is well known that the odds ratio is invariant under permutation of the rows and columns of the fourfold table. This property implies here that θ_j may *also* be derived from the ratio of π_{ij2} and π_{ij1}, which leads to an alternative expression for (9) in terms of the centroids $\mathbf{z}_{ij*}$, defined analogously to $\mathbf{z}_{*jk}$ in (5). We shall have a closer look at this duplication when we illustrate the spatial relationships with an example.

What is the spatial representation of independence? The reader is advised to draw a square with vertices $\mathbf{z}_{1j1}, \mathbf{z}_{1j2}, \mathbf{z}_{2j2}$, and $\mathbf{z}_{2j1}$; when $\mathbf{z}_{*jk}$ and $\mathbf{z}_{ij*}$ are added to this figure, the following relationships are verified easily. If variable A is jointly independent of B and C, we know (see Table 1) that $\theta_j = 1$, so from (9) we derive $d(\mathbf{z}_{1j2}, \mathbf{z}_{*j2})d(\mathbf{z}_{2j1}, \mathbf{z}_{*j1}) = d(\mathbf{z}_{1j1}, \mathbf{z}_{*j1})d(\mathbf{z}_{2j2}, \mathbf{z}_{*j2})$. But we also know, by the construction of the spatial representation, that the interprofile distances are equal, that is, $d(\mathbf{z}_{1j1}, \mathbf{z}_{2j1}) = d(\mathbf{z}_{1j2}, \mathbf{z}_{2j2})$, from which we derive $d(\mathbf{z}_{1j1}, \mathbf{z}_{*j1}) + d(\mathbf{z}_{2j1}, \mathbf{z}_{*j1}) = d(\mathbf{z}_{1j2}, \mathbf{z}_{*j2}) + d(\mathbf{z}_{2j2}, \mathbf{z}_{*j2})$. Taken together, and after some algebraic manipulation, these two equalities imply that *the four distances are equal in opposite pairs:* $d(\mathbf{z}_{1j1}, \mathbf{z}_{*j1}) = d(\mathbf{z}_{1j2}, \mathbf{z}_{*j2})$ and $d(\mathbf{z}_{2j1}, \mathbf{z}_{*j1}) = d(\mathbf{z}_{2j2}, \mathbf{z}_{*j2})$. If we consider the *intersection line* connecting $\mathbf{z}_{*j1}$ with $\mathbf{z}_{*j2}$, it must be parallel to the edges $(\mathbf{z}_{1j1}, \mathbf{z}_{1j2})$ and $(\mathbf{z}_{2j1}, \mathbf{z}_{2j2})$. A similar relation holds for the intersection line between $\mathbf{z}_{1j*}$ and $\mathbf{z}_{2j*}$ with its corresponding edges. So we conclude that independence is a necessary condition for the intersection lines to be parallel to the edges.

It is natural to assign to each centroid $\mathbf{z}_{*jk}$ a mass, π_{+jk}, indicating the proportion of objects that has a profile with j of B and k of C. Similarly, the marginal proportion π_{ij+} will be assigned to $\mathbf{z}_{ij*}$, that is, the sum of the masses of which it is the balancing point. The two intersection lines $(\mathbf{z}_{*j1}, \mathbf{z}_{*j2})$ and $(\mathbf{z}_{1j*}, \mathbf{z}_{2j*})$ themselves intersect in a point $\mathbf{z}_{*j*}$, called the category point (the coordinates of which are called category quantifications) of category j, which is easily shown *also* to be a center of gravity (of all objects in category j, calculated in any of a number of different ways) with mass π_{+j+}. Continuing in this way, the intersection lines connecting the category points, $(\mathbf{z}_{1**}, \mathbf{z}_{2**}), (\mathbf{z}_{*1*}, \mathbf{z}_{*2*})$ and $(\mathbf{z}_{**1}, \mathbf{z}_{**2})$, intersect in $\mathbf{z}_{***}$, the centroid of all objects, with mass 1.

Our high-dimensional spatial representation of the profile frequency table is now complete. For the $2 \times 2 \times 2$ case, it contains the eight original profile points, 3×4 added one-asterisk centroids, three added two-asterisks centroids, and the overall three-asterisks centroid. The masses of these points correspond exactly to all the cell probabilities and the complete set of marginal probabilities of the three-way contingency table. Just as the centroids of the form $\mathbf{z}_{*j*}$ are called category points of B, centroids of the form $\mathbf{z}_{*jk}$ are called the category points (quantifications) of the cross-classification variable BC (similarly, we have category points for AB and AC). So all cells of the bivariate marginal tables can be viewed as categories of some cross-classification variable, which is quantified by centroids located on the edges of the cube of profile points.

As we have seen, lack of interaction implies parallel intersection lines, and this has important further implications for the relationship between the category quantifications of the bivariate marginals with the profile points, on the one hand, and with the univariate marginals on the other hand. We suppose that the origin of the space is chosen as $\mathbf{z}_{***}$. Considering vectors in the face of the cube corresponding to category j of B, which are obtained from the original ones by translation with an amount $-\mathbf{z}_{*j*}$, parallelism implies additivity:

$$(\mathbf{z}_{ijk}-\mathbf{z}_{*j*}) = (\mathbf{z}_{ij*}-\mathbf{z}_{*j*}) + (\mathbf{z}_{*jk}-\mathbf{z}_{*j*}) \tag{10}$$

which follows from the definition of vector addition in terms of the parallelogram formed by the points $\mathbf{z}_{*j*}, \mathbf{z}_{*jk}, \mathbf{z}_{ijk}$, and $\mathbf{z}_{ij*}$ (this is in fact a rectangle, but we want to use only the parallelism, not any properties of the angles). Thus, conditional independence must manifest itself by the fact that one of the three possible pairs of cross-classification variables has additive quantifications when viewed with respect to the univariate centroid, as in (10). Under joint independence, we must have two pairs of cross-classification variables with additive quantifications with respect to their joint univariate centroid. Similarly, it can be shown that, when variables B and C are independent, we have a marginal odds ratio $(\pi_{+11}\pi_{+22})/(\pi_{+12}\pi_{+21}) = 1$, which implies

$$\mathbf{z}_{*jk} = \mathbf{z}_{*j*} + \mathbf{z}_{**k}, \tag{11}$$

that is, the quantifications of the cross-classification variable are equal to the sum of the quantifications of the categories of the original variables. Combining (10) and (11), we obtain the spatial representation of mutual independence:

$$\mathbf{z}_{ijk} = \mathbf{z}_{i**} + \mathbf{z}_{*j*} + \mathbf{z}_{**k}. \tag{12}$$

In this case, the category points of all three cross-classification variables form a parallelogram. So there is a clear one-to-one correspondence between odds structures in the three-way table and additivity structures in the spatial model.

All relationships described so far are exact in the original cube, and we may wonder how well they remain intact in the projected configuration that constitutes the usual result of a homogeneity analysis. Angles and distances are not preserved under projection: squares and rectangles become parallelograms. Projection does preserve parallelism, so (10), (11), and (12) remain *completely valid* in a low-dimensional representation of the profile frequency table.

4 Some Special Properties of Discrimination Measures

We will now propose and illustrate a general procedure for studying interaction in higher way contingency tables that allows us to distinguish the various additivity structures in a low-dimensional representation of the profile frequency table. Our procedure simply amounts to a homogeneity analysis of a profile frequency matrix including all cross-classification variables that can be formed from the original variables in a completely balanced way. If there is reason to expect a three-way interaction (for example, as indicated by a model search in a preliminary log-linear analysis), we include all bivariate and trivariate cross-classification variables. It is essential for our procedure to introduce the additional variables in blocks and not to make some selection among them. First, we will focus on the so-called discrimination measures (Gifi, 1990, sect. 3.8.4), which are quantities that show how well a variable is represented as a group of category points in low-dimensional space.

Let $\mathbf{P}$ be the projection matrix that defines the optimal projection; the sth row of $\mathbf{P}$, which produces the projection on component (or dimension) s, is denoted by $\mathbf{p}_s$. We introduce a different but consistent notation for the projected points to distinguish them from the high-dimensional ones. The projected profile points $\mathbf{x}_{ijk}$ are defined by $\mathbf{x}_{ijk} = \mathbf{P}\mathbf{z}_{ijk}$; the projected centroids are defined by $\mathbf{y}_{i**} = \mathbf{P}\mathbf{z}_{i**}, \mathbf{y}_{ij*} = \mathbf{P}\mathbf{z}_{ij*}$, and so on. The scalar value $y_{ij*}(\mathbf{p}_s) = \mathbf{p}_s^T \mathbf{z}_{ij*}$ is the coordinate of the projection of the centroid for category ij of cross-classification variable AB on the component defined by $\mathbf{p}_s$. Discrimination measures then measure the dispersion of the projected category points as

$$\eta_A^2(\mathbf{p}_s) = \sum_i \pi_{i++}(y_{i**}(\mathbf{p}_s) - y_{***}(\mathbf{p}_s))^2 \tag{13}$$

$$\eta_{AB}^2(\mathbf{p}_s) = \sum_i \sum_j \pi_{ij+}(y_{ij*}(\mathbf{p}_s) - y_{***}(\mathbf{p}_s))^2 \tag{14}$$

for the original variables and cross-classification variables, respectively, where it will be clear how to continue for the higher order interactions. Thus, $\eta_A^2(\mathbf{p}_s)$ is a weighted sum of squares of the category quantifications of variable A with respect to the overall center of gravity along component s. Since the weights (being probabilites) sum to one, it is the *variance* of the quantified categories in dimension $\mathbf{p}_s$ of the spatial model. The average discrimination measure across all variables on component s is denoted by λ_s^2, the *eigenvalue*. We are now ready to look at the results for our example.

The profile frequency matrix including all cross-classification variables is given in Table 2, where the observed count has been supplemented with the expected count under the hypothesis of mutual independence. Five different homogeneity analyses were performed, always in two dimensions. The first analysis pertains to the original set of variables (G, P, E, M); the second uses the two-way cross-classification variables (GP, GE, GM, PE, PM, EM) only. The third analysis includes both the main

Table 2: Profile frequency table for marital status data with all possible cross-classification variables, observed count and expected count under the hypothesis of independence[a]

G	P	E	M	GP	GE	GM	PE	PM	EM	GPE	GPM	GEM	PEM	GPEM	Obs.	Exp.
1	1	1	1	1	1	1	1	1	1	1	1	1	1	1	17	8.76
1	1	1	2	1	1	2	1	2	2	1	2	2	2	2	4	9.61
1	1	2	1	1	2	1	2	1	3	2	1	3	3	3	54	66.23
1	1	2	2	1	2	2	2	2	4	2	2	4	4	4	25	72.66
1	2	1	1	2	1	1	3	3	1	3	3	1	5	5	36	28.89
1	2	1	2	2	1	2	3	4	2	3	4	2	6	6	4	31.70
1	2	2	1	2	2	1	4	3	3	4	3	3	7	7	214	218.47
1	2	2	2	2	2	2	4	4	4	4	4	4	8	8	322	239.69
2	1	1	1	3	3	3	1	1	1	5	5	5	1	9	28	4.66
2	1	1	2	3	3	4	1	2	2	5	6	6	2	10	11	5.12
2	1	2	1	3	4	3	2	1	3	6	5	7	3	11	60	35.27
2	1	2	2	3	4	4	2	2	4	6	6	8	4	12	42	38.70
2	2	1	1	4	3	3	3	3	1	7	7	5	5	13	17	15.39
2	2	1	2	4	3	4	3	4	2	7	8	6	6	14	4	16.88
2	2	2	1	4	4	3	4	3	3	8	7	7	7	15	68	116.34
2	2	2	2	4	4	4	4	4	4	8	8	8	8	16	130	127.65

[a]G, gender (1=female, 2=male); P, premarital sex (1=yes, 2=no); E, extramarital sex (1=yes, 2=no); and M, marital status (1=divorced, 2=married). GP, two-way cross-classification gender × premarital sex (1=female/premarital sex, 2=female/no premarital sex, 3=male/premarital sex, 4=male/no premarital sex), etc. GPE, three-way cross-classification gender × premarital sex × extramarital sex (1=female/premarital sex/extramarital sex, 2=female/premarital sex/no extramarital sex), etc.

286

effect variables and the two-way interactions (G, P, E, M, GP, GE, GM, PE, PM, EM), the fourth analysis adds the four three-factor cross-classification variables and the four-factor interaction to the second analysis, and finally the whole set (G, P, E, M, GP, GE, GM, PE, PM, EM, GPE, GPM, GEM, PEM, $GPEM$) was analyzed. The resulting discrimination measures are given in Table 3, along with the associated eigenvalues.

In the first panel of Table 3, we see that all variables contribute to the first component, premarital sex and extramarital sex being the most important, whereas the second component is determined predominantly by gender and marital status. The second panel reports the analysis with the bivariate cross-classification variables only, and the third panel reports the combined analysis.

Comparing the third with the first and second panels, we see a remarkable similarity between the solutions: the first four discrimination measures of the combined analysis are about equal to those of the analysis with the original variables only, while the last six discrimination measures are about equal to those of the analysis with the

Table 3: Discrimination measures η^2, eigenvalues λ^2, and average discrimination measures for partitions from five different homogeneity analyses with increasing number of cross-classified variables

	Analysis 1		Analysis 2		Analysis 3		Analysis 4		Analysis 5	
	dim 1	dim 2	dim 1	dim 2	dim 1	dim 2	dim 1	dim 2	dim 1	dim 2
G	0.259	0.525			0.254	0.510			0.252	0.501
P	0.538	0.055			0.553	0.059			0.558	0.061
E	0.432	0.092			0.428	0.094			0.425	0.095
M	0.319	0.365			0.314	0.373			0.312	0.377
GP			0.656	0.502	0.650	0.512	0.660	0.494	0.654	0.505
GE			0.610	0.655	0.615	0.664	0.605	0.648	0.611	0.657
GM			0.560	0.883	0.566	0.887	0.558	0.878	0.563	0.883
PE			0.808	0.213	0.804	0.206	0.810	0.220	0.806	0.213
PM			0.733	0.522	0.729	0.513	0.738	0.530	0.733	0.520
EM			0.602	0.415	0.606	0.407	0.599	0.420	0.603	0.412
GPE							0.888	0.680	0.886	0.685
GPM							0.845	0.916	0.843	0.918
GEM							0.789	0.945	0.795	0.948
PEM							0.910	0.593	0.908	0.581
$GPEM$							1.000	1.000	1.000	1.000
λ^2	0.387	0.259	0.662	0.532	0.552	0.422	0.764	0.666	0.663	0.557
$\frac{1}{4}\sum_1^4 \eta^2$	0.387	0.259			0.387	0.259			0.387	0.258
$\frac{1}{6}\sum_5^{10} \eta^2$			0.662	0.532	0.662	0.532	0.662	0.532	0.662	0.532
$\frac{1}{5}\sum_{11}^{15} \eta^2$							0.886	0.827	0.886	0.826
$\sum \eta^2$	1.549	1.037	3.972	3.192	5.520	4.220	8.402	7.323	9.948	8.356

bivariate cross-classification variables only. To be completely clear, we stress that we obtain not perfectly identical results but very similar ones.

Since eigenvalues are averages of discrimination measures, the eigenvalues of the combined analysis are about equal to 0.4 times the eigenvalues of the first panel plus 0.6 times the eigenvalues of the second panel, or, equivalently, the sum over all discrimination measures per dimension in the first and second analyses together is about equal to the sum over all discrimination measures in the third analysis, and so on. The overall similarity between the results for cross-classification variables included or excluded is obtained only if cross-classification variables are included in a completely balanced way. Otherwise, the similarity would be lost.

How do we recognize independence from these tables? We discuss this question in two steps. First, for a precise judgment we need a standard of comparison, because the expected level of a discrimination measure depends on the number of categories. As our standard, we choose the *expected value* of a discrimination measure under the hypothesis of mutual independence (alternatively, the quantities that we call expected value could also be interpreted as the mean discrimination measure across all components). When we consider all higher order cross-classification variables, including the highest one corresponding to a saturated model, starting with m original variables we will have $2^m - 1$ analysis variables. If the total number of categories of the original variables is denoted by $q = n_A + n_B + n_C + \cdots$, then there are $q - m$ nontrivial components to consider. Under the hypothesis of mutual independence, these components will have equal eigenvalues. We derive the expected discrimination measure $\eta_A^2(*)$ for one of the original variables as $(n_A - 1)/(q - m)$, the expected discrimination measure $\eta_{AB}^2(*)$ for one of the two-way analysis variables as $(n_A + n_B - 2)/(q - m)$, the expected discrimination measure $\eta_{ABC}^2(*)$ for one of the three-way analysis variables as $(n_A + n_B + n_C - 3)/(q - m)$, and so on. In our example, where the variables G, P, E, and M have two categories each, under mutual independence $\eta_G^2(\mathbf{p}_s) \cdots \eta_M^2(\mathbf{p}_s)$ will be equal to 0.25, $\eta_{GP}^2(\mathbf{p}_s) \cdots \eta_{EM}^2(\mathbf{p}_s)$ will be equal to 0.50, and $\eta_{GPE}^2(\mathbf{p}_s) \cdots \eta_{PEM}^2(\mathbf{p}_s)$ will be equal to 0.75. The discrimination measure $\eta_{GPEM}^2(\mathbf{p}_s)$ will be equal to 1.00, representing perfect fit, which corresponds to the saturated model in a log-linear analysis.

Second, in the previous section we have seen that independence implies additivity of category quantifications. We will now show that under two-way independence the discrimination measures are additive, too. For instance, the fact that gender and marital status are independent ($\chi_{GM}^2 = 0.031$) is reflected in the discrimination measures $\eta_{GM}^2(\mathbf{p}_s) = (0.563, 0.883)$ being approximately equal to $\eta_G^2(\mathbf{p}_s) = (0.252, 0.501)$ plus $\eta_M^2(\mathbf{p}_s) = (0.312, 0.377)$. Geometrically, departure from independence can be depicted as a distance between two vectors that represent the discrimination measures in two-dimensional space, the first vector (with coordinates 0.563, 0.883) displaying the observed discrimination measure and the second vector (0.565, 0.878) displaying the expected discrimination measure when G and M are independent. For GM, this distance is 0.005; for the other two-way interactions, these distances are 0.167 (GP), 0.090 (GE), 0.186 (PE), 0.161 (PM), and 0.146 (EM), respectively. This pattern shows a very close resemblance to the results from the log-linear analysis with two-way interactions only, as can be seen from Agresti (1990, Table 7.4).

To show that additivity must hold in any component, we first note that additivity in high-dimensional space (e.g., $\mathbf{z}_{ij*} = \mathbf{z}_{i**} + \mathbf{z}_{*j*}$) carries through to low-dimensional space by virtue of the distributive character of projection: $\mathbf{p}_s^T \mathbf{z}_{ij*} = \mathbf{p}_s^T \mathbf{z}_{i**} + \mathbf{p}_s^T \mathbf{z}_{*j*}$. Looking at marginal independence, then by substituting $\pi_{ij+} = \pi_{i++} \pi_{+j+}$ and $y_{ij*}(\mathbf{p}_s) = y_{i**}(\mathbf{p}_s) + y_{*j*}(\mathbf{p}_s)$ into (14), we obtain

$$
\begin{aligned}
\eta_{AB}^2(\mathbf{p}_s) &= \sum_i \sum_j \pi_{i++} \pi_{+j+} \left[(y_{i**}(\mathbf{p}_s) - y_{***}(\mathbf{p}_s)) + (y_{*j*}(\mathbf{p}_s) - y_{***}(\mathbf{p}_s)) \right]^2 \\
&= \sum_i \pi_{i++} (y_{i**}(\mathbf{p}_s) - y_{***}(\mathbf{p}_s))^2 + \sum_j \pi_{+j+} (y_{*j*}(\mathbf{p}_s) - y_{***}(\mathbf{p}_s))^2 \\
&= \eta_A^2(\mathbf{p}_s) + \eta_B^2(\mathbf{p}_s),
\end{aligned}
\tag{15}
$$

where the cross-product vanishes because

$$
\sum_i \pi_{i++}(\mathbf{z}_{i**} - \mathbf{z}_{***}) = \sum_j \pi_{+j+}(\mathbf{z}_{*j*} - \mathbf{z}_{***}) = \mathbf{0}
$$

by definition of $\mathbf{z}_{***}$, and therefore any projected value must be 0. In a similar way we obtain, for the case in which variable A is jointly independent of B and C,

$$
\eta_{ABC}^2(\mathbf{p}_s) = \eta_A^2(\mathbf{p}_s) + \eta_{BC}^2(\mathbf{p}_s)
\tag{16}
$$

and when A and B are conditionally independent given C

$$
\eta_{ABC}^2(\mathbf{p}_s) = \eta_{AC}^2(\mathbf{p}_s) + \eta_{BC}^2(\mathbf{p}_s) - \eta_C^2(\mathbf{p}_s).
\tag{17}
$$

Although these relationships are exact when the stipulated type of independence is exactly fulfilled, for the "no three-way interaction" case, we must do something different. One possible idea would be to settle for an approximation. For instance, using Darroch's (1962) condition of a "perfect" table (which does not exhibit paradoxes), no three-way interaction implies that $\pi_{ijk} = (\pi_{ij} + \pi_{i+k}\pi_{+jk})/(\pi_{i++}\pi_{+j+}\pi_{++k})$, and from this condition we may derive the approximate relationship

$$
\eta_{ABC}^2(\mathbf{p}_s) = \eta_{AB}^2(\mathbf{p}_s) + \eta_{AC}^2(\mathbf{p}_s) + \eta_{BC}^2(\mathbf{p}_s) - \eta_A^2(\mathbf{p}_s) - \eta_B^2(\mathbf{p}_s) - \eta_C^2(\mathbf{p}_s).
\tag{18}
$$

At this point, some experimentation indicated that this is not the way to go; instead, it seems that to test the no three-way interaction case, we should rely on a higher order statistic as well, in contrast to the discrimination measure, which we could argue is a two-way statistic. The suggested diagnostics would then be the category quantifications, and in the next section these will be used to demonstrate that they indeed display three-way interactions.

5   Visual Display of Odds Ratios as Distance Ratios Between Category Points

In Section 3, we have seen how odds are represented as ratios of distances, so that an odds ratio of one corresponds to parallel lines. Two-way association leads to nonparallel lines and three-way association leads to different nonparallel lines, conditional upon one fixed variable. At this point we will display the results from the extended homogeneity analysis in two dimensions, including the first- and second-order cross-classification variables. The category quantifications will be labeled with their level; for instance, g_1 and g_2 denote the female and male categories, respectively. Similarly, the label p_1e_1 denotes respondents who reported both premarital and extramarital sex. In the second-order interactions, $g_1p_1e_2$ denotes the category of women who did report premarital sex but no extramarital sex, and $p_2e_1m_1$ denotes respondents who did not report premarital sex, did report extramarital sex, and are divorced. This notation will also be used in the following equations that give the distance ratios between selected category points to study two particular higher way interactions, that is, the three-way interaction between premarital sex, extramarital sex, and marital status (*PEM*) and between gender, premarital sex, and extramarital sex (*GPE*). We already know that G and M are independent, so all higher order interactions that include *GM* are not very interesting.

According to Agresti (1990, p. 221), the *PEM* interaction seems vital to explaining relationships in the data. To describe this *PEM* interaction, Agresti uses the estimated odds ratios for the log-linear model (*GP*, *GM*, *GE*, *PEM*) and concludes: "Given gender, for those who reported pre-marital sex, the odds of a divorce are estimated to be 1.82 times higher for those who reported extra-marital sex than for those who did not; for those who did not report premarital sex, the odds of a divorce are estimated to be 10.94 times higher for those who reported extramarital sex than for those who did not." We translate this estimated *EM* odds ratio for the two levels of P in a spatial model (see Figure 1); in this figure we have used only the category points relevant for this particular interaction. For the two different levels of P, p_1 and p_2, category points for levels of E and M are connected to form two diamond shapes. Along the edges of each diamond, the distances are given between the four three-way points and their centroids, the two-way points. The closer a two-way point to a three-way point, the more respondents are in that particular three-way point. So we see in Figure 1 that p_1m_2 (premarital sex, married) is closer to $p_1e_2m_2$ (no extramarital sex) than to $p_1e_1m_2$ (extramarital sex) and that p_2e_2 (no premarital sex, no extramarital sex) is closer to $p_2e_2m_2$ (married) than to $p_2e_2m_1$ (divorced). From this we would deduce that extramarital sex is not beneficial to marriage. If we compute the distance ratio for p_1 with respect to the p_1e_1 and p_1e_2 centroids,

$$\frac{d(p_1e_1m_2, p_1e_1)d(p_1e_2m_1, p_1e_2)}{d(p_1e_1m_1, p_1e_1)d(p_1e_2m_2, p_1e_2)} = 1.76 \cong \frac{1.20 \times 0.54}{0.40 \times 0.92} \tag{19}$$

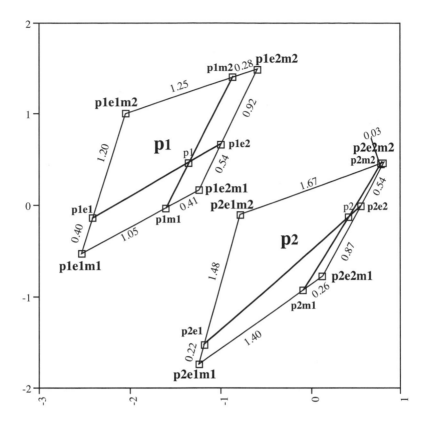

Figure 1: Display of three-way interaction PEM: category points for extramarital sex and marital status connected for the two levels of premarital sex.

(where $\cong$ denotes equality up to rounding errors) and with respect to centroids p_1m_2 and p_1m_1,

$$\frac{d(p_1e_1m_2, p_1m_2)d(p_1e_2m_1, p_1m_1)}{d(p_1e_2m_2, p_1m_2)d(p_1e_1m_1, p_1m_1)} = 1.76 \cong \frac{1.25 \times 0.41}{0.28 \times 1.05} \tag{20}$$

we note that the equality between the two different ratios is preserved. So, in the sequel we need to look at only one of each pair of ratios. We also remark that the estimated odds ratio of 1.82 reported by Agresti (1990) is indeed close to 1.76. If we now inspect the distance ratio for the p_2 category (respondents who did not report premarital sex), we obtain

$$\frac{d(p_2e_1m_2, p_2m_2)d(p_2e_2m_1, p_2m_1)}{d(p_2e_2m_2, p_2m_2)d(p_2e_1m_1, p_2m_1)} = 10.62 \cong \frac{1.67 \times 0.26}{0.03 \times 1.40} \tag{21}$$

and the estimated odds ratio 10.94 reported by Agresti is again very close to this figure.

So the main conclusion of the log-linear analysis, that the effect of extramarital sex on divorce is much greater for respondents who did not report premarital sex, is displayed graphically in the homogeneity analysis solution. As usual, there are two companion pairs of odds ratios; we first look at the distance ratios for the two categories of the variable extramarital sex. The diamonds for levels e_1 and e_2 are to be found in Figure 2, accompanied by the associated distances.

The distance ratio for those who did report extramarital sex is obtained from

$$\frac{d(p_1e_1m_2, e_1m_2)d(p_2e_1m_1, e_1m_1)}{d(p_2e_1m_2, e_1m_2)d(p_1e_1m_1, e_1m_1)} = 0.45 \cong \frac{0.58 \times 0.81}{1.09 \times 0.96}. \tag{22}$$

Agresti gives the estimated *PM* odds ratio for category e_1 as 0.50, so among those who reported extramarital sex, divorce is about two times more likely for respondents with no premarital sex than for those who had premarital sex. For those who did not

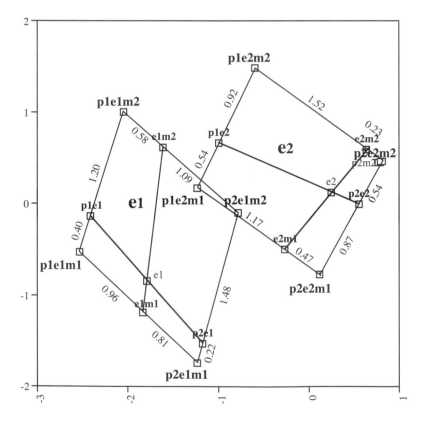

Figure 2: Display of three-way interaction PEM: category points for premarital sex and marital status connected for the two levels of extramarital sex.

report extramarital sex,

$$\frac{d(p_1e_2m_2, e_2m_2)d(p_2e_2m_1, e_2m_1)}{d(p_2e_2m_2, e_2m_2)d(p_1e_2m_1, e_2m_1)} = 2.73 \cong \frac{1.52 \times 0.47}{1.17 \times 0.23}. \tag{23}$$

The estimated *PM* odds ratio for category e_2 is reported as 3.00 by Agresti (1990, p. 222): among those who did not report extramarital sex, divorce is much more likely for respondents who had premarital sex than for those who had no premarital sex. Finally, with respect to the levels m_1 and m_2, Agresti reports the *PE* estimates as 1.82 for divorced and 10.95 for married respondents; from the distances, we recover the odd ratios

$$\frac{d(p_2e_1m_1, p_2m_1)d(p_1e_2m_1, p_1m_1)}{d(p_2e_2m_1, p_2m_1)d(p_1e_1m_1, p_1m_1)} = 2.10 \cong \frac{1.40 \times 0.41}{0.26 \times 1.05} \tag{24}$$

$$\frac{d(p_2e_1m_2, p_2m_2)d(p_1e_2m_2, p_1m_2)}{d(p_2e_2m_2, p_2m_2)d(p_1e_1m_2, p_1m_2)} = 12.65 \cong \frac{1.67 \times 0.28}{0.03 \times 1.25} \tag{25}$$

for m_1 and m_2, respectively. The corresponding diamonds are given in Figure 3.

As a final illustration of this very special property of category quantifications in terms of three-way interactions, we inspect the *GPE* interaction as well, which should be identified, following Agresti, as a "no three-way interaction" case. The *GP* distance ratio was obtained as 0.283 for those who reported extramarital sex and 0.286 for those who did not. So there is only two-way interaction, to the effect that about 3.6 times more men than women had premarital sex. The *GE* distance ratio is 0.695 for those who reported premarital sex and 0.704 for those who did not. Again, there is only two-way interaction: 1.4 times more men than women had extramarital sex. Finally, the *PE* distance ratio is 3.56 for women and 3.61 for men: those who had premarital sex were 3.6 times more likely to have extramarital sex than those who had not, but gender has no effect on the relation between P and E.

6   Discussion

A major point of this chapter is that the use of homogeneity analysis does not need to rely on the assumption that the higher order interactions among the categorical variables are nonsignificant. We first proposed a procedure that uses homogeneity analysis to display the higher order interactions in a $2 \times 2 \times 2 \times 2$ contingency table directly. The multiway contingency table was first transformed into a profile frequency table. Then higher way cross-classification variables were added in a completely balanced way. We demonstrated from the solution of such an extended homogeneity analysis how the higher way interactions are represented in the visual display.

It was shown that the condition "no two-way interaction" could be expressed exactly in terms of the discrimination measures; if two variables are independent, their discrimination measures add up to the discrimination measure of their cross-

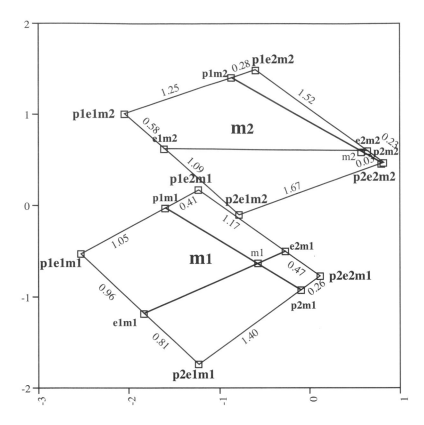

Figure 3: Display of three-way interaction PEM: category points for premarital sex and extramarital sex connected for the two levels of marital status.

classification variable. No two-way interaction is also expressed in ratios between distances between particular category points. If there is no two-way interaction, the latter ratio is equal to 1. These distance ratios were then shown to be the major diagnostic for identifying three-way interaction. If the second-order interaction is significant, the distance ratio based on each pair of two variables will differ substantially for the different levels (categories) of the third variable.

To simplify the computation while using existing software, we would recommend not using standard multiple correspondence analysis or homogeneity analysis programs (as in the SPSS HOMALS procedure). This would amount to an analysis of a much larger matrix than is actually required, because the number of profiles is much smaller than the total number of individual objects.

Instead, simple correspondence analysis could be applied, for example, the SPSS ANACOR procedure. To apply simple correspondence analysis, we would have only to replace each column in the profile matrix by its indicator matrix, collect

Table 4: Weighted indicator supermatrix for marital status data to be used as input for simple correspondence analysis

17	0	17	0	17	0	17	0
4	0	4	0	4	0	0	4
54	0	54	0	0	54	54	0
25	0	25	0	0	25	0	25
36	0	0	36	36	0	36	0
4	0	0	4	4	0	0	4
214	0	0	214	0	214	214	0
322	0	0	322	0	322	0	322
0	28	28	0	28	0	28	0
0	11	11	0	11	0	0	11
0	60	60	0	0	60	60	0
0	42	42	0	0	42	0	42
0	17	0	17	17	0	17	0
0	4	0	4	4	0	0	4
0	68	0	68	0	68	68	0
0	130	0	130	0	130	0	130

the indicator matrices in an indicator supermatrix, and premultiply the latter with a diagonal matrix, containing the corresponding profile frequency on its main diagonal. Table 4 is such a table for the four original variables in our example.

In the empirical cases we have analyzed so far, the row scores in an analysis with only the main effect variables were always very similar to those with all cross-classification variables added. This suggests that in practice we would not need to include all the cross-classification variables, but could derive the higher order category quantifications from a simple analysis, by computing the appropriate centroids of profile points afterward. The resulting visual displays of interaction through the use of diamonds must be very similar as well. In fact, although the diamonds may be slightly different (due to a somewhat different projection from high-dimensional space), the distance ratios they display are identical.

From the row scores of a simple correspondence analysis of a table as shown in Table 4, the higher way centroid for the category point for $g_1p_1e_1$, for example, is obtained as 17 times the first row score (for profile 1111) plus 4 times the second row score (for profile 1112) divided by 21; the category point for $g_1e_1m_1$ is 17 times the first row score (for profile 1111) plus 36 times the fifth row score (for profile 1211) divided by $17 + 36$, and so on. Which combinations should be taken follows from the associated columns in the extended profile frequency matrix in Table 2.

To compare our procedure with already existing ones, the following observations are important. First, our spatial representation is totally different from the usual geometric model used in the theory of log-linear analysis (Fienberg and Gilbert, 1970), which considers the distribution of mass over the cells of the table as one point in a regular polygon. Relationships among the two would be a subject for further study. Second, there are at least two other approaches to the use of cross-classification variables, which are, however, different from ours. In Gifi (1990), it is proposed to replace two of the original (main effect) variables by one cross-classification variable, with the aim of removing "uninteresting" association with respect to the main object of study. Van der Heijden and de Leeuw (1985) use the idea of cross-classification variables with the aim of studying residuals from higher order independence models. They apply simple correspondence analysis to a matrix of order, say, $n_A \times n_{BC}$; as they remark, it is often not obvious which two out of three variables should be cross-classified. We have shown in this chapter that when using our method, no such choice has to be made, and at the same time possible effects can be identified.

Bishop *et al.* (1975, p. 24) remark about the possibility of a linear (additive) model in the cell probabilities instead of their logarithms: "We conclude that the difficulty of relating the additive model to the concept of independence makes it less attractive than the loglinear model." The profile scores from homogeneity analysis are additive combinations of category quantifications, and we have seen that they are related to the concept of independence in a rather simple way. The apparent contradiction is resolved once we realize that the scores from homogeneity analysis do not represent the cell probabilities but the cell itself (the profile). The spatial representation aims at predicting an answer pattern given the profile score, not a probability given the answer pattern.

We have not elaborated on the case in which variables contain more than two categories. Some experimentation has shown that the special properties in terms of discrimination measures are preserved for the multicategory case. With respect to the much more complicated distance ratios, the promising results obtained by applying the general approach proposed in this chapter to the multicategory case are currently being scrutinized.

Chapter 21

Graphical Displays in Nonsymmetrical Correspondence Analysis

Simona Balbi

1   Introduction

The aim of this chapter is to show how a nonsymmetric version of correspondence analysis can be useful for dealing with survey data. The method was first proposed by Lauro and D'Ambra (1984), as an alternative to correspondence analysis, when the dependence structure between two categorical variables has to be analyzed. Coding two qualitative variables as indicator matrices, Lauro and D'Ambra display the distribution of the dependent variable, given the explanatory one, in a suitable factorial subspace. Moreover, the method has been extended to multiway tables (D'Ambra and Lauro, 1989) and developments have been proposed in relation to models (Lauro and Siciliano, 1989; Balbi and Siciliano, 1994) and inferential issues (Siciliano, 1990; Balbi, 1992, 1994).

In this chapter we give special attention to the graphical representations of nonsymmetrical correspondence analysis (NSCA) and stress the conditions in which NSCA is preferred to ordinary correspondence analysis (CA) in exploring data.

2   Outline of the Method

Let **P** be an $I \times J$ correspondence matrix, a cross-tabulation of two categorical variables where all frequencies are divided by the total n of the table. Pearson's mean-squared contingency coefficient Φ^2, called the total inertia, is defined as

$$\Phi^2 = \frac{\chi^2}{n} = \sum_i \sum_j \frac{(p_{ij} - p_{i.}p_{.j})^2}{p_{i.}p_{.j}} \tag{1}$$

where n is the number of observations and $p_{i.} = \sum_j p_{ij}, p_{.j} = \sum_i p_{ij}$.

CA aims at visualizing the association structure among the row and column categories in a suitable subspace by decomposing Φ^2 along principal axes in the style of principal component analysis. However, many other indices, which measure different kinds of departure from the independence hypothesis, have been proposed. Among them, there is the predictability index τ_b:

$$\tau_b = \frac{\sum_i \sum_j \left[(p_{ij} - p_{i.}p_{.j})^2/p_{.j}\right]}{\left(1 - \sum_i p_{i.}^2\right)} = \frac{\sum_j p_{.j} \sum_i \left[(p_{ij}/p_{.j}) - p_{i.}\right]^2}{\left(1 - \sum_i p_{i.}^2\right)} \tag{2}$$

introduced by Goodman and Kruskal (1954), arguing that "measures of association ... should be carefully constructed in a manner appropriate to the problem at hand."

The applicability of τ_b (hereinafter denoted simply by τ) is related to what Goodman and Kruskal call "proportional prediction," that is, the relative decrease in the proportion of incorrect prediction of one categorical variable, given knowledge of the other categorical variable.

Note that the denominator of τ is the heterogeneity measure proposed by Gini (1912). It is a normalizing factor, as it represents the value assumed by the numerator of τ when the knowledge of $p_{.j}$ completely determines $p_{i.}$. Lauro and D'Ambra (1984) show how, dealing with conditional distributions, the numerator of τ can be decomposed along principal axes, just as CA decomposes Φ^2. Thus, when the two ways of a contingency table seem not to be in a symmetric relation (e.g., the row variable depends on the column variable), it could be convenient to visualize the influence of the column variable categories j on the row variable categories, that is, on the empirical conditional distributions $p_{ij}/p_{.j}, i = 1, \ldots, I$, relative to the hypothesis of absence of influence, given by the marginal frequencies $p_{i.}$.

2.1   The Adoption Survey

At present, in Italy there is a wide debate concerning adoption. New laws have been proposed, and interest is focused mainly on the methods for choosing and matching adoptive parents and children. Thus, a sample survey was carried out (Balbi *et al.*, 1995) by interviewing a sample of 100 adoptive parents. The survey was part of a wider collaborative project between judges, psychologists, sociologists, and people working with adoptive families. One of the main goals was to understand when an adoption can be judged successful and why. As a first approximation, the difficulties met by parents, compared with expectations, were considered as a clue. In Italy there is a heated controversy about parents' ages and their importance for the child's fitting into social life. Thus, the first question was, "Can difficulties met by adoptive parents be related to the mother's age?"

A table cross-classifying difficulties (three categories: more than expected, less than expected, equal to expected) and age of mother at the adoption (three categories: less than 36 years old, 36–40 years old, more than 40 years old) was constructed (Table 1).

Performing a CA on such a table might appear superfluous because of its small size. However, it can be useful in showing the reason for choosing NSCA in preference to CA.

2.2 The Method

Let $\tilde{\mathbf{P}}$ be the relative frequency table, centered with respect to the independence hypothesis ($\tilde{p}_{ij} = p_{ij} - p_{i.}p_{.j}$); $\mathbf{D}_c$ is the diagonal matrix of the column marginal frequencies $p_{.j}$. $\tilde{\mathbf{P}}\mathbf{D}_c^{-1}$ is the centered column profile matrix [with general element $(p_{ij}/p_{.j}) - p_{i.}$]. Each row of $\tilde{\mathbf{P}}\mathbf{D}_c^{-1}$ represents the departure from the hypothesis of conditional independence of the corresponding category of I on J. Following Greenacre (1984), NSCA can be seen as a special case of a general analysis based on the (generalized) singular value decomposition: $\tilde{\mathbf{P}}\mathbf{D}_c^{-1} = \mathbf{U}\boldsymbol{\Lambda}^{1/2}\mathbf{V}^\mathsf{T}$, with the orthonormalizing constraints $\mathbf{U}^\mathsf{T}\mathbf{U} = \mathbf{V}^\mathsf{T}\mathbf{D}_c\mathbf{V} = \mathbf{I}$. $\boldsymbol{\Lambda}^{1/2}$ is the diagonal matrix of the square roots of the eigenvalues λ_α of the matrix $\mathbf{A} = \tilde{\mathbf{P}}\mathbf{D}_c^{-1}\tilde{\mathbf{P}}^\mathsf{T}$, sorted in decreasing order. The columns of $\mathbf{U}$ and $\mathbf{V}$ are respectively the left and right (generalized) singular vectors of $\tilde{\mathbf{P}}\mathbf{D}_c^{-1}$.

Table 1: Difficulties met by parents and mother's age at adoption (column percentages in parentheses)

	Less than 36	36–40	More than 40	Total
More than expected	4 (11.1)	8 (21.6)	4 (14.9)	16
Less than expected	18 (50.0)	21 (56.8)	9 (33.3)	48
Equal to expected	14 (38.9)	8 (21.6)	14 (51.8)	36
Total	36 (100)	37 (100)	27(100)	100

Notice that in NSCA the metrics assumed for the row and column spaces are $\mathbf{D}_c$ and $\mathbf{I}$, respectively, whereas in CA they are $\mathbf{D}_c^{-1}$ and $\mathbf{D}_r^{-1}$ (the chi-squared distances). Furthermore, note that:

$$\left(1 - \sum_{i=1}^{I} p_{i.}^2\right) \tau = \sum_{\alpha=1}^{M} \lambda_\alpha$$

where M is the rank of $\tilde{\mathbf{P}}\mathbf{D}_c^{-1}$ $[M = \min(I, J) - 1]$.

The ith row conditional distribution can be decomposed as

$$\frac{p_{ij}}{p_{.j}} = p_{i.} + \sum_{\alpha=1}^{M} \sqrt{\lambda_\alpha} u_{\alpha i} v_{\alpha j} \qquad (j = 1, \ldots, J) \qquad (3)$$

Replacing M with a lower value M^*, a low-rank approximation is obtained. The descriptive index based on the explained inertia $\sum_\alpha^{M^*} = 1\lambda_\alpha / \sum_\alpha^{M} = 1\lambda_\alpha$ is usually adopted in choosing a suitable M^*. A more detailed presentation of the algebra of NSCA can be found in Lauro and Balbi (1995).

2.3   Geometry of NSCA

NSCA looks for the orthonormal basis accounting for the largest part of variability (here in the sense of *predictability*, measured by τ). Let us consider the I-dimensional space, spanned by the columns of $\tilde{\mathbf{P}}\mathbf{D}_c^{-1}$. The origin of this space is at $\mathbf{r}$, the average column profile. The J centered columns of $\tilde{\mathbf{P}}\mathbf{D}_c^{-1}$ are contained in a subspace with at most $I - 1$ dimensions. As in ordinary CA, we are interested in displaying distances between column profiles, but in this case unweighted Euclidean distances, not chi-squared distances:

$$d_{jj'}^2 = \sum_i \left(\frac{p_{ij}}{p_{.j}} - \frac{p_{ij'}}{p_{.j'}}\right)^2 \qquad (4)$$

The projections of the column profiles on the αth principal axis give coordinates:

$$\varphi_\alpha = \mathbf{D}_c^{-1}\tilde{\mathbf{P}}^\mathsf{T}\mathbf{u}_\alpha \qquad (5)$$

Note that, for measuring the predictability in the table, we have to take into account the marginal distribution of the explanatory variable, in the example mother's age. Here the weighting system is defined by $\mathbf{D}_c$.

As in principal component analysis, row points and column points have different geometries. The distance of each row point from the origin is a measure of its expectancy, given the column variable distribution. The distance between two row points indicates their different ways of depending on the explanatory variable:

$$d_{ii'}^2 = \sum_j p_{.j} \left[\left(\frac{p_{ij}}{p_{.j}} - p_{i.}\right) - \left(\frac{p_{i'j}}{p_{.j}} - p_{i'.}\right)\right]^2 \qquad (6)$$

In such a distance it is important to take into account the marginal distribution of the explanatory variable, because its categories contribute differently to the total variability measured by τ. Thus a weighted Euclidean metric, defined by $\mathbf{D_c}$, is adopted in defining the metric structure of the row space. Each row is assigned an equal weight in its contribution to the total variability.

As an additional consequence, the distributional equivalence property in NSCA is preserved only when two categories of the explanatory column variable, having the same profile, are merged into one, with the same profile and weight equal to the sum of the weights of the two merged categories. In such a case, as in CA, distances among row points are not modified. The property does not hold in merging two categories of the dependent variable, although, from practical evidence, Lauro and D'Ambra (1984) assert that there is little consequence in merging row categories with equal profiles.

NSCA row coordinates on the αth principal axis are

$$\boldsymbol{\psi}_\alpha = \tilde{\mathbf{P}}\mathbf{v}_\alpha \tag{7}$$

2.4   Some Rules for Interpreting NSCA Factorial Planes

Coming back to the adoption data, CA has been performed together with NSCA. As in CA, the results of NSCA are presented in maps that show the configuration of points in projection planes formed by the major principal axes.

Table 1 does not present a strong association structure, having a chi-squared value of $\chi^2 = 7.2\,(p = 0.12)$. The value $\tau = 0.04$ means that, knowing the mother's age, we can better predict the difficulties adoptive parents would meet in only 4% of cases (the approximate χ^2-test for τ has a p-value of 0.08). Figure 1 shows the maps obtained by CA and NSCA, both representing 100% of the variability.

In both analyses there is strong evidence that mothers of 40 years and older have a better awareness of adoption difficulties: the first axis (explaining 89% of the total variability in CA and 93% in NSCA) opposes women 36–40 years old (on the left) versus women over 40 (on the right), the latter characterized by difficulties equal to expected. The second axis (explaining 11% of the total variability in CA and 7% in NSCA) opposes the youngest mothers, for whom adoption has been easier than expected, versus the middle age class, who have more difficulties than expected.

The differences between the CA and NSCA displays can be understood by considering the following aspects of the NSCA interpretation:

The scattering of the difficulties category points around the origin displays the dependence strength of difficulties on mother's age; the position of the ith dependent variable category with respect to the origin displays how well the explanatory variable predicts the category (for example, the category MORE is less predictable, being closer to the origin).

The scattering of mother's age categories around the origin also displays the dependence strength of difficulties on mother's age: the position of the jth explanatory

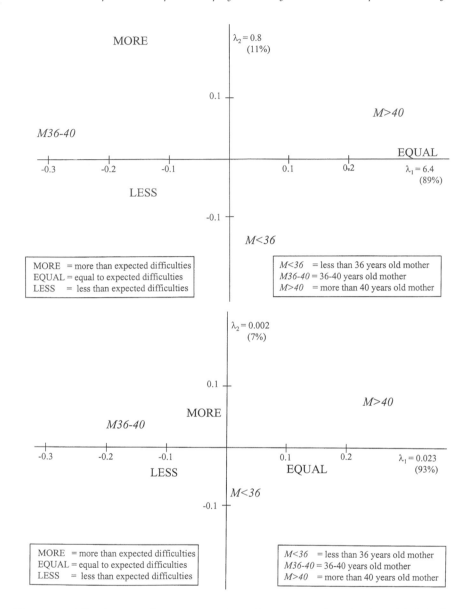

Figure 1: Difficulties met by parents and mother's age at the adoption. (a) C.A. first factorial plane, (b) NSCA, first factorial plane.

variable category with respect to the origin displays how strongly it predicts the behavior of the dependent variable (for example, the age group mother > 40 is a better predictor, being farther from the origin).

If two dependent variable categories are close on an axis (e.g., MORE and LESS, as opposed to EQUAL), their dependence structure with respect to the explanatory variable is similar, according to the feature of that axis.

When two predictor variable categories are close on an axis, they similarly influence prediction of the response categories contrasting on that axis; for example, all mothers 36 years or older have a similar influence on predicting the category MORE.

Keeping these rules in mind, we can interpret the NSCA map for Table 1 (Figure 1b) and compare it with the CA map (Figure 1a). The graphical displays include the complete information in the table, because the dimensionality of the table is equal to 2.

2.5   Aids to the Interpretation of NSCA Maps

Coordinates are not the only elements to be taken into account when one reads NSCA maps. As in principal component analysis, row points' coordinates show the contribution of each category i to the orientation of the relative axis, because the contribution, $\mathbf{CTR}_k(i) = \psi_{\alpha i}^2/\lambda_\alpha$, does not involve weights. For the column contributions, however, the situation is as in CA, where the weighting system must be taken into account: for the jth column, the contribution is $\mathbf{CTR}_k(j) = p_{.j}(\varphi_{\alpha j}^2/\lambda_\alpha)$.

Table 2 shows these contributions, and we can see, for example, that mothers younger than 36 years do not contribute to the first axis. To evaluate how a category is represented on each axis, as in CA, squared factor loadings (or squared cosines) can be computed (Table 3): for the ith row,

$$\cos \alpha(i) = \frac{\psi_{\alpha i}^2}{\sum_j p_{.j}\left((p_{ij}/p_{.j}) - p_{i.}\right)^2}$$

Table 2: Contributions of Table 1 categories to the first two NSCA factorial axes

Contributions (CTR)	First axis ($\times$ 1000)	Second axis ($\times$ 1000)
More than expected	39	628
Less than expected	345	321
Equal to expected	616	51
Less than 36 years old	6	634
36–40 years old	479	151
More than 40 years old	515	215

Table 3: Squared factor loadings of Table 1 categories to the first two NSCA factorial axes

Squared factor loadings (COS)	First axis (× 1000)	Second axis (× 1000)
More than expected	448	552
Less than expected	933	67
Equal to expected	994	6
Less than 36 years old	117	883
36–40 years old	976	24
More than 40 years old	969	31

and for the jth column,

$$\cos \alpha(j) = \frac{\varphi_{\alpha j}^2}{\sum_i \left((p_{ij}/p_{.j}) - p_{i.} \right)^2}$$

Thus, we can see that difficulties equal to expected is almost perfectly represented on the first axis, whereas mothers less than 36 is badly represented on it (as it is an exact two-dimensional representation, squared factor loadings of each category sum to 1 for the two axes). Furthermore, Balbi (1994) proposes evaluating the importance in NSCA of each cell in the table by means of influence functions.

2.6 Joint Plots in NSCA

It is worth noting that an NSCA joint plot is not a true biplot of $\tilde{\mathbf{P}}\mathbf{D}_c^{-1}$, because both rows and columns are expressed in so-called principal coordinates (Greenacre, 1984). As a matter of fact, the centered column profiles matrix is given by

$$\tilde{\mathbf{P}}\mathbf{D}_c^{-1} = \sum_{\alpha=1}^{M} \frac{1}{\sqrt{\lambda_\alpha}} \boldsymbol{\psi}_\alpha \boldsymbol{\varphi}_\alpha^T \mathbf{D}_c^{1/2} \tag{8}$$

As an alternative to the joint plot, Lombardo and Kroonenberg (1993) proposed the use of a "column isometric" biplot (Gabriel, 1971) to enhance asymmetry in displaying points (this is also known as an asymmetric map with column points in principal coordinates). Thus, row coordinates are given by $\boldsymbol{\psi}_\alpha$, while column coordinates are given by $(1/\sqrt{\lambda_\alpha})\mathbf{D}_c^{1/2}\boldsymbol{\varphi}_\alpha$. The different choice is related to different objectives of the joint plot, either displaying the dependence structure or displaying and graphically reconstructing the centered column profile matrix.

3 An Extension to the Analysis of Multiway Tables

In the adoption survey, we were interested in going deeper into the socialization problems of adoptive children. The variable "performance at school" can be chosen as an indicator of their fitting into social life. In our questionnaire, we asked parents to give an opinion on school performance, subjectively using a three-point scale (high, medium, and low). Because there are strong indications that the age at which a child was adopted influences his or her future life, we cross-tabulate school performance with child's age at adoption (Table 4). The additional explanatory variable of mother's age at adoption is also included. To generalize NSCA to three-way or even more complex data sets, multiple NSCA and partial NSCA have been introduced by D'Ambra and Lauro (1989).

Multiple NSCA consists of transforming a multiway table into a suitable two-way table. For example, in the case of a three-way table, let us suppose that the first variable, with I categories, is the dependent variable "performance at school," and the other two variables, with J and K categories, respectively, are the explanatory variables "child's age" and "mother's age" at adoptions.

The proposed flattening procedure consists of combining the JK elements of the two explanatory variables and constructing an $I \times JK$ table. The computation of I and the implementation and interpretation of NSCA are otherwise identical, applied to the flattened table.

In multiple NSCA we deal with a new compound explanatory variable. The dependent variable may be highly conditioned by one of the two explanatory variables, say the second one with K categories. In this case, it can be interesting to analyze a single category K, referring to the conditional independence hypothesis $p_{ijk} = p_{ik}/p_{.k}$. This is called a partial NSCA.

From a geometric viewpoint (Figure 2b), partial NSCA represents in a lower-dimensional space the K clouds of points centered with respect to the respective conditional independence model of the first two variables, given a level k of the third. In multiple NSCA, centering is with respect to independence of the first variable and the combined second and third variables.

Table 4: Performance at school and child's age at the adoption (column percentages in parentheses)

	Less than 1 year old	1–2 year(s) old	3–4 years old	More than 4 years old	Total
High	19 (57.5)	4 (40.0)	7 (35.0)	4 (12.1)	34 (35.4)
Medium	12 (36.3)	4 (40.0)	7 (35.0)	14 (42.4)	37 (38.5)
Low	2 (6.2)	2 (20.0)	6 (30.0)	15 (45.5)	25 (26.1)
Total	33 (100)	10 (100)	20 (100)	33 (100)	96 (100)

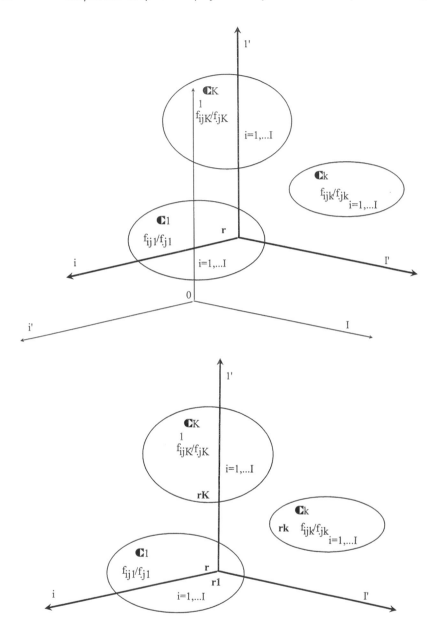

Figure 2: (a) Multiple NSCA, the analysis with respect to the common centroid, (b) Partial NSCA, the analysis with respect to the stratum centroid.

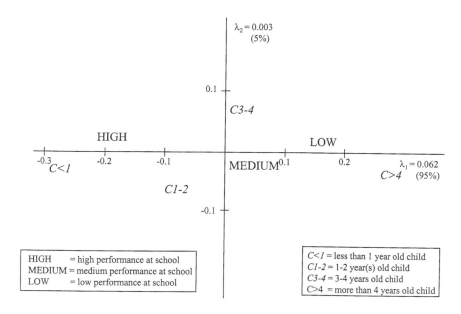

Figure 3: Performance at school and child's age at the adoption: simple NSCA, first factorial plane.

The different roles played by the two variables, cross-classified in Table 4, suggest the use of NSCA ($\tau = 0.10$, with performance dependent on age, $p = 0.003$). The influence of the age at the adoption on school performance was confirmed by NSCA.

The NSCA map in Figure 3 shows children with a low level of performance on the right, predicted by "age at adoption greater than 4 years," opposing children with a high level on the left, predicted by "age at adoption less than 1 year old." We now take into account a second explanatory variable, "mother's age at adoption." To avoid a large number of cells with small frequencies, we aggregate children with ages at adoption of 1–4 years. Table 5 shows the three-way table of frequencies.

Table 5: Performance at school and child's age at the adoption and mother's age at the adoption

	Less than 36 years old			36–40 years old			More than 40 years old			Total
	Less than 1	*1–4*	*More than 4*	*Less than 1*	*1–4*	*More than 4*	*Less than 1*	*1–4*	*More than 4*	
High	13	4	1	4	6	1	2	1	2	34
Medium	6	6	2	4	4	5	2	1	7	37
Low	1	2	1	1	5	3	0	1	11	25
Total	**20**	**12**	**4**	**9**	**15**	**9**	**4**	**3**	**20**	**96**

A multiple NSCA was performed on the two-way table crossing performance at school with the compound explanatory variable child's age (three categories) by mother's age (three categories). The $\tau_{(\text{mother's age.child's age})}$ is equal to 0.127: 83% is accounted for by the first eigenvalue and the residual 17% by the second one.

In Figure 4, the strong relation between age and performance at school observed along the first axis in the previous simple NSCA (Figure 3) is confirmed: performance at school is negatively related, first, with child's age at the adoption ($\tau_{(\text{child's age})}$ = 0.096) and, second, with mother's age at the adoption ($\tau_{(\text{mother's age.})}$ = 0.052). These relationships are effective in the extreme categories: high performance at school is strongly dependent on the combined category "mother less than 35 with child less than 1" (both these points are in the top left quadrant of Figure 4), and a similar positioning relates "mother over 40 with child over 4" and low school performance in the top right quadrant.

Additional information due to the introduction of mother's age is given by comparing configurations: although all variables have ordered categories, the typical "horseshoe" effect is present only for the dependent variable categories and for classes with mother's age 36–40. The marginal distributions of the two explanatory variables can be projected onto the map as supplementary points.

The multiple τ is less than the sum of the two simple τ's, which implies that interactions exist between the two explanatory variables—in Italy, it is almost impossible for parents more than 40 years old to adopt newborn babies.

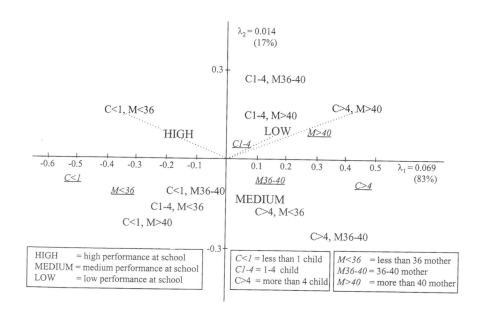

Figure 4: Performance at school and child's age/mother's age at the adoption: multiple NSCA, first factorial plane.

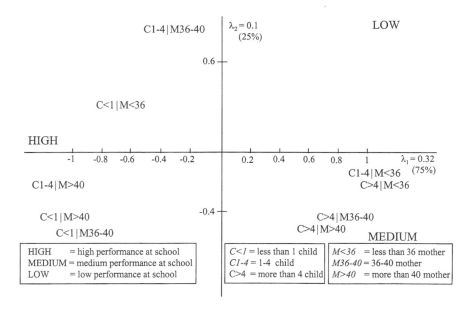

Figure 5: Performance at school and child's age/mother's age at the adoption: partial NSCA, first factorial plane.

Partial NSCA has been performed (Figure 5) by considering the distribution of "performance at school" depending on "child's age" given "mother's age." While the first axis explaining 75% of the total variability shows again how important child's age at the adoption is in the child's school performance, we also see that children adopted at 1–4 years perform relatively well given that the mother's age at adoption is over 40. This second axis shows several combinations at the bottom that lead to medium performance, mainly opposing the combination "mother 36–40 years with child 1–4 years," which results in poor performance relatively often.

Acknowledgments

This work has been supported by a grant from CNR, *Analisi non simmetrica dei dati* (N. C. Lauro). The author thanks N. C. Lauro and V. Esposito for helpful comments.

Chapter 22

Ternary Classification Trees: A Factorial Approach

Roberta Siciliano and Francesco Mola

1   Introduction

Several partitioning procedures have been proposed in the literature to construct decision trees, notably CART (Breiman *et al.*, 1984), RECPAM (Ciampi and Thiffault, 1987), and CHAID (Kass, 1980). In classification trees the data set consists of a large sample on which a categorical response variable and a high number of predictors have been observed. The idea is to use such a sample to "learn" how to predict the response variable from known observations of the predictors. This leads to defining a rule in the form of a *decision tree* in order to classify new cases of unknown class on the basis of observations of the given predictors.

In this chapter we are concerned with the case in which all predictors are categorical, like the response variable. Our main aim is to construct *exploratory trees* in order to emphasize the most significant predictors at each level of the tree. For this purpose a factorial approach is used to grow classification trees, especially the method of nonsymmetrical correspondence analysis (NSCA). Some new insights into the graphical displays of NSCA are presented in order to define a partitioning criterion into three classes. Factorial coordinates and predictability measures are used to distinguish between categories with strong and weak predictive power.

2 Nonsymmetrical Correspondence Analysis

NSCA is a factorial method for the analysis of the dependence in a two-way contingency table (Lauro and D'Ambra, 1984; D'Ambra and Lauro, 1989; Lauro and Siciliano, 1989; Balbi, 1992; Siciliano *et al.*, 1993). In the following we describe the factorial model and we give some insights into the interpretation of graphical displays. These results are used to propose a tree-structured classification via NSCA.

For a two-way contingency table let p_{ij} be the observed proportion such that $\sum_i \sum_j p_{ij} = 1$ ($i = 1, \ldots, I$; $j = 1, \ldots, J$) where the column variable Y depends on the row variable X (notice that in contrast to that of Balbi, Chapter 21, our dependent variable defines the *columns* of the table). NSCA analyzes the centered matrix of row profiles, with general element $(p_{ij}/p_{i.}) - p_{.j}$. The solutions can be obtained by generalized singular value decomposition, which can be written in scalar form as

$$\frac{p_{ij}}{p_{i.}} - p_{.j} = \sum_{k=1}^{K} \alpha_k r_{ik} c_{jk} \tag{1}$$

where $K \le \min(I - 1, J - 1)$, and $\alpha_1 \ge \cdots \ge \alpha_K > 0$. The score parameters of the row categories r_{ik} and of the column categories c_{jk} satisfy the following centering and orthonormality conditions:

$$\sum_i p_{i.} r_{ik} = 0, \qquad \sum_j c_{jk} = 0 \tag{2}$$

$$\sum_i p_{i.} r_{ik} r_{ik*} = \delta_{kk*}, \qquad \sum_j c_{jk} c_{jk*} = \delta_{kk*} \tag{3}$$

where δ_{kk*} is Kronecker's delta, the rows are weighted by the row margins, and the columns are weighted by ones. The objective is approximate the matrix by using a reduced-rank decomposition with a number of factors K^* lower than K, usually a two-dimensional factorial representation. For details on graphical displays and diagnostics for evaluating the quality of reduced-rank factorial representation, see Greenacre and Hastie (1987), Andersen (1995), Le Roux and Rouanet (Chapter 16), and Greenacre (Chapter 17).

In NSCA the coordinates of the row categories are defined by $\alpha_k r_{ik}$ and the coordinates of the column categories are defined by c_{jk}. In this way, we ensure that the graphical display in the reduced space is a *biplot* and can be interpreted in a nonsymmetrical way, namely by using prediction and dependence criteria.

Justification for NSCA lies in the predictability index τ of Goodman and Kruskal, defined as:

$$\tau_{Y|X} = \frac{\sum_i \sum_j \left[(p_{ij}/p_{i.}) - p_{.j}\right]^2 p_{i.}}{(1 - \sum_j p_{.j}^2)} \tag{4}$$

The denominator of the τ index is the index h_Y of total heterogeneity of Y due to Gini, and the numerator of the τ index is the part of total heterogeneity—called explained heterogeneity—due to the predictive power of the predictor categories. The τ index varies between 0 (no predictive power of predictor categories) and 1 (perfect prediction). If $\tau_{Y|X} = 0$, there is independence: $p_{ij}/p_{i.} = p_{.j}$ for all i and j. If $\tau_{Y|X} = 1$, then for each row category i there exists only one category j such that $p_{ij} = p_{i.}$. The index τ can also be interpreted as the relative increase in correct predictions of the response variable when knowledge about the category of the predictor is used.

NSCA decomposes the numerator of the τ index along principal axes and also over the row categories and over the column categories:

$$h_Y \tau_{Y|X} = \sum_k \alpha_k^2 = \sum_i p_{i.} \sum_k (\alpha_k r_{ik})^2 = \sum_j \sum_k (\alpha_k c_{jk})^2 \tag{5}$$

We notice that for $k = 1, \ldots, K$, the component $\sum_k (\alpha_k r_{ik})^2$ is equal to the squared distance of the row category to the origin (where the column marginal distribution is represented), the component $\sum_k (\alpha_k c_{jk})^2$ is equal to the squared distance of the column category to the origin; for $k = 1, 2$ we can approximate such distances in a two-dimensional factorial space by using the first two sets of scores. Using (5), we can define the following predictability measure of the row category R_i:

$$\text{pred}(R_i) = p_{i.} \frac{\sum_k (\alpha_k r_{ik})^2}{\sum_k \alpha_k^2} \tag{6}$$

where $\sum_i \text{pred}(R_i) = 1$. Formula (6) allows one to distinguish which row categories have more predictive power on the response variable (and thus contribute more to the index τ). Similar measures can also be defined for the column categories to understand which response categories are best predicted by the predictor.

Furthermore, the row coordinates $\alpha_k r_{ik}$ are related to the column coordinates c_{jk} by the following transition formula:

$$\sum_j \left(\frac{p_{ij}}{p_{i.}} - p_{.j} \right) c_{jk} = \sum_j \frac{p_{ij}}{p_{i.}} c_{jk} - \sum_j p_{.j} c_{jk} = \alpha_k r_{ik} \tag{7}$$

The left-hand side of (7) consists of two terms: the first term $\sum_j (p_{ij}/p_{i.}) c_{jk}$ gives the weighted average of the column coordinates, where the weights are given by the conditional distribution $p_{ij}/p_{i.}$ (or profile) for row i; the second term subtracts $\sum_j p_{.j} c_{jk}$, a constant term that all rows have in common. Formula (7) shows that the row coordinates are, apart from a constant term, the weighted average of the column coordinates. In the case of perfect prediction, (7) shows that for each column point j there exists at least one row point i with the same coordinates, that is, $\alpha_k r_{ik} = c_{jk}$; thus the predictor category i and the response category j are projected into the same point. Under independence $\alpha_k r_{ik} = 0$ for all the rows, since $p_{ij}/p_{i.} = p_{.j}$ for all i and j. In practice, the predictive power of the row categories is somewhere between the extremes of independence and perfect prediction. Row points far from the origin have high predictive power on the response variable. A row category i with coordinates $\alpha_k r_{ik}$ has high influence in predicting a particular column category j when the column coordinates c_{jk} are high and their signs agree with the signs of the respective row coordinates. In a factorial representation this means that the row point i is close to the column point j and these points are relatively far from the origin.

3 Ternary Trees by NSCA

Consider a data matrix with a categorical response variable (Y) and M categorical predictors $(X_1 \ldots X_M)$ observed on a sample of N cases, often called the learning sample. A classification tree, or classifier, is constructed by recursive partitioning of the cases into two or more subsets, which correspond to the "nodes" of the tree. The partitioning procedure starts with N cases at the "root of the tree" and is performed until the current node is declared to be a "terminal node" according to a stopping rule.

In Classification and Regression Trees (CART) the partitioning procedure is performed to grow the so-called maximal or exploratory tree with the highest number of terminal nodes (i.e., nodes with either a low number of cases or all cases belonging to the same class); then a pruning procedure allows a cutting back of some branches of the tree to provide the final classifier. To each terminal node is assigned the class with the highest proportion of cases. A test sample or a cross-validation procedure can be used to validate the final tree.

Partitioning procedures usually construct binary trees that are simple to interpret. In this case a splitting criterion is defined to divide cases at each node into two disjoint subgroups that are internally as homogeneous as possible. A drawback of tree methods is the time required to grow the tree depending on the number of splits to be tried out at each node (see Mola and Siciliano, 1997, and Aluja-Banet and Nafría, Chapter 5).

CART considers the splitting criterion based on the concept of node impurity, and a commonly used impurity measure is given by the Gini index of heterogeneity for the response variable Y [i.e., the denominator h_Y of (4)]. Among all possible splits of each predictor, the best split in CART is found by maximizing the decrease in impurity when passing from one group to two subgroups.

As an alternative approach, Mola and Siciliano (1994) provide a two-stage splitting criterion with which a reduced number of splits is considered at each node, thus saving some computing time. The basic idea is that a predictor is not merely used as generator of splits but plays a *global role* in the analysis. Thus some variable selection is performed to choose one or more best predictors that generate the set of possible splits at a given node. Two-stage splitting criteria can follow three strategies with respect to the use of a statistical index such as the predictability τ index (Mola and Siciliano, 1994, 1997), the use of a statistical model such as logistic regression (Mola *et al.*, 1996), or the use of a factorial method such as NSCA (Mola and Siciliano, 1996). Following the last strategy, we provide a classification tree procedure to grow ternary trees.

We now describe the main steps of the proposed methodology as applied at the root node to all N cases of the sample. Then the procedure follows recursively for each subsample until the stopping rule declares nodes to be terminal.

3.1   Table Selection

We consider the set of M contingency tables by cross-classifying each predictor with the response variable. Then for each table we calculate the predictability index τ and we select the predictor that provides the highest value.

3.2   Visualizing Dependence

We perform NSCA on the table cross-classifying the response categories with the best predictor identified in the previous step. Using the reduced-rank decomposition of (1) in two dimensions, we make a graphical display of the dependence structure between the response categories and the selected predictor categories. This map is used to explore the dependence structure in the subsample at the current node.

3.3   Partitioning Criterion

We define a classification criterion to find a partition of the N cases into three disjoint subgroups that are internally as homogeneous as possible with respect to their predictability of the predictor categories. The classification criterion is defined by using the predictor (row) coordinates in the first dimension, that is, the principal coordinates $\alpha_1 r_{i1}, i = 1, \ldots, I$. From (2), (3), and (6) these coordinates are centered and the sum of their predictability measures pred(R_i) is equal to 1. The predictor categories having a negative coordinate will predict response categories different from those predictor categories having a positive coordinate, which would lead us to use their sign as a binary splitting criterion.

In practice, however, we can have coordinates close to zero but with different signs, in which case using them to make different predictions does not make sense. Therefore, we use these intermediate predictor categories to define an additional split, leading to the ternary nature of our splitting criterion. To operationalize this idea, we use the predictability measures to distinguish between *strong* and *weak* categories. From (6) we can see that row i will make a proportionately higher or lower contribution to pred(R_i) if $|r_{i1}| \geq 1$ or $|r_{i1}| < 1$, respectively. We say that category i is a *strong* category when $|r_{i1}| \geq 1$, whereas category i is a *weak* category when $|r_{i1}| < 1$. As a result, we distinguish three subsets of predictor categories: (1) row categories such that $r_{i1} \geq 1$ (strong right categories); (2) row categories such that $|r_{i1}| < 1$ (weak categories); (3) row categories such that $r_{i1} \leq -1$ (strong left categories).

The partitioning of the predictor categories induces a partition of the current sample of cases into three subgroups. There can be an empty subgroup of cases in some situations, either when no category belongs to one of the preceding groups or when the predictor itself is dichotomous. The weak or "middle" subgroup includes cases in which the response variable is not strongly characterized by any category of the best predictor. This subgroup probably needs further splitting to improve predictability, whereas the strong subgroups include cases in which the response variable is strongly predicted by some categories—such nodes often include a low number of cases so that strong nodes are often declared terminal nodes.

3.4   Stopping and Assignment Rules

For stopping the recursive partitioning procedure we can use some natural rules, such as to stop when the percentage of cases at a node is below a certain value (e.g., 10%). This approach is recommended when the sample is not so large. We can also consider a stopping rule based on the CATANOVA statistic for the analysis of variation of categorical data, introduced by Light and Margolin (1971; see also Margolin and Light, 1974). Mola and Siciliano (1994) show how to check the strength of the dependence relation of the best predictor on the response variable.

After node splitting is completed, a response category can be assigned to each terminal subgroup of the final partition for prediction purposes. This assignment is usually based on the response category having the highest proportion of cases within the subgroup.

4   An Example

We consider a data set concerning a sample of 286 graduates of the Economy and Commerce Faculty of the University of Naples over the period 1986–1989. The variables shown in Table 1 have been observed by means of a questionnaire. The variable "final score" is assigned by the final committee taking into account the grad-

Table 1: Names and category descriptions of variables

	Categories					
Variables	**1**	**2**	**3**	**4**	**5**	**6**
Final score	Low (L)	Medium–Low (ML)	Medium–high (MH)	High (H)		
Sex	Male	Female				
Origin	Naples	County	Other counties			
Age	-25	26–30	31–35	$+36$		
Diploma	Classical	Scientific	Technical	Magistral	Profes.	
Study plan	Official	Managerial	Economics	Quantitative	Public	Profes.
Time to graduate	4 years	5–6 years	$+7$ years			
Thesis subject	Economy	Law	Quantitative	History	Management	

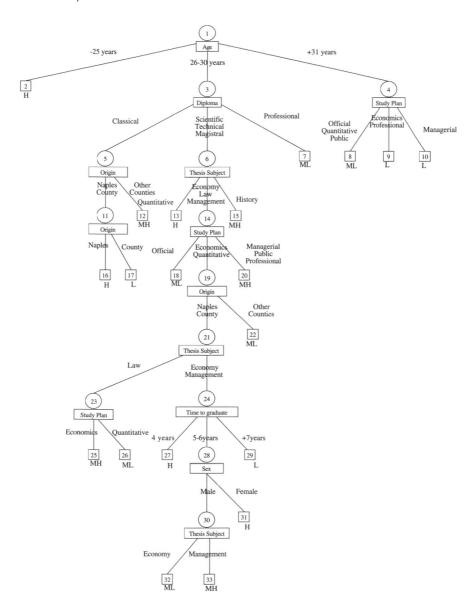

Figure 1: Classification tree where nodes are numbered (circles are for nonterminal nodes and boxes are for terminal nodes). Below a *nonterminal node* the best predictor (in the box) with the partition into strong left, weak, and strong right categories (arcs) is indicated. Below a *terminal node* the assigned response category label is indicated.

uates' average examination scores and the final dissertation. The variable "origin" refers to the place of living, which plays an important role in terms of participation in university activities. A student applying to a university chooses a "study plan," which can be either the official plan suggested by the faculty or any specialized plan such as managerial, economics, quantitative, public, or professional.

We are interested in identifying the variables that best predict the *final score* of each graduate. Figure 1 shows the final ternary tree, where as a stopping rule we have set a minimum node size of about 10% of the cases in the sample. Table 2 describes the partitioning into three subgroups of the categories of the best predictor with the percentage of explained inertia in terms of the index τ retained by the first principal axis. Table 3 describes at each node the distribution of proportions of the response categories: in particular, the first column gives the node number as from Figure 1, the

Table 2: Category partition of the best predictor at each nonterminal node of the tree in Figure 1

Node	Best predictor	Explained $\tau(\%)$	Strong left categories	Weak categories	Strong right categories
1	Age	94	−25 years	26–30 years	31–35 years +36 years
3	Diploma	89	Classical	Scientific Technical Magistral	Professional
4	Study plan	87	Official Quantitative Public	Economics Professional	Managerial
5	Origin	77	—	Naples County	Other counties
6	Thesis subject	54	Quantitative	Economy Law Management	History
11	Origin			Naples	County
14	Study plan	66	Official	Economics Quantitative	Managerial Public Professional
19	Origin	77	—	Naples County	Other counties
21	Thesis subject	99	Law	Economy Management	
23	Study plan	100		Economics	Quantitative
24	Time to graduate	75	4 years	5–6 years	+7 years
28	Sex	100		Male	Female
30	Thesis subject	100	Economy	Management	

Table 3: Response variable distribution of proportions at each nonterminal node of the tree in Figure 1

	% of	% of cases in responses			
Node	cases	L	ML	MH	H
1	100.0	18.9	27.3	28.3	25.5
3	74.1	14.6	25.0	31.6	28.8
4	16.4	44.7	40.4	14.9	0.0
5	14.7	11.9	4.8	7.1	52.4
6	52.1	14.8	27.5	32.9	24.8
11	10.8	9.7	3.2	22.6	64.5
14	42.7	14.8	29.5	29.5	26.2
19	39.2	13.4	30.4	28.5	27.7
21	29.0	8.4	27.7	33.7	30.2
23	10.5	3.3	40.0	40.0	16.7
24	18.5	11.3	20.8	30.2	37.7
28	15.0	9.3	20.9	34.9	34.9
30	11.9	5.9	23.5	41.2	29.4

Table 4: Response variable distribution of proportions at each terminal node with the assigned class label for the tree in Figure 1

	% of	% of cases in responses				Class
Node	cases	L	ML	MH	H	label
2	9.4	7.4	22.2	25.9	44.5	H
7	7.3	19.1	47.6	23.8	9.5	ML
8	5.9	11.8	47.0	41.2	0.0	ML
9	8.4	58.3	41.7	0.0	0.0	L
10	2.1	83.3	16.7	0.0	0.0	L
12	3.9	18.2	9.1	54.5	18.2	MH
13	2.1	0.0	16.7	33.3	50.0	H
15	7.3	19.1	19.1	52.3	9.5	MH
16	8.0	0.0	0.0	26.1	73.9	H
17	2.8	50.0	12.5	12.5	25.0	L
18	0.7	0.0	50.0	0.0	50.0	ML
20	2.8	37.5	12.5	50.0	0.0	NH
22	10.1	27.6	37.9	13.8	20.7	ML
25	8.7	4.0	36.0	48.0	12.0	MH
26	1.8	0.0	60.0	0.0	40.0	ML
27	1.8	0.0	20.0	0.0	80.0	H
29	1.8	40.0	20.0	20.0	20.0	L
31	3.2	22.2	11.1	11.1	55.6	H
32	3.9	9.0	36.4	27.3	27.3	ML
33	8.0	4.4	17.4	47.8	30.4	MH

second column gives the percentage of sample cases that fall in each node, and the remaining columns row by row give the distribution of proportions for the response variable at each node of the tree. Table 4 shows the terminal node information: for each terminal node we give the percentage of sample cases, the distribution of proportions of the response variable, and the assigned response category.

The first table selected according the proposed methodology cross-classifies the response variable final score with the predictor age. In Figure 2 we present the two-dimensional map of NSCA applied to this table. The first principal axis explains a very high percentage of the τ index (94%). We notice an opposition between older and younger graduates corresponding to an opposition between low and high scores, respectively. The age category 26–30 years is a weak category, close to the origin, generating a "middle" subgroup of 74.1% of the cases. The strong category, -25 years, splits off a subgroup of 9.4% of the cases, which form a terminal node predicting a high score. The other strong categories, 31–35 and $+36$ years, split off 16.4% of the cases, associated with low scores, but will be split still further. Usually, the negative correlation between age and study performance is due to the fact that

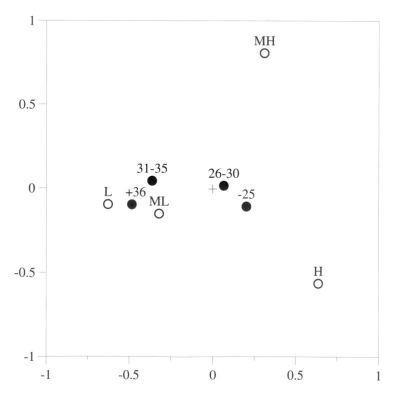

Figure 2: Nonsymmetrical correspondence analysis of cross-classification of age versus final score at node 1.

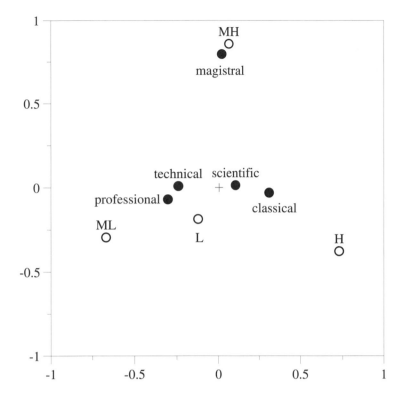

Figure 3: Nonsymmetrical correspondence analysis of cross-classification of diploma versus final score at node 3.

most of the old graduates are busy with working activity and aim to take the degree with any final score value.

The middle subgroup at node 3 is split by the predictor "diploma" with a distinction between classical, associated with high final scores, and professional, associated with low and middle–low scores. Figure 3 shows the NSCA map at node 3.

Node 4 is split by the variable "study plan" and Figure 4 shows the corresponding map. We notice in particular the opposition between official, quantitative, and public, associated with medium–high final scores, and managerial, associated with low scores. Notice that in node 4 there are no graduates with high final scores.

For brevity, we do not illustrate the factorial representations of the remaining nodes and we refer to Table 2 for further interpretation of the partitioning sequence in Figure 1.

In Table 5 we show the misclassification matrix when we use the final tree as a classification rule: each proportion in the diagonal of the table gives the probability of correct classification of each response category.

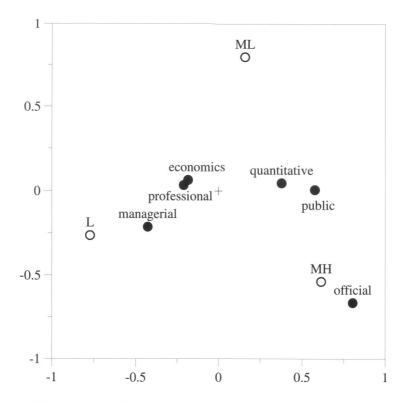

Figure 4: Nonsymmetrical correspondence analysis of cross-classification of study plan versus final score at node 4.

Table 5: Misclassification matrix for the response variable using the classification rule of Figure 1

	Predicted class			
True class	L	ML	MH	H
L	46%	27%	20%	7%
ML	17%	47%	24%	12%
MH	2%	23%	55%	20%
H	4%	19%	19%	58%

Conclusion

This chapter provides a methodology for growing ternary trees via an exploratory multidimensional method, NSCA. The proposed approach can be particularly convenient when the sample is very large and many predictors are considered. We organize the analysis as a sequence of NSCAs that leads to the construction of a decision tree to classify new cases with unknown responses as well as to explore the data set by constructing maps at each node of the classification tree where splitting occurs. When the sample is very large, we can consider splitting nodes according to the category combinations of a pair of variables, in the style of multiple NSCA (see Balbi, Chapter 21, and Lauro and Siciliano, 1989).

Acknowledgments

This research was supported by CNR research funds number 95.02041.CT10.

PART III

Multidimensional Scaling and Biplot

Although they originated as specific methods and algorithms, multidimensional scaling (MDS) and the biplot are now generic terms for a wide variety of techniques for data visualization. In general, data suitable for MDS are in the form of a square symmetric matrix and are measures of similarity or dissimilarity between pairs of objects, for example, a correlation matrix between variables, paired comparison data, or the number of co-occurrences of responses in a multiple-choice test. MDS assumes that these data are related to distances between objects in pairs: if objects A and B are observed, or measured, to be more similar or "closer" to each other than objects C and D, for example, then A and B should be displayed closer in the map than C and D. The term MDS has its origin in the so-called nonmetric multidimensional scaling of Shepard and Kruskal in the early 1960s, in which a map of a set of objects is achieved by approximating the rank ordering of observed distances between the objects rather than the distances themselves. MDS includes metric scaling, a much older idea, in which a stronger condition is imposed on the map that the distances themselves be approximated, not just their ordering. Whichever variation of MDS is performed, the resulting map is interpreted in terms of the distances between points, which is the most intuitive way of interpreting a display of this kind.

The term "biplot" was proposed by Gabriel in 1971 to describe methods that treat a rectangular matrix of metric data as if it contained scalar products between the row and column objects. The aim is to represent each row and column by a point so that the displayed row–column scalar products approximate the data values. Usually the data are centered and normalized before being "biplotted." The interpretation of the biplot in terms of scalar products might not seem straightforward at first, but becomes clearer when one thinks of a scalar product between points R and C as the projection of R onto the direction defined by C, multiplied by the length of C. Thus the projections of all the rows R onto the direction vector defined by C will give the set of approximations for the (transformed) data of column C. When the columns are variables and the rows cases, then the column points define directions that can be thought of as axes, which can be calibrated in units of the variable. The row points can be projected onto this "biplot axis" to read off the approximate values directly.

The following six chapters are a varied collection that take these ideas into the realm of categorical data analysis. Chapter 23, by Kimball Romney, Carmella Moore, and Timothy Brazill, first illustrates the use of CA as a metric MDS method for general similarity data, not necessarily frequencies. Their reason for using CA rather than MDS itself is that similarity matrices obtained from different individuals can be stacked and analyzed jointly with CA, rather than needing to use a different and more time-consuming methodology such as individual differences scaling or generalized Procrustes analysis. The empirical results presented here open up a new area of application of CA and demonstrate again that through appropriate coding CA can bring out the structure of interest in the data.

In Chapter 24, Ingwer Borg and Patrick Groenen explain a way of interpreting an MDS solution, called facet theory, devised by Louis Guttman. Rather than the dimensional interpretation of the map, this approach concentrates on regions of the MDS solution called "facets." Facets are used to confront hypotheses from social science theory about the similarities between the variables of interest with the MDS display. The authors use an example of eight intelligence test items: according to psychological theories, they expect a subdivision along "language" (numeric versus geometric) as well as along "requirement" (application versus inference). This theory implies a certain pattern of points in the MDS display, in the form of patterns of concentric circular bands. This is a more structured approach to the use and interpretation of MDS in which the geometry actually reflects the underlying substantive theory.

Chapter 25, by Carles Cuadras and Josep Fortiana, deals with a particular situation in which one has two sets of distance matrices for the same set of objects, for example, distances among a set of researchers in terms of their areas of interest as well as information related to their coauthorship of papers. Clearly, one can perform separate MDS analyses on these two sources of information and then compare them qualitatively, but the authors strive to achieve a common analysis in which these sources of information are related. The advantage of their approach is that two quite different types of data can be related in one analysis, via the distance matrices each one generates.

Chapter 26 contains another application to electoral data, the third in this book. Magda Vuylsteke-Wauters, Jaak Billiet, Hans de Witte, and Frans Symons look at the voting and political attitudes in the Flemish region of Belgium. Their data consist of several attitude scales for each respondent as well as a discrete observation in the form of their voting behavior in the 1991 elections. The authors show how the attitudes and voting preferences can be related and mapped using what is called a canonical correlation biplot, where the political parties are represented as points and the attitude scales as vectors. The result is a compact expression of the data where it is possible to distinguish the voter's attitudes that separate the political parties.

In Chapter 27, Ruben Gabriel, Purificación Galindo, and José Luis Vicente-Villardón describe the use of the biplot to diagnose various forms of independence in contingency tables. For rectangular matrices, certain simple models result in certain patterns of points in the biplot map; for example, rows and columns in two perpendicular straight lines diagnose an additive model for the table. Applied to the logarithms of the frequencies in a two-way contingency tables, an additive model is equivalent to independence between the rows and columns. Moving to three-way contingency tables implies various types of independence, again diagnosed by straight line patterns. An important application is to diagnose independence in part of the three-way table by identifying subsets of points that fall on straight lines.

Chapter 28, by John Gower and Simon Harding, treats the display of multivariate categorical data. In multiple correspondence analysis (MCA) all the cases and all the categories are represented in a joint display. As described in Chapter 17, in asymmetric maps we can look at distances (or scalar products) between row and column points to predict the categories to which each case belongs. This is not the only way to obtain a joint display. Gower and Harding's approach uses the extended matching coefficient to measure distance between cases, rather than the chi-squared distance. Instead of looking at row–column distances, they partition the full space of the points into regions according to the categories of each variable. By construction, each region predicts the corresponding category perfectly. When these prediction regions intersect with our low-dimensional display of the cases, then we have areas of the map where we can predict the categories accurately.

Chapter 23

Correspondence Analysis as a Multidimensional Scaling Technique for Nonfrequency Similarity Matrices

A. Kimball Romney, Carmella C. Moore, and Timothy J. Brazill

1 Introduction

This chapter is an empirical investigation of the validity of two somewhat novel generalizations of correspondence analysis (CA). The chapter uses results from empirical data sets to explore the following issues: (1), the appropriateness of using CA as a general multidimensional scaling (MDS) technique for nonfrequency similarity data; and (2), the appropriateness of using CA to analyze stacked similarity matrices to obtain a common spatial representation of many individual matrices simultaneously.

The first question we address is the appropriateness of using CA in a purely descriptive manner to obtain spatial representations of the interrelations among points from similarity data regardless of the level of measurement. For example, nonmetric MDS has been widely used since the 1960s to obtain Euclidean representations of square symmetric data matrices. If it could be shown that CA could be generalized for use in such contexts, it would make possible many applications that cannot be easily accomplished with nonmetric MDS, for example, the analysis of asymmetric similarity data or the comparison of multiple configurations by analyzing stacked similarity matrices (both symmetric and asymmetric).

The second question we address is the application of CA to stacked similarity matrices, where each matrix in the stack consists of a square symmetric similarity matrix obtained from a single individual. The aim would be to produce representations of several individuals (or cases) in the same space for comparisons among individuals and subgroupings of individuals. Current methods for comparisons among individual configurations [that do not, like INDSCAL (e.g., Carroll and Chang, 1970), impose a common group configuration] require a separate analysis of each case followed by some sort of rotation into a common orientation. The current method of choice (see, e.g., Borg, 1977; Borg and Lingoes, 1987) for such analysis of multiple representations of fundamentally similar configurations is provided by generalized Procrustes analysis (Gower, 1975). We will demonstrate that CA of stacked similarity matrices provides a much simplified approach to such comparisons and gives representations virtually identical to those obtained using an implementation of generalized Procrustes analysis called PINDIS (Borg, 1977; Lingoes, 1987).

2   Historical Background of CA and Nonmetric MDS

CA was originally derived for the analysis of contingency tables containing cross-tabulated frequency data. For example, in one of the earliest papers on CA, Fisher (1940) used a contingency table in which individuals were cross-classified on two categories, eye color and hair color, to illustrate the scaling of categorical variables. In that paper CA was treated in the context of discriminant analysis. In another early example, Guttman (1941) derived the method from the perspective of the indicator matrix form. The full application of many of the concepts in CA, such as the partitioning of chi-square, is based on underlying statistical theory and valid only when based on random sampling from multinomial populations. Detailed historical summaries of CA may be found in Nishisato (1980), Greenacre (1984), and Gifi (1990). An elementary introduction to CA may be found in Weller and Romney (1990).

Nonmetric MDS techniques developed in a very different methodological environment from CA. In the 1960s nonmetric MDS techniques began to replace metric MDS (e.g., Torgerson, 1958) and became available on mainframe computers. These methods were immediately put to use in a descriptive manner to represent in Euclidean (or other Minkowski) space the similarity (or distance) among a wide variety of observed "proximity" measures among a set of objects. The key to the early work in nonmetric MDS was to compute a configuration of points in which the interpoint distances "closely" approximated the experimentally observed ranked "proximities." The breakthrough consisted of optimizing a monotonic goodness-of-fit function, called "stress," which measured the discrepancy between input "proximities" and displayed "distances." The early papers by Kruskal (1964a, 1964b) and Shepard (1962a, 1962b) describe this work in detail, and a broader prospective may be obtained from Shiffman *et al.* (1981), Young and Hamer (1987), Borg and Lingoes (1987), Shepard *et al.* (1972), and Romney *et al.* (1972).

In the nonmetric MDS literature, there are examples of a large variety of kinds of "proximity" data. Examples include confusion data on the Morse code, airline miles among cities, and ratings of similarities among nations. Other than the assumption that the proximities represent some sort of similarity or distance among the points, there were no apparent limits on what sort of data might be used. Almost no attention was paid to the level or scale of measurement of the input data such as nominal (categorical), ordinal, interval, or ratio as outlined by Stevens (1968). In fact, the only input to a nonmetric MDS program was the rank order of the experimentally obtained pairwise proximities; thus the output was invariant under arbitrary monotonic transformation of the data.

In the beginning period of nonmetric MDS, there was genuine skepticism about whether it "really worked." There were extensive tests against known patterns, usually Euclidean (see Borg and Lingoes 1987; p. 10, for example), and standard presentations of the so-called Shepard diagram consisting of a scatterplot between the fitted distances and the input proximities. Simple examples of this strategy may be found in Kruskal and Wish (1978). Other criteria were used, such as interpretability of the dimensions or the overall configuration and replicability.

In the spirit of the earlier justification of nonmetric MDS, we present in this chapter simple empirical demonstrations of how CA can be used to obtain descriptive MDS representations of similarity data. In the first example we check the validity of the results by comparison with the known "correct" answer. We also demonstrate that CA applied to stacked similarity matrices provides virtually the same descriptive results as those obtained by separate analysis of each case followed by generalized Procrustes analysis. The chapter does not provide mathematical theorems, although we refer to sources that contain them when relevant.

3 CA as Descriptive MDS

The model for CA that we consider here is the simple case for contingency tables, where an $I \times J$ contingency table $\mathbf{N}$ is approximated by weighted least-squares by a matrix of lower dimensionality, using the low-rank approximation properties of the singular value decomposition (e.g., see Gifi, 1990, p. 276 ff.). The results of CA are the matrices $\mathbf{X}$ and $\mathbf{Y}$ of standard coordinates for the rows and columns, from which maps are constructed by scaling one or both of these coordinate matrices by some function of the singular values.

To simplify the discussion in the remainder of the chapter we define "similarity" data to include both distance (or dissimilarities) and similarities. These empirical measures were referred to as proximities in the nonmetric MDS literature. It is understood that similarities may be derived from distances by subtracting each distance from some constant equal to or larger than the largest distance in the matrix (Gifi,

1990, p. 281). Weller and Romney (1990, pp. 70–76) discuss some precautions that apply when using CA on nonfrequency data.

It might be noted that the attempt to recover distances among cities was used early in the development of nonmetric MDS to demonstrate that the interpoint rankings of similarities were sufficient to recover the actual Euclidean distances in a satisfactory way, (e.g., Kruskal and Wish, 1978; Borg and Lingoes, 1987). Gifi (1990, p. 280) used the analysis of a matrix of similarities derived from physical distances among 23 cities in The Netherlands to illustrate that the application of CA "is not restricted to frequency data and to show its potential as an MDS technique." We depend heavily upon Gifi's arguments and refer the reader to their mathematical and statistical defense for the use of CA for the analysis of similarity matrices.

Despite these examples, however, there has been some concern that CA does not return actual Euclidean distances when they are input to the analysis. Researchers would usually choose metric scaling (Gower, 1966), which does recover Euclidean distances exactly. This shortcoming of CA has been addressed in a paper by Carroll *et al.* (1997) in which it is proved that a special variant of CA recovers Euclidean distances and in fact yields solutions equivalent up to a similarity transformation to those of classical MDS.

We will not repeat the proof but will review the procedure of Carroll *et al.* (1997). They begin with a matrix $\mathbf{D} = \{d_{ij}\}$, where d_{ij} are the distances among points in a known Euclidean space of r dimensions. The matrix $\mathbf{S} = \{s_{ij}\}$ is then calculated, where

$$\mathbf{S} = k - d_{ij}^2, \qquad i, j = 1, 2, \ldots, n \qquad (1)$$

and k is a number several hundred times greater than the maximum d_{ij}^2. The matrix $\mathbf{S}$ is then analyzed with standard CA. At the singular value decomposition step the product, $\mathbf{U\Lambda U}^\mathsf{T}$ is obtained. Note that the input matrix is symmetric so that $\mathbf{U} = \mathbf{V}$ and is a diagonal matrix of eigenvalues.

Then the vector $\mathbf{U}$ is rescaled as follows:

$$\mathbf{X} = \sqrt{\Lambda}\,\mathbf{U}\sqrt{\frac{s_{..}}{s_{i.}}} \qquad (2)$$

which is the usual transformation of standard to principal coordinates.

A final rescaling of the principal coordinates, $\mathbf{X}$, is then performed:

$$\mathbf{X}^* = \sqrt{\frac{1}{2k}}\mathbf{X} \qquad (3)$$

Distances recovered in the CA solution are thus

$$d_{ij}^* = \left[\sum_{r=1}^{R}(x_{ir}^* - x_{jr}^*)^2\right]^{1/2} \qquad (4)$$

As k approaches infinity the distances d_{ij}^* tend to the true original distances d_{ij}. R is the number of positive, nonzero, nontrivial eigenvalues and corresponds to the number of dimensions of the known original Euclidean distances, again depending on an appropriate choice of k.

We want to emphasize that when we talk about using CA as a general tool for MDS we stress that we do not carry over any of the theoretical or inferential aspects of the CA model based on frequency data. Whereas nonmetric MDS optimizes a monotonic function, the procedures in CA optimize a bilinear function. The major question is whether a bilinear function is robust with respect to a number of possible distortions almost always encountered with empirical similarity data.

The descriptive use of traditional concepts usually used in a more formal way is not limited to our extension of these concepts in applying CA to nonfrequency data. For example, in discussing the generalized Procrustes program PINDIS, Borg (1977, p. 620) comments:

> The reader might be somewhat surprised that we conceived of a configuration in terms of total variance and variance components, *i.e.*, explained, common, unexplained, *etc.*, variance. Of course, this is a purely formalistic use of these terms: algebraically, the coordinate matrices (...) are indistinguishable from data matrices associated with some random vector; furthermore, since least-squares minimization is the standard procedure in estimating statistical parameters—as it is in the fitting problems in *PINDIS*—we use this familiar terminology in an entirely *descriptive* way.

We should also note that we are not the first to use CA in the context of descriptive MDS applications. For example, both Meulman (1986) and Meyer (1992) considered similarities between CA and various multivariate data analysis techniques by presenting these techniques within a general MDS framework. In the next section we show how CA can be used descriptively on various forms of similarity matrices derived from interpoint airline distances among selected U.S. cities.

4 Correspondence Analysis and Non-Linear Transformations

One of the concerns about CA as a general method for the analysis of similarity data is how robust the method is with respect to reasonably severe nonlinear transformations of the raw data. In this section we compare the results of the effects of a series of nonlinear transformations on data from Kruskal and Wish (1978, p. 8), who provide a matrix of airline distances among 10 U.S. cities. We have arbitrarily constructed six different 10×10 similarity matrices from the data by the application of distinct transformations as described in the following. We then analyze each of these six matrices separately with CA. After presenting the results, we stack the six matrices into a single 60×10 matrix and apply CA to the stacked data. This provides us with a comparison, to a known standard, of each of the six separate analyses as well as with results from the joint analyses using the stacked representation.

The values for each cell value, s_{ij}, for the six similarity matrices were constructed as follows: (1) A standard similarity transformation was applied by subtracting the airline distances from the largest distance in the matrix plus one. (2) A squared transformation was produced by subtracting the squared airline distance from 10^7 and then dividing by 10^4. This emulates the procedure of Carroll *et al.* (1997) for obtaining Euclidean distances where k is given the value of 10^7. The division by 10^4 is simply to reduce the values for numerical comparison—results from descriptive CA are invariant under multiplication of the raw data by a constant. (3) A natural log transformation was computed by subtracting the natural log of the airline distances from the largest value in the log matrix. (4) A square root transformation was computed by subtracting the square root of the airline distances from the largest value in the square root matrix. (5) A rank order transformation was calculated from all off-diagonal values of the airline distances, where the highest rank was assigned to the two cities closest (most similar) to each other and entered on the diagonal. (6) A Lickert-type rating on a scale of one to nine was constructed, with the value of nine being assigned to cities closest to each other and entered on the diagonal.

Note that when CA is used on square symmetric similarity matrices, where a single object is represented by both a row and column score, we enter on the diagonal a number at least as large as the largest similarity score in any off-diagonal cell. From a commonsense point of view, this simply represents the fact that an object is at least as similar to itself as to any other object.

The six transformations just outlined may be viewed as a kind of test of how robust CA is over a wide variety of monotonic transformations. From this perspective any differences among the results are seen as "accidental" or error variance in the sense that we would hope that the method is robust to such perturbations. Note that there is a wide disparity in both the absolute magnitude and the variability of the individually derived matrices. We first present a summary of the results obtained by separately analyzing each of the data sets produced by the various transformations.

The pertinent statistics for each individual data set and the selected results from individual CA of each set are shown in Table 1 (information in the table on results for stacked matrices will be discussed later). The statistics on the raw data sets were computed on the values produced by the transformations described earlier and are based on the off-diagonal lower half-matrix. The mean, standard deviation, and coefficient of variation were computed for each set. The table also includes information on the first two nontrivial singular values and the inertia associated with the first two nontrivial singular values, the squared correlation between the reconstructed similarity values from the two-dimensional CA solution with (1) the original airline distances between every pair of cities and (2) the appropriate transformed matrix of similarities described earlier. When using CA for descriptive MDS we suggest that this last figure may be taken as a least-squares linear approximation of goodness-of-fit that plays the role of "stress" in nonmetric MDS goodness-of-fit measures.

Table 1: Basic statistics on individual analysis of each of the six transformations

Statistic	Transformation					
	Simil.	**Square**	**Log**	**Root**	**Rank**	**Rating**
Mean	1317.87	750.28	6.13	16.90	45.50	5.00
Standard deviation	707.16	218.30	0.59	9.76	26.27	2.61
Coefficient of variation	0.54	0.29	0.10	0.58	0.58	0.52
1st singular value	0.51	0.26	0.10	0.55	0.53	0.47
2nd singular value	0.12	0.05	0.04	0.17	0.14	0.12
% cumulative inertia, 1-d	0.92	0.97	0.69	0.84	0.89	0.90
% cumulative inertia, 2-d	0.98	0.99	0.82	0.92	0.96	0.96
r^2 separate[a]	0.99	0.96	0.89	0.97	0.98	0.97
r^2 stacked[b]	0.99	0.95	0.69	0.97	0.99	0.99
r^2 separate[c]	0.99	1.00	0.94	0.98	0.98	0.98
r^2 stacked[d]	0.99	0.99	0.57	0.97	0.97	0.95

[a] Squared correlation coefficient between original airline distances and two-dimensional reconstructed similarities from correspondence analysis for the six separate analyses.

[b] Squared correlation coefficient between original airline distances and two-dimensional reconstructed similarities from correspondence analysis for the stacked analysis.

[c] Squared correlation coefficient between transformed input data matrices and two-dimensional reconstructed similarities from correspondence analysis for the six separate analyses.

[d] Squared correlation coefficient between transformed input data matrices and two-dimensional reconstructed similarities from correspondence analysis for the stacked analysis.

The following conclusions and observations may be drawn from the results in Table 1. The extent to which the reconstructed values from CA correlate with the input data reveals that for each of the six transformations the recovery is remarkably good. The worst squared correlation, 0.94, for the natural log data is still quite high.

One interesting observation is that the singular values vary widely. The implication of this is that they are not of much use in diagnosing a good solution from a bad solution as they depend on "accidental" features of the input data. The cumulative percentages of inertia are quite good in all cases, although they are lowest for the natural log data. We conclude from these results that in this example CA gives a very accurate representation of the data for the original cities across the six transformations.

Before presenting the results for the comparison among the stacked data sets, it will be a useful check on stacking to compare the summary column scores with those obtained from an aggregated summary of the data. Historically, nonmetric MDS was normally performed on aggregate data. Individual data were summed across subjects, resulting in a single square symmetric matrix for analysis. When analyzing a matrix obtained by stacking several replications, or unknown transformations, of what is assumed to be similar underlying data, we would expect to find that the column scores from the stacked matrix are a very close approximation of the scores resulting from a single square aggregated data matrix. Figure 1 shows the comparison; the circles represent the results from the aggregated data and the triangles represent the results from the column scores of the stacked data. This close similarity simply confirms that the two ways of summarizing a single aggregate representation are giving similar results.

We will now illustrate the use of CA on the six transformed and stacked similarity matrices. When CA is applied to this 60×10 matrix, the first 10 row scores represent the configuration given by the first (similarity) transformation, the second

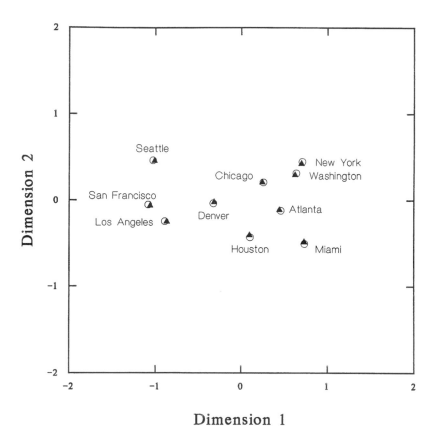

Figure 1: Comparison, based on the cities data, between (1) the overall results from scaling a single aggregated matrix obtained by adding six matrices and (2) the column scores from the stacked correspondence analysis of the same six matrices (the filled triangles represent the stacked correspondence analysis).

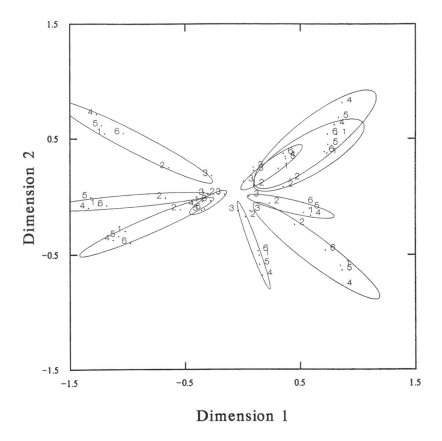

Dimension 1

Figure 2: Plot of the unstandardized row scores for each of the cities from the stacked correspondence analysis with each of the six transformed data sets identified by number and with a .95 confidence ellipse of the mean of each city.

10 row scores represent the configuration given by the second (squared) transformation, and so on through the six transformations. We can then compare the resulting representation with those from the six individual analyses.

Figure 2 shows the results from the stacked procedure where the six configurations for the cities are represented by a number for each transformation as defined earlier. In this and succeeding plots the ellipses are .95 Gaussian bivariate confidence regions on the centroid of the points (cities in this case). According to Wilkinson (1989, p. 214), each "ellipse is centered on the sample means of the X and Y variables. Its major axes are determined by the unbiased sample standard deviations of X and Y, and its orientation is determined by the sample covariance between X and Y". In Figure 2 the ellipses are elongated toward the midpoint of the picture. This is because the picture for the natural log transformation (points labeled "3"), for example, forms a dramatically "smaller" representation than the other points. The representation for the squared transformation (points labeled "2") forms an intermediate-sized representation.

This difference in size produced by different transformations has nothing to do with the original magnitude of the values in the respective matrices. For example, the values in the squared transformation are the next to largest of the six transformations, and the logged values are near the smallest. The difference is related to the value of the coefficient of variability of the matrices. The larger the coefficient of variability of each matrix, the larger the overall configuration of the joint representation. This is not a special feature of stacked matrices but is a general characteristic of CA. In general, when examining the points in a cloud of row scores, for example, the points most distant from the centroid are those with the highest coefficients of variation. Table 1 shows that the natural log data has a coefficient of variation about one fifth, and the squared data about one half, of the size of the remaining four data sets.

We view the differences in size among the representations of the different transformations as artifacts of the transformations. These differences are without substantive interpretation beyond the difference in the coefficient of variation. In order to compare the configurations we need to correct for these sorts of differences. We recognize that there is no standard practice with respect to this situation. One solution that we have used is to standardize the x_i scores for each subject to zero mean and variance equal to the singular values or the square root of the singular values. These standardization procedures worked very well in studies by Kumbasar *et. al.* (1994) and Romney *et al.* (1996). Note that programs such as PINDIS (Borg, 1977) do such rescaling as a standard routine inside the program.

When applied to the cities data, such a standardization produces the configuration shown in Figure 3. In Figure 3 the points representing the six transformations are clustered into tight sets around the location for each city. We have plotted the 95% confidence ellipses about the means of the clouds for the cities. The size of each ellipse gives an idea of the "resolving power" of the method given the various sorts of transformations that enter into the calculations. In practice, this would be the resolving power of the measurements based on some particular sample or subsample of subjects.

We can compare the results of the stacking procedure with the individual analyses presented earlier by reference to Table 1. The squared correlations between reconstructed distances based on the stacked results are compared with the original distances as well as each of the six transformed similarity matrices. The figures show that results are comparable to those of the individual analysis. The fact that in most cases the results are somewhat attenuated compared with the individual analyses demonstrates that the fit is not an artifact of stacking.

5   Correspondence Analysis of Longitudinal Ranking Data

In this section we present the results of an analysis of the classic Newcomb (1956, 1961) fraternity data. The data were collected in 1955–56 for 17 previously unac-

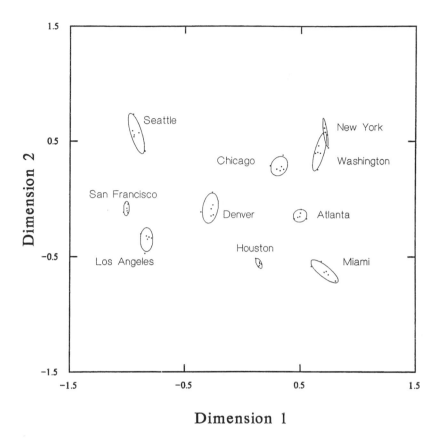

Figure 3: Plot of the standardized row scores for each of the cities from the stacked correspondence analysis with a .95 confidence ellipse of the mean of each city.

quainted students. The male subjects were provided room and board in a fraternity house in return for serving as subjects in an experiment on friendship. The data we report here were collected once a week for 15 weeks and consist of Newcomb's rankings (originally collected as ratings on a 100-point scale) of how well each subject liked each other subject. Each of the 15 weekly 17 × 17 matrices contains a row of rankings supplied by each subject indicating how well he likes each of the other subjects. Sample results of the study were published by Newcomb (1956, 1961), and the actual data are published by Nordlie (1958). The data have been reanalyzed many times, most recently by Nakao and Romney (1993).

Nakao and Romney's research included a Procrustes analysis of the data for the 15 weeks. They performed a nonmetric MDS on the data from each week based on a subject-by-subject correlation matrix. They then used a simple Procrustes procedure to rotate each week into the same orientation as the 14th week. The 14th week was selected as the target week because it was deemed the most representative configuration. We note that the new analysis of the Newcomb data, using CA of the

15 similarity matrices, presented in the following, is consistent with the nonmetric MDS results of Nakao and Romney (1993) in all aspects.

As with the cities data, we carry out an individual analysis of each week as well as a single analysis of all 15 weeks treated simultaneously with the stacking procedure. We then compare these results with those of a PINDIS analysis in which results from individual correspondence analyses of each week are used as input. These procedures will facilitate a controlled comparison of the stacking procedure with the generalized Procrustes analysis.

Our first step in the analysis is to transform the original data into similarity form. This is done by reversing the ranking and placing values equal to one plus the largest rank on the diagonal (Weller and Romney, 1990, p. 71). This means that the "closest" friend gets a rank of 16 (rather than 1) and 17 is entered on each diagonal. The overall results of the stacked data are given in Figure 4.

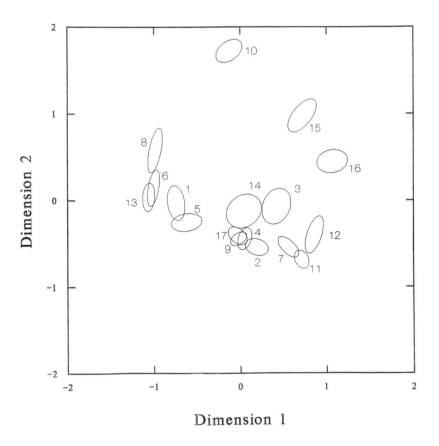

Figure 4: Plot of the 17 Newcomb subjects over the 15 weeks produced with row scores from stacked correspondence analysis and with a .95 confidence ellipse around the mean location of each of the subjects.

Figure 4 shows the scaled data with the 15 weekly positions of each subject summarized as a single .95 confidence ellipse of the mean position of that subject. The ellipses for the subjects with less well-defined social positions (i.e., subjects whose positions are more variable from week to week) are larger and represent a genuine phenomenon of the data. They tend to be social outliers. In the Nakao and Romney (1993) study, subjects 3, 10, 14, 15, and 16 were identified as outliers. It can be seen that these are the subjects with the largest ellipses. In addition to outliers, the earlier study distinguished two groups, namely numbers 1, 5, 6, 8, and 13 versus numbers 2, 4, 7, 9, 11, 12, and 17 (Nakao and Romney, 1993, p. 119). It can be seen that these subgroupings are fairly well defined in Figure 4.

In the PINDIS analysis we used coordinates from CA of each individual week done separately for each of the 15 weeks. Figure 5 shows the picture produced by the generalized Procrustes procedure for comparison with Figure 4 based on CA of the

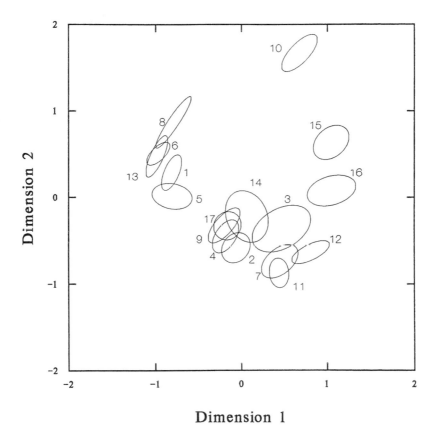

Figure 5: Plot corresponding to Figure 4 except that scores are obtained with generalized Procrustes analysis (implemented with PINDIS) using 15 separate analyses of the week-by-week data.

stacked data. One can see that the pattern is very similar to that observed in Figure 4. The resolution is not quite as sharp as in Figure 4, but clearly virtually identical (with slightly different orientations) configurations are visible.

A comparison of the "centroid" (Borg, 1977) and "consensus" (Gower, 1975) summary configuration from generalized Procrustes analysis with the column scores summary from the stacked CA illustrates just how similar the two procedures are. Figure 6 shows the two configurations with the results from the stacked CA shown with solid triangles and the results from the generalized Procrustes analysis shown with open circles. The conclusions drawn from the two figures would have to be identical; there are no differences of any practical import.

The question arises of whether adding a large number to the diagonal of each matrix biases the results of the stacking procedure. One would expect on an a priori basis that large numbers on the diagonal would, by themselves, produce clustering

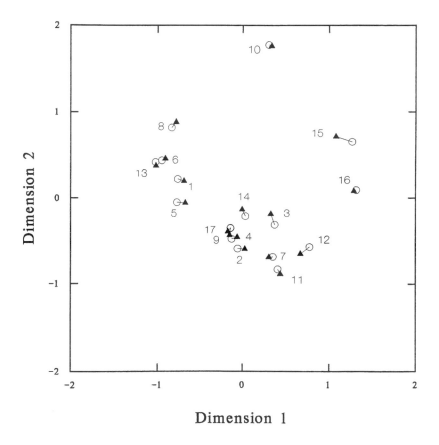

Figure 6: Plot comparing the summary location of each of the 17 Newcomb subjects obtained from the column scores of correspondence analysis (filled triangles) and the centroid (or consensus) configuration from PINDIS (open circles).

among the items. In order to study this effect, we performed some simulations using similarity matrices of random data with a large constant on the diagonal. Indeed, large numbers on the diagonal do produce a measurable clustering in otherwise random data. The possible biasing effects deserve further study. The practical question is whether, in the case of empirical data where there is a strong signal, such as the Newcomb data, the stacking procedure biases the results when used to compare various individuals or subgroups within the stacked data.

The most impressive evidence that the stacking procedure does not bias the data comes in a week-by-week comparison of the pictures produced by the stacked data compared with individual analysis of the week-by-week data. There is not sufficient space to show the stacked and unstacked results for each week here. However, a very careful examination of the week-to-week configurations shows virtually no differences in any of the comparisons. In each case the patterns are very similar and exhibit differences on the order of magnitude of the different transformations of the city data. In other words, the differences are in the range of individual measurement and sampling variability. We present a comparison figure for week 1, one of the most variable weeks. Figure 7 compares the results from the CA of week 1 only with those for the same week from the stacked data.

In the figure the circles are the results of an individual CA and the filled triangles represent the positions given from the stacked data. The lines connect the same subject in the two representations. One can see that an investigator would draw the same conclusions from either picture. The positions of a given individual are within reasonable sampling variability of each other. Note that this week is a comparatively "bad" example of fit. Later weeks show much closer correspondence between the two configurations.

6 Discussion

We have illustrated the possible utility of CA as a general MDS technique for application to a variety of types of similarity data. Some have suggested extreme caution in terms of the level of measurement that is appropriate for generalizing CA beyond frequency data. In converting the airline distances into a variety of similarity matrices, we included examples, such as ranking and Lickert rating, that were not ratio scale measurements. Our own feeling is that any data that are appropriate for nonmetric MDS are legitimate for analysis with CA. The resulting representations should be just as valid as those resulting from nonmetric MDS.

We have also demonstrated that the results from the stacking procedure can be used as a simple and accurate way to compare a series of individual configurations. The configuration for any single individual obtained from the stacked CA is, within sampling variability limits, the same as the configuration obtained from the analysis of that individual data matrix by itself. By extension, because generalized Procrustes analysis is limited to rotation, translation, and scaling transformations, the configu-

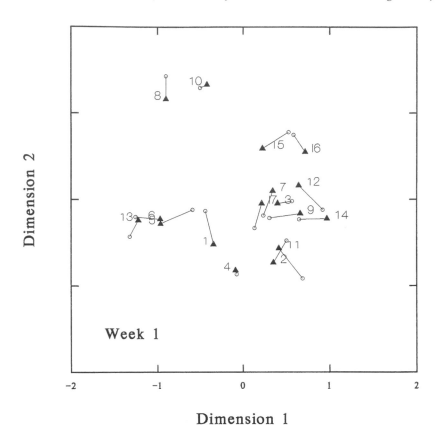

Figure 7: Plot of the Newcomb data for week 1 comparing the position of each subject obtained from the stacked correspondence analysis (filled triangles) with that obtained by the correspondence analysis of the week 1 data taken by themselves (open circles).

rations produced by the stacked CA would also be similar to results from generalized Procrustes.

The stacked similarity matrices approach is of considerable practical utility. For example, Romney *et al.* (1996) analyzed 732 stacked 15 × 15 symmetric similarity matrices representing similarity judgments derived from triadic comparisons for 15 English kinship terms. The purpose was to compare the effects of gender and linguistic background factors on the cognitive structure of the kinship terms. The use of generalized Procrustes analysis would have required 732 separate scalings to provide input coordinates for the analysis. The stacking procedure does all of this in a single step. We feel that this practical difficulty has discouraged the use of generalized Procrustes analysis for comparisons among individuals in large data sets. For example, the largest example we have found in the literature is that of the 41 subjects analyzed originally by Green and Rao (1972) and reanalyzed with PINDIS by Borg (1977, pp. 643–647) and Borg and Lingoes (1987, pp. 337–340). It also should be recognized that the fit statistics that are computed in PINDIS are very useful. Similar statistics need to be developed for the stacked CA results.

There is still a great need for further research and discussion about the two main issues that we have introduced in this chapter. With respect to the use of CA as a general MDS tool in a purely descriptive manner, there is a need for the development of a consensus about the appropriate goodness-of-fit measures to describe the results of a single analysis. What is the appropriate analogue of the nonmetric MDS measure of stress?

The very close correspondence between the centroid (or consensus) solution from generalized Procrustes and the column scores from the stacked CA shown in Figure 6 suggests that there may be a way to derive an analytic description of the relationship between the two (perhaps with limits on the discrepancy).

Acknowledgments

This research was supported by National Science Foundation grant SES-9210009 to A. Kimball Romney and William H. Batchelder.

Chapter 24

Regional Interpretations in Multidimensional Scaling

Ingwer Borg and Patrick J.F. Groenen

1 Dimensional and Regional Interpretations in MDS

Interpreting solutions in multidimensional scaling (Kruskal and Wish, 1978; Borg and Lingoes, 1987; Borg and Groenen, 1997) means linking geometric properties of the MDS configuration to substantive (physical, psychological, semantic, logical) properties of the represented objects. Many geometric properties could be considered, but one particular aspect of the point configuration, the points' coordinates on Cartesian coordinate axes, dominates the MDS literature. Indeed, the very name multidimensional scaling suggests that such "dimensions" play more than just a technical role in MDS. This has historical reasons, because MDS was originally meant to generalize one-dimensional scaling to several dimensions. Although this introduced many new possibilities to relate the model representation to the physical parameters of the stimuli, the search for meaning remained exclusively focused on dimensions (see, e.g., Torgerson, 1952).

Guttman (1977, p. 101) was among the first to note that this perspective turned things upside down, putting coordinate systems before the geometry that they are supposed to coordinate: "Euclidean space can be defined without a coordinate system. Indeed, this is how Euclid did it. Descartes came centuries later." What Guttman stressed is that the focus on dimensions entails a shift of attention away from the geometric properties of the MDS configuration ("figure") and toward one particular coordination of its points. Geometric properties of a figure, according to Klein (1872), are just those properties that remain invariant under certain transformations, such as rigid motions or similarity transforms. The point coordinates do not satisfy this criterion.

In most applications today, MDS is not used in a context in which the given objects can be coordinated in terms of physical or psychological dimensions. Rather, MDS is typically used for visually exploring the structure of similarities of "vaguely" described objects such as politicians, intelligence test items, or facial expressions. But although this vague prior knowledge does not lead to coordinations on which a meaningful distance function can be based, it typically allows a cross-classification of the stimuli. Politicians, for example, belong to party X or Y; are single, married, divorced, or widowed; have blue, brown, green, or other eye color; and so forth. Intelligence test items may require verbal, arithmetic, visual, or other abilities; they may require the subject to find a rule or use a known rule; they can be presented in writing or verbally; and so on. Facial expressions can be classified, a priori, into a whole range of impressions such as joy, disgust, friendly smiles, and anger; they can also be rated, for example, on a dimension of no to high ego involvement. Such cross-classifications are, in a sense, coordinations of the objects, but they do not have metric properties in general, because the criteria used for classification purposes (facets) are typically only qualitative or ordinal ones. The question, in any case, is how such a prior-knowledge facet system relates to empirically observed dissimilarities among the elements of its classes. The classes result from conceptually partitioning a monolithic domain of interest (such as politicians, intelligence items, or facial expressions) in different ways, and hence it seems natural to ask whether the MDS representation of the corresponding dissimilarity scores reflects this classification system in a geometric way.

A general approach to formulating this question is to ask whether the MDS configuration can be partitioned into regions, facet by facet, so that all points in one region are equivalent on that facet. In a plane, a region is defined as a connected set of points such as the inside of a rectangle or a circle. More generally, a set of points is connected if each pair of its points can be joined by a curve all of whose points are in the set. Partitioning a set of points into regions means to split the set into classes such that each point belongs to exactly one class. For example, do all politicians with brown eyes lie in one region of the MDS space, all those with blue eyes in another region, and all those with another eye color in a third region? If such a regional correspondence holds, one may ask further questions about the particular shape of the regions. For example, the regions may be such that they cut an MDS plane into essentially parallel stripes. This obviously comes close to the notion of a dimension. But, of course, there are other regional patterns, such as a system of concentric bands around a common origin, that are not related to Cartesian dimensions. Thus, the regional approach is more general than the dimensional approach, including the latter as a special case.

2 Partitioning MDS Spaces Using Facet Diagrams

Although regional interpretations are possible without prior studies on the objects of interest, in practice regionalizations are usually "confirmatory" ones. They are based on classification systems, mostly those using a facet-theoretical framework (Borg and Shye, 1995; Shye and Elizur, 1994). Consider the following example by Galinat and Borg (1987).

In experimental investigations a number of properties of a situation have been shown, one by one, to have an effect on judgments of time duration. The following mapping sentence shows four of these properties within a design meant to measure symbolic duration judgments, that is, duration judgments on hypothetical situations:

$$
\text{Person } \{p\} \text{ believes that the } \left\{ \begin{array}{l} A = \underline{\text{affective tone}} \\ 1 = \text{pleasant} \\ 2 = \text{neutral} \\ 3 = \text{unpleasant} \end{array} \right\} \text{ situation with}
$$

$$
\begin{array}{cc}
N = \underline{\text{number}} & V = \underline{\text{variability}} \\
\left\{ \begin{array}{l} 1 = \text{many} \\ 2 = \text{few} \end{array} \right\} & \left\{ \begin{array}{l} 1 = \text{monotonous} \\ 2 = \text{variable} \end{array} \right\} \text{ events that are}
\end{array}
$$

$$
\begin{array}{l}
D = \underline{\text{difficulty}} \\
\left\{ \begin{array}{l} 1 = \text{difficult} \\ 2 = \text{easy} \end{array} \right\} \text{ to handle is felt as } \rightarrow \left\{ \begin{array}{c} \text{reaction} \\ \text{very short in duration} \\ \text{to} \\ \text{very long in duration} \end{array} \right\}
\end{array}
$$

In each particular way of reading the mapping sentence, one element from the population $\{p\}$ is picked and crossed with one particular combination of the elements of the content facets. The content facets distinguish among different situations by considering four properties of its events: affective tone, number, variability, and difficulty. Altogether, the mapping sentence defines 24 different situation types (Table 1).

Table 1: Twenty-four situation types with structuples and mean empirical duration ratings; greater value indicates longer duration (Galinat and Borg, 1987)

No.	Structuple	Mean duration	No.	Structuple	Mean duration	No.	Structuple	Mean duration
	ANVD			ANVD			ANVD	
1	1212	3.29	9	1121	4.37	17	2221	4.66
2	2112	3.54	10	1211	4.41	18	3112	4.70
3	1221	3.87	11	2211	4.42	19	3211	4.93
4	1112	3.90	12	1222	4.43	20	3221	4.94
5	1122	3.95	13	3111	4.46	21	3122	5.00
6	1111	4.00	14	2111	4.54	22	2222	5.08
7	2212	4.03	15	2122	4.57	23	3212	5.15
8	2121	4.05	16	3121	4.57	24	3222	5.67

For example, a situation with the "structuple" (3222) is defined to be a generally unpleasant one, where few but different things are happening and where one has no problems in coping with what is happening.

How do persons judge the duration of these situation types? For each of the $3 \times 2 \times 2 \times 2 = 24$ structuples, one vignette was constructed to illustrate a situation of that type. For example, the vignette used for the "pleasant-many-variable-easy" situation was this: "You are playing a simple card game with your children. It is quite easy for you to win this game because your kids are no serious opponents. The game requires you to exchange many different cards. The game is fun throughout the 3 minutes that it lasts." This description is supplemented by the question: "What do you think, how long would this card game seem to last? Would it seem longer or shorter than 3 minutes?"

The respondents were asked to rate each such vignette with respect to its likely subjective duration. The intercorrelations of these ratings are mapped into a four-dimensional MDS space (stress = 0.13). Figure 1 shows the plane spanned by the first two principal axes of the MDS configuration. The points in this plane are labeled by the item numbers of Table 1.

We now ask whether this MDS configuration mirrors any of the design facets in the sense that points representing different types of situations fall into different

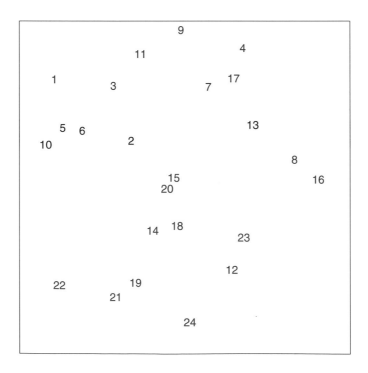

Figure 1: MDS plane of first two principal axes for duration data.

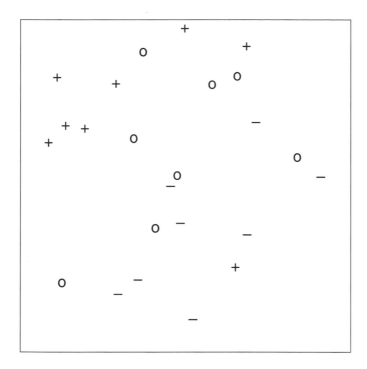

Figure 2: Facet diagram over Figure 1 for facet "affective tone."

regions of the space. Let us begin with the affective-tone facet and try to partition the space into regions containing "pleasant," "neutral," and "unpleasant" points, respectively. This task is greatly facilitated by an appropriate facet diagram. A facet diagram is simply a reproduction of an MDS configuration plot where the points are labeled by their structuples or by their codings on a particular facet. Figure 2 shows the facet diagram for the affective-tone facet, plotted over the two-dimensional MDS plane from Figure 1. The points in Figure 2 are labeled − if they represent situations defined as unpleasant, + for pleasant, and ∘ for neutral.

The diagram shows that the three types of points are not distributed randomly. Rather, the plane can be partitioned into regions so that each region contains only or almost only points of one particular type. Figure 3 shows such a partitioning. It contains two minor errors: the two solid arrows indicate where these points should lie to be in the appropriate regions. Obviously, they are not far from these regions. There is also one major error, a "pleasant" point located in the negative region. The dashed arrow attached to this point indicates the direction of required shifting.

Figure 4 represents an alternative partitioning that is error free. This partitioning depends, however, very much on the position of point 12 (marked by an arrow) and, thus, may be less reliable in further replications.

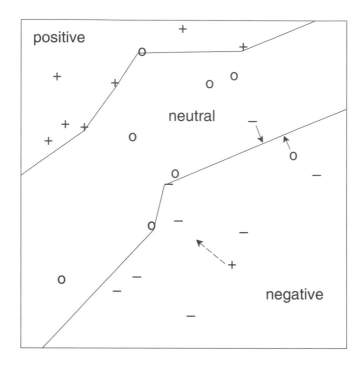

Figure 3: Facet diagram, for facet "affective tone" over Figure 1, with axial partition-ing.

The two partitionings, moreover, imply different things. The concentric regions of Figure 4 predict that duration ratings on unpleasant situations correlate higher among one other, on the average, than those for pleasant situations. The parallel re-gions of Figure 3 do not restrict the correlations. Both partitions are similar, however, in splitting the plane into ordered regions, where the neutral region lies in between the pleasant and the unpleasant ones. Hence, the regions are ordered as the affective-tone facet itself. Neither the spatial organization induced by the straight lines nor that in-duced by concentric circular lines would therefore have problems in accommodating an affective-tone facet that distinguishes more than just three levels.

We thus see that the affective-tone facet is reflected in the structure of the duration ratings. The decision on which of the two partitionings is ultimately correct requires further data.

The plane spanned by the third and fourth principal components (not shown here) can be partitioned by the facet number—without error—and also by variability—with two errors. The partitioning lines are almost straight and orthogonal to each other (Borg and Shye, 1995). The facet difficulty, on the other hand, does not show up in the MDS configuration; that is, the points representing easy and difficult situations, respectively, seem to be so scrambled that they cannot be discriminated by any but the most irregular partitionings.

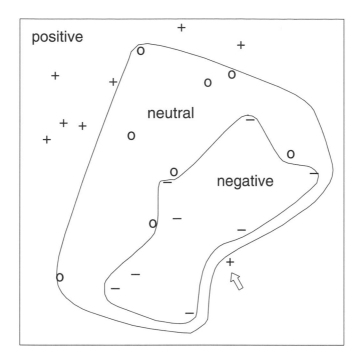

Figure 4: Facet diagram, for facet "affective tone" over Figure 1, with modular partitioning.

3 Facet Theory and Regions in MDS Spaces

Partitionings are based on substantive classifications of the represented objects. The facets may be predicted to play a particular role in partitioning the MDS space, but in no case is a particular dimensional system (such as a Cartesian one) chosen a priori and then interpretationally forced onto the content. Rather, the opposite is true: the content leads to the approximation of a particular dimension system.

To see this more clearly, consider a classic case, the cylindrex of intelligence items. The items in paper-and-pencil intelligence test batteries require the subject to find verbal analogies, solve arithmetic problems, or identify patterns that complete series of figures, for example. Hence, they can be classified by the facet "language of presentation" into numerical, verbal, and geometrical ones. At the same time, such tests relate to different abilities, which gives rise to a second facet, "required mental operation." It classifies tests into those in which the testee has to infer, apply, or learn a rule, respectively (Guttman and Levy, 1991). In combination, these two facets distinguish nine types of intelligence.

Table 2 shows the intercorrelations of eight intelligence test items. For example, item 1 in Table 2 is coded as numeric (re language) and as application (re requirement), whereas item 5 is geometrical and inference.

Table 2: Intercorrelations of eight intelligence test items, together with content codings on the facets "language" = {N = numerical, G = geometrical} and "requirement" = {A = application, I = inference}

Language	Requirement		1	2	3	4	5	6	7	8
	Structuple									
N	A	1	1.00	.67	.40	.19	.12	.25	.26	.39
N	A	2	.67	1.00	.50	.26	.20	.28	.26	.38
N	I	3	.40	.50	1.00	.52	.39	.31	.18	.24
G	I	4	.19	.26	.52	1.00	.55	.49	.25	.22
G	I	5	.12	.20	.39	.55	1.00	.46	.29	.14
G	A	6	.25	.28	.31	.49	.46	1.00	.42	.38
G	A	7	.26	.26	.18	.25	.29	.42	1.00	.40
G	A	8	.39	.38	.24	.22	.14	.38	.40	1.00

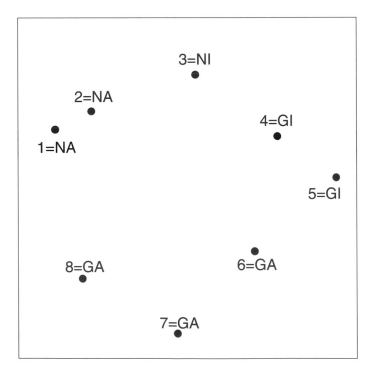

Figure 5: Two-dimensional MDS of correlations in Table 2.

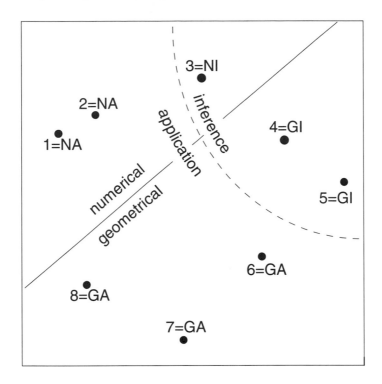

Figure 6: MDS space with four regions resulting from G versus N, and A versus I distinctions.

The correlations in Table 2 can be represented with stress = 0.015 in a two-dimensional MDS space (Figure 5). Figure 6 demonstrates that the MDS configuration can be cut such that each partitioning line splits it into two regions containing only points of one type: points of the N type lie above the solid line and points of the G type below that line. The dashed line separates I-type points from A-type points.

One notes, however, that there is considerable leeway in choosing the partitioning lines. Why, then, was a curved line chosen for separating I-type points from A-type points? The reason is that this line yields a structure that looks like a slice from the *universe* of all possible item types discriminated by the given two facets. If items of all nine types (including "learning" and "verbal") had been observed, one can predict that the MDS configuration would form a pattern similar to a dartboard, a radex, shown schematically in Figure 7. If, in addition, one added another facet ("communication") that distinguishes among oral, manual, and paper-and-pencil items, one would obtain the three-dimensional cylindrex shown in Figure 8. In the cylindrex, "communication" plays the role of an axis along which the radexes for items using a fixed form of communication are stacked on top of one other.

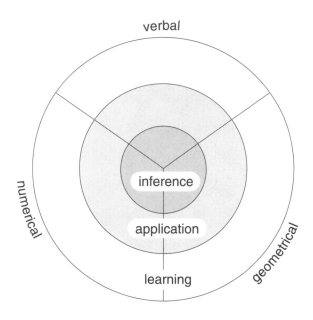

Figure 7: Schematic radex of intelligence items.

4 Regional Laws

The cylindrex structure has been confirmed so often for intelligence test items that now it is considered a regional law (Guttman and Levy, 1991). What Figure 6 shows, therefore, is a partial replication of the cylindrex law.

What does such a regional law mean? First of all, it reflects regularities of the data. For example, restricting oneself to items formulated in a particular language (such as paper-and-pencil tests) and, thus, to a radex as in Figure 7, one notes that inference items generally correlate higher among each other than application items, and learning items are least correlated. Thus, knowing that some person performs well on a given inference item allows one to predict that he or she will most likely also perform well on other inference items, whereas good performance on a given learning item says little about the performance on other learning items. One can improve the predictions, however, if one constrains them to learning tasks that use a particular language of presentation.

One notes, moreover, that the MDS regions for inference, application, and learning are ordered. This order cannot be predicted or explained from the properties of the qualitative facet "required mental operation." Nevertheless, it reliably shows up in hundreds of replications (Guttman and Levy, 1991) and, thus, asks for an expla-

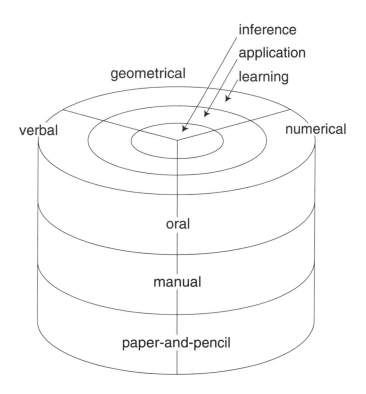

Figure 8: Cylindrex of intelligence items (after Guttman and Levy, 1991).

nation. Snow *et al.* (1984) reported a factor analysis that shows that items which relate to points in the center of the radex (i.e., inference tasks) are "complex" items and those represented at the periphery (such as learning tasks) are "specific" items. This repeats, to some extent, what the radex says: items whose points are closer to the origin of the radex tend to be more highly correlated with other items. Snow *et al.* (1984) add, however, that more complex tasks show "increased involvement of one or more centrally important components." Hence, their explanation for the inference–application–learning order seems to be that these facet elements are, in fact, discrete semantic simplifications of a smooth gradient of complexity.

One can ask the complexity question in a different way, and define a task t_1 as more complex than t_2 if "it requires everything t_1 does, and more" (Guttman, 1954, p. 269). Formally, this implies an interlocking of content structuples, analogous to the perfect Guttman scale. Specifying such structuples requires one to identify basic content facets with a common range, where the concepts inference, application, and learning then become only global labels for comparable (hence ordered) content structuples of these underlying facets. For a fixed element of the language facet, such a system would allow one to predict a particular order of regions (simplex).

But, then, this leads to the question of what pulls the different simplexes—one for each type of required mental operation, that is, one for items that require application, learning, or inference of an objective rule, respectively—to a common origin? To explain this empirical structure requires an additional pattern in the structuples. Formally, for the three directions of the intelligence radex, it would suffice to have an additional coding of the items in terms of the extent to which they require each of the three mental operations.

In any case, with many points and/or differentiated facets, a simple correspondence between regions and structuples is a remarkable finding. Arbitrary assignments of structuples to the points do, in general, not lead to such lawfulness. Partitionings with relatively smooth cutting lines are generally also more reliable. Moreover, they help clarify the roles the various facets play with respect to the data. Such roles are reflected in the particular ways in which they cut the space.

5   Alternative Facets

A given object of interest can always be facetized in more than one way. Every new facet offers a new alternative. But does each new facet also have a new statistical effect? Consider an example. Work value items require the respondent to assess the importance of different outcomes of his or her work. Conceptually, two different kinds of facets have been proposed for organizing such items: one facet distinguishes the work outcomes in terms of the need they satisfy, and the other facet is concerned with the allocation criterion for rewarding such outcomes. Consider Table 3, where 12 common work value items are coded in terms of seven facets. The facets and the structuples were taken from the literature on organizational behavior (Borg and Staufenbiel, 1993).

Figure 9 shows a two-dimensional MDS representation for the correlations of the 13 work value items assessed in a representative German sample. The radex partitioning is based on the facets M (solid radial lines), R (dashed radial lines), and L (concentric lines). It is easy to verify that the other facets also induce perfect and simple partitionings of this configuration. These partitionings are, moreover, quite similar: the respective regions turn out to be essentially congruent, with more or with fewer subdivisions. Differences of the various wedgelike partitionings are primarily related to the outcome advancement, which is most ambiguous in terms of the need that it satisfies. Hence, one can conclude that all these theories are structurally quite similar in terms of item intercorrelations. This suggests, for example, that Herzberg's motivation and hygiene factors correspond empirically to Elizur's cognitive and affective/instrumental values, respectively.

We note, moreover, that the similar partitioning of the MDS space into wedgelike regions, induced by different facets that are formally not equivalent, gives rise to a partial order of the induced sectors. The interlocking of the Herzberg and the

Table 3: Work value items with various facet codings[a]

No.	H	M	A	E	R	L	B	Work value
1	m	a	g	k	i	g	3	Interesting work
2	m	a	g	k	i	g	3	Independence in work
3	m	a	g	k	i	g	3	Work that requires much responsibility
4	m	a	g	k	i	n	4	Job that is meaningful and sensible
5	m	r	g	k	e	i	1	Good chances for advancement
6	m	r	r	a	s	i	1	Job that is recognized and respected
7	h	b	r	a	s	n	4	Job where one can help others
8	h	b	r	a	s	n	4	Job useful for society
9	h	b	r	a	s	n	4	Job with much contact with other people
10	h	s	e	i	e	i	2	Secure position
11	h	s	e	i	e	i	1	High income
12	h	p	e	i	e	n	4	Job that leaves much spare time
13	h	p	e	i	e	n	4	Safe and healthy working conditions

[a]H(erzberg) = {h = hygiene, m = motivators}; M(aslow) = {p = physiological, s = security, b = belongingness, r = recognition, a = self-actualization }; A(lderfer) = {e = existence, r = relations, g = growth}; E(lizur) = {i = instrumental–material, k = cognitive, a = affective–social}; R(osenberg) = {e = extrinsic, i = intrinsic, s = social}; L(evy-Guttman) = {i = independent of individual performance, g = depends on group performance, n = not performance dependent}; B(org-Elizur) = {1 = depends much on individual performance, 2 = depends more on individual performance than on system, 3 = depends both on individual performance and on system, 4 = depends on system only}.

Maslow facets implies, for example, that the hygiene region contains the subregions physiological, security, and belongingness, while the motivators region contains the subregions recognition and self-actualization. Hence, the subregions are forced into a certain neighborhood relation that would not be required without the hierarchical nesting.

Elizur *et al.* (1991) report further studies on work values, conducted in different countries, which show essentially the same radex lawfulness. Note that this does not imply similarity of MDS configurations in the Procrustean sense in which configurations can be brought, by admissible transformations, to a complete match, point by point. Rather, what is meant here is that several configurations—which do not even have to have the same number of points—exhibit the same law of formation: they can all be partitioned in essentially the same way (i.e., in the sense of a radex) by just one fixed coding of the items, thus showing similar contiguity patterns.

6 Prototypical Roles of Facets

The axial partitioning shown in Figure 3 can be seen as a primitive Cartesian coordinate axis. With more and more ordered categories in the affective-tone facet, there

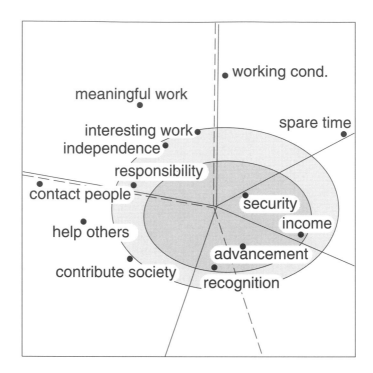

Figure 9: Radex partitionings of 13 work value items.

should be correspondingly more parallel regions and thereby an ever closer approximation to a Cartesian axis. Moreover, in the complete four-dimensional MDS space, the plane orthogonal to the one shown in Figure 3 is partitioned into four quadrants by the facets "number" and "variability." This further strengthens the hypothesis that these facets essentially suggest the usual (Cartesian) dimension system.

Another coordinate system is suggested if we accept the circular partitioning shown in Figure 4. In this case, the three effective facets give rise to a cylindrical coordinate system.

Mathematically, it is immaterial in which way a multidimensional figure is coordinated. In the given example, however, the coordination was not chosen arbitrarily. Rather, it was based on a priori distinctions of content. We stress this point here because the data determine only the distances among the points, not any "dimensions." Dimensions are imposed on a distance geometry for different reasons. One reason is computational and serves the purpose of being able to replace ruler-and-compass construction methods by computation. The other reason is interpretational and builds on imposing content onto the geometry.

The content facets often play one of three prototypical roles in this context. This is shown in the three panels of Figure 10. The panels exhibit schematic facet diagrams, whose points are labeled *a*, *b*, and *c*. In the panel on the left-hand side,

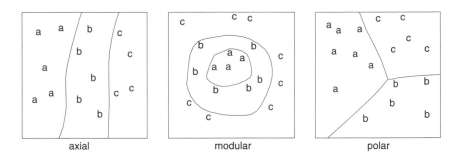

Figure 10: Three prototypical roles of facets in partitioning a facet diagram: axial (left), modular (center), and polar (right).

the space is partitioned in an axial way. The panel in the center shows a modular partitioning. The panel on the right-hand side shows a polar facet. An axial facet is one that corresponds to the usual linear dimension, cutting the space into parallel stripes (axial simplex of regions). A modular facet leads to a pattern that looks like a set of concentric bands (radial simplex of regions). Finally, a polar facet cuts the space, by rays emanating from a common origin, into sectors, similar to cutting a pie into pieces (circumplex of regions).

A number of particular combinations of facets that play such roles lead to structures that were given special names because they are encountered frequently in practice. For example, the combination of an angular facet and a radial facet in a plane, having a common center, constitutes a radex. Adding an axial facet in the third dimension renders a cylindrex. Another interesting structure is a multiplex, a conjunction of at least two axial partitionings. Special cases of the multiplex are called duplex (two axial facets), triplex (three axial facets), and so on. The multiplex corresponds to the Cartesian coordinate system as a special case if the facets are (densely) ordered and the partitioning lines are straight, parallel, and orthogonal to each other.

There are also structures that are found less frequently in practice, such as the spherex (polar facets in three-dimensional space) and the conex (similar to the cylindrex, but with radexes that shrink as one moves along the axial facet).

7   Regions, Clusters, and Factors

The notion of a region is quite general. Clusters are just special cases of regions. Lingoes (1981) defined a cluster as a particular region whose points are all closer to each other than to any point in some other region. This makes the points in a cluster look relatively densely packed, with empty space around the cluster. For regions, such a requirement is generally not relevant. All they require is a rule that allows one to decide if a point lies within or outside the region. The points 5 and 6 in Figure 5 are in different regions, but complete linkage clustering, for example, puts them into one cluster together with point 4, while it assigns points 7 and 8 to another cluster. For regions, the distance of two points—on which clustering is based—does not matter. Indeed, two points can be very close and still be in different regions. Conversely, two points may be far apart and yet belong to the same region. Moreover, clusters are usually identified on purely formal criteria, whereas regions are always based on substantive codings of the represented objects. Guttman (1977) commented therefore as follows: "theories about non-physical spaces ... generally call for continuity, with no vacuum or no clear separation between regions.... The varied data analysis techniques going under the name of cluster analysis generally have no rationale as to why systematic clusters should be expected at all .... The term cluster is often used when region is more appropriate, requiring an outside criterion for delineation of boundaries" (p. 105).

Factors from factor analyses are not directly related to regions or to clusters. However, it is often asked in practice what one would have found if one had analyzed a correlation matrix by a factor analysis rather than by MDS. Factor analysis, like cluster analysis, is a procedure that is substantively "blind" or that, if used in a confirmatory way, forces a preconceived formal structure onto the data representation, namely factors. The factors are (rectilinear) dimensions that are run through point clusters, usually under the additional constraint of mutual orthogonality. For Table 2, a factor analysis yields three factors with eigenvalues greater than 1. After Varimax rotation, one finds that these factors correspond to three clusters in Figure 5, $\{1, 2, 3\}$, $\{4, 5, 6\}$, and $\{6, 7, 8\}$. Hence, in a way, the factors correspond to a polar partitioning of the MDS configuration in the given case, with three factors or regions in a two-dimensional MDS space. With positive correlation matrices, this finding is rather typical; that is, one can expect $m + 1$ factor-induced regions in an m-dimensional MDS space. The reason for this is that positive correlations are conceived of in factor analysis as a vector bundle that lies in the positive hyperoctant of the Cartesian representation space, whereas MDS—which does not fix the origin of the space— looks only at the surface that contains the vector endpoints. Thus, Figure 5 roughly shows the surface of a section of the sphere whose origin lies somewhere in the center of the points but behind (or above) the plane. The factors, then, correspond to a tripod fixed to the origin and rotated such that its axes lie as close as possible to the points. Hence, one notes that the location of this dimension system depends very much on the distribution of the points in space, while this is irrelevant for regions, although, of course, a very uneven distribution of the points in space will influence the MDS solution through the stress criterion.

8   Discussion

Partitionings of geometric configurations that consist of only a few points are easy to find but they leave the exact shape of the partitioning lines quite indeterminate. More determinacy and greater falsifiability are brought in by increasing the number of items. Another principle for restricting the choice of partitioning lines is to think beyond the sample. In Figure 6, the partitioning lines were chosen, in part, by considering the universe of all intelligence items, a cylindrex (Figure 8).

The system of partitioning lines should, in any case, not attend too much to the particular sample. "Simple" partitionings with relatively smooth cutting lines are typically more robust. But what is simple? Surely, a regionalization consisting of simply connected regions as in an axial or an angular system is simple, but so are the concentric bands of a circumplex. Hence, simple means, above all, that the partitioning is simple to characterize in terms of the roles of the facets that induce the regions. Naturally, if one admits greater local flexibility for the partitioning lines, then the number of errors of classification can generally be reduced. However, irregular *ad hoc* partitionings also reduce the likelihood to find similar structures in replications and in the universe of items.

Admitting very irregular lines also makes it difficult to reject a regional hypothesis. Generally, partitionings become more unlikely to result from chance the more points they classify correctly, the more differentiated the system of facets is, the simpler the partitioning lines are, and the greater the stability of the pattern is over replications.

In addition, the pattern of regions should make sense. Irregular lines are difficult to characterize and make it hard to formulate the role of the respective facet. For the intelligence items, in contrast, the radial order of inference, application, and learning is not only simple and replicable but also seems to point to an ordered facet "complexity," where inference is the most complex task (see earlier). If application items, then, come to lie in the radex center, such further search for substantive meaning is thwarted.

It would be desirable to have an MDS procedure that not only represents the similarity data optimally by distances of an MDS space but also enforces certain regionalities onto the MDS solution. For axial facets one can enforce an appropriate regionality by linear constraints (de Leeuw and Heiser, 1980; Borg and Groenen, 1997). A general solution, however, is not known.

A correspondence between data and content categories can also be established a posteriori. One may recognize certain groupings or clusters in the points and then think about a rationale afterward to formulate new hypotheses. When the definitional framework is complex, one typically does not predict a full-fledged regional system (like a cylindrex) unless past experience leads one to expect such a system. Rather, one uses a more modest strategy with exploratory characteristics and simply tries to partition the space, facet by facet, with minimum error and simple partitioning lines. Even more liberal and exploratory is the attempt to identify space partitions according to new content facets, not conceived in advance. The stability of such partitions is then tested in replications.

Establishing a regional correspondence is one thing, but researchers typically also want to "understand" such regularities. Why, for example, are work values organized in a radex? An answer to this question can be derived, in part, from reasoning in Schwarz and Bilsky (1987). These authors studied general values. One of the facets they used was "motivational domain" = {achievement, self-direction, security, enjoyment, . . . }. These distinctions were considered nominal ones, but there was an additional notion of substantive opposition. Four such oppositions were discussed, for example, achievement versus security: "To strive for success by using one's skills usually entails both causing some change in the social or physical environment and taking some risks that may be personally or socially unsettling. This contradicts the concern for preserving the status quo and for remaining psychologically and physically secure that is inherent in placing high priority on security values" (p. 554). Hence, the region of achievement values was predicted to lie opposite to the security region. If we use this kind of reasoning post hoc on the work value radex of Figure 9, we can explain the opposite position of the sectors "r" and "a" (in Maslow's sense) by a certain notion of "contrast" of striving for self-actualization and for recognition, respectively.

To predict regional patterns requires one to clarify the expected roles of the facets in the definitional framework. This involves, first of all, classifying the scale level of each facet. For ordered facets, one predicts a regional structure whose regions are also ordered in some way, so that the statement that some region R comes "before" another region R' has meaning. The order of the regions should correspond to the order specified for the elements of the corresponding facet. For qualitative facets, any kind of simple partitionability of the point configuration into regions, each of whose points share the same facet element, is interesting. The distinction of facets into qualitative and ordinal ones represents a "role assignment" (Velleman and Wilkinson, 1994) that is "not governed by something inherent in the data, but by interrelations between the data and some substantive problem" (Guttman, 1971, p. 339), that is, by certain correspondence hypotheses linking the observations and the definitional system. Hence, if one can see a conceptual order among the facet's elements and hypothesize that this order is mirrored in the observations collected on corresponding items, then the facet "is" ordered—for testing the hypothesis. Scale level, thus, remains context related.

Consider as an example the facet "color" = {red, yellow, green, blue, purple}. One would be tempted to say, at first, that color "is" a nominal facet. Yet, with respect to similarity judgments on colors, "color" has been shown to be ordered empirically in a circular way. Furthermore, with respect to physical wavelengths of colors, "color" is linearly ordered.

Acknowledgments

The second author is supported by The Netherlands Organization for Scientific Research (NWO) by grant nr. 575-67-053 for the PIONEER project "Subject Orientated Multivariate Analysis."

Chapter 25

Visualizing Categorical Data with Related Metric Scaling

Carles M. Cuadras and Josep Fortiana

1 Prelude: Metric Scaling

Measuring straight line distances on a map with a ruler is an easy task. From the map in Figure 1, showing the locations of four European cities, we obtain the given distances. Such arrays of distances are common in road maps. Data that can be likened to distances are common in multivariate statistics, where they are often called dissimilarities. A dissimilarity matrix is a square, symmetric matrix of nonnegative data and has zeros on its diagonal.

Let us now consider the following question: given a dissimilarity matrix, such as the one in Figure 1, how can we reconstruct from it the map on which it is based?

Metric scaling, also called principal coordinate analysis (although as a rule the first term is used in a more general sense), is a technique that allows us to construct a map, or Euclidean configuration, from a matrix of dissimilarities. Sometimes this construction is not possible: a necessary condition for it is that the dissimilarities must obey the triangle inequality, in which case they are called distances.

Because the same set of distances can be obtained from several Euclidean configurations of points, one of them is selected as the usual metric scaling solution. The criterion used for this selection is explained in the following. In our example of four European cities, the solution is given in Figure 2.

The main advantage of metric scaling becomes apparent when we process a dissimilarity matrix that has not been obtained from actual measurements from a map. For instance, our "dissimilarities" could be "time spent traveling by car from one city to another" or "number of daily flights between two cities."

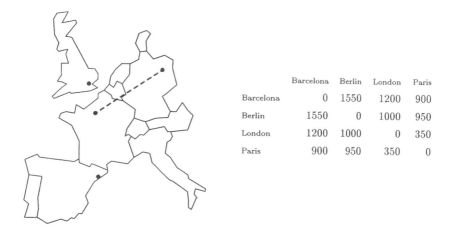

Figure 1: Geographical interdistances between four European cities. We can measure distances on a map ....

	Barcelona	Berlin	London	Paris
Barcelona	0	1550	1200	900
Berlin	1550	0	1000	950
London	1200	1000	0	350
Paris	900	950	350	0

Barcelona	-824.6	255.8
Berlin	720.9	373.0
London	148.5	-445.9
Paris	-44.76	-182.9

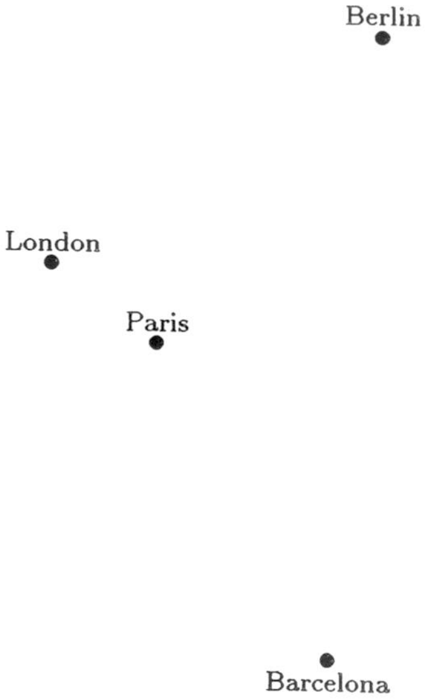

Berlin

London

Paris

Barcelona

Figure 2: ... or we can try to reconstruct a map from the distances between a set of objects. Euclidean configuration for four European cities obtained by metric scaling from the set of interdistances. In the left-hand side matrix of coordinates, the first column corresponds approximately to the N-S direction and the second column to the E-W direction; hence in the right hand diagram, the first and second coordinate have been plotted along the vertical and horizontal axes, respectively.

An exact Euclidean configuration for these general dissimilarities may require more than two dimensions. However, since a representation on a plane is still useful, from all exact Euclidean configurations we will choose one such that its first two coordinates give a best approximation to the original dissimilarities. More generally it is known that, if an exact Euclidean configuration requires p coordinates, the metric scaling configuration is characterized by the property that for each k, $1 \leq k \leq p$, its first k coordinates give the best k-dimensional approximation to the true distances.

2 Introducing Related Metric Scaling

Data often contain information that is duplicated in some sense: for example, (1) opinion polls before and after a political event, (2) results of elections classified by cities and geographical distances between the same cities, or (3) preferred leisure time activities of married couples, questioning husband and wife separately. For 1 we have different observations obtained at different times for the *same* objects and variables. For 2 we have the same objects but a *different* kind of distance matrix. For 3 we have the same variables observed for *paired* individuals.

Table 1 shows an artificial data set, consisting of the answers of six married couples A–a, B–b, C–c, D–d, E–e, F–f, to a survey on preferred leisure time activities.

Using metric scaling, we can obtain graphical representations of men (Figure 3, left-hand diagram) and women (Figure 3, right-hand diagram). We can observe that C and E share the same set of preferences whereas A and D differ widely, that for women a and b share the same preferences, and so on.

How can we represent the set of couples so that the information for husbands and wives is in the same display? A straightforward method is to join the left and

Table 1: Leisure time activity preferences expressed by six married couples (1 = "Yes, I enjoy," 0 = "I dislike/try to avoid").[a]

	Husbands				Wives		
	Traveling	Home	Sports		Traveling	Home	Sports
A	1	1	0	*a*	1	1	0
B	1	0	1	*b*	1	1	0
C	0	1	0	*c*	0	0	1
D	0	1	1	*d*	0	1	1
E	0	1	0	*e*	1	0	0
F	1	1	1	*f*	1	0	1

[a]Each row represents a married couple. Labels reflect this relationship, for example, the wife of *A* is labeled *a*.

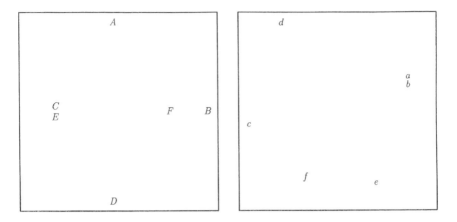

Figure 3: Metric scaling graphical representations of Table 1. The diagram for men appears on the left (percentage of variance 86.7%), and the diagram for women appears on the right (percentage of variance 92.8%.)

right halves of Table 1 and to perform a metric scaling with the resulting 6 × 6 data matrix containing all the data, yielding Figure 4a. For example, *Aa* now refers to the row of six elements corresponding to the first couple.

Another possibility is to use related metric scaling, an extension of metric scaling. Its aim is to analyze two distance matrices together, taking into consideration the

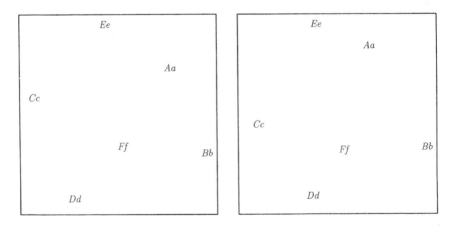

Figure 4: Metric scaling graphical representations of Table 1. The diagram for the whole data, using ordinary scaling (percentage of variance 72.3%), appears on the left, and the diagram obtained with related metric scaling, joining the two distance matrices (percentage of variance 65.7%), appears on the right.

Table 2: Average number of days per week on which pairs of wives meet

	a	b	c	d	e	f
a	7	6	1	2	4	3
b	6	7	1	2	4	3
c	1	1	7	2	3	4
d	2	2	2	7	2	3
e	4	4	3	2	7	2
f	3	3	4	3	2	7

possibility of redundant information. This leads to a related distance matrix, which can be represented using ordinary metric scaling.

Figure 4b shows the result of performing related metric scaling from the two distance matrices obtained from the two halves of Table 1. It is not surprising that this diagram is very similar to the one on its left, because both are obtained from the same whole data set.

To distinguish the differences, we note that the metric scaling representation of $Aa, \ldots, Ff$ according to the whole data set is equivalent to considering the six columns of Table 1 as if they were associated with preferences in six different activities, which is not the case, because they are associated in pairs: activities 1 and 4 (traveling), 2 and 5 (home), 3 and 6 (sports). For instance, it seems contradictory to admit that, for example, the Ff couple simultaneously prefers to stay at home and not to stay at home.

There are circumstances in which a related metric scaling is not only advisable but also the only possibility. Suppose that we prefer to relate the husbands' leisure activities to the information about the wives in Table 2: the average number of days per week on which each pair of women meet each other. Taking these figures as measuring similarities, for example, the similarity of a and b is $s(a, b) = 6$, we can easily convert them into dissimilarities by subtracting them from $7, d(a, b) = 7 - s(a, b)$, giving Table 3. Now the data for the husbands and wives are of different types, which prevents us from performing ordinary metric scaling.

By using related metric scaling, however, we can still obtain a representation of Table 3, as shown in Figure 5. More generally, related metric scaling can display any data consisting of, or convertible to, an associated pair of dissimilarity matrices. Since these two matrices are not independent of each other, we would like to relate them in the graphical display.

3 Description of Methodology

Metric scaling, or "classic scaling," originated in Schoenberg (1935), Young and Householder (1938), and Torgerson (1952, 1958) and was extended and related to other multivariate techniques by Rao (1964) and Gower (1966). Since then, it has

Table 3: Preferences expressed by six husbands for leisure time activities (1= "Yes, I enjoy," 0= "I dislike/try to avoid") and distance matrix between their wives, as explained in the text

| | Husbands | | | | Wives | | | | | |
	Traveling	Home	Sports		a	b	c	d	e	f
A	1	1	0	a	0	1	6	5	3	4
B	1	0	1	b	1	0	6	5	3	4
C	0	1	0	c	6	6	0	5	4	3
D	0	1	1	d	5	5	5	0	5	4
E	0	1	0	e	3	3	4	5	0	5
F	1	1	1	f	4	4	3	4	5	0

been widely applied in many disciplines and is considered a useful complement to cluster analysis. Descriptions of the method and its properties can be found in standard textbooks on multivariate analysis (Mardia *et al.*, 1979, p. 397; Seber, 1984, p. 235) and monographs (Davison, 1983; Cox and Cox, 1994).

Given n objects, $\{1, 2, \ldots , n\}$, say, and a distance matrix between them, $\mathbf{\Delta} = [\delta_{ij}]$, the aim of metric scaling is to find, for each object i, a set of coordinates in m

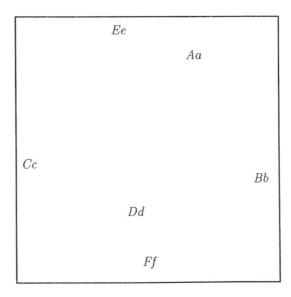

Figure 5: Related metric scaling graphical representation of Table 3 (percentage of variance 70.9%).

dimensions

$$\mathbf{x}_i = (x_{i1}, x_{i2}, \ldots, x_{im})^\mathsf{T} \tag{1}$$

such that δ_{ij} is equal to or closely approximated by

$$d_{ij} = \sqrt{\sum_{k=1}^m (x_{ik} - x_{jk})^2} \tag{2}$$

which is the Euclidean distance between $\mathbf{x}_i$ and $\mathbf{x}_j$. In practice, a two-dimensional graphical display is often used. The $n \times m$ matrix of coordinates $\mathbf{X} = [x_{ij}]$ is chosen in such a way that the first two coordinates give the best fit to the initial squared distance

$$\delta_{ij}^2 \approx d_{ij}(2)^2 = (x_{i1} - x_{j1})^2 + (x_{i2} - x_{j2})^2 \tag{3}$$

where $d_{ij}(2)$ is the Euclidean two-dimensional distance.

To derive the formulas to compute $\mathbf{X}$, with rows $\mathbf{x}_1^\mathsf{T}, \ldots, \mathbf{x}_n^\mathsf{T}$, let us write (2) as

$$d_{ij}^2 = ||\mathbf{x}_i - \mathbf{x}_j||^2 = \mathbf{x}_i^\mathsf{T}\mathbf{x}_i + \mathbf{x}_j^\mathsf{T}\mathbf{x}_j - 2\mathbf{x}_i^\mathsf{T}\mathbf{x}_j. \tag{4}$$

The $n \times n$ matrix $\mathbf{S} = [s_{ij}]$, with $s_{ij} = \mathbf{x}_i^\mathsf{T}\mathbf{x}_j$, is called the inner product matrix associated with $\mathbf{\Delta} = (\delta_{ij})$. With this notation, (4) becomes

$$d_{ij}^2 = s_{ii} + s_{jj} - 2s_{ij} \tag{5}$$

If we impose on $\mathbf{X}$ the condition $\sum_{i=1}^n \mathbf{x}_i = \mathbf{0}$, in order to obtain a centred configuration, we have the equality

$$\sum_{i=1}^n s_{ij} = \sum_{j=1}^n s_{ij} = 0$$

which allows us to solve (5) for s_{ij} by taking row, column, and overall averages, as is the usual procedure for analogous equations found in classical ANOVA and log-linear models. The result is

$$s_{ij} = -\frac{1}{2}\left(\delta_{ij}^2 - \overline{\delta^2}_{i\cdot} - \overline{\delta^2}_{\cdot j} + \overline{\delta^2}_{\cdot\cdot}\right)$$

where $\overline{\delta^2}_{i\cdot}$, $\overline{\delta^2}_{\cdot j}$, and $\overline{\delta^2}_{\cdot\cdot}$ are the row, column, and overall averages of the two-way table $[\delta_{ij}^2]$, respectively. Thus, s_{ij} is computed directly from the distances δ_{ij}.

The next step in metric scaling is to find the spectral decomposition $\mathbf{S} = \mathbf{U}\mathbf{\Lambda}\mathbf{U}^\mathsf{T}$, where $\mathbf{U}$ is the $n \times m$ matrix of orthonormal eigenvectors of the symmetric matrix $\mathbf{S}$ that correspond to the first m eigenvalues, ordered as $\lambda_1 \geq \lambda_2 \geq \cdots \geq \lambda_m > 0$, $m \leq n - 1$, contained in the diagonal matrix $\mathbf{\Lambda}$. The metric scaling solution is the matrix

$$\mathbf{X} = \mathbf{U}\mathbf{\Lambda}^{1/2}, \quad \text{where} \quad \mathbf{\Lambda}^{1/2} = \text{diag}(\sqrt{\lambda_1}, \sqrt{\lambda_2}, \ldots, \sqrt{\lambda_m}) \tag{6}$$

Note that $\mathbf{S}=\mathbf{X}\mathbf{X}^\mathsf{T}$ is the matrix of scalar products $s_{ij} = \mathbf{x}_i^\mathsf{T}\mathbf{x}_j$, implying that the set $\{\mathbf{x}_i^\mathsf{T}\}$ of rows of $\mathbf{X}$ satisfies (4) and, equivalently, (2).

On the other hand, each column $\mathbf{X}_j$ of $\mathbf{X}$ can be understood as a "variable," which takes the value x_{ij} on the set of individuals $i, i = 1,\ldots,n$. The condition imposed, that the sum of rows of $\mathbf{X}$ is null, is equivalent to the column means being zero: $\overline{\mathbf{X}}_j = 0, j = 1,\ldots,m$. In addition, taking into account the definition (6) and the orthonormality of the columns of $\mathbf{U}$, we obtain the variances and covariances

$$\text{var}(\mathbf{X}_j) = \mathbf{X}_j^\mathsf{T}\mathbf{X}_j/n = \lambda_j/n, \qquad j = 1,\ldots,m$$

$$\text{cov}(\mathbf{X}_j, \mathbf{X}_k) = \mathbf{X}_j^\mathsf{T}\mathbf{X}_k/n = 0, \qquad j,k = 1,\ldots,m, \qquad j \neq k$$

which allow us to interpret the variables $\mathbf{X}_j$ as principal components.

Since the first two columns $(\mathbf{X}_1, \mathbf{X}_2)$ of the metric scaling solution (6) are associated with the two largest eigenvalues (i.e., the two largest variances), they give the best two-dimensional approximation, as required in (3). A measure of the quality of the approximation is given by the percentage of variance: $100(\lambda_1 + \lambda_2)/(\lambda_1 + \cdots + \lambda_m)$. For instance, in Figure 2 the percentage of variance is 99.5%, hence the picture is an accurate representation of the set of four European cities.

Figures 2 to 4a were obtained by the methods described up to this point. The distance used in Figure 3 is deduced from the matching coefficient between individuals; that is, the squared distance equals the number of variables minus the number of coincident values of their coordinates, computed from Table 1. For example, the distance between A and B is $\delta(A, B) = \sqrt{3 - 1} = \sqrt{2}$, since there are three coordinates, and A and B agree in one of them. Similarly, using the right-hand part of Table 1, $\delta(a, b) = \sqrt{3 - 3} = 0$, and for Figure 4a, using the six columns of Table 1, $\delta(Aa, Bb) = \sqrt{6 - 4} = \sqrt{2}$.

Suppose now that we have two $n \times n$ distance matrices $\mathbf{\Delta}_A = (\delta_A(i, j))$, $\mathbf{\Delta}_B = (\delta_B(i, j))$, which are defined either on the same finite set or on two different sets with the same number n of objects, paired between them. The two preceding examples cover both possibilities. Our objective here is to construct a joint $n \times n$ distance matrix $\mathbf{\Delta}_{AB} = (\delta_{AB}(i, j))$, which allows us to represent the n objects in a single graphic display, relating the displays obtained from $\mathbf{\Delta}_A$ and $\mathbf{\Delta}_B$.

The problem of constructing $\mathbf{\Delta}_{AB}$ is similar to that of constructing a joint probability distribution given its marginals. These constructions must follow some compatibility rules and often a dependence structure is imposed (Cuadras, 1992). Another example of this type of construction is the iterative proportional fitting procedure for adjusting a multivariate contingency table by maximum likelihood to a hierarchical log-linear model, where the set of marginals is determined by the given model and their actual values are computed from the observed table (see, e.g., Bishop et al., 1975, sect. 3.5).

We propose the following properties for δ_{AB}, with marginal distances δ_A and δ_B:

1. If $\delta_A = 0$ then $\delta_{AB} = \delta_B$; if $\delta_B = 0$ then $\delta_{AB} = \delta_A$.
   Comment: if all the objects are identical under δ_A, then this distance has no influence on the joint distance.

2. If $\delta_A = \delta_B$, then $\delta_{AB} = \delta_A = \delta_B$.
   Comment: if the distances are the same under δ_A and δ_B, then the joint distance must maintain these values.
3. If the principal coordinates obtained from δ_A and those obtained from δ_B are orthogonal, then $\delta_{AB}^2 = \delta_A^2 + \delta_B^2$.
   Comment: this is Pythagoras' theorem. If $\mathbf{X}_A$ is obtained from δ_A and $\mathbf{X}_B$ is obtained from δ_B, the orthogonality condition is $\mathbf{X}_A^\mathsf{T}\mathbf{X}_B = \mathbf{0}$.

There are many joint distances satisfying these conditions. Here we propose one. Let $\mathbf{S}_A$, $\mathbf{S}_B$ be the inner-product matrices associated with $\mathbf{\Delta}_A$ and $\mathbf{\Delta}_B$, respectively. Then the matrices of principal coordinates $\mathbf{X}_A$ and $\mathbf{X}_B$ satisfy $\mathbf{S}_\alpha = \mathbf{X}_\alpha\mathbf{X}_\alpha^\mathsf{T} = \mathbf{U}_\alpha\mathbf{\Lambda}_\alpha\mathbf{U}_\alpha^\mathsf{T}$, $\alpha = A, B$. We define the joint distance δ_{AB} between two objects i and j, whose coordinates are $\mathbf{x}_i$ and $\mathbf{x}_j$ with respect to δ_A and $\mathbf{y}_i$ and $\mathbf{y}_j$ with respect to δ_B, by

$$\delta_{AB}^2(i, j) = \delta_A^2(i, j) + \delta_B^2(i, j) - \tau_{AB}(i, j) \tag{7}$$

where

$$\tau_{AB}(i, j) = (\mathbf{x}_i - \mathbf{x}_j)^\mathsf{T}\mathbf{\Lambda}_A^{-1/2}\mathbf{X}_A^\mathsf{T}\mathbf{X}_B\mathbf{\Lambda}_B^{-1/2}(\mathbf{y}_i - \mathbf{y}_j) \tag{8}$$

encapsulates the dependence between the A and the B variables.

It can be proved that the joint distance defined by (7) satisfies properties 1, 2, and 3, provided that $\mathbf{\Delta}_A$ and $\mathbf{\Delta}_B$ have the same geometric variability, that is, if

$$\frac{1}{n^2}\sum_{i=1}^n\sum_{j=1}^n\delta_A^2(i, j) = \frac{1}{n^2}\sum_{i=1}^n\sum_{j=1}^n\delta_B^2(i, j)$$

(Cuadras and Fortiana, 1995a). Note that this condition can always be assumed to hold, because multiplying one of the marginal distances by an appropriate constant amounts to a change of measurement unit. In the illustrations of this chapter, the geometric variability of the second distance matrix has been equaled to that of the first distance matrix.

In addition, the inner product matrix $\mathbf{S}_{AB}$ associated with the matrix $\mathbf{\Delta}_{AB}$ of joint distances is given by

$$\mathbf{S}_{AB} = \mathbf{S}_A + \mathbf{S}_B - \frac{1}{2}\left(\mathbf{S}_A^{1/2}\mathbf{S}_B^{1/2} + \mathbf{S}_B^{1/2}\mathbf{S}_A^{1/2}\right) \tag{9}$$

where $\mathbf{S}_\alpha^{1/2} = \mathbf{U}_\alpha\mathbf{\Lambda}_\alpha^{1/2}\mathbf{U}_\alpha^\mathsf{T} = \mathbf{X}_\alpha\mathbf{\Lambda}_\alpha^{-1/2}\mathbf{X}_\alpha^\mathsf{T}$, $\alpha = A, B$. Finally, the related metric scaling solution $\mathbf{X}_{AB}$ is computed from the spectral decomposition of $\mathbf{S}_{AB}$.

4 An Empirical Application

We applied related metric scaling to a subset of data from a study about statistical research in Spain. The data matrix in Table 4 contains the number of papers published

Table 4: Number of papers published by 11 Spanish authors classified into 11 subjects of statistics.[a]

	ber	cua	gim	gip	gom	mor	oll	par	pen	sal	sat	Total
GE	4	0	0	0	0	0	0	0	3	0	2	9
PT	9	1	0	0	0	5	0	0	0	0	0	15
PD	1	6	0	0	0	2	0	0	0	0	3	12
ID	0	0	21	11	0	0	0	33	0	13	0	78
FS	0	0	16	9	0	0	0	5	0	0	0	30
SI	3	2	0	0	11	5	2	0	2	0	0	25
BS	27	0	0	2	0	8	0	8	8	0	0	53
MA	5	6	0	0	0	0	0	0	0	1	12	24
MS	0	6	0	0	0	0	9	0	0	0	0	15
RE	2	6	0	0	5	0	1	0	8	0	0	22
TS	0	0	0	0	1	0	0	0	15	0	0	16
Total	51	27	37	22	17	20	12	46	36	14	17	299

[a]Abbreviations for authors: ber = J. M. Bernardo, cua = C. M. Cuadras, gim = M. A. Gil, gip = P. Gil, gom = W. González–Manteiga, mor = E. Moreno, oll = J. M. Oller, par = L. Pardo, pen = D. Peña, sal = M. Salicrú, sat = A. Satorra. Abbreviations for subjects: GE = Mathematical methods, sampling, applications, general, PT = Probability theory, PD = Probability distributions, SI = Statistical inference, BS = Bayesian statistics, ID = Statistical information and divergences, FS = Fuzzy sets, MS = Multidimensional scaling and statistical distances, MA = Multivariate analysis, classification, RE = Regression, ANOVA, experimental designs, TS = Time series, modeling processes.

by 11 representative authors (columns) on 11 subjects (rows). The data were collected from the Extended Current Index of Statistics (CIS) Database (Thisted, 1994).

Figure 6a is the graphic display of the authors, obtained by ordinary metric scaling, from Table 4. To obtain a distance matrix, we computed first the *profile* of each author, that is, the proportion of papers on each of the subjects considered. For example, from Table 4 the profiles of ber and cua are

	GE	PT	PD	ID	FS	SI	BS	MA	MS	RE	TS
ber: =	[.078	.176	.020	.000	.000	.059	.529	.098	.000	.039	.000]
cua: =	[.000	.037	.222	.000	.000	.074	.000	.222	.222	.222	.000]

The distance between authors i and j, with profiles $\mathbf{p}_i = [p_{i1}, \ldots, p_{im}]^\mathsf{T}$, and $\mathbf{p}_j = [p_{j1}, \ldots, p_{jm}]^\mathsf{T}$, where $m = 11$, can be computed, for example, using the Hellinger distance:

$$\delta_H(i, j) = \sqrt{\sum_{k=1}^{m} \left(\sqrt{p_{ik}} - \sqrt{p_{jk}} \right)^2}$$

For example, $\delta_H(\text{ber, cua}) = \sqrt{1.092} = 1.045$.

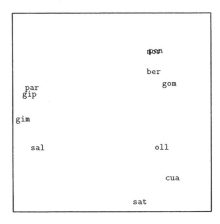 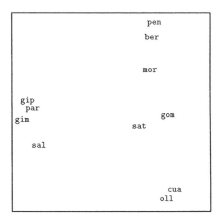

Figure 6: Two–dimensional Euclidean representations of authors: (a) using only the first (Hellinger's) distance matrix (percentage of variance 60.5%) and (b) related metric scaling representation (percentage of variance 43.1%).

The Hellinger distance is one of the classic distances between probability distributions. Rao (1995) proposed its application to represent graphically a set of rows or columns of a frequency table. Principal coordinate analysis of the matrix of Hellinger distances between a set of profiles is an alternative to correspondence analysis. Correspondence analysis can be defined as the metric scaling of the chi-squared distances between profiles, where each profile is weighted (Greenacre, 1984). Hellinger distance is a sensible choice when, as is the case in Table 4, the frequency table has a product–multinomial structure, that is, each column contains measurements on a different individual or case.

In Figure 6a, three clusters are apparent: center left, right top, and right bottom. These clusters can be associated with the subjects (ID,FS), BS and (MS,MA), respectively. However, this display is not a faithful representation, because several authors have published joint papers, the information on them is not independent, and we should correct for this fact. Therefore, in addition to Table 4, we consider, for each pair of authors the number of papers published jointly.

Only six authors in the selected set have written joint papers, as shown in the lower triangle of Table 5. The raw information contained in the lower triangle and diagonal of Table 5 can easily be converted into dissimilarity data. For instance, we can define

$$\delta(i, j) = 1 - a_{ij} / \min\{a_{ii}, a_{jj}\} \tag{10}$$

where a_{ij} is the number of joint papers by authors i and j and a_{ii} is the number of individual papers by author i. For example, for $i = 2$ and $j = 7$ (authors cua and oll), we have $\delta(2, 7) = 1 - 5/\min\{27, 12\} = 1 - 0.417 = 0.583$. The upper triangle of Table 5 contains the resulting dissimilarity matrix.

Table 5: Number of joint papers by 11 authors (lower triangle and diagonal) and distance matrix (upper triangle) computed from (10); total number of papers by each author on the diagonal

	ber	cua	gim	gip	gom	mor	oll	par	pen	sal	sat
ber	51	1	1	1	1	1	1	1	1	1	1
cua	0	27	1	1	1	1	0.583	1	1	0.929	1
gim	0	0	37	0.546	1	1	1	1	1	1	1
gip	0	0	10	22	1	1	1	1	1	1	1
gom	0	0	0	0	17	1	1	1	1	1	1
mor	0	0	0	0	0	20	1	1	1	1	1
oll	0	5	0	0	0	0	12	1	1	1	1
par	0	0	0	0	0	0	0	46	1	0.571	1
pen	0	0	0	0	0	0	0	0	36	1	1
sal	0	1	0	0	0	0	0	6	0	14	1
sat	0	0	0	0	0	0	0	0	0	0	17

Again, as in the artificial example in Section 2, we have two sources of information on the given set of individuals: Table 4 and the upper triangle of Table 5. These data are of two different types: Table 4 contains information on individuals, and Table 5 contains information on pairs of individuals. Related metric scaling provides a way to mix these two types of information, taking into account possible redundancies.

Figure 6b is the graphical representation of the set of authors by related metric scaling. We can appreciate that the pairs (cua, oll) and (gim, gip) are now slightly closer, as they have jointly authored papers, and gom now occupies a more isolated position, consistent with the fact that this author has produced no joint papers with the remaining authors in the analyzed set. The apparently larger displacement of sat, who also has no joint papers with other authors, and his approximation to gom are due to the loss of variability incurred when projecting on a plane. To see the isolation of gom and sat, more dimensions would have to be visualized.

Chapter 26

Contrasting the Electorates of Eight Political Parties: A Visual Presentation Using the Biplot

Magda Vuylsteke-Wauters, Jaak Billiet, Hans de Witte, and Frans Symons[1]

1 Introduction

Belgian election studies of the Flemish voter's perceptions and attitudes revealed that the ecologist (green) party "Agalev" and the radical right-wing party "Vlaams Blok" were each other's antipodes. In the 1991 general election in Flanders, these two electorates were polarized by their attitudes toward immigrants, materialism and postmaterialism, economic conservatism, and Flemish nationalism (Billiet and de Witte, 1995). These findings partially support both the thesis about the emergence of two new cleavages (universalism/particularism and postmaterialism/materialism) and the finding of a new right–left cleavage. What happened to the "old" cleavages that divided Belgian society so sharply in the past (Lorwin, 1971)? Do the ideological conflicts between church and state, beween labor and capital, and between the linguistic communities (Dutch speakers and the Francophones) no longer play a role? Are these old cleavages no longer relevant and are the values on which they were built completely replaced by new value orientations (see van Deth 1995)? In order to determine the relevance and dominance of the hypothesized new and old cleav-

---

[1]Frans Symons passed away in July 1997.

ages, we will analyze the distributions of the electorates with respect to 14 attitude scales expressing value orientations that are related to the old and new cleavages. We are using the term "cleavages" in the sense of value cleavages, because we will confine ourselves to attitudes and value orientations, neglecting social background variables and institutional ties (see Lipset and Rokkan, 1967; Knutsen and Scarbrough, 1995, p. 497). How well can these 14 attitude scales cleave the voters into the eight electorates, and how can we arrive at a clear visualization of the large amount of information? In a previous study, Billiet and de Witte (1995) used logistic regression in order to predict the odds ratios of voting for each party. They used the whole set of attitude scales as predictors. With logistic regression, however, the visualization of the results remains a real problem. In this chapter, we will offer a biplot representation based on an original view of canonical correlation analysis as a combination of projection and rotation methods (Vuylsteke-Wauters, 1994). It is demonstrated that this approach is capable of displaying the complexity of the data containing two sets of variables.

2 Data and Measurements in the Flemish Voters' Study

In late 1991 and early 1992 a national survey was conducted of voting and political attitudes in Belgium involving 2691 interviews in the Flemish region. The sample was constructed with the equal probability method and was representative of all adults 18–74 years old (Carton *et al.*, 1993). A two-stage sample with equal probabilities was used. In the first stage, the municipalities were selected at random. About 120 of the 316 Flemish communities were included in the sample. In the second stage, a random sample of respondents was selected from the national population registers. The response rate was 64%. The interviewers were trained in an approved experimental training program developed by the research group.

The 14 attitude scales that are used here are based on sets of items selected via tests of measurement models with confirmatory factor analysis (see, for example, Bollen, 1989). The confirmatory factor analysis was preceded by an exploratory factor analysis based on another part of the sample. The complete set of items and scales, as well as the measurement models, are documented by Billiet and de Witte (1995).

The voters' study contained an index of religious identification and church attendance, which is related to a set of three attitude scales that can be identified as indicators of the concept "sociocultural conservatism" (Middendorp, 1991). These attitudes are the rejection of the liberalization of abortion (abbreviated henceforth as ABORTION), the preference for clearly distinct roles between men and women in society (SEXROLES), and an aversion to the free expression of opinions in public (NOFREEOP). Because of their relationship with church involvement, they can be used as attitudinal expressions of the value orientations behind the first old cleavage (Middendorp, 1991).

The second old cleavage refers to the conflict between labor and capital. This dimension was operationalized by four attitude items that constitute one scale regarding socioeconomic conservatism (ECONCONS) in the sense of the rejection of socioeconomic equality and the wish to limit the influence of labor unions and the government.

The linguistic conflict, the third old cleavage, was operationalized by a scale in which the respondents had to indicate whether they wanted Flanders (or Belgium) to decide everything itself (FLABELG). This scale was strongly related to a set of items that measured other aspects of Flemish nationalism but that were not presented to all respondents.

The postmaterialist value orientation (POSTMAT) was measured by the number of postmaterialistic objectives the respondents chose from a list of 12 materialistic and postmaterialistic political objectives (see Inglehart, 1987, 1990). A particular aspect of postmaterialism is readiness to make social and financial sacrifices for the preservation of the sociophysical environment. This orientation was measured by a set of six items (MILIEU). Together with the postmaterialist value orientation, this ecological orientation is considered to be an operationalization of the first new cleavage: "postmaterialism versus materialism."

The potential second new cleavage was operationalized by four attitude scales: a negative attitude toward immigrants in the sense of feeling threatened by the presence of immigrants in the acquisition of scarce goods such as jobs, social security benefits, housing, and culture (OUTGROUP); biological racism, or the idea that the white race is superior and has to be kept pure (SUPRACE); the emphasis on traditional values and principles, such as authority and respect for law and order (AUTHORIT); the idea that one is insufficiently protected against petty criminality (PETTYCRI). The authoritarianism scale is a short version of the California F-scale (Adorno *et al.*, 1950) designed to measure "potential fascism" (see, for example, Meloen *et al.*, 1994). Previous research confirmed the relationship of biological racism with all other aspects of the extreme right-wing ideology.

The last relevant but more or less hybrid dimension refers to social and political indifference and distrust. Three scales were selected to operationalize this dimension. One was a scale measuring political inefficacy (Campbell *et al.*, 1954, p. 187) and expressing feelings of powerlessness in the domain of politics (POWERLES). The second was a scale measuring utilitarian individualism (Bellah *et al.*, 1985): being driven purely by self-interest and personal material success are core elements of this kind of individualism (INDIVID). This attitude reflects a pessimistic and even misanthropic world view. The third was a scale measuring feelings of social isolation (SOCISO). This aspect reflects the experience of disintegration of the traditional social networks. Each of these three scales is related to Srole's concept of "anomia" (Srole, 1956) and may be associated with protest voting. All scales were transformed into 11-point scales ranging from 0 to 10, with zero indicating the highest level of disapproval with the scale content and 10 indicating the highest level of agreement.

The second set of variables deals with preferences, that is, the respondents' voting behavior in the 1991 general elections, which is recoded in a set of dummy variables. In the Flemish part of Belgium, there were seven major political parties in 1991: the Christian Democrats (CVP), the Liberal Party (PVV), the Socialist Party (SP), the traditional Flemish nationalists of the "Volksunie" (VU), the ecology party (Agalev), the extreme right-wing party the "Vlaams Blok," and the Libertarians (Rossem). Those who voted blank or turned in invalid votes can be considered as a specific electorate as well. Voting is compulsory in Belgium, but a number of voters (about 7% in 1991) cast their ballots without filling them in (i.e., blank vote) or rendered them invalid by writing comments on them.

The Christian Democratic Party (CVP) is the heir of the Catholic Party that originated from the ideological conflict between church and state at the beginning of the 19th century. The degree of church involvement and membership of Christian organizations are still important predictors of voting for this party.

The conflict between labor and capital at the turn of the 19th century forced the liberal party at that time to split into two parties: the socialists and the liberals. The Socialist Party (SP) converted its ideology from Marxism to a broader progressive stand on socioeconomic issues in which the basic value of social equality is stressed. Its voters tend to be nonbelievers or marginal Catholics who participate in socialist organizations. The liberal Party for Freedom and Progress (PVV) takes a liberal–conservative stand on socioeconomic issues, with a focus on socioeconomic freedom and on the restriction of the influence of the state in this domain. This party appeals to the higher strata and the more highly educated.

The Volksunie (VU), which promotes the interests of the Flemish, is a result of the third cleavage: the linguistic or communitarian conflict in Belgium between the Dutch-speaking and the French-speaking parts of the country.

The ecological or "green" party Agalev is perceived as the result of the rise of a new value orientation in Belgian society that stresses postmaterialism (see, for example, Inglehart 1987, 1990). The electorate of this party is rather young and non-Catholic and originates from the higher strata and educational levels.

Also, the rise of the extreme right-wing party the Vlaams Blok has often been attributed to a new cleavage. More than 50% of the Vlaams Blok voters mentioned their aversion to immigrants as the main reason for their electoral choice (Swyngedouw, 1992). In explaining the rise of the Vlaams Blok, some studies refer to feelings of political inefficacy as part of a broader, more encompassing new alignment of attitudes and value orientations such as individualism, ethnocentrism, and authoritarianism.

Finally, the rise of a clear protest party, such as the Libertarians, Rossem, and the rather large number of blank or invalid votes could reflect feelings of political indifference and distrust in politics. These attitudes and feelings, of course, are not a new value orientation and certainly not a new political cleavage. Nevertheless, it is important to consider them here, as they may be of relevance in determining the voting behavior of certain electorates.

3 Attitudinal Differences Between the Eight Electorates

Let us start the analysis by looking at a univariate comparison of the mean scores for each attitude for all the parties. In Table 1, the scales are classified according to the cleavages that were distinguished in the previous section. In an overall comparison of the mean scores for all the political parties, we generally find statistically significant ($P < 0.001$) differences between parties, the only nonsignificant difference being for "social isolation" ($P = 0.015$). A nonparametric Kruskal–Wallis test was performed because none of the attitudes had normally distributed data (Siegel and Castellan, 1988, pp. 206–216). The number of tests performed was limited by comparing only the two most extreme parties with all the others. Because we had 13 tests for each attitude, the significance level for rejecting the null hypothesis for each test was set at .001, which corresponds roughly to an overall probability of .05 of a type I error. In Table 1, the parties that differed significantly from all the other parties because they had high or low mean scores are printed in boldface type.

As was expected, the electorate of the Christian Democratic Party (CVP) scored on average significantly higher on the scales used as indicators for the first old cleavage; however, this is true only for two of the three scales. The second old cleavage is clearly built on the opposition between the Liberals (PVV) and the Socialists (SP). The indicator of the third old cleavage is capable of distinguishing the Volksunie significantly from the Vlaams Blok and both of these parties from all the others.

Finally, Agalev is clearly the emanation of the new cleavages in its two possible components, but this electorate is only the antipode of the Vlaams Blok voters in the dimension of three extreme right-wing orientations, especially ethnocentrism. The dimension of negative feelings concerning political powerlessness and utilitarian individualism contrasts the electorate of Agalev with those who voted blank or turned in invalid votes.

The deficiencies of this way of looking at the data are manifold. First, we need a method that is capable of displaying the distribution for each electorate on all the attitudes without resulting in an overwhelming number of tables. Second, we also require an approach in which the attitudes in the electorates are analyzed simultaneously. Third, the method should differentiate between the attitudes according to their relevance and dominance in creating cleavages into the electorate. This means that the method should be capable of stating the net contribution of each attitude and value orientation for joining each electorate. Fourth, it would be appropriate if the analytical method resulted in a visualization of the data, displaying which attitudes are capable of cleaving the electorate along the lines of the political parties. An approach that satisfied these requirements is a combination of projection and rotation methods based on the biplot and canonical correlation analysis (Vuylsteke-Wauters, 1994).

Table 1: Mean scores on the 14 attitude scales in the different electorates[a]

Attitudes	Party							
	Agalev	**CVP**	**PVV**	**SP**	**Vl.Blok**	**VU**	**Rossem**	**Null**
First "old" cleavage								
No liberalization of abortion								
(ABORTION)	3.28	**5.40**	3.93	2.93	3.37	4.39	2.65	4.09
Differences in sex roles								
(SEXROLES)	**1.59**	2.73	2.48	2.55	2.54	2.16	2.71	2.91
No freedom of opinion								
(NOFREEOP)	2.73	**3.21**	2.69	2.54	2.62	2.65	2.77	2.64
Second "old" cleavage								
Economic conservatism								
(ECONCONS)	3.39	3.68	**4.79**	2.92	3.48	4.03	3.94	3.21
Third "old"' cleavage								
Flanders must decide								
(FLABELG)	4.19	4.16	4.00	3.86	**5.35**	**6.86**	4.98	3.21
First "new" cleavage								
Post-materialism[b]								
(POSTMAT)	**6.64**	4.25	4.29	4.88	4.50	5.09	4.92	4.40
Sacrifices for milieu[b]								
(MILIEU)	**5.43**	3.99	3.79	3.74	3.74	4.24	3.67	3.53
Second "new" cleavage								
Insufficient protection								
(PETTYCRIM)	**6.51**	7.43	7.43	7.44	**8.09**	7.35	7.70	7.50
Negative toward migrants								
(OUTGROUP)	**3.76**	5.37	5.70	5.32	**7.09**	5.16	5.69	5.90
Superiority of white race								
(SUPRACE)	**2.17**	4.31	4.05	3.79	**5.00**	3.44	3.81	3.94
Authoritarianism								
(AUTHORIT)	**5.17**	6.87	6.55	6.58	6.69	6.60	6.00	6.83
Protest voting								
Political powerlessness								
(POWERLES)	4.71	5.25	5.26	5.45	5.83	5.12	5.87	**7.54**
Utilitarian individualism								
(INDIVID)	**2.53**	3.54	3.72	3.92	4.01	3.10	3.62	**4.23**
Social isolation[c]								
(SOCISOL)	2.28	2.54	2.32	2.60	2.87	2.35	2.25	2.98
Total (N)	**225**	**696**	**491**	**401**	**227**	**207**	**78**	**124**

[a]The parties are the ecologists (Agalev), the Christian Democrats (CVP), the Liberal Party (PVV), the Socialists (SP), the right-wing party (Vl. Blok), the traditional Flemish nationalists (VU), the Libertarians (Rossem), and blank or null votes (null). A number of voters are not included in the analysis: those who did not report their vote (133), those who did not vote for several reasons (51), and those who voted for other parties (26).

[b]Original scales (materialism, no sacrifices) are reversed.

[c]Not significant at the .001 level. The associations between political party and all the other scales are significant on this level.

4   A Biplot Presentation by Means of Canonical Discriminant Analysis

Visualizing the group structure in the different electorates with respect to the 14 attitudes can be done by various methods. A well-known graphical display of the group structure in the multivariate data is the scatterplot of the data produced by the canonical variables. A more informative plot would be obtained if we could characterize on the same plot which attitudes are most responsable for the discrimination of the groups. This leads us to vizualise not only the voters—in their group structure—but also the attitudes and their contribution to the separation between the electorates. This kind of graphical display is called a biplot (Gabriel, 1971), where both observations (voters) and variables (attitudes) are plotted.

4.1   Canonical Correlation Analysis

Canonical correlation analysis allows the researcher to examine patterns of relationships between sets of variables. With this technique we are able to study the differences between the eight electorates with respect to the 14 attitudes simultaneously. As a data reduction technique, canonical correlation analysis in many ways subsumes factor analysis. Rather than concentrate on the relationships within a single set of variables, the analysis tries to find pairs of unobserved latent variables underlying two sets of variables. In our case, the first set of variables is the categorical variable "party choice" coded as a set of eight dummy variables with data value 0 or 1. The second set of variables consists of the 14 attitude scales, measured on a quasi-interval scale.

Notice that our objective is not to perform statistical tests on the relationship between the two sets of variables, for which various statistical assumptions on the data would be necessary, but rather to arrive at a visualization of the group structure in the different electorates with respect to the full spectrum of attitudes.

Given two sets of centered and standardized variables and a sample of n observations, called the X variables and Y variables, respectively, denote the first data matrix with p variables by $\mathbf{X}$ $(n \times p)$ and the second one with q variables by $\mathbf{Y}$ $(n \times q)$. The aim of canonical correlation analysis is to find two new sets of uncorrelated variables, called the U variables and the V variables, each with m variables, that are linear combinations of the original X and Y variables, respectively. The U variables and V variables are such that the first pair of variables, say U_1 and V_1, have maximum correlation; then the second pair, U_2 and V_2, uncorrelated with U_1 and V_1, have maximum correlation, and so on (see, for example, Gittins, 1985).

The solution is given by the following pair of generalized eigenvalue decompositions:

$$\mathbf{R}_{11}^{-1}\mathbf{R}_{12}\mathbf{R}_{22}^{-1}\mathbf{R}_{21}\mathbf{R}_{11}^{-1} = \mathbf{A}\mathbf{R}^2\mathbf{A}^\mathsf{T} \qquad \text{where } \mathbf{A}^\mathsf{T}\mathbf{R}_{11}\mathbf{A} = \mathbf{I}$$

$$\mathbf{R}_{22}^{-1}\mathbf{R}_{21}\mathbf{R}_{11}^{-1}\mathbf{R}_{12}\mathbf{R}_{22}^{-1} = \mathbf{B}\mathbf{R}^2\mathbf{B}^\mathsf{T} \qquad \text{where } \mathbf{B}^\mathsf{T}\mathbf{R}_{22}\mathbf{B} = \mathbf{I}$$

or equivalently by the following generalized singular value decomposition:

$$\mathbf{R}_{11}^{-1}\mathbf{R}_{12}\mathbf{R}_{22}^{-1} = \mathbf{ARB}^{\mathsf{T}} \qquad \text{where } \mathbf{A}^{\mathsf{T}}\mathbf{R}_{11}\mathbf{A} = \mathbf{B}^{\mathsf{T}}\mathbf{R}_{22}\mathbf{B} = \mathbf{I}$$

where $\mathbf{R}_{11}$ and $\mathbf{R}_{22}$ are the correlation matrices of the X and Y variables, respectively, $\mathbf{R}_{12}$ is the $p \times q$ matrix of correlations between the two sets, and $\mathbf{R}_{21} = \mathbf{R}_{12}^{\mathsf{T}}$. The singular values in the diagonal matrix $\mathbf{R}$ are the maximized correlations, called canonical correlations, arranged in decreasing order. The U and V variables themselves are obtained by linear transformations using the eigenvectors: $\mathbf{U} = \mathbf{XA}$ and $\mathbf{V} = \mathbf{YB}$.

Linear discriminant analysis, on the other hand, is a method designed to discriminate between groups on the basis of sets of multivariate observations observed on their group members [this method is often described as canonical variate analysis or multivariate analysis of variance (MANOVA) in textbooks on multivariate analysis]. This is indeed the situation we have here, where we have eight electorates and 14 scale variables observed on the sample from each electorate.

Vuylsteke-Wauters (1994) showed that linear discriminant analysis can be seen as a special case of canonical correlation analysis. In this special case when one of the sets of variables, say $\mathbf{X}$, consists of dummy variables, the correlation matrix $\mathbf{R}_{11}$ is singular, and a generalized matrix inverse has to be used. The resulting canonical variables have the property that they are uncorrelated as well as standardized to have variance 1; in other words, the new variables are orthonormal and the correlation between the U variables and V variables is simply $\mathbf{U}^{\mathsf{T}}\mathbf{V}$. Gabriel (1981a) gives applications of the biplot in the context of MANOVA. The fact that canonical correlation analysis admits canonical variate analysis as a special case is also described by Gower (1989a). The biplot properties of canonical correlation analysis are discussed in detail by ter Braak (1990).

The original X and Y variables and the new U and V variables can be related by computing correlation coefficients analogous to factor loadings, which in geometrical terms can be expressed as the projection of the unit vectors for the original sets of variables onto the unit vectors defining a new basis in the respective spaces: $\mathbf{X}^{\mathsf{T}}\mathbf{U}$ and $\mathbf{Y}^{\mathsf{T}}\mathbf{V}$. When $\mathbf{X}$ is the (centered, standardized) indicator matrix, then it is the second set of loadings $\mathbf{Y}^{\mathsf{T}}\mathbf{V}$ that gives the canonical coefficients which allow us to interpret the canonical variables which maximally discriminate between the groups.

4.2   Correlation Biplots

We use the results of canonical correlation analysis applied to the indicator variables and attitude scale variables as a basis for a correlation biplot display of the political parties and the attitude variables. In a two-dimensional biplot, we use the first two columns of $\mathbf{V}$, denoted by $\mathbf{V}_{(2)}$, as a basis for the subspace. The projections of the attitude variables in $\mathbf{Y}$ are then $\mathbf{G} = \mathbf{Y}^{\mathsf{T}}\mathbf{V}_{(2)}$. The rows of $\mathbf{G}$ are used to depict the attitude variables in the biplot, usually drawn as lines emanating from the origin of the display to the points (Symons *et al.*, 1983).

Individuals are displayed by the rows of $\mathbf{F} = \mathbf{V}_{(2)}$ and are thus displayed by their standardized canonical scores. In order to depict the eight political parties, the corresponding subsets of individual points are averaged to find the mean points, or centroids, of each party. This process of averaging is the same as projecting the dummy variables in $\mathbf{X}$ onto the basis $\mathbf{V}_{(2)}$.

In interpreting the canonical correlation biplot, we can draw a unit circle, as is frequently done in the usual correlation biplot of a single set of variables. Variables whose points extend as far as the unit circle are very well reconstructed in the display, which means that they are important in explaining group differences. Short vectors, however, indicate poor display in the sense that they do not contribute to the explanation as far as the first two canonical dimensions are concerned. Otherwise, the interpretation is very much like that of the regular biplot; we look for large scalar products between party points and attitude scale points, which indicate that people voting for that party have high values on those attitude scales.

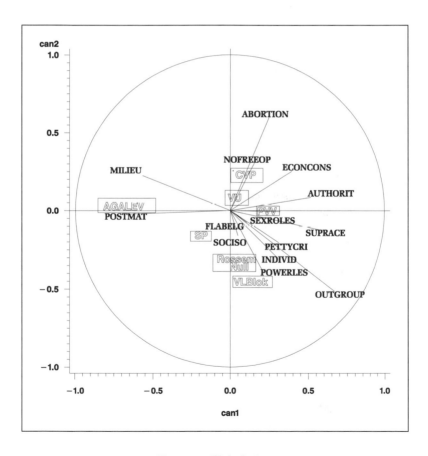

Figure 1: Global view

5   Interpreting the Biplot Presentations

The biplot for the two first canonical axes (Figure 1) accounts for 53% of the separation between the eight electorates based on Pillai's trace. The attitudes with the largest distances from the origin are OUTGROUP, POSTMAT, MILIEU, and ABORTION. Agalev is the most outlying party and has the highest (absolute) mean scores on the canonical variable that represents POSTMAT and MILIEU. For the representation of the OUTGROUP variable, the Vlaams Blok has the highest score. These results confirm the relevance and dominance of two new cleavages. One of the old cleavages, CVP, is best discriminated from other voters by the attitude toward ABORTION. This finding is in line with the important role of the church (Billiet and de Witte, 1995). This figure also shows that the other attitudes used as indicators for the three cleavages that we have found are not adequately represented in the first pair of axes and seem of minor significance for separating the electorates.

Another way of looking at the biplot is obtained by drawing a line through the origin and a particular group mean and projecting the group means on this line in order to obtain a ranking of the groups. This projection onto a biplot axis is illustrated for AGALEV in Figure 2. The projections of the parties approximate their order on the postmaterialist/materialist scale in the reduced spacee. We can see that AGALEV, for example, has a much higher value of POSTMAT than all other parties.

The other two old cleavages, built around economic conservatism (ECONCONS) and linguistic conflicts (FLABELG), seem to be of little relevance in the reconstruction displayed in Figure 1. We can find these attitudes in the display of the third and the fourth canonical axes (see Figure 3), which account together for another 30% of the separation between the groups. In Figure 1 CVP and VU are close together, whereas in Figure 3 VU separates out in the direction of FLABELG.

Another attitude well displayed in Figure 3 is economic conservatism (ECONCONS), which was intended to measure the labor–capital (or old right–left) cleavage. It is apparent from this figure that the voters of the PVV are somewhat separated on that value orientation.

The plots in the third and the fourth canonical axes sustain Elchardus' (1994) view about the old and the new right–left cleavages, but with the restriction that the

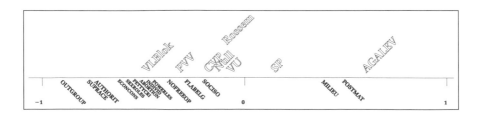

Figure 2: Focus on AGALEV

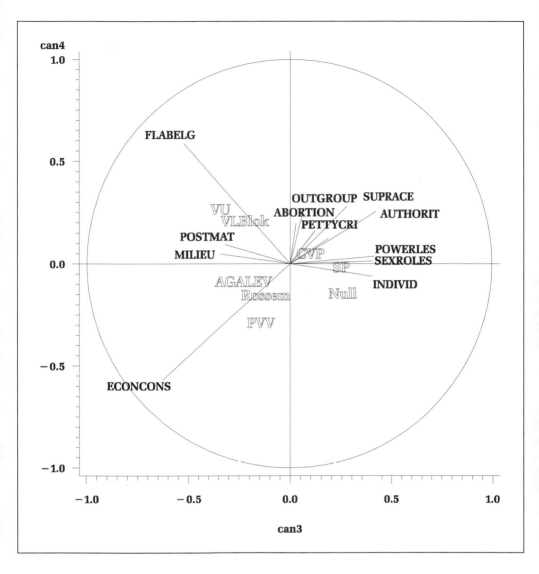

Figure 3: Biplot with the 3rd and 4th canonical axes

social–economic cleavage is not as dominant as he stated. The Agalev and the SP voters belong to different segments of the so-called progressive value orientations.

Finally, political inefficacy or powerlessness (POWERLES) is best represented by the fifth axis (not shown here), which accounts for 7% of the separation between the groups. As was expected, those who voted blank or turned in invalid votes are separated from the other voters on this dimension.

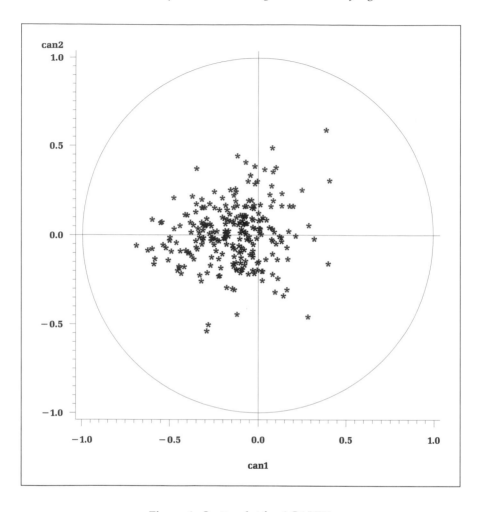

Figure 4: Scatterplot for AGALEV

The group means on the normalized canonical variables do not provide information about the position of the individual voters. The idea of cleavages assumes that the group members are more or less concentrated around the means: the more they are concentrated, the more legitimate the use of the term, at least when the other conditions for cleavages are met. Figure 4 shows the positions of individual voters of Agalev on the first two canonical axes.

6 Conclusions and Discussion

Our graphic presentation of correlation biplots sustains the position that the "new" value orientations have not completely replaced the "old" value orientations as cleaving forces in the political landscape in Flanders. The image is rather one of fragmentation or pluralization as the new values are added to existing orientations (van Deth, 1995, p.3). In the space shaped by the first and the second canonical axes, we could identify two new cleavages and one old cleavage, which were represented by attitude scales as indicators for value orientations. The Agalev voters were most clearly separated from all the others not only because they scored high on postmaterialism but also because they scored low on ethnocentrism and on traditional value orientations. From other studies we know that the more educated professionals with jobs in education, welfare, and culture are strongly overrepresented among the Agalev electorate. As postmaterialists, the Agalev voters are not located in the traditional class structure but mainly take positions among the new middle class (Knutsen and Scarbrough, 1995, p. 496; Inglehart, 1990, p. 332).

The attitude toward immigrants was able to separate adequately the Vlaams Blok voters (10%) from the electorates of most other parties. At least one attitude, authoritarianism, seems ambiguous as an indicator of the value orientation that is represented by the attitude toward immigrants. Its association with the traditional value orientation and with cultural conservatism may be responsible for this. Ethnocentrism seems an expression of feelings of being threatened, related to distrust of politics and feelings of being insufficiently protected by the authorities (see Figure 1).

The traditional values are still relevant for a substantial part of the electorate. This was already displayed in the space of the first and second axes in which the CVP electorate (27% of the vote) was separated from the others by the traditional religious value orientation expressed by the attitude toward abortion. From our other studies, we know that the older part of the electorate (over 50 years old) is mainly responsible for the ongoing relevance of this old cleavage and that its dominance has been slowly declining over the years. Nevertheless, the orientation to traditional values, expressed here by the attitude toward abortion, is still an important factor. The other two old cleavages were discovered in the space of the third and second axes (Figure 3). The old right–left cleavage was expressed by a value orientation called "economic conservatism," and it separated most obviously the voters of the SP from the PVV electorate. The third old cleavage, which is linked to the conflict between the linguistic communities and the demand for more autonomy for Flanders, also appeared in the space of the third and fourth axes (Figure 3). Both the Vlaams Blok and the Volksunie are perceived as Flemish nationalist parties but it is the Volksunie (9% of the vote in 1991) that grouped the largest number of Flemish nationalist voters.

The canonical correlation biplot was used to study the differences between the eight electorates with respect to the 14 attitudes in a simultaneous display. This visualization provides an understanding that fits both the theoretical considerations and the empirical evidence from other studies using attitudes and social-background variables.

Software Note

The biplots were programmed as a macro in SAS release 6.11, using only SAS Base and SAS/GRAPH. Input data can be any pair of orthogonal variables, such as principal components and canonical variables. This macro is available from the first author.

Acknowledgments

The Inter-University Centre for Politicial Opinion research consists of J. Billiet, M. Swyngedouw, A. Carton, and R. Beerten, with offices at the Katholieke Universiteit Leuven. The project is part of the National Social Research Programme and is directed by the Federal Services of Scientific, Technical, and Cultural Affairs.

Chapter 27

Use of Biplots to Diagnose Independence Models in Three-Way Contingency Tables

K. Ruben Gabriel, M. Purificación Galindo, and José Luis Vicente-Villardón

1 Introduction

An essential part of the analysis of contingency tables is testing for independence of classifications. In two-way tables this is straightforward, because there is a single hypothesis of independence. In three-way tables there are many possible independence hypotheses, each of which may be tested, but consistent inferences must take into account the implication relations between them (Roy and Mitra, 1956; Agresti, 1990). Analyses of three-way contingency tables are therefore often difficult to interpret, especially because they produce too many acceptable models. That creates a need for methods that simplify the appraisal of the data and reduce the profusion of acceptable models (Whittaker, 1990). This chapter illustrates how this may be done by visualizing biplots of logarithms of frequencies of contingency tables. Because independence models for contingency tables become additive models for the logarithms of the frequencies and rules for visual diagnosis of additivity on biplots are known (Bradu and Gabriel, 1978), the corresponding rules can be applied to biplots of logarithms of frequencies to diagnose various types of independence in contingency tables.

Biplots (Gabriel 1971, 1981a, 1981b) are visual displays of matrices by vectors for each row and each column which are constructed so the inner product of vector $\mathbf{a}_i$ for row i and vector $\mathbf{b}_j$ for column j approximates the matrix element in cell (i, j). In the present discussion of contingency tables, it is the matrix of logarithms of cell frequencies that is biplotted and the approximation is by least squares weighted by the frequencies, as is appropriate if the frequencies are Poisson variables (see the description of software at the end of the chapter). Actually, the logarithms are centered before they are fitted and displayed, because this focuses the display on differences rather than on the general magnitude of the frequencies. An analogous centering of nonfrequency data has been explained by Bradu and Gabriel (1978).

2   Diagnosis in Two-Way Tables

In a two-way contingency table with frequencies f_{ij} $(i = 1, \ldots, I; j = 1, \ldots, J)$, independence is defined as

$$i \perp\!\!\!\perp j : \quad \frac{p_{ij}}{p_{i'j}} = \frac{p_{ij'}}{p_{i'j'}} \quad \text{for all } i, i'; j, j'$$

where $\perp\!\!\!\perp$ symbolizes independence (Dawid, 1979), i and j indicate the row and column classifications, respectively, i, i' and j, j' their categories, and p_{ij} $(i = 1, \ldots, I; j = 1, \ldots, J)$ the probability of cell (i, j). Equivalently, by taking logarithms, one can write this definition as

$$i \perp\!\!\!\perp j : \quad \lambda_{ii',jj'} = 0 \quad \text{for all } i, i'; j, j'$$

where

$$\lambda_{ii',jj'} = \log(p_{ij}) - \log(p_{i'j}) - \log(p_{ij'}) + \log(p_{i'j'})$$

These tetrad differences have estimates

$$\hat{\lambda}_{ii',jj'} = \log(f_{ij}) - \log(f_{i'j}) - \log(f_{ij'}) + \log(f_{i'j'})$$

and independence may be inferred if they are small.

In our biplots of contingency tables it will be understood that we always approximate centered logarithms of frequencies, not the frequencies as such. The biplot markers $\mathbf{a}_i$ for the rows $(i = 1, \ldots, I)$ and $\mathbf{b}_j$ for the columns $(j = 1, \ldots, J)$ have vector inner products that satisfy

$$\log(f_{ij}) - \overline{\log(f)} \cong \mathbf{a}_i^{\mathrm{T}} \mathbf{b}_j \quad (i = 1, \ldots, I; j = 1, \ldots, J)$$

where $\cong$ means "is approximated by" and $\overline{\log(f)}$ stands for an average of the logarithms of the frequencies. (Centering serves only to focus the display on differences and does not affect the diagnostic rules. Any convenient "average" can therefore be used for centering.) The tetrad difference $\hat{\lambda}_{ii',jj'}$ is visually approximated by

$$\mathbf{a}_i^{\mathrm{T}} \mathbf{b}_j - \mathbf{a}_{i'}^{\mathrm{T}} \mathbf{b}_j - \mathbf{a}_i^{\mathrm{T}} \mathbf{b}_{j'} + \mathbf{a}_{i'}^{\mathrm{T}} \mathbf{b}_{j'} = (\mathbf{a}_i - \mathbf{a}_{i'})^{\mathrm{T}} (\mathbf{b}_j - \mathbf{b}_{j'})$$

and so

$$\hat{\lambda}_{ii',jj'} \cong (\mathbf{a}_i - \mathbf{a}_{i'})^{\mathrm{T}}(\mathbf{b}_j - \mathbf{b}_{j'})$$

Hence the biplot criterion for diagnosing independence $i \perp\!\!\!\perp j$ is whether or not the inner products $(\mathbf{a}_i - \mathbf{a}_{i'})^{\mathrm{T}}(\mathbf{b}_j - \mathbf{b}_{j'})$ are close to zero for all i, i' and j, j'. But two vectors have a zero inner product if, and only if, they are orthogonal, so this criterion is equivalent to $(\mathbf{a}_i - \mathbf{a}_{i'})$ being approximately orthogonal to $(\mathbf{b}_j - \mathbf{b}_{j'})$ for all i, i' and j, j'. Clearly, this can occur if, and only if, the $\mathbf{a}_i$'s and $\mathbf{b}_j$'s are close to perpendicular straight lines. The visual diagnostic rule is therefore:

> "Diagnose $i \perp\!\!\!\perp j$ if the $\mathbf{a}_i$'s and $\mathbf{b}_j$'s are close to perpendicular straight lines," and this will be referred to as the *perpendicularity rule*.

This argumentation applies equally to any subtable of some of the rows and columns of the contingency table. The rule for such a subtable is to diagnose independence if the $\mathbf{a}_i$'s and $\mathbf{b}_j$'s for the subtable's rows and columns are on perpendicular lines or planes. These diagnostics are stated in terms of planes, rather than just lines, because more than two dimensions may be needed for closely fitting a biplot to a complete table in which there is no overall independence.

This is illustrated in Figure 1 for a 3×4 table of large frequencies generated from Poisson distributions such that the first three columns are independent of the rows, but the fourth column is not. For these frequencies the biplot has the $\mathbf{a}_1, \mathbf{a}_2, \mathbf{a}_3$ markers for the rows roughly collinear and on a line that is perpendicular to another line which is close to markers $\mathbf{b}_1, \mathbf{b}_2, \mathbf{b}_3$ for the first three columns; marker $\mathbf{b}_4$, on the other hand, deviates noticeably from the latter line. By the foregoing rule, this biplot pattern diagnoses independence only in the subtable of the first three columns, which is the correct diagnosis.

A three-dimensional biplot provides a perfect fit, and interactive computer displays (for example, SAS, 1994) can be used to rotate this and find a planar projection that has $\mathbf{a}_1, \mathbf{a}_2, \mathbf{a}_3$ and $\mathbf{b}_1, \mathbf{b}_2, \mathbf{b}_3$ very close to perpendicular straight lines (Figure 2).

Subset independence may arise for a variety of reasons related to the properties of the classifications. It can also appear because unusually high or low counts occur in a few cells, and these are best treated as outliers from the general pattern (see Bradu and Gabriel, 1978, p. 48, for an analogous discussion).

3   Diagnosis in Three-Way Tables

Biplots display markers for the rows and the columns of a matrix, so that each matrix element (logarithm of frequency, centered on an average) is represented by the inner product of the corresponding row and column marker vectors. This does not readily generalize to three-way tables because no mathematically tractable "product" of three vectors is available. Biplot displays of a three-way table can, however, be constructed

	Poisson Parameters		
10000	15000	20000	20000
8000	12000	16000	20000
12000	18000	24000	25000

	Frequencies		
9930	15096	19818	19970
7902	12094	15688	20057
11959	17865	24201	25063

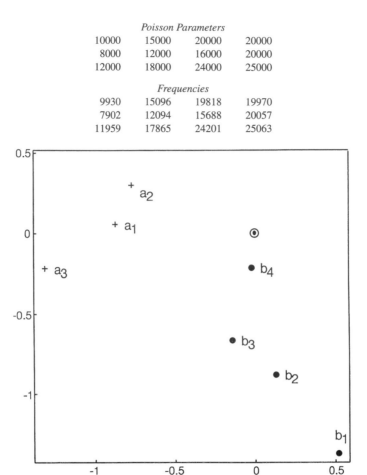

Figure 1: Biplot of centered logarithms of frequencies of a table generated from Poisson distributions with parameters corresponding to independence except in the last column.

by first combining two of the table's classifications and thus reducing it to a two-way table, as shown next.

A three-way contingency table with frequencies f_{ijk} and probabilities p_{ijk} can be displayed by a biplot with markers $\mathbf{a}_i$ ($i = 1, \ldots, I$) and $\mathbf{b}_{j,k}$ ($j = 1, \ldots, J; k = 1, \ldots, K$) that satisfy

$$\log(f_{ijk}) - \overline{\log(f)} \cong \mathbf{a}_i^{\mathrm{T}} \mathbf{b}_{jk} \quad (i = 1, \ldots, I; j = 1, \ldots, J; k = 1, \ldots, K)$$

for an average $\overline{\log(f)}$. This biplot is analogous to that of Section 2, except that it uses the combination of the **j** and **k** classifications [with categories $(j, k) = (1, 1), (1, 2), \ldots, (J, K)$], where the earlier biplot simply had classification **j** (with

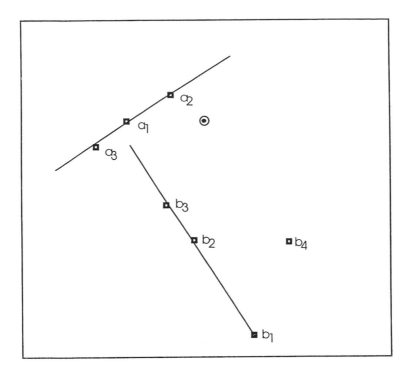

Figure 2: Projection of 3D biplot of logarithms of frequencies of the table given in Figure 1, showing collinearity of markers for first three rows and markers for first three columns and perpendicularity of the two marker lines. (Lines joining markers are drawn to help examine patterns used for diagnosis.)

categories $j = 1, \dots, J$). By analogy, it leads to the multiple perpendicularity rule:

> "Diagnose $i \perp\!\!\!\perp (j,k)$ if the $\mathbf{a}_i$'s and $\mathbf{b}_{jk}$'s are close to perpendicular straight lines." That is

$$i \perp\!\!\!\perp (j,k) : \frac{p_{ijk}}{p_{i'jk}} = \frac{p_{ij'k'}}{p_{i'j'k'}} \quad \text{for all } i, i'; j, j'; k, k'$$

which is known as *multiple independence* of i and (j,k).

It was noted earlier that independence can be diagnosed for a subtable if the markers for its rows and columns satisfy the perpendicularity criterion. Applying this to the $\mathbf{b}_{jk}$'s for a particular category k of classification $\mathbf{k}$ leads to the conditional perpendicularity rule:

> "Diagnose $i \perp\!\!\!\perp j/k$ if for a given k the $\mathbf{a}_i$'s and $\mathbf{b}_{jk}$'s are close to perpendicular straight lines," where

$$i \perp\!\!\!\perp j/k : \quad \frac{p_{ijk}}{p_{i'jk}} = \frac{p_{ij'k}}{p_{i'j'k}} \quad \text{for all } i, i'; j, j'; \text{ and given } k$$

which is known as *conditional independence* of *i* and *j*, *given category k of classification k*.

More generally, if the preceding criterion applies to *each k*, it leads to the partial perpendicularity rule:

"Diagnose $i \perp\!\!\!\perp j/k$ if, for each k, the $\mathbf{a}_i$'s and $\mathbf{b}_{jk}$'s are close to perpendicular straight lines," that is

$$i \perp\!\!\!\perp j/k : \quad \frac{p_{ijk}}{p_{i'jk}} = \frac{p_{ij'k}}{p_{i'j'k}} \quad \text{for all } i, i'; j, j'; \text{ and all } k$$

which is known as *conditional independence* of *i* and *j*, *given classification k*.

These diagnoses are illustrated in Figure 3 for a $3 \times 3 \times 3$ table of large frequencies generated from Poisson distributions such that $i \perp\!\!\!\perp j/k$ for $k = 1$ and 2 but not for $k = 3$. Conditional independence $i \perp\!\!\!\perp j/k$ therefore holds only in the subtable that excludes $k = 3$. On the biplot, the $\mathbf{a}_i$'s are roughly collinear and both the $\mathbf{b}_{j1}$'s and the $\mathbf{b}_{j2}$'s are very close to lines perpendicular to the **a** line: The $\mathbf{b}_{j3}$'s, however, do not lie near any straight line. The visual rules therefore indicate $i \perp\!\!\!\perp j/1$ and $i \perp\!\!\!\perp j/2$, but not $i \perp\!\!\!\perp j/3$, and thus correctly diagnose the independence structure.

Analogous visual criteria for conditional independence of *i* and *k*, given category *j* or classification *j*, can be obtained by permuting the roles of *j* and *k* in the preceding rules.

Returning to the example in Figure 3, one may check $i \perp\!\!\!\perp k/1$ by means of the markers $\mathbf{b}_{11}$, $\mathbf{b}_{12}$, and $\mathbf{b}_{13}$. These are not near a line perpendicular to the **a** line, so conditional independence $i \perp\!\!\!\perp k/1$ is not diagnosed. The same applies to $i \perp\!\!\!\perp k/2$ and to $i \perp\!\!\!\perp k/3$ because neither $\mathbf{b}_{21}$, $\mathbf{b}_{22}$, and $\mathbf{b}_{23}$ nor $\mathbf{b}_{31}$, $\mathbf{b}_{32}$, and $\mathbf{b}_{33}$ is anywhere near lines perpendicular to the **a** line.

The consistency of these criteria may be noted by recalling that multiple independence $i \perp\!\!\!\perp (j, k)$ holds if and only if conditional independences $i \perp\!\!\!\perp j/k$ and $i \perp\!\!\!\perp k/j$ both hold. This is reflected by the geometric equivalence of their diagnostic criteria. It is readily verified that occurrence of "the $\mathbf{a}_i$'s and $\mathbf{b}_{jk}$'s are close to perpendicular straight lines" is equivalent to simultaneous occurrence of both conditional perpendicularity criteria, that is, "for each k, the $\mathbf{a}_i$'s and $\mathbf{b}_{jk}$'s are close to perpendicular straight lines" and "for each j the $\mathbf{a}_i$'s and $\mathbf{b}_{jk}$'s are close to perpendicular straight lines."

The foregoing perpendicularity rules for diagnosis relate to independence of the *i* classification, which is represented by the biplot $\mathbf{a}_i$ markers, from the *j* and/or *k* classifications, which are combined for representation by the biplot $\mathbf{b}_{jk}$ markers. Different geometric considerations are used for visual diagnosis of independence of

Poisson parameters

i	k=1			k=2			k=3		
	j=1	j=2	j=3	j=1	j=2	j=3	j=1	j=2	j=3
1	10000	8000	12000	30000	40000	50000	25000	35000	10000
2	15000	12000	18000	15000	20000	25000	20000	45000	20000
3	20000	16000	24000	15000	20000	25000	20000	55000	30000

Frequencies

i	k=1			k=2			k=3		
	j=1	j=2	j=3	j=1	j=2	j=3	j=1	j=2	j=3
1	9926	7775	11993	29993	39919	49524	24816	34795	9991
2	15039	11950	18140	15193	20210	24973	20299	45248	19815
3	19926	16130	24841	15090	20053	25221	20143	55042	29799

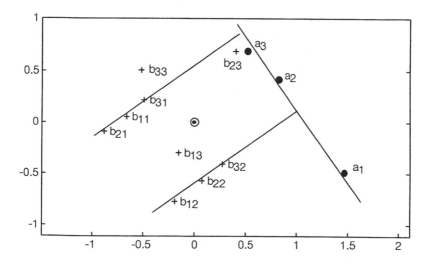

Figure 3: Biplot of centered logarithms of frequencies of a 3×3×3 table generated from Poisson distributions with parameters corresponding to conditional independence of **i** and **j** given $k = 1$ and 2, but not $k = 3$. (Lines joining markers are drawn to help examine patterns used for diagnosis.)

the *j* and *k* classifications. Thus, for *conditional independence* of **j** and **k**, given **i**, which is

$$j \perp\!\!\!\perp k/i: \quad \frac{p_{ijk}}{p_{ij'k}} = \frac{p_{ijk'}}{p_{ij'k'}} \quad \text{for all } j, j'; k, k'; \text{ and all } i$$

the rule is

"Diagnose $j \perp\!\!\!\perp k/i$ if for each $j, j'; k, k'$ the $(\mathbf{b}_{jk}, \mathbf{b}_{j'k}, \mathbf{b}_{jk'}, \mathbf{b}_{j'k'})$ are close to a parallelogram, that is, if $\mathbf{b}_{jk} + \mathbf{b}_{j'k'}$ is close to $\mathbf{b}_{jk'} + \mathbf{b}_{j'k}$." This will be referred to as the parallelogram rule.

The parallelogram rule differs from the perpendicularity rules in that it tests independence of the two classifications that are combined in the biplot display. It is not an application of the two-way perpendicularity diagnostic but can be understood by rewriting the hypothesis as

$$j \perp\!\!\!\perp k/i : \quad \lambda_{i,jj',kk'} = 0 \quad \text{for all } i; j, j'; k, k'$$

where $\lambda_{i,jj',kk'} = \log(p_{ijk}) - \log(p_{ij'k}) - \log(p_{ijk'}) + \log(p_{ij'k'})$. The latter are estimated by

$$\hat{\lambda}_{i,jj',kk'} = \log(f_{ijk}) - \log(f_{ij'k}) - \log(f_{ijk'}) + \log(f_{ij'k'})$$

and approximated in the biplot by $\hat{\lambda}_{i,jj',kk'} \cong \mathbf{a}_i^T(\mathbf{b}_{jk} - \mathbf{b}_{j'k} - \mathbf{b}_{jk'} + \mathbf{b}_{j'k'})$. This quantity is small for all i if $(\mathbf{b}_{jk} - \mathbf{b}_{j'k} - \mathbf{b}_{jk'} + \mathbf{b}_{j'k'})$ is close to zero, that is if $\mathbf{b}_{jk} - \mathbf{b}_{j'k}$ is close to $\mathbf{b}_{jk'} - \mathbf{b}_{j'k'}$, which in turn corresponds to the figure formed by $\mathbf{b}_{jk}, \mathbf{b}_{j'k}, \mathbf{b}_{jk'}, \mathbf{b}_{j'k'}$, in that order, being close to a parallelogram.

As in the case of a two-way layout, the three-way table diagnostics can also be applied to subsets of the categories of one or more classifications. No diagnostic rule is provided for conditional independence $j \perp\!\!\!\perp k/i$, that is, of j and k, given category i of classification i, since that is not easily visualized on the biplot.

To illustrate the parallelogram rule for conditional independence $j \perp\!\!\!\perp k/i$, consider again the three-way table of large Poisson frequencies, but interchange the i and k classifications, so that $j \perp\!\!\!\perp k/1$ and $j \perp\!\!\!\perp k/2$, but not $j \perp\!\!\!\perp k/3$, and hence not $j \perp\!\!\!\perp k/i$. Figure 4 shows the $2 \times 3 \times 3$ subtable of frequencies excluding $i = 3$ and its biplot marker. Consider the $\mathbf{b}_{jk}$ markers for any given k: $\mathbf{b}_{1k}$ is slightly above and to the right of $\mathbf{b}_{2k}$, and $\mathbf{b}_{3k}$ is well below $\mathbf{b}_{2k}$. And the same pattern holds for every k. It is readily seen that this entails that $\mathbf{b}_{1k} - \mathbf{b}_{2k}$ is the same for all k, $\mathbf{b}_{2k} - \mathbf{b}_{3k}$ is the same for all k, and $\mathbf{b}_{3k} - \mathbf{b}_{1k}$ is the same for all k, so the required tetrads of $\mathbf{b}_{jk}$'s form parallelograms. The parallelogram rule correctly diagnoses $j \perp\!\!\!\perp k/i$ for this subtable. Figure 5 shows the entire $3 \times 3 \times 3$ table of frequencies and its biplot. Here the $\mathbf{b}_{jk}$ markers do not display the same pattern for every k, and hence the rule would lead to the diagnosis that $j \perp\!\!\!\perp k/i$ does not hold for this table. Again, that is the correct diagnosis.

On a planar biplot, visualization of the diagnoses is quite straightforward, but it requires some care when the $\mathbf{a}$'s and $\mathbf{b}$'s are in a three-dimensional space. Any two-dimensional view will reveal a line in 3-D as a line or point, but a plane in three dimensions cannot be revealed by a single two-dimensional view. It requires two views of a 3D biplot to ascertain whether a set of points is close to a line or to a plane or to assess a parallelogram pattern.

The preceding discussion is of biplots of three-way $i \times j \times k$ tables in which the j and k classifications are combined. Alternative biplots can be constructed for the combination of classifications i and j, or of i and k, and all the above results apply after suitable permutation of the indices. Each of the three possible biplots can analyze all the types of independence except for conditional independence of the combined classifications given a category of the other classification. The rule of

Poisson parameters

i	k=1			k=2			k=3		
	j=1	j=2	j=3	j=1	j=2	j=3	j=1	j=2	j=3
1	10000	8000	12000	15000	12000	18000	20000	16000	24000
2	30000	40000	50000	15000	20000	25000	15000	20000	25000

Frequencies

i	k=1			k=2			k=3		
	j=1	j=2	j=3	j=1	j=2	j=3	j=1	j=2	j=3
1	9926	7775	11993	15039	11950	18140	19926	16130	24841
2	29993	39919	49524	15193	20210	24973	15090	20053	25221

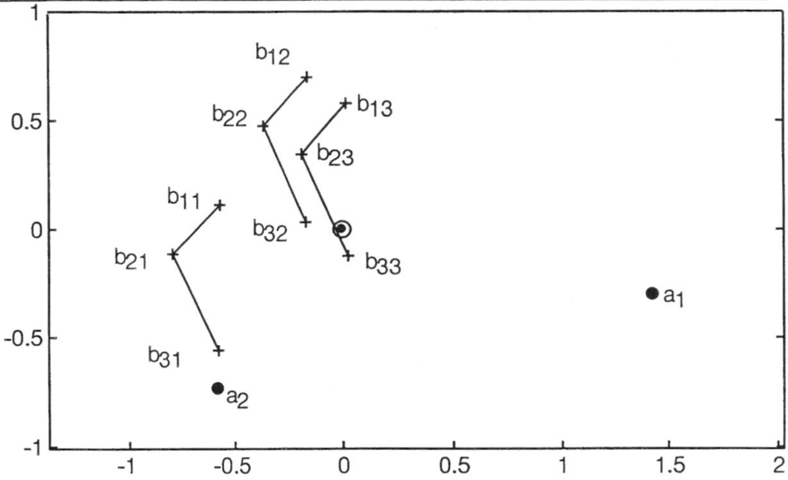

Figure 4: Biplot of centered logarithms of frequencies of a $2 \times 3 \times 3$ table generated from Poisson distributions with parameters corresponding to $j \perp\!\!\!\perp k/i$. (Lines joining markers are drawn to help examine patterns used for diagnosis.)

diagnosis applied to any particular type of independence will, however, depend on what biplot is used. Thus, $i \perp\!\!\!\perp k/j$ is diagnosed by the conditional perpendicularity criterion on the $(\mathbf{a}_i, \mathbf{b}_{jk})$ biplot but by the parallelogram criterion on the $(\mathbf{a}_j, \mathbf{b}_{ik})$ biplot.

4   An Applicaton—Danish Reemployment Data

In a study of the determinants of reemployment of workers in Denmark, Andersen (1994) considered data for laid-off employees on length of employment (L in six categories), cause of layoff (K, two categories: closure of the company or replacement of the employee), and whether or not they had been reemployed (E, two categories)— see Table 1. Hypotheses of independence of these three classifications were examined

Poisson parameters

i	k=1			k=2			k=3		
	j=1	j=2	j=3	j=1	j=2	j=3	j=1	j=2	j=3
1	10000	8000	12000	15000	12000	18000	20000	16000	24000
2	30000	40000	50000	15000	20000	25000	15000	20000	25000
3	25000	35000	10000	20000	45000	20000	20000	55000	30000

Frequencies

i	k=1			k=2			k=3		
	j=1	j=2	j=3	j=1	j=2	j=3	j=1	j=2	j=3
1	9926	7775	11993	15039	11950	18140	19926	16130	24841
2	29993	39919	49524	15193	20210	24973	15090	20053	25221
3	24816	34795	9991	20299	45248	19815	20143	55042	29799

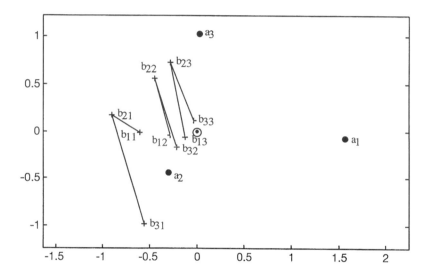

Figure 5: Biplot of centered logarithms of frequencies of a 3 × 3 × 3 table generated from Poisson distributions with parameters corresponding to $j \perp\!\!\!\perp k/i$ independence for $i = 1, 2$, but not for $i = 3$. (Lines joining markers are drawn to help examine patterns used for diagnosis.)

with the help of biplots for two of the combinations of classifications, as shown in Figures 6 and 7.

The ($\mathbf{a}_E$, $\mathbf{b}_{K,L}$) biplot of Figure 6 fits the data perfectly because it represents the log frequencies of a matrix with only two rows (and 6 × 2 columns). The $\mathbf{a}_E$ line through the "no" and "yes" $\mathbf{a}_E$ markers for reemployment is shown on the plot, so the diagnoses of conditional independence can proceed by checking the relation of the $\mathbf{b}_{K,L}$ markers to this line. What is evident is that for workers laid off for closure the $\mathbf{b}_{K,L}$ markers are close to a line (shown in Figure 6) that is perpendicular

Table 1: Survey of Danish workers who had been laid off

Length of employment	Cause of layoff		Reemployment E	
L	K	Code	Yes	No
Less than 1 month	Closure	C<1 m	8	10
	Replacement	R<1 m	40	24
1 month to less than 3	Closure	C 1 m	35	42
	Replacement	R 1 m	85	42
3 months to less than a year	Closure	C 3 m	70	86
	Replacement	R 3 m	181	41
1–2 years	Closure	C 1 yr	62	80
	Replacement	R 1 yr	85	16
2–5 years	Closure	C 2 yr	56	67
	Replacement	R 2 yr	118	27
More than 5 years	Closure	C>5 yr	38	35
	Replacement	R>5 yr	56	10

to the **a** line. That leads to the diagnosis $E \perp\!\!\!\perp L/C$ of conditional independence of reemployment from length for workers laid off for closure (C). For employees laid off for replacement (R), the situation is not so simple since the $\mathbf{b}_{K,L}$ markers are on two separate lines, each of which is perpendicular to the $\mathbf{a}_E$ line (both lines are shown in Figure 6). Hence $E \perp\!\!\!\perp L/R$ cannot be diagnosed for all lengths of employment, but one may diagnose $E \perp\!\!\!\perp L_{[<3m]}/R$ as well as $E \perp\!\!\!\perp L_{[\geq3m]}/R$: In other words, for replaced workers the association between length and reemployment depends on whether their employment was less than 3 months or at least 3 months but is independent of the exact length.

In addition to this pattern, one may observe that the $\mathbf{b}_{K,L}$'s for all workers laid off for closure project onto the $\mathbf{a}_E$ line closest to the "no" marker , and the $\mathbf{b}_{K,L}$'s for workers replaced after more than 3 months project onto the $\mathbf{a}_E$ line farthest toward the "yes" marker. Thus, the chance of reemployment is least for workers laid off for closure and greatest for workers replaced after more than 3 months.

Considering all the $\mathbf{b}_{K,L}$ markers for each of the two length of employment groups, one could also say that reemployment is independent of length of employment both within the short employment group (less than 3 months) and within the long employment group (3 months or more), but not overall.

Inspection of this biplot did not lead to a simple diagnosis of conditional independence of reemployment and length, given the cause of the layoff, but showed that for replaced employees the situation was more complex. Although an independence model did not fit the entire table, it did fit separate subtables and was useful for exploring the data.

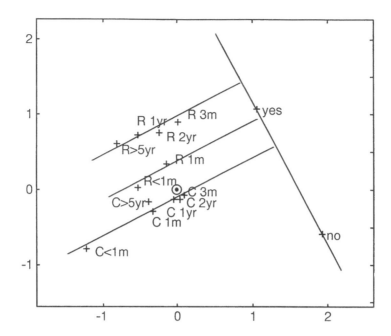

Figure 6: (a_E $b_{K,L}$) biplot of centered logarithms of frequencies of data for Danish workers. (Lines joining markers are drawn to help examine patterns used for diagnosis.)

The same three-way contingency table can be biplotted for different combinations of two classifications. In Figure 7 reemployment is combined with length of employment, so the visual representation is by a (a_K, $b_{E,L}$) biplot, the a_K line that goes through the C marker (for closure) and the R markers (for replacement) is not perpendicular to any particular sets of $b_{E,L}$ markers, and therefore one cannot diagnose independence of cause K from either reemployment E conditional on length L or length L conditional on reemployment E. However, the difference between the $b_{E,L}$ markers for "yes, < 1m" and "no, < 1m" is pretty much the same as the difference between the $b_{E,L}$ markers for "yes, 1m" and "no, 1m"—both differences are marked by lines in Figure 7—and so the tetrad ("yes, < 1m," "yes, 1m," "no, 1m," "no, < 1m") of $b_{E,L}$ markers is close to a parallelogram. It then follows from the parallelogram rule that, for lengths of employment below 3 months, there is conditional independence of reemployment and length given cause, that is, $E \perp\!\!\!\perp L_{[\leq 3m]}/K$.

Similarly, the "yes, length" and "no, < length" differences—also indicated in Figure 7 by lines—are pretty much the same for all lengths from 3 months up, so the corresponding parallelograms exist and lead to the diagnosis $E \perp\!\!\!\perp L_{[\geq 3m]}/K$. The "yes, length" and "no, < length" differences are not, however, the same for all six length categories, and therefore one may not diagnose conditional independence $E \perp\!\!\!\perp L/K$, that is, conditional independence for all lengths.

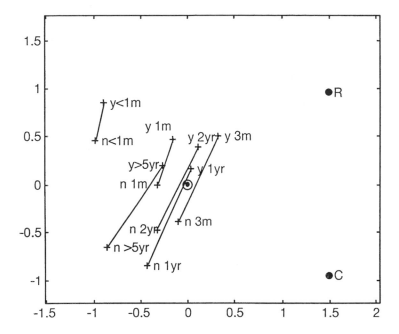

Figure 7: (a_K $b_{E,L}$) biplot of centered logarithms of frequencies of data for Danish workers. (Lines joining markers are drawn to help examine patterns used for diagnosis.)

The same diagnoses were obtained from the biplot of Figure 7 as from that of Figure 6, even though different diagnostic rules were used. A third biplot, in which length of unemployment is in six rows of the contingency table and the combination of cause and reemployment is in four columns, cannot be fitted as closely in the plane, or in three dimensions, and has not been found as helpful for diagnosis.

5 Some Comments

This chapter proposes rules for visual diagnosis of independence that are based on patterns on biplots. Ideally, they allow diagnosis of one model for an entire table, but in practice they often indicate models for subtables, as in the unemployment application. That illustrates an important feature of visualization: it allows the eye to pick out unexpected patterns, such as those in various subtables, and to reveal features that would not be tested by standard methods because they were not anticipated.

Random variation may hamper the identification of patterns on biplots and the consequent diagnosis of models. The schematic illustrative examples in Sections 2 and 3 used very large samples to sidestep this difficulty, but for contingency tables based on empirical data it may be difficult to judge whether a pattern holds or the deviations from it are significant. Such inferences have to be checked by formal tests of significance, although in some situations one may use an alternative type of biplot that incorporates approximate visual tests of independence (Gabriel, 1995).

Identification of diagnoses from biplots is fairly straightforward in two dimensions, since the rules use only straight lines, right angles, and repetitive patterns. It becomes more difficult when a three-dimensional display is used for improved fit to the data, because lines and angles and patterns in space may need to be visualized by real-time rotation of the three-dimensional biplot. The usefulness of the methods proposed here depends on how well the logarithms of the frequencies are approximated and how easily users are able to discern and interpret patterns on biplots. The authors' experience suggests that this is not difficult to acquire and well worth the effort.

Software Description

The computational algorithm used is an adaptation of criss-cross regression lower rank fitting (Gabriel and Zamir, 1979). It is applied to the matrix of the $\log(f_{ij}) - \overline{\log(f)}$ values with weights f_{ij}. A simple way to initialize the iteration is by using the Householder–Young (1938) rank 2 approximation to the matrix, or, if there are zero entries, doing so after substitution of $f_{ij} + 1/2$ for f_{ij}. If singularities arise, they may be circumvented by making small changes in the weighting, analogous to what is done in ridge regression (Hoerl and Kennard, 1970).

Computations were carried out by means of iterative routines programmed in MATLAB (Mathworks, 1995) and are available from the authors at Salamanca; an adaptation to SAS is also being prepared. Animated display and rotation have been carried out with JMP software (SAS, 1994).

Chapter 28

Prediction Regions for Categorical Variables

John C. Gower and Simon A. Harding

1 Introduction

Quantitative information on two or more variables is often represented relative to coordinate axes. In this chapter we show how the familiar concepts associated with quantitative axes may be extended to categorical variables. We shall be concerned with low-dimensional approximations to high-dimensional representations relative to coordinate axes, and to understand the properties of the approximation we must first recapitulate the familiar properties of Cartesian coordinate axes. Figure 1 shows two coordinate axes referring to variables x_1 and x_2, marked with scales.

The position of a sample with value two units of the first variable and one unit of the second variable is at the point P of Figure 1a and is obtained as a vector sum, as shown. Of course, many people prefer to think of this as moving two units in an eastward direction followed by one unit north, but the vector-sum terminology embodies the mathematical concept that extends directly to cope with any number of coordinate axes. We term the operation of positioning a point with known sample values *interpolation*. Figure 1b shows the inverse operation of associating the values of the variables that pertain to the point P. This is done by projecting from P onto the two axes and reading off the nearest scale value. Again, the notion of projection easily extends to any number of variables. Although we express this operation in terms of projection, or more precisely orthogonal projection, it is simpler to regard it as finding the nearest scale marker to P on each axis. We term the operation of determining the values of the variables to be associated with a given point *prediction*. The operations

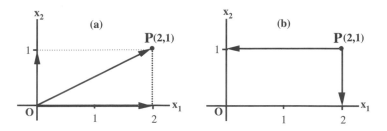

Figure 1: (a) Positioning of the sample (2,1) at the point P. (b) The values of the two variables associated with the point P.

of interpolation and prediction are inverses of each other and are consistent, in the sense that the values predicted for an interpolated point are those given initially. The reason for this terminology, which may seem perverse for exact representations, will become clearer when we discuss approximations.

Coordinate systems are very often at the basis of the visualization of data. With n samples and just two variables, we get a scatter of n points whose coordinate values are the sample values. Indeed, this is how a scatterplot is defined. Visual interpretation consists of inspection for patterns, such as straight lines or other curves, clusters of points, and, when there are sufficient data, inspection for varying densities of sample points throughout the plot. Another important interpretive tool is the visualization of differences between a pair of samples as the distance between the corresponding plotted points. There are many mathematical ways to define distance, but throughout this chapter we shall use ordinary Euclidean distance as measured with a ruler. Even this is less straightforward than it might seem, for it is evident that by changing the scales on the two axes, perhaps merely to reflect changes in the units with which the variables are measured, we shall change distances between plotted points. A brief discussion of how scaling manifests itself with categorical variables is given in our concluding remarks.

Two-dimensional scatterplots are familiar and their usefulness for initial data examination is recognized by all. With more than two variables we may examine all pairs of variables in a series of scatterplots. With p variables this gives $\frac{1}{2}p(p-1)$ scatterplots. Thus with 4 variables we have 6 scatterplots and with 10 variables we have 45 scatterplots. As the number of variables grows, the greatly increased number of scatterplots becomes hard to assimilate, but for modest values of p all the scatterplots may be displayed as a $p \times p$ array. A useful supplement, often available in commercial software, is painting or brushing (see, e.g., Cleveland, 1985), whereby the points associated with specified values of a third variable, possibly categorical, are highlighted or colored. For example, if we plot height against weight, the points referring to male samples may be colored black and those referring to female samples white. In the height/weight scatter the generally taller and heavier males would show up in the appropriate region, whereas in an age/family-size scatter any such relationship would be unlikely. Despite such refinements, multiple scatterplots are not ideal for visualizing relationships in multivariate data.

Multidimensional scaling (MDS) (see, e.g., Cox and Cox, 1994) offers an alternative generalization of the scatterplot. MDS starts by defining the distances between all pairs of samples. These distances may be observed values, but in the forms of MDS considered here, the distances are calculated as some simple function of the values taken by the p variables for each pair of samples. Then a set of points, one for each sample, is sought that generates the calculated distances. If this can be done at all, usually many dimensions are required for an exact representation, so MDS finds a representation in a few dimensions that approximates the given distances. Usually, few is two and will be so taken in the following. This is not a restriction on the methodology but reflects the difficulty of visualizing more than two dimensions except, perhaps, for three-dimensional models and with the aid of interactive graphics. In the two-way scatterplot arising from all forms of MDS, often called a "map," the axes are mathematical constructs that relate to the original variables in complicated ways. In contrast, biplots (see, e.g., Gower and Hand, 1996) relate the scatter of points representing samples directly to the values of the variables associated with those samples. Biplots are MDS maps representing the samples supplemented by information related to the variables.

2 Visual Representations of Quantitative and Categorical Variables

In this section we begin by briefly describing principal components analysis (PCA), which is the simplest form of MDS for quantitative variables. Similar methods may be used to represent a multivariate sample with p categorical variables. We describe a simple way to handle categorical variables based on what is known as the extended matching coefficient and close with the better known, but related, method of multiple correspondence analysis. Gower and Hand (1996) give all the technical details needed to construct all these biplots. In Section 3 we demonstrate the methodology using data from a survey of British sugar beet production.

2.1 Principal Components Analysis

In PCA, the plane of approximation is a subspace of an exact representation in p dimensions. The two-dimensional PCA approximation is especially useful when supplemented by nonorthogonal linear biplot axes (Gabriel, 1971) that represent the variables. Indeed, these biplot axes are the projections of the original coordinate axes onto the plane of approximation. When the biplot axes are endowed with scales, a practice that seems to be gaining acceptance (see Gower and Harding, 1988; Gabriel and Odoroff, 1990; Greenacre, 1991), then, although nonorthogonal, they may be used like familiar coordinate axes. The scales on these axes may be used to interpolate new samples by evaluating vector sums as in Figure 1a. This justifies the term interpolation because new samples may be interpolated into the map determined by the old samples; of course, the old samples interpolate into their correct positions in the map. The same axes, but with different scales, may be used for prediction. Thus, the position of any point in the PCA display may be orthogonally projected onto each biplot axis in turn and the corresponding p scale markers predict the values of the original variables that are to be associated with the sample represented by the point. It can be shown that this two-dimensional procedure gives the same results as projecting the point onto each of the original p-dimensional axes representing the variables, as in Figure 1b, where $p = 2$. This justifies the term prediction because the values of the variables to be associated with any point in the map may be predicted. Reading scale markers by projecting onto the prediction axes is equivalent to evaluating an inner product and gives a graphical way of predicting the best two-dimensional approximation to the observed sample values. Here "best" is used in the sense of Eckart and Young (1936) as the set of predictions that minimize the sum of squares of the differences between the observed and predicted values of all the variables. The inner product interpretation figures largely in the literature on biplots, but we think it unnecessarily obfuscates what is essentially the familiar process of referring to coordinate axes. The directions of the horizontal, vertical, and any higher dimensional orthogonal principal axes that are used to construct the PCA display are often interpreted. The directions of the biplot axes offer an alternative basis for interpretation and, we believe, one that is better. Therefore, we recommend the retention of only the planar approximation with its linear biplot axes, discarding all other axes.

In approximations the two sets of scale markers are inversely related, as in exact representations, but now interpolation and prediction are not consistent operations.

2.2 Reference Systems for Categorical Variables

First, let us see how the equivalent of coordinate axes may be defined for categorical variables. Because categorical variables take only a finite number of levels, they cannot be represented by an axis with a continuous scale. Instead, each categorical variable is represented by a set of points, one for each category level as shown in Figure 2.

In Figure 2a we show three points representing the levels *green, red, blue* of a categorical variable "color." These points are at the vertices of an equilateral triangle and are known as category level points (abbreviated to CLPs). The set of CLPs for a categorical variable corresponds to a linear axis that represents a quantitative variable and the labels attached to the CLPs correspond to the markers that give the numerical values of a quantitative variable. Notice that this representation of a three-level categorical variable requires two dimensions as well as the origin. In general, a categorical variable with L_k levels is represented by L_k CLPs at the vertices of a regular simplex and therefore occupies $L_k - 1$ dimensions. Thus the representation of two or more categorical variables tends to require many dimensions, which are difficult to show in the two dimensions that suffice in Figure 1 for two quantitative variables. The simplest case is for two variables each at two levels, which gives four

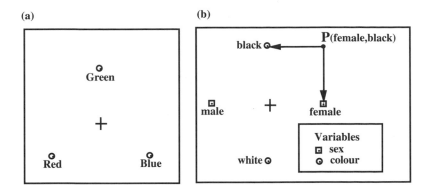

Figure 2: (a) Category level points for a single categorical variable, "color," with three levels—"green," "red," and "blue." (b) The prediction for the point P of the levels of two categorical variables, "sex," and "color," each with two levels shown in two dimensions.

CLPs generally requiring three dimensions (one for each variable and one for the origin). However, to give an idea of things in two dimensions, we may show each variable as two CLPs on each of two orthogonal lines intersecting at an origin. This is shown in Figure 2b, where the variables are "sex" with levels "female", "male," and "color" with levels "black," "white." Interpolation proceeds as before by vector sums but now there are only $2 \times 2 = 4$ possibilities, which occur at the vertices of a square. The vertex corresponding to (female, black) is shown at the point P. Prediction cannot be obtained by projection because there are no axes on which to project. However, the more fundamental concept of finding the *nearest* point survives and predictions are given by finding the nearest CLPs as shown in Figure 2b. Thus, we have the correspondences between coordinate representations of quantitative and categorical variables that are shown in Table 1.

2.3 Placing the CLPs (the Extended Matching Coefficient)

So far we have not seriously discussed the relative positions of the CLPs for different variables. There are several possibilities, but we shall mention only the two most important. The simplest definition of the CLPs is to place them on orthogonal axes a unit distance from an origin. Thus, the equilateral triangle of Figure 2a is obtained as in Figure 3. CLPs for other variables are obtained by extending this system by unit points on as many axes as there are category levels—L_k for the kth variable. This gives a total of $L = L_1 + L_2 + \cdots + L_p$ axes and L unit points as CLPs.

The word "axes" is used here merely for verbal convenience; in fact, only the simplices in mutually orthogonal spaces are essential and even the origin may differ for each set of CLPs. The vector sum method places a sample at one of $L_1 L_2 \ldots L_p$ points, giving a method for positioning any sample described by categorical variables.

Table 1: Representation of quantitative and categorical variables

Quantitative variables	Categorical variables
Each variable is represented by a linear axis.	Each variable is represented by a set of CLPs.
A scale is marked on each axis and each mark is labeled with a numerical value.	The CLPs are labeled with the names of the category levels.
The position of a point relative to the axes is obtained as the vector sum of the labels giving the values of the variables.	The position of a point relative to the set of all CLPs is obtained as the vector sum of the labels giving the relevant levels of the variables.
The values of the variables to be associated with a given point are obtained by orthogonal projection onto the axes and reading off the values given by the labels. This is the same as finding the nearest label to the given point on each axis.	The values of the variables to be associated with a given point are obtained by finding the nearest label in each set of CLPs.

Thus, coordinate axes are associated with continuous variables, and the discrete sets of CLPs are associated with categorical variables. To refer to both kinds of representation, Gower and Hand (1996) propose the term "reference system" and give examples of reference systems that combine continuous and categorical variables.

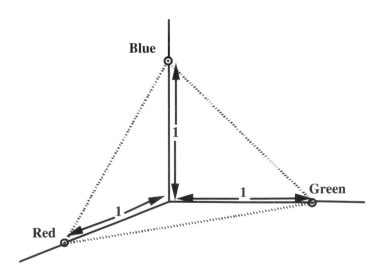

Figure 3: A simple configuration of CLPs for a categorical variable color.

The squared distance between two samples is given by the number of category levels in which they differ; which is $p - m$, where m is the number of matching category levels; m/p is known as the extended matching coefficient (EMC), which for two-level categorical variables (i.e., $L_1 = L_2 = \cdots = L_p = 2$) becomes the well-known simple matching coefficient. The EMC has the usual property of similarity coefficients that it is nonnegative, has value zero when no category levels are common to the two samples, has value one when the two samples share the same category levels, and otherwise lies in the zero–one interval.

Configurations of points generating the EMC may be approximated in two dimensions by PCA or, equivalently in this case, by classical scaling/principal coordinates analysis. As with PCA, we may project the "axes" to give biplot axes, but now there is only one scale point, corresponding to a category level, on each axis. Thus, projection gives L biplot axes, each with one marker corresponding to a CLP; other points on the axes have no direct meaning as they do for quantitative variables. These projected CLPs may be used for interpolating new points but there are difficulties in using them for the more important operation of prediction. This is because each variable is represented by several axes, three in the case of color in Figure 3. One could project onto each of these axes and predict the color corresponding to the projected CLP that was nearest a point of projection. Unfortunately, this does not necessarily give the one that is nearest the true CLP. In the exact representation, all points that correspond to a color, say red, are nearest the CLP for red. Therefore we can imagine the whole configuration as partitioned into regions containing the points nearest the CLP for red and similarly for green and blue. Such regions are called *neighbor regions*. The points in the approximation space that are in the neighbor region labeled red form a prediction region for the color red and similarly for the other colors and for all the other categorical variables. The geometry is illustrated in Plate 6.

The right-hand side of the figure shows the CLPs in high-dimensional space and the left-hand side shows the low-dimensional approximation space. To reduce confusion, the two parts are shown as well separated, although in reality they may be more intermingled. The neighbor regions are completely determined by the CLPs but, generally, there is no set of points in the plane of approximation that determine the prediction regions. In particular, within the plane of approximation there is no set of points for which the prediction regions are neighbor regions. This is unfortunate, for if there were, we could make the great simplification of representing the prediction regions by their generating points. To determine in which prediction region a point lies, each CLP could be projected onto the plane of approximation, also recording the (squared) distance of each CLP from the plane. Then, in principle, the distance of any point in the plane of approximation from each CLP could be calculated and hence its membership of a neighbor region/prediction region determined, but this is too cumbersome for ordinary use. In practice, it is better to show the boundaries between the prediction regions in full and with one diagram for each categorical variable. The prediction regions associated with the kth categorical variable are analogous to the kth linear biplot of PCA and are used in a similar way. In PCA one finds the nearest point on the axis and this involves orthogonal projection; with categorical variables, one finds the nearest CLP and this involves assignment to a prediction region. The prediction region diagrams may be superimposed, and occasionally this is a useful thing to do, but usually the picture becomes too complicated. To produce p separate diagrams diverges from the PCA biplot for quantitative variables, where all the linear axes may be shown on one diagram.

Fixing attention on the kth categorical variable, the whole of space may be divided into L_k neighbor regions, each of which is an $L_k - 1$ dimensional region extended orthogonally to the space of CLPs into the L-dimensional space. The approximation in few dimensions is a subspace of the full space, and where the subspace intersects the neighbor regions gives prediction regions, each labeled with a category name. In Plate 6 there are only three categories, so there are three prediction regions. With L_k categories there will be L_k prediction regions, some of which may be closed polygons and some may be hidden behind other regions, so the associated category level is never predicted.

Plate 7 is an example of the end product of an analysis, showing the prediction regions for a categorical variable with four levels, together with the positions of 10 samples, which have been numbered. The figure shows the convexity of the prediction regions and also shows a closed region for red. Prediction is obvious: sample number 4 is predicted as being green, while number 9 is predicted as red, and similarly for the other colors. Some predictions may be wrong, but in a good approximation most will be correct. Predictions for points close to a boundary, as with sample number 9, are especially likely to be uncertain. A table can be made of correct predictions versus actual predictions and the percentage of correct predictions may be used as measure of the quality of the approximation to give a criterion analogous to the least-squares criterion of the Eckart–Young theorem. Multidimensional scaling (MDS) offers a global method for positioning all the samples irrespective of the prediction regions for particular categorical variables that may be shown, as in Plate 7. For some purposes it may be desirable to show the samples in positions that predict better for particular variables than for others, but this would require new forms of MDS.

2.4 Multiple Correspondence Analysis

A more usual multivariate display for categorical variables is given by multiple correspondence analysis (MCA). There are many ways in which the methodology of MCA may be developed but here we adopt the approach used by Gower and Hand (1996), which emphasises the close relationship with PCA. It turns out that this is almost identical to the preceding approach. Indeed, the only difference is in the definition of the CLPs, which in MCA are not equidistant from the origin. If in a sample of 169 individuals, 9 are red, 16 are blue, and 144 are green, then, apart from a scaling factor, the corresponding CLPs are distant $1/3$, $1/4$, and $1/12$ from the origin—these are the inverses of the square roots of the frequencies. The coordinates of the CLPs are modified from the unit values that pertain to the EMC by replacing each unit by the corresponding inverse square root. Thus, as in Figure 3, the CLPs for color still form a triangle but it is not equilateral. This applies in general. The difference is analogous to using a quantitative scale with unequally spaced scale markers. With the MCA settings of the CLPs, the squared distance between a pair of samples is known as the chi-squared distance; the use of inverse square roots of frequencies gives greater weight to rare category levels than to common category levels. Chi-squared distance derives from the ordinary correspondence analysis of a two-way contingency table, where it has considerable justification (see, e.g., Greenacre, 1984), but its use in MCA has been criticized by Greenacre (1991). Certainly, the characteristic of chi-squared distance that gives rare categories greater weight than common categories is not always what is required. In ordinary correspondence analysis two sets of chi-squared distance, one between the rows and the other between the columns of the contingency table, have equal status. In MCA, the relevant contingency table is binary (termed the indicator matrix), containing only units and zeros giving, for each sample, the presence and absence of the category levels. Then, intercolumn chi-squared distance gives a measure of the distance between different levels of the same or two different categorical variables. We agree with Greenacre that this has little interest. By working in terms of prediction regions and having only the row points, which represent the samples, our approach avoids difficulties of these kinds.

Gower and Hand (1996) show that the extended matching coefficient (EMC) is monotonically related to many other distances (but not including chi-squared distance) that may be derived from comparisons among the rows of an indicator matrix. It follows that in the context of nonmetric MDS (see, e.g., Cox and Cox, 1994) all such coefficients are equivalent. In metric MDS this equivalence vanishes but lends support to using the simplest form, the EMC itself, unless there are strong reasons for adopting some more complicated definitions of distance. If one insists on using a more complicated form, then prediction regions may still be constructed but the methodology is less straightforward (Gower and Hand, 1996).

2.5 Computation of Prediction Regions

As we have seen, prediction regions are the intersection of the plane of approximation with the neighbor regions determined by the CLPs. It follows that in two-dimensional approximations we may proceed for the kth variable as follows. Consider the plane of approximation as being made up of pixels, as it would be on a computer screen. For each pixel, compute its distance from the L_k CLPs and color the pixel according to the category level of the nearest CLP. When all pixels have been colored, the plane of approximation will be partitioned into the neighbor regions for the kth categorical variable. Although our example in Figure 3 is couched in terms of a variable "color," and hence the coloring of pixels is particularly apt, the method will work for any categorical variable. Indeed, the coloring is needed only to determine the linear boundaries between the prediction regions. Once the boundaries have been found, they may be shown as lines and the regions they enclose labeled as in Plate 7.

The pixel coloring algorithm is simple but it is not very efficient. Gower (1993) has described the basis for an efficient algorithm that gives insight into the geometry of prediction regions. In principle, this algorithm will work for any value of L_k and for any number of dimensions, but it is yet to be implemented.

3   Prediction Regions for the Sugar Beet Data

The data for this example are drawn from the British Sugar Crop Survey (1993). This survey is the principal means for collecting information on the UK sugar beet crop. For each farm, the survey collects data on the factory where the crop is processed, crop, sowing, fertilizer usage, disease, and pest control. The illustrations of prediction regions given in this section concern a subset of 53 farms, selected from the total of 580 farms, and the following eight categorical variables:

Region	North, West, East
Factory	N1, N2, N3, W1, W2, E1, E2. E3, E4, E5
Drill	D1, D2, D3, D4, D5, D6, D7, D8, D9
Soil type	Sands, Sandy-loam, Silty-loam, Clay
Variety	Regina, Amethyst, Hilma, Gala, Matador, Rex, Planet, Saxon, Celt, Giselle, Triumph, Zulu, Aztec, Cordelia
Stubble cultivation	No, Yes
Straw disposal	Left, Removed, Incorporated, No straw
Sugar content	A quantitative variable grouped into an ordered categorical variable with three levels (low, medium, and high)

In this list, the factory names are given in coded form, with the initial letter indicating the geographical region containing the factory; drill types are also coded.

We do not attempt an exhaustive analysis here. The two-dimensional approximation for the EMC accounts for only 16.7% of the total variation, which, in this case, occupies 39 dimensions (obtained as a total of 49 levels, less 8 variables less 2 because region can be deduced from "factory"). However, the percentage of correct predictions is 64.6%. The two-dimensional fit of the MCA at 32.0% was rather better when judged as percentage of the total variation, but the percentage of correct predictions at 62.3% was a little worse. Both sets of solutions were visually similar and we present diagrams only for the EMC.

Plate 8 shows the two-dimensional plot for the variable "regions." The numbers refer to the positions of the farms. The usual projections of the CLPs are not shown, but when they are, they may be used for vector sum interpolation rather than for prediction. The three prediction regions shown correspond to the three geographical regions (North, West, East). The farms tend to cluster into three groups that are enclosed within prediction regions, which, with the exception of farm 34, correspond precisely with the geographical regions. Plate 9 shows similar plots on a reduced scale for regions and for the other variables. For factories we see that the prediction regions are very similar to those of Plate 8, reflecting that the factories occupy the same geographical space as the farms. However, there are 10 factories but only three prediction regions appear. These are for the factories numbered E1, W2, and N3, which are those with the highest frequencies. Inevitably, predictions are wrong for the farms that process their beet at the less popular factories and the correct and incorrect predictions are indicated in the figure by open and black circles, respectively. Comparison with Plate 8 verifies that, despite incorrect predictions, at least every farm is allocated to a factory in its own region. The CLPs for the factories that are not shown are farther away from all the pixels than the CLPs for the three factories that are shown and hence their prediction regions are hidden.

With other variables, the picture is less clear-cut. For example, in Plate 9, for "straw disposal" there are only two prediction regions, corresponding to "straw removed" (37 farms) and "straw incorporated" (14 farms). The bulk of the 51 farms are not correctly classified, and in the figure the black circles denote the incorrect classifications. The remaining two farms—number 1, which is said to have "no straw," reasonably falls into the "straw removed" region, and number 38, recording "straw left," which is certainly not "straw removed" and is akin to "straw incorporated," as predicted. Recall that the open and black circles merely record correct and incorrect predictions; we could have colored every point by the color that correctly gives its recorded category level—correct predictions would superimpose a colored dot on the same background color and would be recognized as open circles, as in Plate 9. More in keeping with the painting and brushing of scatterplots would be to color the dots according to the category levels of some other variable. For example, if the dots representing farms in the plot for geographical region were colored according to their method of straw disposal, it would be seen that the Western region almost universally prefers straw removal (farm number 4 is an anomaly), as does the Eastern region, with a few exceptions; the Northern region would be seen to be divided in its methods for straw disposal. Similar remarks apply to the other variables, but the visualizations say everything that needs to be said.

The linear biplot axes associated with the classical analysis of numerical variables are usually displayed on a single diagram. The corresponding plot for categorical variables would require the superimposition of all the prediction regions. Thus, with the present example, all eight components of Plate 9 would have to be superimposed. Clearly, the resulting display would be highly confusing, so we do not attempt it; Gower and Hand (1996) give an example with four variables where superimposition is feasible. Superimposition can be helpful when there are only a few variables and when the different prediction regions tend to overlap, indicating association between the corresponding variables. The practical exploitation of such possibilities calls for interactive graphical facilities that permit one to modify a current set of superimpositions by adding or removing the prediction region for a nominated variable. This example shows that prediction regions give a visual representation of categorical variables that can focus attention on the main features of a sample.

The results for the EMC and MCA may be compared numerically by tabulating the prediction error rates as in Table 2. For EMC there are 274 correct predictions and 150 false predictions, slightly better than the corresponding figures for MCA of

Table 2: Error rates for MCA and EMC

	Multiple correspondence analysis			Extended matching coefficient		
	Predicted			Predicted		

REGION

True	N	W	E	N	W	E
N	18	0	0	16	0	2
W	0	17	0	0	17	0
E	0	0	18	1	0	17

FACTORIES

	E1	W2	N3	E1	W2	N3
E1	7	0	0	7	0	0
W2	0	10	0	0	10	0
N3	0	0	12	0	0	12
W1	0	7	0	0	7	0
others	9	1	7	8	2	7

DRILLS

	D3	D6		D3	D6	
D3	8	3		9	2	
D6	0	25		0	25	
Others	8	9		14	3	

SOIL TEXTURE

	Sandy Loam	Silts/Loams		Sandy Loam	Silts/Loams	Clays
Sands	3	1		1	1	2
Sandy Loam	17	6		17	6	0
Silts/Loams	11	9		9	10	1
Clays	3	3		0	5	1

Multiple correspondence analysis

VARIETIES

True	Predicted	
	Saxon	Celt
Saxon	17	0
Celt	9	0
Regina	3	0
Others	22	2

STUBBLE CULTIVATION

	No	Yes
No	21	7
Yes	3	22

STRAW DISPOSAL

	Removed	Incorporated
Removed	35	2
Incorporated	12	2
Left	1	0
No straw	0	1

SUGAR CONTENT

	Low	Medium	High
Low	6	9	3
Medium	4	16	1
High	6	4	4

Extended matching coefficient

VARIETIES

	Predicted		
	Saxon	Celt	Regina
Saxon	12	5	0
Celt	5	4	0
Regina	1	1	1
Others	15	8	1

STUBBLE CULTIVATION

	No	Yes
No	23	5
Yes	1	24

STRAW DISPOSAL

	Removed	Incorporated
Removed	29	8
Incorporated	1	13
Left	1	0
No straw	0	1

SUGAR CONTENT

	Low	Medium	High
Low	6	6	6
Medium	4	11	6
High	1	3	10

264 and 160. As with linear biplots, some variables are approximated better than others. Table 2 shows that the variables "region," "drills," "stubble cultivation," and "straw disposal" give better than 60% correct predictions for both EMC and MCA, so are reasonably well represented, whereas "varieties" are very poorly represented. However, for "straw disposal" MCA gets 35 correct predictions for the "removed" category with only 2 for "incorporated," while the corresponding figures for EMC are 29 and 13. Note that these numbers are computed values that, for EMC, may be compared with Plate 8. What may seem small discrepancies arise from the occasional coincidence of pairs of points in the figures and some ambiguity when a point lies on a line separating prediction regions.

4   Conclusion

The methods discussed here may be set within developments of biplot theory discussed by Gower and Hand (1996), where the concept of biplot axes is extended to most forms of MDS. In general, quantitative variables require different sets of axes for interpolation and for prediction and, rather than linear axes, we may require nonlinear axes, called trajectories. We have seen how biplot axes marked with scales for continuous variables correspond to prediction regions labeled with names for categorical variables. Gower and Hand (1996) show how both types of variables may be exhibited simultaneously. Ordered categorical variables may be treated by using linear axes with irregularly marked scales, which give prediction regions that are parallel bands of differing widths. Between the general disposition of CLPs in $L_k - 1$ dimensions, as described earlier, and the unidimensional CLPs for ordered categorical variables, there is the possibility of dispositions of CLPs in intermediate numbers of dimensions; one approach that gives suitable coordinates for CLPs can be found in the multiple solutions given by the homogeneity analysis program HOMALS described by Gifi (1990) and available in SPSS. CLPs, once found, define neighbor regions and the methods described earlier remain valid for deriving the boundaries of prediction regions. The variant of MCA termed joint correspondence analysis (Greenacre, 1988b, and Chapter 17 in this volume) provides its own set of CLPs, to which Gower and Hand (1996) show how to add the sample points.

Finally, we return to the introduction, where we drew attention to the effect of scaling quantitative variables on the distances in scatter plots and MDS. With categorical variables this problem manifests itself in the choice of coordinate positions for the CLPs. We have discussed two possibilities: (1) where with the EMC every CLP is at a vertex of a unit simplex and (2) where MCA places the CLPs at vertices whose positions depend on the category frequencies in the data. There is an analogy between (1) choosing equal scales for a set of quantitative variables and (2) scaling quantitative variables by data-derived quantities such as standard errors or ranges.

We have seen how error rates are associated with prediction regions. In Plate 9 it is clear that a small change in the boundary between the East and North regions would remove the one incorrect classification. This verifies that the plane of approximation, determined by least-squares fits to the samples positioned as vector sums of CLPs as we have used earlier, does not minimize the error rate; it is an interesting research problem to find what plane does. Further, can the positions of the CLPs be determined to minimize error rates while maintaining acceptable definitions of inter sample distance?

PART IV

Visualization and Modeling

In the fourth part of the book we turn our attention to statistical models for categorical data and how visualization can assist the modeling process and the interpretation of results.

A traditional approach in the social sciences is to formulate a number of hypotheses about the relationships between the variables of interest and to test them using a statistical model. Well-known models for categorical data include latent class models and log-linear and log-bilinear models such as Goodman's RC model. The combination of modeling and visualization ideas is the aim of the chapters in this part. Either the authors show how to visualize the residuals from a fitted model to search for structures in these residuals for improving the model, or they discuss similarities between modeling approaches and visualization techniques. Visualization techniques can also be used in situations in which models contain many parameters, where simple display methods can assist in the interpretation of the results.

Chapter 29, by Clifford C. Clogg, Tamás Rudas, and Stephen Matthews, introduces a new idea in contingency table modeling, based on a mixture model. The main idea is to split cell probabilities into two parts: one part that can be attributed to that part of the population where the model of interest holds and another part that can be attributed to that part of the population where the model does not hold. The model of interest can be any suitable one, for example, the independence model or the quasi-independence model, belonging to any kind of contingency table. Instead of applying the model to the whole population, which is the usual statistical approach, the model of interest is applied only to the respondents for whom this model holds, with all other persons belonging to the alternative model. It follows that all values in the alternative model, which can be treated as residuals, are positive. The visualization part in the chapter refers to different ways of displaying those residuals.

In Chapter 30, Yoshio Takane gives an introduction to the visualization in ideal point discriminant analysis (IPDA) of contingency tables. IPDA is a model of the conditional probability of row i given column j in terms of the distances d_{ij} between row and column points in a low-dimensional Euclidean space. IPDA is one of the best examples of the integration of visualization and modeling ideas, where the data are modeled directly as a function of the graphical elements, in this case, interpoint distances. Takane gives three applications of IPDA to illustrate its usefulness.

Chapter 31, by Ulf Böckenholt, shows shifts in ideal points over time. According to the unfolding model, persons evaluate choice alternatives by comparing them to their ideal alternatives; they select their most preferred options, the ones that are closest (or least dissimilar) to their ideal option. An important constraint of unfolding theory is that, although persons may differ in terms of their preferences for the choice options, they agree on the similarity relationship among them. Thus, in the unidimensional case the choice options' positions along a common (latent) continuum are perceived homogeneously by all persons. The author uses an extension of this model to explain shifts in ideal points over time. Using two empirical data sets from marketing research Böckenholt demonstrates the possibilities of visualizing these shifts.

In Chapter 32, Allan L. McCutcheon demonstrates how to use correspondence analysis complementary to latent class analysis (LCA) in comparative social research. Using data from the General Social Survey Program, the author introduces several LCA models for answering different research questions. In the given examples, LCA is used to account for the observed heterogeneity in a multiway cross-tabulation by characterizing a set of unobserved, internally latent classes; for example, McCutcheon examines a five-class model conducted from questions on religious beliefs in seven nations. The classes mirror the latent "levels of belief." One aim of LCA is to determine the proportion of respondents in each country belonging to each class. The solution is a matrix of positive numbers that can be visualized by applying CA. Thus, CA gives a quick understanding of the solutions of the LCA models.

Chapter 33, by L. Andries van der Ark and Peter G. M. van der Heijden, discusses the visualization of latent class analysis and latent budget analysis (LBA), with special reference to correspondence analysis. Since the latent class model studies the joint probabilities of each cell of the contingency table to be analyzed, this approach should be used if the row variable and the column variable are both response variables or, in other words, if there is no causal interpretation in the table. The response variables are presumed independent given the latent classes. On the other hand, the latent budget model should be used if one of the variables is an explanatory variable and the other is a response variable. However, the authors show that LCA and LBA are equivalent techniques and that the parameters from the one model can be obtained from the parameters of the other. Van der Ark and van der Heijden also show how to project latent budgets onto a CA solution, thus providing a new way of interpreting these latent values.

In Chapter 34, Jay Magidson proposes the use of general ordinal logit displays for the visualization of the effects in categorical outcome data. In the traditional approach of log-linear modeling with an ordinal variable, the solution consists of long lists of parameter estimates and related statistics, which are often difficult to interpret. Using examples in which the categorical outcome is either dichotomous or ordinal and the predictor variables are either nominal or ordinal, the author demonstrates the power of graphical displays for interpreting the data. Together with the traditional statistics, which reflect the model fit and the significance of effects, the displays show which effects are minor and which are major. Thus, one can conclude from the display where there are possibilities for improving the log-linear model.

Chapter 35, by Antoine de Falguerolles, discusses the visualization of the residuals in log-linear models. The general idea of this kind of model is that the model formula for the predictor consists of a linear term and an additional bilinear term of reduced rank that models the interaction between the rows and columns. The advantage of modeling the interaction in this way is that the fitted row and column parameters of the bilinear term can be plotted on orthogonal axes and interpreted as a biplot. The author illustrates the methodology with two examples, first a three-way contingency table of suicide behavior, treated as a two-way table where "causes of death" is cross-tabulated against the combined variable of sex and age group. The second example is a square mobility table illustrating the quasi-symmetric model with a bilinear term, which also leads to a biplot display of the fitted model parameters.

Chapter 29

Analysis of Contingency Tables Using Graphical Displays Based on the Mixture Index of Fit

Clifford C. Clogg, Tamás Rudas, and Stephen Matthews

1 Introduction

We present here a new approach to the visualization of structure in categorical data. This approach assumes that a simple model is considered in order to define the structure of interest. We consider methods for the analysis of two-way contingency tables, and as an example of data of this kind we shall analyze the occupational mobility table given in Table 1. This contingency table is taken from the famous study by Blau and Duncan (1967), as condensed by Knoke and Burke (1980). This table cross-classifies American men in 1962 according to their current occupation category and their fathers' occupation category. The approach presented here is not limited to mobility, or other two-way tables, or to the models considered. It could be applied to any of the several models that have been suggested for the analysis of social mobility and related two-way tables (see Goodman, 1984; Goodman and Clogg, 1992; Clogg and Shihadeh, 1994; Luijkx, 1994). The method can also be generalized to higher dimensional contingency tables.

Our approach relies on the following logic. First, a model H is proposed either as a baseline model or as a structural model. Second, this model is embedded in a special two-point mixture model. Third, the special residuals from this mixture representation

Table 1: A 5 × 5 occupational mobility table

				Son			
Father		1	2	3	4	5
1	Professional and managerial	152	66	33	39	4
2	Clerical and sales	201	159	73	80	8
3	Craftsmen	138	125	184	172	7
4	Operatives and laborers	143	161	209	378	17
5	Farmers	98	146	207	371	226

Source: Blau and Duncan (1967, p. 496), as condensed by Knoke and Burke (1980, p. 67). The cell frequencies in the Blau–Duncan table, as reported by Knoke and Burke and used here, are actually population estimates divided by ten thousand; the given sample size ($n = 3396$) is not the actual sample size used to estimate the population totals.

are examined. These residuals summarize structure or unmodeled structure in relation to model H. The mixture-model residuals are very different from ordinary residuals in two important ways: they are always valid and always nonnegative. The new residuals may be described by tabular and graphical displays.

When applied to the mobility data, this approach leads to splitting the cell frequencies or probabilities into two parts: one part that can be attributed to the part of the population where the model of interest holds and another part that can be attributed to the part of the population where the model does not hold. For example, it will be shown that the model of quasi-uniform association can describe the mobility process in about 95% of the population, and nearly half of the remaining 5% of the population is concentrated into one cell of the mobility table.

The new approach offered here is one way to use models as a guide for the visualization of structure in categorical data. The approach specifies what can be compactly summarized with parameter values from some simple models and what needs to be summarized with graphical or tabular displays. Our methods can serve as a kind of rapprochement between modeling and graphical techniques for the analysis of categorical data.

2   Some Models

For the two-way contingency table cross-classifying variables R and C, let (i, j) denote a given cell and let f_{ij} and F_{ij} denote the observed and expected frequencies, respectively, for $i = 1, \ldots, I, j = 1, \ldots, J$. Let $p_{ij} = f_{ij}/n$ and $P_{ij} = F_{ij}/n$ denote the observed (or empirical) cell proportions and the expected (or theoretical) cell proportions.

Our approach starts with specifying a model H for the table. Three models will be considered for illustrative purposes. These are the *independence (I), quasi-independence (QI)*, and *quasi-uniform (QU) association* models. These three models form useful baselines in many cases in which two-way tables are considered. Written as log-linear models, these are

$$I : \log(F_{ij}) = \lambda + \lambda_{R(i)} + \lambda_{C(j)} \tag{1}$$

$$QI : \log(F_{ij}) = \lambda + \lambda_{R(i)} + \lambda_{C(j)} + \lambda_{RC(i,i)} \tag{2}$$

$$QU : \log(F_{ij}) = \lambda + \lambda_{R(i)} + \lambda_{C(j)} + \lambda_{RC(i,i)} + \phi\, i\, j \tag{3}$$

The QI model can be obtained from the I model by adding (on the logarithmic scale) special parameters to the cells on the main diagonal, supposing these are not structural zeros. The QU model is a special form of quasi-symmetry and is useful when the variables have ordered categories (Goodman, 1984). This model assumes essentially that, except for the cells on the main diagonal, all the local odds ratios $(F_{i,j}F_{i+1,j+1})/(F_{i,j+1}F_{i+1,j})$ have the same value.

Now consider the sufficient statistics for the above models (see, for example, Agresti, 1990). These are

$$I : \{n, (f_{i+}), (f_{+j})\} \tag{4}$$

$$QI : \{n, (f_{i+}), (f_{+j}), (f_{ii})\} \tag{5}$$

and

$$QU : \left\{ n, (f_{i+}), (f_{+j}), (f_{ii}), \sum_i \sum_j f_{ij} i\, j \right\} \tag{6}$$

where the ranges of the subscripts for sets of statistics have been suppressed for convenience. Note that these sets of sufficient statistics are not minimal, but they summarize all the relevant sample information for the respective model, assuming the model is true for the entire population. A more precise formulation of the sets of sufficient statistics would require the specification of the actual sampling scheme. Depending on this, the sample size n may or may not be a statistic observed from the data, and even if it is, its value could be easily obtained from the other statistics listed. The sample size is included here to facilitate generalizations later in this chapter.

When any of the models I, QI, QU is analyzed by maximum likelihood procedures, the values of the sufficient statistics are fitted, or carried over from the data to the estimate. For the I model, these quantities are the sample size, row marginal distribution, and column marginal distribution. For the QI model, the same quantities plus the cell entries on the main diagonal are fitted, and for the QU model there is one additional quantity, the observed cross product, using row scores i and column scores j.

When a model is not true for the entire population, the "sufficient" statistics are not sufficient, for they do not capture all the information relevant in the data. Actually, the values of the preceding statistics in this case may be completely irrelevant. Traditional approaches to model fitting and testing assume that the model of interest is either true for the entire population or not and assess the likelihood of obtaining the given sample under the former assumption. In the next section we shall show how the idea that a model may be true for a part of the population can be used to measure the fit of the model. In this approach, the preceding sets of "sufficient" statistics will be sufficient for only that part of the population where the model holds.

The I and QI models are often taken as baseline models, but the QU model might be viewed as an approximation for the model that generated the data in several cases. Using these models as a guide, we will present an approach that summarizes model misfit, either via tabular displays or via graphical displays in terms of the sufficient statistics for the models as well as in terms of other quantities.

3   The π^* Index of Structure

The material in this section follows Rudas *et al.* (1994) and Clogg *et al.* (1995), where proofs of the assertions made here can be found.

Consider a model H for the contingency table, for example, any of the models described earlier. Model H is embedded into the following model H_π

$$H_\pi : P_{ij} = (1 - \pi)\Pi_{1(ij)} + \pi\Pi_{2(ij)}; \quad \Pi_1 \in H, \Pi_2 \text{ unspecified} \tag{7}$$

Model (7) can be given the following interpretation: A dichotomous latent variable, say X, is posited with $\Pr(X = 1) = 1 - \pi$ and $\Pr(X = 2) = \pi$, that is, with latent class proportions, or mixing weights $1 - \pi$ and π. Within the tth latent class ($t = 1, 2$), $\Pi_{t(ij)} = \Pr(\text{cell}(i, j) \mid X = t)$; that is, the $\Pi_{t(ij)}$ denote the conditional probabilities of interest.

Model (7) is different from the ordinary two-class latent structure model in two respects. First, the model H originally specified applies only to the first latent class. The second latent class is not modeled, or is unrestricted, or is estimated nonparametrically. The usual latent class model assumes independence or "local independence" in both latent classes. Second, model H need not be independence or some restricted version of independence. Any restricted or unsaturated model can be used to define H and hence the mixture model H_π.

For a given model H, the mixture representation H_π in (7) defines a class of models as π varies between zero and one. Note that $H_0 = H$; that is, for $\pi = 0$ (7) is equivalent to H. The specific model with $\pi = 0$ says that model H is completely congruent with the data, or that model H applies to the entire population. When $\pi = 1$, the data, or the true distribution that generated the data, would be said to be completely outside the model formulated initially. Also, H_1 does not imply any restriction as to the true distribution.

The class of models in (7) has the following nesting property: for $\pi < \pi', H_\pi \subset H_{\pi'}$. This means that if $H = H_0$ does not contain the true distribution, then a value of π with $0 < \pi \leq 1$ can always be found so that the model in (7) fits the distribution perfectly, since H_1 is not restricted and therefore contains any distribution. Or, if H_π does not fit the distribution perfectly for a given value of π, then a value of π' with $\pi < \pi' \leq 1$ can always be found so that model $H_{\pi'}$ contains the distribution. Because of this property, the class of models in (7), with π considered as a parameter, can be used to represent the lack of fit of model H. There is always a value of π for which H_π fits the underlying distribution perfectly; therefore this model can be used to quantify the lack of fit of H, that is, to summarize the structure in the true distribution in relation to H and to pinpoint cells where this lack of fit arises. For cases where model H serves as a definition of structure, the magnitude of the structure can be quantified in terms of π, and the local structure can be investigated using the entries in the table of the conditional probabilities for the second latent class (i.e., $\Pi_{2(ij)}$).

For a specified model H, we define our index of structure (or of lack of fit) π^* as the smallest value of π for which the model in (7) contains (fits) the true distribution P. That is, the functional $\pi^*(P)$ is defined as

$$\pi^*(P) = \inf\{\pi : P = (1 - \pi)\Pi_1 + \pi\Pi_2; \ \Pi_1 \in H, \Pi_2 \ \text{unspecified}\} \qquad (8)$$

where the cell subscripts have been suppressed. Because of the nesting property, H_π for any $\pi^* < \pi \leq 1$ will also fit P. It is important to recognize that the parameter π^* is the minimum value of the mixing proportion π for which H_π describes the distribution. The value of π^* is unique, because it is a minimum. Also, the maximum likelihood estimator (MLE) of π^* is obtained by substituting $\hat{P} = \{\hat{p}_{ij}\} = \{f_{ij}/n\}$ for P in the functional defined in (8). Iterative calculation of the MLE of π^* is described by Rudas *et al.* (1994). Basically, the algorithmic problem is to find the smallest value of π for which the mixture representation in (7) produces perfect fit for the observed distribution. For other algorithms see Xi and Lindsay (1997).

For a given model H, such as independence or "perfect mobility," suppose now that π^* or $\hat{\pi}^*$ is available. These quantities have the following simple interpretation: π^* is the minimum fraction of the population that is "outside" model H, and $\hat{\pi}^*$ is the MLE of that fraction.

Note that whereas in the traditional hypothesis-testing approach the null and alternative hypotheses may both be false, the representation in (7) is always possible with an appropriate value of π, and inference based on it is always valid.

4 The π^* Index Applied to the Mobility Data

The framework outlined will now be applied to the data in Table 1. Standard chi-squared fit statistics as well as the MLE of π^* appear in Table 2. The chi-squared values indicate that each model is unacceptable, although the QU model provides a relatively good fit. The $\hat{\pi}^*$ values for the three models are 0.310, 0.147, and 0.052, re-

Table 2: The usual fit statistics and the mixture index of structure π^* for three models applied to the data in Table 1[a]

Model	X^2	L^2	df	$\hat{\pi}^*$	$\hat{\pi}_L$
I	875.10	830.98	16	0.310	0.282
QI	269.07	255.14	11	0.147	0.123
QU	30.78	27.82	10	0.052	NA

[a] X^2, L^2 refer to the Pearson and the likelihood-ratio chi-squared statistics; df to the degrees of freedom; $\hat{\pi}_L$ denotes an approximate 95% lower confidence bound for π^*. NA means that the confidence interval contains zero. See text for a description of the models.

spectively. These quantities have the following interpretation. With the independence model we estimate that 31% of the population is outside the model, and this quantity can be taken as the amount of structure to be explained either by other models or by graphical summary. In other words, we estimate that 69% of the population can be described by independence and no special visualization techniques are required for this part of the population. For the *QI* model, nearly 15% of the population is estimated to be outside the model, or about 85% of the population can be described by quasi-independence and, again, no special visualization techniques are required for this part of the population. Alternatively, about half of the structure not described by the independence model, in terms of the fraction of the population where it is present, is accounted for by including special parameters for the cells on the main diagonal. Finally, for the *QU* model, only about 5% of the population is estimated to be outside the model, and this amount seems to be quite small. That is, in spite of the fact that the *QU* model does not fit the data well using a strict interpretation of the chi-squared statistic, the model fits well in the sense that it is able to describe about 95% of the population.

It is well known that values of the chi-squared statistics are proportional to the sample size for sets of data with the same observed distribution $\hat{P}$. Given that for the present set of data the cell frequencies are merely population estimates in ten thousands (the actual frequencies used to estimate these are not available in the source), it is not valid to rely on the chi-squared values as strict test statistics. Depending on the actual sample size, they may be substantially smaller or substantially greater than the values reported in Table 2. The $\hat{\pi}^*$ values, on the other hand, do not depend on the sample size in the preceding sense; for a given model, these values depend on $\hat{P}$ only. Therefore, in this case, the inference based on the mixture index of fit π^* appears to be more valid than inference based on the chi-squared values.

The last column of Table 2 reports 95% lower confidence bounds for π^*. For the *QU* model the resulting confidence interval contains the value zero. This implies that the hypothesis that the true value of π^* is equal to zero, that is, the model describes the entire population, cannot be rejected at the 5% level. This test relies on the assumption that the true sample size is the one reported in Table 1.

Another interpretation of the quantity $\hat{\pi}^*$ can be obtained by linking its value to the sufficient statistics. All of the models (1), (2), and (3) include the constant term λ. The MLE of λ for a table of fixed size under any of the models can be computed using the sample size n only from among the sufficient statistics listed under (4), (5), and (6), respectively. The sample size is sufficient for the constant term under any of the models (1), (2), or (3) (in fact, under any log-linear model assumed to hold for the entire population). In these cases, the constant term λ in the model is a scaling factor making the estimated frequencies sum to the sample size (or, consequently, the estimated probabilities sum to 1).

Consider now the following reparameterization of the mixture model (7) for the $H = I$ model in (1):

$$F_{ij} = \exp\left(\kappa_1 + \kappa_{1R(i)} + \kappa_{1C(j)}\right) + \exp\left(\kappa_2 + \kappa_{2R(i)} + \kappa_{2C(j)} + \kappa_{2RC(ij)}\right) \quad (9)$$

with the usual assumption that the κ_1 and κ_2 parameters summed in any of their arguments give zero. Then subject model (9) to the requirement that

$$\sum_i \sum_j \exp\left(\kappa_1 + \kappa_{1R(i)} + \kappa_{1C(j)}\right) \text{ is maximal} \quad (10)$$

Model (9) is the same mixture representation that appears in (7), but the independent and nonrestricted parts are now in log-linear representation, with the $(1 - \pi)$ and π mixing weights absorbed. Condition (10) means that the weight of the independent part $[(1 - \pi)$ in (7)] should be as large as possible.

Then $n\hat{\pi}^*$ is sufficient for the constant term κ_2 in model (9)–(10), and $n(1 - \hat{\pi}^*)$ is sufficient for κ_1. That is, under (1) $n (= n1)$ is sufficient for the constant term, while under (9)–(10), $n(1 - \hat{\pi}^*)$ is sufficient for the constant term for the part of the population where (1) is supposed to hold. This reflects the fact that instead of the fraction 1 (i.e., the entire population), model (1) is supposed to be true only in a smaller fraction of size $1 - \pi^*$.

This argument generalizes directly to any model that includes a constant term, not just to models (2) and (3), but to any log-linear model as well.

5 Residuals Based on the Mixture Representation of a Model

We now consider some properties of the Π_2 matrix, which is the unrestricted part of the mixture model (7). Suppose that H is the model of independence with $(I-1)(J-1)$ degrees of freedom or, equivalently, $I + J - 1$ estimated parameters, or constraints on the fitted frequencies [with $IJ - (I + J - 1) = (I-1)(J-1)$]. The table of conditional probabilities for the unstructured latent class (Π_2) will have at least $I + J - 1$ zeros. Consequently, the number of nonzero entries in this matrix will be no greater than $(I-1)(J-1)$, which coincides with the number of degrees of freedom for the model I. In general, if model H has d degrees of freedom, then the number of nonzero entries in Π_2 will be less than or equal to d, and there will be at least $IJ - d$ zeros (see Xi, 1996).

The nonzero entries in Π_2 define residuals in the context of the mixture representation. The pattern of nonzero entries summarizes the local misfit or local structure. The pattern and size of nonzero entries in this matrix may suggest ways to modify the original model. These entries can be examined using conventional tabular displays or by using simple graphical displays. Note that H might be either a model proposed as an "explanation" of the data, in which case π^* summarizes lack of fit of this model and the pattern of nonzero entries in Π_2 describes cell-by-cell lack of fit of this model, or H might be a model used primarily to define structure (H is some standard baseline model), in which case π^* measures the amount of structure not described by H and the pattern of nonzero entries in Π_2 describes the local structure not described by H. By considering the local misfit or local structure as summarized in the unrestricted matrix of probabilities Π_2, a new and fundamentally different kind of residual analysis is possible.

To facilitate comparison, the entries in $\hat{\Pi}_2$ can be multiplied by $\hat{\pi}^*$; when this is done, the decomposition corresponding to (8) or to (9) and (10) can be written as

$$p_{ij} = q_{ij} + r_{ij} \qquad (11)$$

where $p_{ij} = f_{ij}/n$ [the observed proportion in cell (i, j)], $q_{ij} = (1 - \hat{\pi}^*)\hat{\Pi}_{1(ij)}$ is the component associated with the part of the population estimated to lie in H, and $r_{ij} = \hat{\pi}^*\hat{\Pi}_{2(ij)}$ is the component associated with the part of the population estimated to be outside H. We shall call the r_{ij} quantities the mixture-model residuals, or MMRs for short.

The MMRs are different from ordinary residuals in two very important aspects. First, the MMRs are always valid, in the sense that the representation (7) from which they are derived is always valid (for some value of π) in contrast to the usual residuals that are based on the assumption that model H is true for the entire population, which may or may not be correct. Moreover, when model H is not true, the meaning of the ordinary residuals is somewhat dubious. Second, the MMRs are always nonnegative, have the straightforward interpretation of being the distribution in the part of the population where H does not hold true, and therefore can be analyzed by methods similar to those that can be used to analyze the distribution of the estimates under model H.

The decomposition in (11) leads to an analysis of local structure that is consistent with the index of overall structure in the following sense. Summing over i and j gives

$$1 = \sum_i \sum_j p_{ij} = \sum_i \sum_j q_{ij} + \sum_i \sum_j r_{ij} = (1 - \hat{\pi}^*) + \hat{\pi}^*$$

The MMRs sum to the "overall residual" $\hat{\pi}^*$. The relative contribution of a given MMR to the overall residual is $r_{ij}/\hat{\pi}^*$.

In some cases the MMRs may be used to identify the parts of the population where H does not fit or does not fit well (Rudas and Zwick, 1997).

6 Analyzing Lack of Fit and Local Structure Using Tabular Displays

Tables 3, 4, and 5 give the r_{ij} MMRs for the I, QI, and QU models, respectively. For such a simple table (i.e., a 5×5 table), these tabular displays go a long way toward summarizing the lack of fit of the models. They also describe the structure in the data not captured by the given model. Note that the sizes of the residuals decrease as we go from one model to the next.

For the independence model, the residuals in Table 3 admit the following simple interpretation. Nonindependence arises from the upper left 3×3 subtable and the lower right 1×1 subtable, that is, cell (5, 5). Contributions from cells (1, 1), (2, 1), (2, 2), and (5, 5) are dominant. The other entries in this table are virtually zero or

Table 3: Mixture-model residuals for the independence model applied to the data in Table 1[a]

| | Son | | | | |
Father	1	2	3	4	5
1	0.042	0.015	0.003	0	0.001
2	0.053	0.038	0.008	0	0.001
3	0.027	0.017	0.026	0	0
4	0.013	0.004	0	0	0.001
5	0	0	0.001	0	0.062

[a]There are 9 zeros and $25 - 9 = 16$ nonzero entries, corresponding to the number of parameters (or degrees of freedom) for the independence model. The values sum to the value of $\hat{\pi}^*$ (i.e., to 0.310), except for rounding error. See Table 1 for the definition of the categories.

Table 4: Mixture-model residuals for the quasi-independence model applied to the data in Table 1[a]

| | Son | | | | |
Father	1	2	3	4	5
1	0	0.012	0	0.001	0.001
2	0.048	0	0	0	0.001
3	0.017	0.002	0	0	0
4	0.010	0	0	0	0.002
5	0	0	0.005	0.047	0

[a]There are 14 zeros and $25 - 14 = 11$ nonzero entries, corresponding to the number of parameters (or degrees of freedom) associated with the quasi-independence model. Apart from rounding, the entries sum to $\hat{\pi}^*$. See Table 1 for the definition of categories.

Table 5: Mixture-model residuals for the quasi-uniform association model applied to the data in Table 1[a]

	Son				
Father	**1**	**2**	**3**	**4**	**5**
1	0	0.005	0	0.004	0.001
2	0.025	0	0	0.003	0.002
3	0.001	0	0	0.006	0
4	0	0	0.003	0	0
5	0	0.003	0	0	0

[a]There are 15 zeros and $25 - 15 = 10$ nonzero entries, corresponding to the number of parameters (or degrees of freedom) associated with the quasi-uniform association model. See Table 1 for the definition of categories.

identically zero. The upper right 4×2 subtable and the lower left 2×4 subtable contribute very little to the lack of fit; that is, independence almost characterizes these subtables.

Because the quasi-independence model fits the frequencies on the main diagonal, $r_{ii} = 0$ for each cell on the main diagonal for this model in Table 4. The interesting local structure is thus confined to the off-diagonal cells. The (upward) mobility in the $(2, 1)$ and the (upward) mobility in the $(5, 4)$ cells are the most important contributors to the lack of fit here. These two cells account for nearly 65% of the total lack of fit of the QI model.

The apparent patterns seen in the first two displays disappear once the quasi-uniform association model is considered, with one important exception. As seen in Table 5, $r_{21} = 0.025$, so this cell denoting upward mobility from the next-to-highest origin category to the highest destination category is not described well by the QU model. This cell is the only one having an MMR deserving much further comment, and it accounts for approximately half of the overall lack of fit.

7 Some Elementary Graphical Displays

A number of simple graphical displays suggest themselves as useful representations of the preceding analyses. Figures 1, 2, 3, and 4 present four of these for the models I, QI, and QU.

Note that the analyses based on Figures 1, 2, 3, and 4 would be the same whether carried out in terms of probabilities or in terms of frequencies. The sample size, of course, plays an important role in effecting the lower confidence bounds $\hat{\pi}_L$ for π^* reported in Table 2. But the estimates themselves do not depend on the sample size.

Figure 1 shows the decomposition (11) for each of the models, giving for every cell the fraction of the probability or of the frequency that is estimated to have come

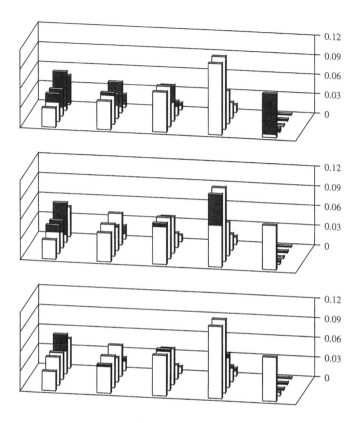

Figure 1: The *I*, *QI*, and *QU* models applied to the Blau–Duncan data. Decomposition into model part (q_{ij} — white) and residual part (r_{ij}—black), as in (11).

from the part of the population where model *H* holds (q_{ij}) and the fraction that is coming from the other part (r_{ij}). The model part is white and the residual part is black. Improved model fit, as one moves from the *I* model to the *QI* model and then to the *QU* model, is reflected by the reduced heights of the black bars, that is, of the residuals. The first panel of the figure indicates that there is a substantial amount of observations on the main diagonal of the table (that is, immobile population), which is not accounted for by the *I* model. The *QI* model reproduces the main diagonal entirely, and residuals are found only in the off-diagonal cells. With the *QU* model, the only substantial black (that is, residual) part is seen in cell (2, 1).

The MMRs (r_{ij}) are presented in Figure 2; this figure contains the same information as Tables 3, 4, and 5. In other words, this figure gives the black bars from Figure 1. Residuals, or equivalently misfit, are concentrated to the upper left corner of the table when the *I* model is used; that is, except for the stronger than expected persistence in the farmer category, the *I* model fails to describe the data mostly because of status persistence in the upper strata of the society. The same status persistence is not seen

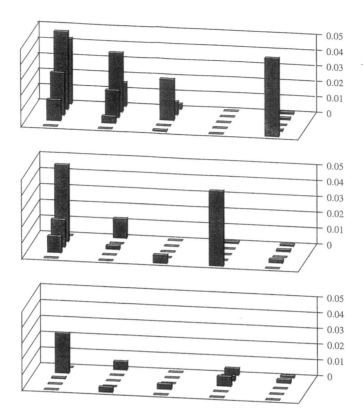

Figure 2: The I, QI, and QU models applied to the Blau–Duncan data. Mixture-model residuals (r_{ij}).

in the lower strata of the society, except for the farmers. The third panel of the figure indicates that the QU model gives a fairly good description of the data, except that it fails to account for a part of the mobility of the sons of fathers in the clerical and sales category into the professional and managerial category. Whether or not this part is substantial, can be investigated using the next figure.

A comparison of the parts not explained by model H in the π^* approach and of the parts explained by the model is facilitated by plotting r_{ij}/q_{ij} in Figure 3. The panels of this figure show for every model how the "residual" probability compares to the "explained" probability. This plot gives an impression about misfit, which is very different from the impression that one can gain from Figure 2. For the QI model, Figure 2 suggests that the two main sources (or locations) of misfit are cells $(2, 1)$ and $(5, 4)$. Figure 3 suggests that from among these, only cell $(2, 1)$ is important, if the ratio of the residual to the explained part is considered. Similarly, for the QU model, the only major absolute source of misfit is cell $(2, 1)$, but relative misfit here is much smaller than in the cells $(1, 5)$ and $(2, 5)$. That is, in "absolute terms," the fit of the

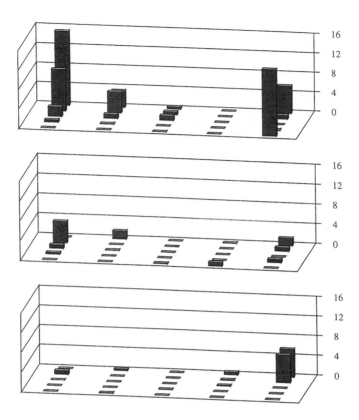

Figure 3: The I, QI, and QU models applied to the Blau–Duncan data relative. Residuals (r_{ij}/q_{ij}).

QU model is poorest for the upward mobility of sons of fathers in the clerical and sales category; however, the part unexplained is not too big compared with the part explained in this category. In the upper right corner of the table the model strongly underestimates the frequency, relative to the observed frequency. This applies to cells with downward mobility: sons in the farmer category whose fathers are in the professional and managerial category or in the clerical and sales category. The small observed frequencies in these categories indicate that this type of downward mobility is rarely found in the population; however, the QU model still strongly underestimates its probability. The statistical reason for this is that the overall fit of a model depends more strongly on the fit in the cells with large observed frequencies than on the fit in the cells with small observed frequencies. Therefore, the relative fit is poorest in the cells with small observed frequencies.

Finally, Figure 4 shows the relative contributions of the cells to the overall lack of fit of the models by plotting $r_{ij}/\hat{\pi}^*$. This figure is a rescaled version of Figure 2. For the QU model, nearly half of the total residual (that is, of $\hat{\pi}^*$) arises in cell (2, 1)

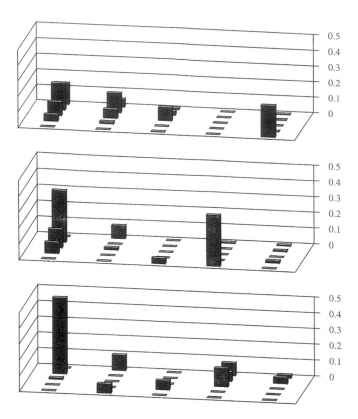

Figure 4: The I, QI, and QU models applied to the Blau–Duncan data. Contribution of the cells to Lack of Fit $(r_{ij}/\hat{\pi}^*)$.

representing the stronger than expected upward mobility from the clerical and sales category into the professional and managerial category.

Further analysis of the MMRs including the application of various visualization techniques could be performed taking into consideration the special structure of the residual matrix. In fact, the residuals in the π^* approach are always nonnegative and can be given a probability distribution interpretation and therefore any technique, computational or visual, can be used for the analysis of residuals that can be used to analyze *any* probability distribution.

Acknowledgments

Clogg's research was supported in part by grant SBR-9310101 from the National Science Foundation and in part by the Population Research Institute, Pennsylvania

State University, with core support from the National Institute of Child Health and Human Development (grant 1-HD28263-01). Rudas's research was supported in part by grant OTKA T-016032 from the Hungarian National Science Foundation. The authors are indebted to A. Löw for producing Figures 1, 2, 3, and 4.

Chapter 30

Visualization in Ideal Point Discriminant Analysis

Yoshio Takane

1   Introduction

Ideal point discriminant analysis (IPDA) was originally proposed as a technique for discriminant analysis with mixed measurement level predictor variables (Takane, 1986; Takane *et al.*, 1987). However, it was soon realized that it could also be used as a technique for analysis of contingency tables (Takane, 1987). It has also been extended to cover a wider range of data types (Takane, 1989a, b). In this chapter we focus on the second use of IPDA, demonstrating its advantages (and disadvantages) in the analysis of contingency tables.

   IPDA allows spatial representations of rows and columns of contingency tables. Specifically, it represents rows and columns of contingency tables as points in a multidimensional Euclidean space. By looking at distances between them, we immediately know which rows and columns are closely related to each other. IPDA also allows incorporating external information about the rows of contingency tables. This information serves as predictor variables for discriminating columns of the tables. Based on the statistical inference capabilities of IPDA, we can decide which predictor variables are useful for the discrimination. This in turn provides information about which attributes of the rows are important for them to have closer relationships with certain columns.

   As an example, let us look at the example data set in Table 1. This is a 16 × 2 contingency table obtained in a clinical study involving cancer patients. Column categories pertain to whether the patients lived less than 10 years or longer than 10 years

Table 1: The data from Madsen (1976)

Stage	Type of operation	Radiation	Pathology	Survival <10 yr	Survival >10 yr
1. Low	1. Extensive	1. No radiation	1. Localized	1	21
			2. Spread	9	20
		2. Radiation	1. Localized	0	23
			2. Spread	17	41
	2. Not extensive	1. No radiation	1. Localized	0	4
			2. Spread	1	9
		2. Radiation	1. Localized	1	2
			2. Spread	2	7
2. High	1. Extensive	1. No radiation	1. Localized	1	3
			2. Spread	37	3
		2. Radiation	1. Localized	1	4
			2. Spread	63	7
	2. Not extensive	1. No radiation	1. Localized	0	0
			2. Spread	3	1
		2. Radiation	1. Localized	0	1
			2. Spread	13	4

after surgery. Rows of the table represent descriptions of the patients according to all possible combinations of four binary variables. The patients were cross-tabulated by the length of survival and their 16 combined categories, which we call "profiles." In analyzing a table like this, we are typically interested in finding out how the patient profiles are related to the survival rate and which variables or combinations of variables are specifically related to the survival rate. IPDA answers both of these questions. By spatial representation it will be immediately clear which profiles are more closely related to which survival categories. By variable selection it will become clear which main effects and/or interaction effects among the predictor variables are important for survival of the patients.

In this chapter we demonstrate the use of IPDA through the analysis of this and two other data sets, emphasizing how effectively various visualization techniques can be used with IPDA for more comprehensive accounts of contingency tables.

2 Outline of the Method

Let us briefly explain what IPDA does. In IPDA, we represent both rows and columns of contingency tables as points in a multidimensional Euclidean space. We let x_{ia} and y_{ja} denote the coordinates of row point i and column point j, respectively, on dimension a. The Euclidean distance between the two points is then given by

$$d_{ij} = \sqrt{\sum_a (x_{ia} - y_{ja})^2} \tag{1}$$

Let n_{ij} denote the observed frequency of column j given row i. We assume, for the sake of parsimony and numerical stability, that the coordinate of the column point, y_{ja}, is given by a weighted average of coordinates of row points, namely

$$y_{ja} = \sum_i n_{ij} x_{ia} / n_{\cdot j} \tag{2}$$

where $n_{\cdot j} = \sum_i n_{ij}$. We posit that the conditional probability of row i given column j is proportional to $\exp(-d_{ij}^2)$, that is,

$$p_{i|j} \propto \exp(-d_{ij}^2) \tag{3}$$

It is important to realize that this probability is a decreasing function of the distance between row i and column j. That is, the closer they are located, the higher is the probability that row i arises from column j, and the further apart they are, the less is the chance that row i arises from column j. Let p_j denote the prior probability of column j. The joint probability of row i and column j is then given by

$$p_{ij} = p_j \exp(-d_{ij}^2)/C \tag{4}$$

where

$$C = \sum_k \sum_l p_l \exp(-d_{kl}^2) \tag{5}$$

and the conditional probability of column j given row i by

$$p_{j|i} = p_j \exp(-d_{ij}^2)/C_i \tag{6}$$

where $C_i = \sum_k p_k \exp(-d_{ik}^2)$. This conditional probability is fitted to n_{ij} so as to maximize the log-likelihood function,

$$\ln L = \sum_i \sum_j n_{ij} \ln p_{j|i} + \text{constant} \tag{7}$$

with respect to the coordinates of row points, x_{ia}. The coordinates of column points, y_{ja}, are simply calculated by (2), once x_{ia}'s are obtained.

The model allows spatial representations of rows and columns of contingency tables. It also allows incorporating external information in the representations. For example, we may represent locations of row points as linear combinations of known predictor variables. Let $\mathbf{X}$ denote the matrix of x_{ia}'s and $\mathbf{G}$ a matrix of predictor variables. Then

$$\mathbf{X} = \mathbf{GB} \tag{8}$$

where $\mathbf{B}$ is the matrix of weights. We estimate $\mathbf{B}$ directly, from which $\mathbf{X}$ is calculated from (8).

We may compare goodness of fit of various models and specifications, including the dimensionality of the representation space and various specifications of $\mathbf{G}$, and choose the best fitting model. Akaike's information criterion (AIC; Akaike, 1974), defined by

$$AIC_\pi = -2 \ln L_\pi^* + 2n_\pi \tag{9}$$

where π is a specific model fitted, $\ln L_\pi^*$ is the log likelihood of model π maximized over its parameters, and n_π is the number of parameters in model π. The model that yields the smallest value of AIC is considered the best fitting model. AIC penalizes the maximum likelihood by the number of parameters used in a model to maximize its future predictability.

The matrix of the negative expected Hessian (second-order derivatives of the log-likelihood function with repect to the parameter vector, $\boldsymbol{\theta}$),

$$I = -E\left(\frac{\partial^2 \ln L}{\partial\boldsymbol{\theta}\partial\boldsymbol{\theta}^T}\right) \tag{10}$$

evaluated at the maximum likelihood estimates of $\boldsymbol{\theta}$, is called the information matrix. The inverse of the information matrix provides variance–covariance estimates of the parameter estimates, which may be used to draw confidence regions or bands around the estimates to indicate the degree of their stability.

The spatial representation in IPDA facilitates a holistic understanding of the relationship between rows and columns of a contingency table in a manner similar to correspondence analysis. Statistical evaluation of various constraints, on the other hand, facilitates an analytic understanding of structures in contingency tables much the same way as in the log-linear analysis of contingency tables. Confidence regions indicating the degree of stability of estimates of point locations can also be drawn, as already discussed, based on the asymptotic properties of maximum likelihood estimators. The conditional probability surface can be plotted for each criterion group (column) as a function of coordinates (locations) in the space. In this chapter, we demonstrate the use of these visualization techniques through the analysis of three example data sets from studies on ovarian cancer, merit distribution, and psychiatric symptoms.

3   Analysis of Ovarian Cancer Data

The data are a five-way contingency table (Table 1), binary each way, pertaining to the survival of patients who had surgical operations to remove ovarian cancer

(Madsen, 1976). Columns of the table represent criterion groups, representing two categories of survival, survival of less than 10 years (column 1) and survival of longer than 10 years (column 2) after the operation. The other four binary variables are deemed to be predictor variables: stage of cancer: $1 = $ low, $2 = $ high; type of operation: $1 = $ extensive, $2 = $ not extensive; radiation: $1 = $ radiation treatment, $2 = $ no radiation treatment; pathology: $1 = $ localized, $2 = $ spread. By factorially combining the four binary variables, we obtain 16 profiles of patients corresponding to the rows of the table. In analyzing this data set we are specifically interested in finding out which profiles have prospects of longer survival and which predictor variables or combinations of the variables are closely related to the length of survival.

We have tried a number of possible combinations of the four predictor variables and interactions among them to find the best representation of the row points. The representation is necessarily unidimensional because there are only two column categories. Results of fitting various models are given in Table 2. Eleven models were fitted, including the saturated model and the independence model. The saturated model takes observed $p_{j|i}$, that is, $n_{ij}/n_{i.}$, as the estimate of true $p_{j|i}$. The independence model, on the other hand, assumes that there is no relationship between rows and colums of the table. These two models serve as benchmark models. The saturated model represents the most general model conceivable and the independence model the opposite extreme. Other models listed in the table are labeled by numbers representing predictor variables included in the models. For example, $1, 2, 3, 4$ in the first row indicates a model in which all four predictor variables (all of them are main effects) are included, and $1, 4, 2 \times 4$ in the fifth row designates a model in which variables $1, 4$ (both are main effects) and the two-way interaction between variables 2 and 4 are included. The minimum AIC indicates that the model with the main effect of stage of cancer (variable 1), the main effect of pathology (variable 4), and the

Table 2: Selection of predictor variables for Madsen's data

Model	Predictors	AIC (# of parameters)	
1	$1, 2, 3, 4$	252.9	(5)
2	$1, 3, 4$	254.3	(4)
3	$1, 2, 4$	251.0	(4)
4	$1, 4$	252.3	(3)
5	$1, 4, 2 \times 4$	252.6	(4)
6	$1, 2, 4, 2 \times 4$	250.9	(5)
7	$1, 4, 2 \times 4$	*249.0	(4)
8	$4, 2 \times 4$	349.7	(3)
9	$1, 2 \times 4$	260.8	(3)
10	Saturated model	263.5	(15)
11	Independence model	416.8	(1)

* Best fitting model.

interaction between the type of operation and pathology (variables 2 and 4) is the best fitting model (indicated by a star in the table). Note that in coding the interaction effect between variables 2 and 4, the nonextensive operation for localized cancer and the extensive operation for spread cancer were taken as category 1. The nonextensive operation for spread cancer and the extensive operation for localized cancer, on the other hand, constitute category 2. (There is one degree of freedom for this interaction, which is defined by the difference between the two categories.) The radiation variable (variable 3) did not have any significant effects. Compare models 1 and 3, where the difference between the two models lies in the absence of variable 3 in model 1. Model 3 is found to be a better fitting model according to the minimum AIC, implying that the contribution of variable 3 is significant. The main effect of the type of operation seemed significant in the absence of the interaction effect between this variable and pathology (compare models 3 and 6), but once the interaction effect was included, it was no longer significant (compare models 6 and 7). These results should be taken with some caution, however. The sample size ($N = 299$) in this data set is a bit too small to rely completely on the asymptotic properties of maximum likelihood estimators. Preferably, we would like to have three times as many observations.

Estimates of parameters and their standard errors are presented in Table 3. Categories associated with positive weights are supposed to have favorable effects on survival. The patient with a low stage and localized cancer has a better prospect of longer survival. A nonextensive operation is better when the cancer is still localized, but an extensive operation is better when the cancer has already spread. Locations of patient profiles are also given in the table. Because the radiation variable has no significant effects, we collapsed the original 16 profiles into eight patterns by ignoring this variable. Thus, for example, profile 111 indicates low stage (variable 1), extensive operation (variable 2), and localized cancer (variable 4). Figure 1 displays the locations of the eight patient profiles and the two column categories. The patient profiles are indicated at the top of the figure, and the column categories labeled c_1 and c_2 are indicated at the bottom. The profiles located more to the right have better prospects of longer survival. Profile 111 has the highest prospect of longer survival, followed by 121, while profile 212 has the lowest prospect among all. Two curves labeled c_1 and c_2 indicate the conditional probabilities of column 1 (survival of less than 10 years) and column 2 (survival of more than 10 years) as functions of coordinates of patient profiles along the horizontal axis. These curves are similar to item characteristics in item response theory. They happen to be either monotonically increasing or decreasing in the entire range of the coordinate values. For example, the c_2 curve is monotonically increasing even in the outside range of the column category 2, where the distance between a point and the column category point increases as the point moves away to the right. However, this is the way it should be according to model (6), because what determines the conditional probability is the difference between the squared Euclidean distances from a point, say x, to the two column categories, that is, $d_{x1}^2 - d_{x2}^2$, but as the point moves away from c_2 to the right, the former becomes larger more quickly than the latter. Dotted curves enclosing the conditional probability curves indicate the 95% confidence bands for these curves.

Table 3: Estimates of parameters from Madsen's data

Variable		Estimate of category weight	Standard error
1 Stage	1. Low	.58	(.02)
	2. High	−.65	(.05)
4 Pathology	1. Localized	.66	(.18)
	2. Spread	−.17	(.05)
2 × 4 (Operation type	1.[a]	.28	(.12)
× pathology)	2.[b]	−.13	(.05)

Pattern number	Predictor pattern (1, 2, 4)	Coordinate of row point	Standard error
1.	111	1.53	(.13)
2.	112	.28	(.07)
3.	121	1.11	(.20)
4.	122	.69	(.16)
5.	211	.29	(.17)
6.	212	−.95	(.06)
7.	221	−.12	(.23)
8.	222	−.54	(.15)

Length of survival	Coordinate of column point	Standard error
Less than 10 years	−.61	(.03)
Longer than 10 years	.61	(.03)

[a]Extensive operation for spread cancer or nonextensive operation for localized cancer.
[b]Extensive operation for localized cancer or nonextensive operation for spread cancer.

The true conditional probability curves lie within the bands with probability .95. The confidence bands were drawn from variance–covariance estimates of conditional probabilities as functions of point coordinates that can easily be derived from those of parameter estimates based on the delta rule (Rao, 1973, pp. 388–389).

Although it is the difference between squared Euclidean distances that determines the conditional probabilities of column categories, certain between-row-and-column distances are still interpretable in probabilistic terms. The following three statements summarize the relationships between d_{ij} and various kinds of probabilities: $p_{i|j}$ (the conditional probability of row i given column j), p_{ij} (the joint probability of row i and column j), and $p_{j|i}$ (the conditional probability of column j given row i).

1. According to (3), $p_{i|j}$ is inversely monotonic with d_{ij} within column j, so that $d_{ij} > d_{i'j} \iff p_{i|j} < p_{i'|j}$. Row i' is more probable than row i in column j. For

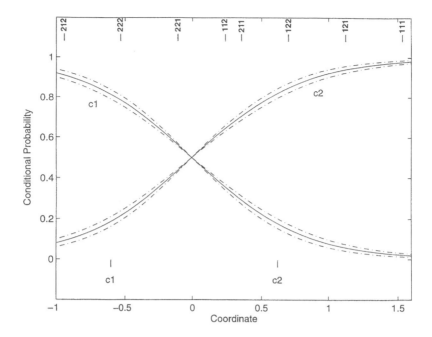

Figure 1: Ovarian cancer data: conditional probability curves.

example, the probability of profile 222 is greater than the probability of profile 111 for $j = 1$ in Madsen's data, because the distance between profile 222 and column category 1 (c_1) is smaller than that between profile 111 and c_1.

2. According to (4), p_{ij} is not necessarily inversely monotonic with d_{ij} unless p_j is constant across j. However, this is approximately true in Madsen's data, so that the joint probability of profile 222 and c_1 is greater than that of profile 111 and c_1.

3. According to (6), $p_{j|i}$ is inversely monotonic with d_{ij} within row i for different columns (j's) only if p_j is constant across j. That is, $d_{ij} > d_{ij'} \iff p_{j|i} < p_{j'|i}$. For example, profile 111 is more likely to belong to c_2 than to c_1 in Madsen's data. (The conditional probability of c_2 is greater than that of c_1 given profile 111.) However, $p_{j|i}$ is not inversely monotonic with d_{ij} within column j for different rows. For example, $d_{12} > d_{32}$, but $p_{2|1} > p_{2|3}$ in Madsen's data. Although the distance between c_2 and profile 111 is greater than that between c_2 and profile 121, the conditional probability of c_2 given profile 111 is greater than that of c_2 given profile 121.

Interpretations of between-row-and-column distances are thus rather intricate, and care should be exercised when they are interpreted in probabilistic terms.

4 Analysis of Merit Distribution Data

Table 4 shows the merit distribution data at McGill University in 1987. There are four merit categories in the amount of salary increase in that year: $2400, $1650, $750, and $0, which constitute columns of the table. The numbers of faculty members who received particular merit increases are tabulated within each of 14 faculties. In the faculty of science, for example, 64 professors received the $2400 merit increase, 74 professors the $1650 increase, 43 professors the $750 increase, and 21 professors no merit increase. McGill University introduced the merit salary system for the first time in 1987, and high-rank university officials were interested in finding out whether the merit allocations were fair across different faculties.

There are four criterion groups in this data set, so that up to three-dimensional solutions can be obtained. The data were analyzed by IPDA with dimensionality varied from one to three. Since there were no obvious predictor variables that "structure" the rows of the table, each faculty was treated as distinct without any particular relationships assumed among them. The unidimensional solution was found to be the best fitting solution according to the minimum AIC (Takane, 1989a). This solution yielded the AIC value of 3196.4 with 16 parameters estimated, while the two-dimensional counterpart yielded 3204.3 with 28 parameters. Two benchmark models resulted in AIC values of 3248.3 (the independence model) and 3225.6 (the saturated model) with 3 and 32 parameters, respectively. This indicates that there are indeed significant differences in the distribution of merit allocations across different faculties.

Table 4: Merit distribution data across faculties of McGill University in 1987

	Faculty	(1) $2400	(2) $1650	(3) $750	(4) $0
1.	Agriculture	13	27	19	15
2.	Arts	56	81	68	13
3.	Dentistry	7	9	3	1
4.	Education	30	32	27	11
5.	Engineering	36	42	32	15
6.	Graduate Studies	13	11	11	5
7.	Law	13	10	6	2
8.	Management	20	13	13	8
9.	Medicine	24	46	44	52
10.	Music	9	9	11	9
11.	Religious Studies	7	4	3	3
12.	Science	64	74	43	21
13.	Libraries	18	27	19	4
14.	Others	9	13	13	15

Estimates of parameters in the best fitting solution are provided in numerical form in Table 5 along with their standard errors. They are also presented graphically in Figure 2. In this figure, column categories are indicated at the bottom by numbers from 1 to 4. Despite the fact that no order constraints were explicitly imposed, they turned out to be in the order of the monetary value of the merit categories. The right-hand side of the figure points to more favorable evaluation. The row points representing the 14 faculties are indicated by numbers from 1 to 14 at the top of the figure. Dentistry (3) was found to be the faculty most lenient to its members, and the Medical School (9) was the most stringent. In fact, Dentistry was on the verge of being closed a few years ago due to lack of research productivity, while the Medical School at McGill is considered one of the top medical schools in the world. The vertical axis represents the conditional probabilities of particular merit categories. These conditional probabilities are functions of how lenient or stringent faculties were in their allocations of merit. The four conditional probability curves depict how they change as functions of the point coordinate along the horizontal axis. They are analogous to the category characteristic curves in mutiple-choice

Table 5: Estimates of parameters for the merit distribution data

	Faculty	Coordinate of row point	Standard error
1.	Agriculture	−.63	(.27)
2.	Arts	.54	(.19)
3.	Dentistry	1.42	(.97)
4.	Education	.18	(.30)
5.	Engineering	.10	(.26)
6.	Graduate Studies	.08	(.46)
7.	Law	1.21	(.72)
8.	Management	.01	(.39)
9.	Medicine	−1.23	(.13)
10.	Music	−.79	(.37)
11.	Religious Studies	−.06	(.69)
12.	Science	.64	(.41)
13.	Libraries	−1.13	(.30)
14.	Others	.38	(.20)

	Merit category	Coordinate of column point	Standard error
1.	$2400	.15	(.02)
2.	$1650	.07	(.01)
3.	$750	−.01	(.01)
4.	$0	−.42	(.02)

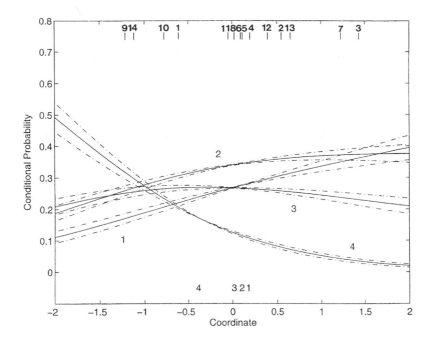

Figure 2: Merit distribution data: conditional probability curves.

item response models. They are either monotonically increasing or decreasing for two extreme categories but are unimodal for intermediate categories. (Two column categories in the two-column case, as in the previous example, correspond to the two extreme categories in the multicolumn case.) As before, dotted curves enclosing the conditional probability curves indicate the 95% confidence bands.

5   Analysis of Psychiatric Symptoms Data

Both examples discussed so far involved only unidimensional spaces. The next example involves a two-dimensional space. Table 6 presents Maxwell's (1961) data, in which there are three criterion groups, SC (schizophrenia), MD (manic-depressive), and AX (anxiety state), constituting columns of the table, and four binary predictor variables, each indicating the presence (2) or absence (1) of a certain symptom: A, whether the patient is anxious; S, whether the patient is suspicious; T, whether the patient has the schizophrenic type of thought disorder; G, whether the patient has delusions of guilt. By factorially combining the four binary variables, we obtain 16 symptom patterns, representing the rows of the table.

IPDA was applied to the table with a number of possible structures for the rows and with varying dimensionalities. It turned out that the model with the main

Table 6: The data from Maxwell (1961)

Pattern number	A	S	T	G	SC	MD	AX

Pattern number	Predictor pattern				Observed frequency in groups		
	A	S	T	G	SC	MD	AX
1	1	1	1	1	38	69	6
2	1	1	1	2	4	36	0
3	1	1	2	1	29	0	0
4	1	1	2	2	9	0	0
5	1	2	1	1	22	8	1
6	1	2	1	2	5	9	0
7	1	2	2	1	35	0	0
8	1	2	2	2	8	2	0
9	2	1	1	1	14	80	92
10	2	1	1	2	3	45	3
11	2	1	2	1	11	1	0
12	2	1	2	2	2	2	0
13	2	2	1	1	9	10	14
14	2	2	1	2	6	16	1
15	2	2	2	1	19	0	0
16	2	2	2	2	10	1	0
Total					224	279	117

effects of the four predictor variables yielded the best representation of the data in a two-dimensional space (see Takane, 1987, for more detailed comparisons among various model specifications). This solution is displayed in Figure 3. In this figure, the points corresponding to the three criterion groups are marked SC, MD, and AX, and those corresponding to the 16 symptom patterns are numbered from 1 to 16. The ellipses surrounding the points indicate the 95% asymptotic confidence regions. They were drawn under the assumption of asymptotic normality of maximum likelihood estimators whose variance–covariance estimates are obtained by the Moore–Penrose inverse of the information matrix (10), which happened to be singular due to the translational and rotational indeterminacies in the Euclidean space. These confidence regions indicate that the point locations are estimated fairly accurately in the solution. However, when the rows of the table are treated as completely distinct without any relationships assumed among them, they are estimated very poorly, as can be seen in Figure 4. Large confidence ellipses in Figure 4 indicate that point locations in this solution can vary rather drastically without impairing the overall goodness of fit of the model, showing the importance of proper constraints in deriving a reliable configuration, particularly when the data are weak in the sense that the sample size is small.

Estimates of weights applied to the four predictor variables and coordinates of the 16 symptom profiles and the three criterion groups are given in Table 7 along

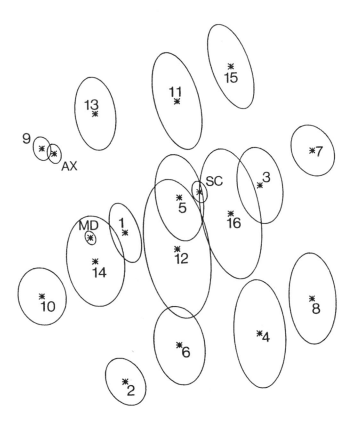

Figure 3: Two-dimensional configuration derived from Maxwell's data with 95% asymptotic confidence regions when the main effects of the four symptom variables are used as row constraints.

with their standard errors. Figure 5 shows conditional probability surfaces for the three criterion groups. They are $p_{SC|x}$, $p_{MD|x}$, and $p_{AX|x}$ as functions of coordinate vector **x**. The functions are evaluated at equally spaced grid points of **x**. These surfaces are two-dimensional analogues of category characteristic curves in unidimensional multiple-choice item response models. The surface for $p_{MD|x}$ is high up in the front but goes down toward the back, although this may be a bit difficult to see in the figure.

Figure 6a, b, and c display the same conditional probability surfaces in the form of three-dimensional isoprobability contour plots. As before, the vertical axis represents the conditional probability. The 16 symptom profiles are indicated by numbers from 1 to 16. From these figures we can deduce the propensities of the three criterion groups given the symptom patterns. Regions (the set of **x**) in which $p_{j|x} = \max_k(p_{k|x})$ are indicated by + in the figures. We can immediately see which symptom patterns are likely to belong to which criterion groups. Patterns 3, 4, 5, 7, 8, 11, 12, 15, and 16 are most likely to belong to SC. All these patterns except one (pattern 5) have

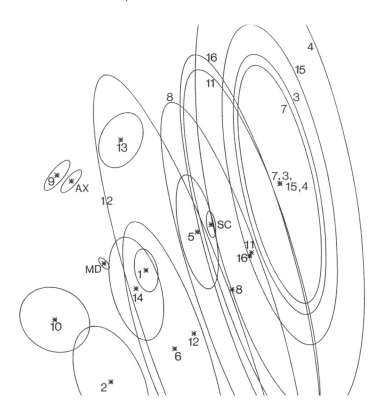

Figure 4: Two-dimensional configuration derived from Maxwell's data with 95% confidence regions: unconstrained case.

thought disorder (T)—see Table 6. It seems that T is a rather decisive indicator of SC. Suspiciousness, on the other hand, is a weak indicator of SC. It is an indicator of SC (pattern 5) when no other symptoms suggest other categories. There are three other patterns (6, 13, and 14) that are characterized by suspicion but do not belong to SC. Patterns 6 and 14 are more strongly affected by delusions of guilt (G) and are classified into MD, while pattern 13 is affected by anxiety (A) and classified into AX. Patterns 1, 2, 6, 10, and 14 are likely to belong to MD. All these patterns except one (pattern 1) have delusions of guilt (G), indicating that this variable is a fairly good indicator of MD. However, thought disorder (T) is a stronger variable. Whenever T is also present, symptom patterns 4, 8, 12, and 16 tend to be classified into SC. Pattern 1 does not have any of the four symptoms. It seems that this pattern is classified into MD simply because this category has the largest prior probability. Finally, only two patterns, 9 and 13, are likely to belong to AX. Anxiety (A) seems to indicate AX only when no other symptoms indicate otherwise.

Table 7: Estimates of parameters for Maxwell's data

| Variable | Category | | Estimates of weights | | |
		Dim. 1	Standard error	Dim. 2	Standard error
A	1	.47	(.04)	−.48	(.06)
	2	−.39	(.03)	.40	(.05)
S	1	−.16	(.03)	−.10	(.05)
	2	.40	(.07)	.25	(.12)
T	1	−.29	(.02)	−.10	(.04)
	2	1.11	(.08)	.37	(.17)
G	1	.00	(.03)	.40	(.04)
	2	−.01	(.08)	−1.13	(.11)

| Predictor Pattern | | Coordinates of row points | | | |
		Dim. 1	Standard error	Dim. 2	Standard error
1	1111	.02	(.07)	−.28	(.12)
2	1112	.01	(.08)	−1.81	(.10)
3	1121	1.42	(.10)	.18	(.16)
4	1122	1.40	(.11)	−1.34	(.23)
5	1211	.58	(.10)	.07	(.18)
6	1212	.56	(.11)	−1.45	(.16)
7	1221	1.98	(.09)	.54	(.11)
8	1222	1.96	(.19)	−.99	(.04)
9	2111	−.83	(.05)	.60	(.10)
10	2112	−.85	(.12)	−.92	(.11)
11	2121	.57	(.21)	1.07	(.14)
12	2122	.55	(.29)	−.46	(.09)
13	2211	−.27	(.09)	.95	(.16)
14	2212	−.29	(.12)	−.57	(.19)
15	2221	1.13	(.10)	1.42	(.18)
16	2222	1.11	(.13)	−.10	(.27)

| Psychiatric category | Coordinates of column point | | | |
	Dim. 1	Standard error	Dim. 2	Standard error
SC	.79	(.03)	.13	(.05)
MD	−.34	(.02)	−.33	(.03)
AX	−.71	(.03)	.55	(.04)

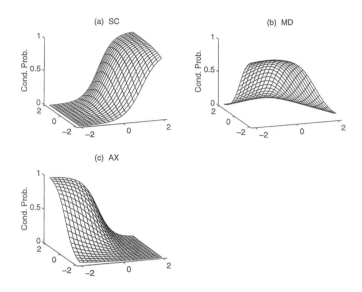

Figure 5: Conditional probability surfaces for the three criterion groups.

6 Concluding Remarks

We have seen three examples of IPDA analysis in which visualization plays a significant role. Interpretations of between-row-and-column distances in terms of various probabilities (conditional and joint), however, require special care, as has been discussed in Section 3. The between-row-and-column distances are inversely monotonically related to $p_{i|j}$ within column j and to $p_{j|i}$ only within row i only if p_j is constant across all j's. Although consistent with the general Bayesian framework, this may be considered as a weakness of the model in IPDA. In order to make all the between-row-and-column distances unconditionally inversely monotonic with the corresponding probabilities it may be necessary to fit the following model:

$$p_{ij} = \exp(-d_{ij}^2) / \sum_k \sum_l \exp(-d_{kl}^2) \tag{11}$$

This is similar to (4), but (11) does not have p_j's.

There are two other kinds of distances, within-row distances and within-column distances, that one may be tempted to interpret. These distances are not directly fitted and consequently do not allow any probabilistic interpretations. Still, the distances are comparable within each set. We may say, for example, that in Maxwell's data symptom patterns 1 and 14 are more similar to each other than symptom patterns 7 and 14 in relation to the three criterion groups and that AX and MD are more similar to each other than SC and AX (see Figure 4). The similarity between criterion groups

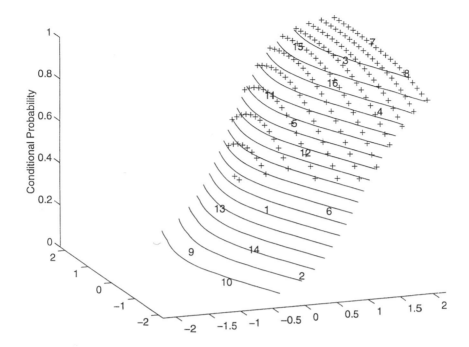

Figure 6(a): Isoprobability contours for SC.

is reflected in the probabilities of misclassification between them, although no formal relationship can be established between the two.

The distances (including the between-row-and-column distances) are not comparable across different sets. This is due to the restriction (2), which makes the variance among the coordinates of column points generally smaller than that of the row point coordinates. This is due to the regression effect. One may venture lifting this restriction to make all three kinds of distances strictly comparable, but that may incur a cost of less numerical stability in estimates of point locations in IPDA.

Software Notes

A Fortran program for IPDA and the program write-up can be obtained from the author. Figures in this chapter were drawn using MATLAB (Math Works, Inc., Natick, MA).

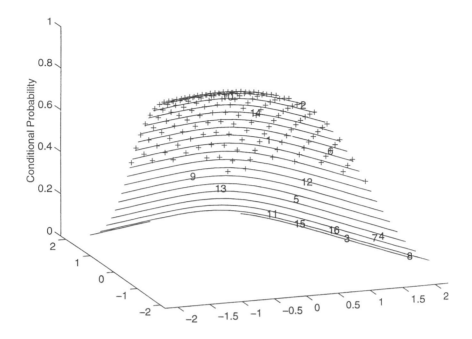

Figure 6(b): Isoprobability contours for MD.

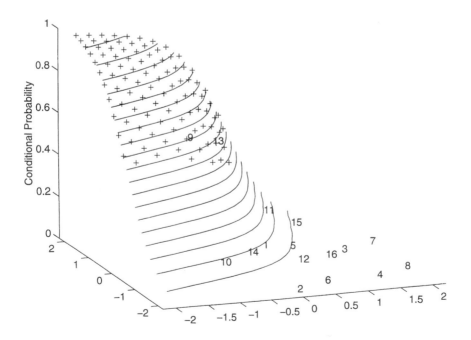

Figure 6(c): Isoprobability contours for AX.

Acknowledgments

The work reported in this chapter has been supported by grant A6394 to the author from the Natural Sciences and Engineering Research Council of Canada.

Chapter 31

Modeling Time-Dependent Preferences: Drifts in Ideal Points

Ulf Böckenholt

1 Introduction

An important objective in the analysis of preferential or attitudinal data is to obtain a graphical representation of the similarity structure that underlies the choice options while taking into account individual differences. Coombs' (1964) unfolding theory provides a conceptually simple yet powerful approach for accomplishing this goal. According to unfolding theory, persons evaluate choice alternatives by comparing them with their ideal alternatives. When asked to pick, for example, the most preferred option, individuals select the one that is closest or least dissimilar to their ideal option. A crucial constraint of Coombs' unfolding approach is that, although persons may differ in terms of their preferences for the choice options, they agree on the similarity relationships among them. Thus, in the unidimensional case the choice options' positions along a common (latent) continuum are perceived homogeneously by all persons; however, the positions of the individual ideal options may vary from person to person.

Because it is frequently the case that individuals differ in their preferences but not in their perception of choice options, there are numerous applications of unfolding approaches that yield parsimonious and easily interpretable graphical representations of choice or attitudinal data (for example, see Bossuyt, 1990; Carroll and Pruzansky, 1980; Heiser, 1981; van Schuur, 1984). However, despite its usefulness in these

461

and other studies, the potential and generality of the unfolding approach have not been fully explored. In particular, the great majority of unfolding studies focused exclusively on the analysis of cross-sectional data. This is unfortunate because the unfolding model provides a promising framework for forecasting and modeling stability and change in preferences over time by distinguishing between time-dependent changes in the similarity relationships of items and in ideal point positions. An appealing feature of this framework for applied settings is its potential in predicting choice behavior when new options are introduced or existing ones are modified. For example, in marketing studies unfolding analysis can be an important methodological tool for the development of new products that are close to most of the consumers' ideal points (Horstman and Slivinski, 1985; Hubert and Busk, 1976).

To some extent, the paucity of unfolding applications that focus on the analysis of longitudinal choice data may be attributed to lack of a comprehensive methodological approach. This chapter addresses this problem by demonstrating that straightforward extensions of latent-class unfolding models proposed for the analysis of cross-sectional data by Böckenholt and Böckenholt (1990a, 1991) may also prove useful for the analysis of time-dependent data. A critical feature of the proposed approach is its ability to take into account that individuals may differ both in their preferences and in the way their preferences change over time. Two applications with different data types are presented to illustrate the usefulness of unfolding models for the analysis of time-dependent data. One application investigates effects of an information campaign on perceived environmental threats of car usage, and another one assesses the impact of a new product on preferences for established brands. In both applications it is shown that preferential or attitudinal changes over time can be explained solely by shifts in ideal-point positions.

2   Individual-Level Unfolding Models for Pick any/*m* and Paired Comparison Data

This section discusses unfolding models for two well-known data collection techniques, the pick any/*m* and paired comparison methods. The former method seems closest to actual choice behavior by asking respondents to select preferred items from a set of *m* items. In applications of the second technique, respondents are presented with two items at a time and are asked to pick the preferred one. The paired comparison technique may be the method of choice in laboratory settings when an experimenter wants to impose minimal constraints on the response behavior of a respondent.

When Coombs (1964) introduced the pick any/*m* procedure, he posited that persons select the items that are closest to the position of their ideal points. More formally, let the positions of item i and of the ath person's ideal point be δ_i and β_a, respectively. A response of person a to item i is denoted by the binary variable X_{ai}. Item i is selected when its distance to the ideal point is smaller than some threshold τ,

$$X_{ai} = 1 \quad \text{when} \quad |\delta_i - \beta_a| \leq \tau \tag{1}$$

and is not selected otherwise,

$$X_{ai} = 0 \quad \text{when} \quad |\delta_i - \beta_a| > \tau$$

In the case of paired comparison data, the response of person a is denoted by the binary variable Y_{aij}, which takes on the value 1 when item i is closer to the person's ideal point and 0 otherwise,

$$Y_{aij} = 1 \quad \text{when} \quad |\delta_i - \beta_a| \leq |\delta_j - \beta_a| \tag{2}$$

Because choice behavior is frequently inconsistent, it is necessary to formulate probabilistic versions of (1) and (2). Hoijtink (1990) proposed the following response function under the premise that a probabilistic unfolding model should reduce to its deterministic counterpart as a boundary case

$$\Pr(X_{ai} = 1) = p_{ai} = \frac{1}{1 + |\delta_i - \beta_a|^{\gamma}} \tag{3}$$

where γ ($\gamma \geq 1$) moderates the strength of the proximity relation on the choice probability whenever the distance between the ideal point and the item position differs from 0 or 1. This model predicts that an item is chosen with certainty when its position coincides with the one of the ideal point.

An analogous model for paired comparison data is derived by Böckenholt and Böckenholt (1990a). According to their approach, the probability that item i is preferred to item j by person a is given by

$$\Pr(Y_{aij} = 1) = p_{aij} = \frac{|\beta_a - \delta_j|^{\gamma}}{|\beta_a - \delta_i|^{\gamma} + |\beta_a - \delta_j|^{\gamma}} \tag{4}$$

This representation makes the strong prediction that item i is preferred to item j with certainty regardless of the position of item j if the position of item i coincides with that of the ideal point. Note that the unfolding paired comparison model has the same structure as the well-known Bradley–Terry–Luce (BTL) model (Luce, 1959),

$$\Pr(Y_{aij} = 1) = \frac{\omega_{ia}}{\omega_{ia} + \omega_{ja}}$$

where ω_{ia} represents the utility of item i for person a, $\sum_i \omega_{ia} = 1$ and $0 < \omega_{ia} < 1$. Setting $\omega_{ia} = |\beta_a - \delta_i|^{-\gamma}$, we obtain the paired comparison model in (4). This decomposition of an item's (dis)utility into an ideal point and an item parameter is nontrivial in the sense that the unfolding paired comparison model assumes that individual differences in the evaluation of the items can be explained solely by differences in the ideal point positions. Thus, whereas the BTL model allows for $m!$ possible rank orders of the items in the population, the item scale estimated from the unidimensional unfolding model is consistent with only $[m(m-1)/2 + 1]$ of those. This shows that the unfolding model may be considerably more restrictive than the BTL model in an analysis of group-level paired comparison data because it requires individuals to be homogeneous in their perception of the items. It is useful to fit both the BTL and unfolding models because the former model provides an important benchmark for the latter one.

3 Modeling Stability and Change in Choice Data

3.1 A Latent Class Representation

This section presents a group-level representation of choice data by formulating latent-class versions of models (3) and (4). Instead of estimating a different ideal point position for each individual, we assign individuals to classes such that members of a latent class are indistinguishable in their predicted response behavior. The approach is less restrictive than it may seem initially. The number of different responses, which typically is much smaller than the sample size, limits possible distinctions that can be made among the individuals (see Lindsay *et al.*, 1991, for a related situation).

The unobserved classes are determined by invoking the principle of local independence, which states that the latent-class membership variable accounts completely for any relationship among the observed responses (Lazarsfeld and Henry, 1968). Consequently, the probability of observing pick any/*m* responses given that person *a* is a member of latent class *s* is

$$\Pr(x_a \mid a \in s) = \prod_{i=1}^{m} p_{si}^{x_{ai}} (1 - p_{si})^{1 - x_{ai}}$$

and $x_a = (x_{a1}, x_{a2}, \ldots, x_{am})$. Each item *i* has a certain probability p_{si} of being selected by members of latent class *s*. The marginal probability of observing the responses of a randomly picked person in a pick any/*m* task is

$$\Pr(x_a) = \sum_{s=1}^{S} \pi_s \prod_{i=1}^{m} p_{si}^{x_{ai}} (1 - p_{si})^{1 - x_{ai}} \tag{5}$$

where π_s represents the relative size or proportion of class *s* and $\sum_s \pi_s = 1$.

Equation (5) corresponds to the unconstrained latent class models. Although this representation does not provide any information about the underlying scale of the items and the positions of the ideal points, this information can be extracted from the data by constraining the class-specific probabilities, p_{si}, to be a function of an unfolding model,

$$p_{si} = \Pr(X_{ai} = 1 \mid a \in s) = \frac{1}{1 + |\delta_i - \beta_s|^\gamma}$$

and every member of latent class *s* is characterized by the class' ideal point position, β_s.

Similar results can be derived for the paired comparison data. In this case, we need to replace p_{si} by p_{sij} and x_{ai} by y_{aij} in (5),

$$\Pr(\mathbf{y}_a) = \sum_{s=1}^{S} \pi_s \prod_{i=1}^{m-1} \prod_{j=i+1}^{m} p_{sij}^{y_{aij}} (1 - p_{sij})^{1 - y_{aij}}$$

where $\mathbf{y}_a = (y_{a12}, y_{a13}, \ldots, y_{a(m-1)m})$. The constrained version of p_{sij} for the BTL model is

$$p_{sij} = \Pr(Y_{aij} = 1 \mid a \in s) = \frac{\omega_{is}}{\omega_{is} + \omega_{js}}$$

and for the unfolding model

$$p_{sij} = \frac{|\beta_s - \delta_j|^\gamma}{|\beta_s - \delta_i|^\gamma + |\beta_s - \delta_j|^\gamma}$$

Incorporating the response mechanism for the pick any/m and paired comparison tasks in latent class models has several advantages. First, by displaying individual differences and similarities among the items along a joint continuum, the interpretation of the results is greatly simplified. Second, by estimating ideal points on the class level, a parsimonious description (and uncluttered display) of individual differences is obtained even when the number of respondents is large. Third, for a given number of latent classes we can compare the fit of the unconstrained and the constrained latent class models to determine whether the postulated response mechanism is consistent with the data. If the differences between both models are nonsignificant, the constrained model provides a more informative and parsimonious representation of the data. For example, instead of estimating mS class-specific probabilities in the case of pick any/m data and $m(m - 1)S/2$ class-specific probabilities in the case of paired comparison data, we need only estimate S ideal points, $m - 1$ item parameters, and one power parameter, γ. The number of item parameters is $m - 1$ as opposed to m because the origin of the item scale is arbitrary. Note that for the paired comparison data, one can also compare the fit of the unfolding model with the fit of the BTL model. This comparison is likely to be more powerful because the latent class BTL model is more parsimonious than the unconstrained latent class model. Fourth, a decision regarding the number of latent classes, which is usually unknown, is not dependent on the specification of the response mechanism. Instead, the decision may be based on the results of the unrestricted latent class analysis.

3.2 Shifts in Ideal Points

For the investigation of stability and change in preferential or attitudinal data, consider the situation of N individuals measured at T time points. If preferences or attitudes are stable, both item and ideal point parameters are time homogeneous. In contrast, systematic response differences at different time point positions may indicate changes in the perception of (some of) the items and/or in the ideal points. Both hypotheses about the locus of change in choice data are of interest. However, because in the reported applications the time-dependent response variability can be accounted for by shifts in ideal point positions, the following discussion is restricted to this case.

As in the previous section, we distinguish between unconstrained and constrained latent class models. According to unconstrained latent class models, individuals may switch among latent classes over time; however, the class-specific item probabilities are time homogeneous (Hagenaars, 1990; Langeheine and van der Pol, 1990; Poulsen, 1990). The same holds for constrained latent class models with the additional restriction that the class-specific probabilities are a function of the ideal point models. As a result, shifts in class memberships correspond to changes in ideal point positions. An important implication of this constraint is that a change in ideal point positions affects relative preferences for all items. Thus, in contrast to the BTL model, preference change is global in the sense that the utilities of all items are affected by an ideal point shift.

Depending on the duration of a longitudinal study, it seems likely that some but not all members of a latent class change their ideal point positions. For example, when observing consumers over time, it is frequently useful to distinguish between loyal consumers and switchers (Böckenholt and Langeheine, 1996). The former group does not change its ideal point position, but the latter group may vary its ideal point position in systematic or unsystematic ways. In general, shifts in ideal point positions may follow certain patterns that can be incorporated in the latent class model. For example, pick any/*m* data observed at three equidistant time points may be modeled as

$$\Pr(x_a^{(t_1)}, x_a^{(t_2)}, x_a^{(t_3)}) = \sum_q \sum_r \sum_s \pi_{qrs} \Pr(x_a^{(t_1)} \mid a \in q)$$

$$\times \Pr(x_a^{(t_2)} \mid a \in r) \Pr(x_a^{(t_3)} \mid a \in s) \tag{6}$$

where the responses of person a at time point t are denoted by $x_a^{(t)}$, and π_{qrs} refers to the probability of belonging to classes q, r, and s at time points t_1, t_2, and t_3, respectively. A change in ideal point positions occurs whenever $\pi_{qrs} > 0$ provided $q \neq r \neq s$. A similar representation for paired comparison data is obtained when replacing $x_a^{(t)}$ by $y_a^{(t)}$ in (6).

Many useful special cases of (6) can be derived by imposing constraints on π_{qrs} to describe switches among the latent class or, more specifically, among the ideal point positions. For example, an important special case is the first-order Markov chain with stationary transition probabilities. This restriction was originally proposed by Wiggins (1973)

$$\pi_{qrs} = \pi_q \, \pi_{r|q} \, \pi_{s|r}$$

where $\pi_{r|q}$ refers to the probability of being a member of class r given membership in class q during the previous time period (see also Langeheine and van de Pol, 1990). The no-change ideal point model is obtained by setting $\pi_{r|q} = 0$ and $\pi_{s|r} = 0$ when

$r \neq q$ and $s \neq r$, respectively. Other constraints can be derived by imposing a log-linear structure on π_{qrs} (Bishop et al. 1975; Böckenholt, 1997). An obvious special case of (6) is the random change model with $\pi_{qrs} = \pi_q \pi_r \pi_s$, which describes changes over time as a function of the class sizes, which, in turn, are constrained to be equal over time.

4 Applications

4.1 Attitude Toward Car Use and Environment

In this questionnaire study conducted by Doosje and Siero (1991) two independent samples of $N = 300$ respondents were asked about their attitudes toward car use and the environment before and after a proenvironment information campaign. To illustrate applications of the unfolding model, three procar and three proenvironment items were selected from this questionnaire:

A. Instead of environmental protection measures with respect to car use, the road system should be extended.
B. It is better to deal with other forms of environmental pollution than car driving.
C. Considering the environmental problems, everybody should decide for themselves how often to use the car.
D. A cleaner environment demands sacrifices such as decreasing car usage.
E. Car users should have to pay taxes per mile driven.
F. Putting a somewhat higher tax burden on car driving is a step in the direction of a cleaner environment.

Because the items are binary, a total of $2^6 = 64$ response patterns can be observed for the pre- and postinformation campaign data. Table 1 contains the likelihood ratio (LR) tests obtained from the unconstrained latent class model fitted to both samples. The LR test is computed as

$$G^2 = 2 \sum_{v=1}^{V} f_v \ln(f_v / \hat{f}_v)$$

Table 1: Goodness-of-fit statistics of unconstrained latent class models

	Precampaign		Postcampaign		Pre=post	
Number of classes	G^2	df	G^2	df	ΔG^2	df
1	295.5	57	271.8	57	21.5	6
2	56.2	50	82.2	50	35.0	13
3	35.7	43	61.5	43	44.3	20
4	28.0	36	44.2	36	57.8	27

Table 2: Parameter estimates of unconstrained three-class model

Class	A	B	C	D	E	F	$\hat{\pi}_s^{(1)}$	$\hat{\pi}_s^{(2)}$
1	.75	.98	.91	.33	.14	.16	.43	.25
2	.22	.74	.77	.84	.34	.32	.34	.51
3	.06	.40	.37	.98	.91	.85	.23	.24

where f_v and $\hat{f}_v$ denote the observed and expected frequencies of the vth response pattern, respectively. Provided standard regularity and identifiability conditions are satisfied, the LR statistic follows asymptotically a χ^2-distribution with degrees of freedom (df) equal to the difference between the number of response patterns (V) and the number of estimated parameters and sampling constraints. In this application the df for the latent class models are $[64 - S(m + 1)]$.

According to Table 1, three classes are required for a satisfactory fit of the pre- and postcampaign data. However, the fit of the latent class model seems to be better for the pre- than for the postcampaign data. The last two columns of Table 1 contain LR statistics and their corresponding df under the hypothesis that the latent class parameters (class-specific probabilities and class sizes) are equal for both studies. For all latent class solutions this hypothesis can be rejected, which indicates that the information campaign had some effect on the respondents' attitudes toward car use and environment.

The effect of the information campaign can be examined by testing which subsets of the latent class parameters, the class-specific probabilities or the class sizes, differ significantly between samples. Although not presented in detail here, these partitions indicate that the main reason for the differences between the pre- and postcampaign data is related to changes in the class size estimates.

Table 2 contains the estimates of the three-class solution with class-specific probabilities constrained to be equal for the pre- and postcampaign data but different class sizes. With some minor exceptions, the latent class probabilities display a single-peaked structure. As a result, a more parsimonious and informative description of the data may be obtained by applying an unfolding model. This observation is confirmed by Table 3, which contains the parameter estimates of the three-class unfolding model.

Table 3: Parameter estimates of three-class unfolding model

Class	A	B	C	D	E	F	$\hat{\beta}_s$	$\hat{\pi}_s^{(1)}$	$\hat{\pi}_s^{(2)}$
1	.72	.95	.92	.29	.13	.12	−.96	.44	.26
2	.22	.75	.80	.88	.40	.37	.00	.34	.52
3	.10	.31	.34	.97	.88	.85	.72	.22	.22
$\hat{\delta}_i$	−1.65	−.64	−.58	.46	1.18	1.23			

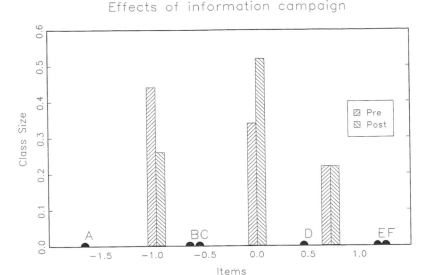

Figure 1: Joint item and latent class ideal point scale with histograms depicting the estimated class sizes before and after the information campaign.

- Clearly, the class-specific probabilities differ little from the unconstrained latent class solution in Table 2.

The results of the unfolding model are depicted in Figure 1 with the horizontal axis being the joint item and latent class ideal point scale and the vertical axis serving to depict the relative size of each latent class in the solution. The items are ordered along a continuum ranging from the position that "car use does not pose an environmental problem" to the position that "car use damages the environment and should be restricted by some governmental interventions." In both the pre- and postcampaign data, about 22% of the respondents (located between items D and E) favor a strong proenvironment position. Before the campaign a substantial number of the respondents (44% located between items A and B) did not consider car use to be an environmental problem. One effect of the campaign was to reduce the size of this group and to increase the number of respondents who acknowledge the negative influence of car usage on the environment (from 34% to 51%). However, the campaign did not increase the number of respondents favoring a tax increase as a means of reducing car usage.

4.2 Product Introduction Study

The second example is taken from a national marketing study. A sample of 211 cigarette smokers were asked to compare a leading brand (A), a competitive brand (B), and a new brand (C) (developed by the manufacturer of the B brand) in a pretest market study. The goal of the study was to investigate the reactions of A and B consumers to the new brand. It was hypothesized that the new brand C may appeal to some of the A consumers and that, as a result, the group of A consumers would split into two parts, one smaller group preferring the new brand over A and another larger group disliking the new brand.

The study was conducted over three waves, which were roughly 4 weeks apart. In wave 1 respondents were asked to compare A and B and in waves 2 and 3 the same respondents compared the three pairs of brands (A, B), (A, C), and (B, C). Of the 211 respondents who were initially recruited, a total of 83 participated in all waves of the study. Thus, in the analysis to follow we will refer to the 83 respondents who participated in all waves as the complete data (C) and to the responses of the 128 respondents who participated in only the first two waves as the incomplete data (IC).

The brands were compared by instructing participants to allocate 11 chips according to their preferences between two options. Because the number of chips is odd, participants could not express indifference. To reduce the sparseness of the original table, adjacent response categories were collapsed such that the direction but not the degree of preference was preserved. The data for the first two waves that were used in the analyses are displayed in Table 4.

When applying the ideal point model, it seems plausible to assume that the new brand C falls somewhere on a continuum between the A and B brands. Because C was designed to appeal to consumers of the A brand, its position may be closer to the A than to the B brand with the result that more A than B consumers may prefer

Table 4: Paired comparison data of product introduction study

Wave 1	Wave 2				
Y_{AB}	Y_{AB}	Y_{AC}	Y_{BC}	$N^{(C)}$	$N^{(IC)}$
0	0	0	0	12	4
0	0	0	1	8	22
0	0	1	0	2	1
0	0	1	1	7	13
0	1	0	0	10	8
0	1	0	1	2	2
0	1	1	0	3	2
0	1	1	1	2	3
1	0	0	0	7	6
1	0	0	1	1	4
1	0	1	0	1	2
1	0	1	1	2	0
1	1	0	0	14	3
1	1	0	1	2	3
1	1	1	0	8	41
1	1	1	1	2	14

the new brand. The attraction effect of the new brand C can be further formalized in two ways. First, we can assume that at wave 1 there are two ideal point positions, one close to the A and another one close to the B brand. As a result of the new brand introduction, some of A and B consumers change their ideal point positions and move it closer to the position of the new brand.

Alternatively, we can assume that initially there are at least three ideal point positions, two close to the established brands and a third one between the two brands corresponding to consumers who feel less strongly about their preferred brand. Appropriately positioned, the new brand may be of more appeal to this third group than any of the established brands. Thus, the first scenario postulates a switch in ideal point positions and the second scenario postulates a more heterogenous ideal point distribution. Because it is not possible to distinguish empirically between the heterogeneity and the switching hypothesis solely on the basis of the paired comparison data, the results of the latent class unfolding models are presented for both scenarios. It is shown that both approaches provide complementary views of the data.

The following analyses also investigate possible differences between the complete and incomplete data. There is little reason to believe that in these types of studies attrition occurs completely at random. Instead, respondents who drop out of a study are likely to do so for specific reasons; for example, respondents may dislike the new product and instead of voicing their dislike, refuse to continue to participate in the study. Similarly, it seems reasonable to expect that respondents who are favorably disposed to the new product are likely to participate in all waves. To test this hypothesis, the data were analyzed under the constraint that the item positions are the same in the complete and incomplete data but the relative class sizes associated with the ideal points were left unconstrained. This parsimonious representation facilitates testing the hypothesis that differences between the complete and incomplete data are not related to the perception of the brands but a result of positive or negative reactions toward the new product.

Heterogeneity Hypothesis   Provided there is an attraction effect of the new brand, we expect at least three latent classes under the heterogeneity hypothesis where each class is characterized by an ideal point position that is close to one of the three products. This hypothesis was tested by first fitting the BTL model with one to three classes and then constraining the (dis)utilities of the best fitting BTL model to be a function of the ideal point and brand positions. The resulting LR statistics of the one-class, two-class, and three-class models are $G^2 = 195.05$ ($df = 28$), 78.1 ($df = 24$), and 19.9 ($df = 20$), respectively. As expected, only the three-class BTL model provides a satisfactory fit of the data and the model's utility estimates are given in Table 5. Moreover, setting $\gamma = 1$, we obtain $G^2 = 20.1$ ($df = 21$) for the three-class unfolding model. Clearly, the decomposition of the (dis)utilities into an absolute difference between an ideal point and an item position is consistent with the data. The corresponding parameter estimates are given in Table 6. Note that according

Table 5: Class-specific utility estimates of three-class BTL model

Class	A	C	B	$\hat{\pi}^{(IC)}$	$\hat{\pi}^{(C)}$
1	.98	.01	.01	.42	.08
2	.17	.69	.14	.25	.71
3	.07	.08	.85	.33	.21

Table 6: Parameter estimates of three-class unfolding model

Class	A	C	B	$\hat{\beta}_a$	$\hat{\pi}^{(IC)}$	$\hat{\pi}^{(C)}$
1	.99	.01	.00	1.01	.42	.08
2	.18	.67	.15	−.02	.26	.72
3	.06	.09	.85	−1.43	.33	.20
$\hat{\delta}_i$	1.00	.25	−1.26			

to both the BTL and the unfolding model the introduction of the new brand did not affect the relative preferences for the A and B brands.

Depicting the parameter estimates of the unfolding model, the left panel of Figure 2 contains the brand positions and the corresponding ideal points with their relative sizes for the complete and incomplete data. We note that an ideal point is close to each brand and that the position of the new brand is closer to A than to B. Moreover, the relative class sizes indicate clearly that the main difference between the complete and incomplete data sets is related to the group with an ideal point closest to the new brand. This group has the smallest size in the complete data and the largest size in the incomplete data, which indicates that attrition may be strongly related to preferences for the new brand.

Ideal Point Switching Hypothesis   According to this hypothesis, there are two ideal points at wave 1, one corresponding to each brand, and when exposed to brand C some members of both classes move their ideal point toward the position of the new brand. Thus, we need to estimate the class sizes at wave 1 and a transition matrix that contains the proportion of A and B consumers who move toward brand C. Because no switching is expected between the A and B brands, the corresponding transition probabilities are set equal to 0. Within this framework it is straightforward to investigate whether A consumers are more attracted to the new brand than B consumers. This question can be tested by an LR statistic obtained from the difference between paired comparison models with unconstrained and equal transition probabilities.

The G^2 statistics of the BTL and the corresponding unfolding model are 18.7 ($df = 18$) and 18.9 ($df = 19$). When constraining the transition probabilities to be equal for both classes, we observe a nonsignificant increase in the G^2 statistics for

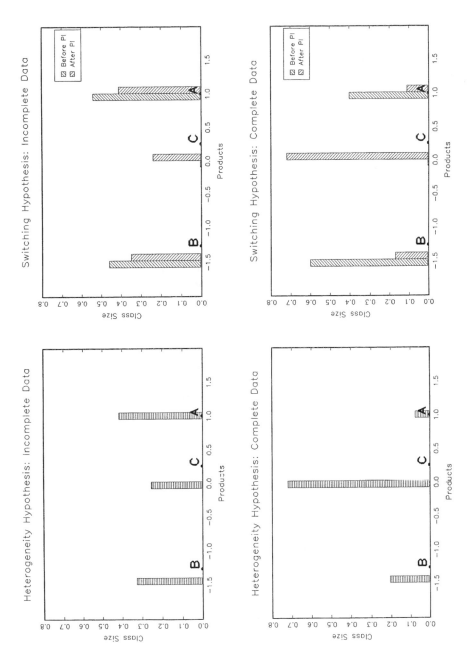

Figure 2: Joint item and latent class ideal point scale with histograms depicting the estimated class sizes.

Table 7: Class sizes and transition probabilities of unfolding model

Class	Wave 1	Wave 2		
s	$\hat{\pi}_s^{(IC)}$	$\hat{\pi}_{1\|s}^{(IC)}$	$\hat{\pi}_{2\|s}^{(IC)}$	$\hat{\pi}_{3\|s}^{(IC)}$
1	.54	.76	.00	.24
2	.46	.00	.76	.24
s	$\hat{\pi}_s^{(C)}$	$\hat{\pi}_{1\|s}^{(C)}$	$\hat{\pi}_{2\|s}^{(C)}$	$\hat{\pi}_{3\|s}^{(C)}$
1	.40	.29	.00	.71
2	.60	.00	.29	.71

the BTL model $G^2 = 20.5$ ($df = 20$), and for the unfolding model $G^2 = 20.7$ ($df = 21$). Because the differences between both models are very minor, we present only the unfolding model's estimates of the initial class sizes and transition probabilities of the complete and incomplete data in Table 7. About 76% and 29% of the A and B consumers are brand loyal in the incomplete and complete data sets, respectively. However, because the initial class sizes differ for the two data sets, a larger percentage of the A consumers is brand loyal and a larger percentage of the B consumers is attracted to the new brand. Thus, this analysis shows that although both A and B consumers have the same transition probabilities, more B than A consumers prefer the new product.

This result is depicted graphically in the right panel of Figure 2. The bars with positive slope lines correspond to the ideal points' class sizes before the product introduction and the bars with negative slope lines correspond to the ideal points' class sizes after the product introduction. We note that slightly more A than B consumers switch to the new brand in the case of the incomplete data. However, for the complete data, a much larger percentage of B than A consumers is attracted to the new brand. In conclusion, these results indicate that the new brand may attract, as intended, some consumers of the A brand but at the price of a substantial cannibalization effect.

5   Discussion

One major problem in the analysis of cross-sectional choice data is to account for heterogeneity caused by individual taste differences. It is well known in the fields of attitude and preference analyses that individuals may perceive and evaluate the alternatives among which they choose in very different ways. Latent class models seem to be well suited to account for these taste differences by allowing for different subpopulations with distinctly different preferences (Böckenholt and Böckenholt, 1991; Croon, 1990; DeSarbo *et al.*, 1994; Formann, 1992). Moreover, the synthesis of unfolding and latent class models provides a parsimonious and versatile framework for determining spatial representations of both individual taste differences and the similarity structure of items at a particular point in time.

However, many decision problems are faced not once but repeatedly. For instance, the choice of a residential location, the selection of a travel mode for a trip to work, and the purchase of consumer brands within a product class are recurrent choice situations. Clearly, by investigating intertemporal choices we obtain valuable information about decision making and the inherently dynamic nature of choice processes that is not available when modeling individual choices at a point in time. A crucial feature of the presented framework is that it takes into account not only that individuals may differ in their preferences but also that they may differ in the way they change their preferences. Consequently, we can test a rich set of hypotheses about the loci of change in latent preferences as a result of an intervention. Modeling shifts in ideal positions may prove instrumental in testing psychological theories about systematic variations in the relationship between perception and choice over time (Loewenstein and Elster, 1992).

By implementing Coombs' unfolding approach we obtain graphical displays that yield a concise summary of the data and are easy to interpret. In the reported applications it proved sufficient to represent the items unidimensionally. However, in other settings multidimensional spatial representations of the items may be more appropriate. Although not discussed in this chapter, it is straightforward to develop multidimensional counterparts of the presented unidimensional unfolding models (Böckenholt and Böckenholt, 1991). These extensions combined with the notion of shifting ideal points yield an important set of methodological tools for graphical representations of the separate effects of taste heterogeneity and preference changes over time.

Acknowledgments

This research was partially supported by National Science Foundation grant SBR-9409531. The author is grateful to William R. Dillon for helpful comments on this research.

Chapter 32

Correspondence Analysis Used Complementary to Latent Class Analysis in Comparative Social Research

Allan L. McCutcheon

1   Introduction

Latent class analysis and correspondence analysis share a common goal of data reduction for cross-tabulated data. One of the principal goals of latent class analysis (LCA) is to account for the observed heterogeneity in a multiway cross-tabulation by characterizing a set of unobserved, internally homogeneous classes. One of the principal goals of correspondence analysis (CA) is to represent graphically a multiway cross-tabulation in a reduced-dimensional space. In this chapter, we examine some of the uses of CA to display graphically the results of LCA, especially when these results are obtained for data from several groups simultaneously. The groups may be different nations, states, regions, cultural groups if the research focuses on cross-cultural comparisons, or separate samples drawn from the same population at two or more time points when the research focuses on social change (see, e.g., McCutcheon, 1987b; Hagenaars, 1990). Indeed, the groups may be any mutually exclusive set of observations on which identical variables are measured.

In the sections that follow, we first briefly examine the simultaneous LCA in which identical measures have been collected in cross-sectional samples from several populations. The increasing availability of surveys with identical indicator variables (e.g., the International Social Survey Program, the European Values Studies, the Eurobarometer studies), as well as the increasing availability of trend studies with repeated indicator variables (e.g., the American, German, and Polish General Social Surveys, the British Social Attitudes Survey), now make it possible to explore the latent structures of many nations' values at several points in time. As an example of religious beliefs from the 1991 International Social Survey Program (ISSP) we will show, however, that communicating the findings from the simultaneous LCA for many samples may prove difficult. The graphical representation provided by CA offers an attractive alternative for presenting the research findings from simultaneous LCA and provides valuable insights for comparative social research.

2   Correspondence Analysis Used Complementary to Multi-Sample Latent Class Analysis

One of the principal goals of latent class analysis is data reduction for categorical data (McCutcheon, 1987b; Hagenaars, 1990). When several categorical variables are available for measuring an unobserved phenomenon, the information available in the associations among the observed indicators may be used to characterize the latent variable. Consider, for example, the responses to five questions regarding traditional (Christian) religious beliefs about "the afterlife" asked of respondents from several nations in the 1991 ISSP:

A. Which best describes your beliefs about God?

    I don't believe in God now and I never have.

    I don't believe in God now, but I used to.

    I believe in God now, but I didn't use to.

    I believe in God now and I always have.

B. Do you believe in life after death?

C. Do you believe in the Devil?

D. Do you believe in Heaven?

E. Do you believe in Hell?

Questions B–E allowed four response categories ("yes, definitely," "yes, probably," "no, probably not," and "no, definitely not"). Each of these five questions is dichotomized into those who report (current) belief or nonbelief in God, life after death, the Devil, Heaven, and Hell. The cross-classification of k dichotomous items yields 2^k response patterns ranging from those who respond "yes" to all k items to those who respond "no" to all k items. In the case of these five dichotomies, there are (2^5) 32 possible response patterns.

    In comparative social research, the difficulties posed by the multiple combinations of indicator responses are multiplied by the number of nations (groups) from which data are collected. In our first example, we examine the responses to the five questions for seven Western nations: Britain, Germany (old Federal states only), Ireland, Italy, The Netherlands, Norway, the United States. Thus, the five dichotomous indicator items give 32 response patterns in each of the seven nations.

LCA allows us to explore the question of whether the 32 categories produced by all possible combinations of response patterns can be represented by some lesser number of categories, without loss of information. As with other latent variable approaches, LCA employs the axiom of *local independence* as the key principle to solve the data reduction problem (Goodman, 1974; McCutcheon, 1987b; Hagenaars, 1990). In LCA, the axiom of local independence imposes the condition that observed indicators are statistically independent of one another within a set of latent (categorical) classes. When this condition holds, the latent classes represent internally homogeneous types. The formal representation of an LCA model with five indicator variables may be expressed as:

$$\pi_{ijklm} = \sum_t \pi_t \, \pi_{i|t} \, \pi_{j|t} \, \pi_{k|t} \, \pi_{l|t} \, \pi_{m|t} \tag{1}$$

where the expected probability for each of the cells of the observed cross-tabulation (π_{ijklm}) is the product of the latent class probability (π_t) for the T latent classes of the latent variable X and the corresponding conditional probabilities for each of the indicator variables for the T latent classes (e.g., $\pi_{i|t}$).

In the usual LCA model, the latent classes are characterized by analyzing the associations among the indicator variables for observations from a single population. In comparative research, we often have identical indicator variables collected in each of the populations. If we define this grouping (population) variable as G and let S represent the number of populations from which independent samples are drawn, we can express the simultaneous (or multisample) latent class model (SLCM) as

$$\pi_{ijklms} = \sum_t \pi_{ts} \, \pi_{i|ts} \, \pi_{j|ts} \, \pi_{k|ts} \, \pi_{l|ts} \, \pi_{m|ts} \tag{2}$$

where $\pi_{i|ts}$, $\pi_{j|ts}$, $\pi_{k|ts}$, $\pi_{l|ts}$, and $\pi_{m|ts}$ represent the conditional probabilities relating each of the indicator variables to the relevant latent class (t) in each of the S populations and π_{ts} represents the joint distribution of the T latent classes and the S populations.

When we engage in comparative (multisample) analysis, one of our first concerns focuses on the issue of *model invariance*, that is, the degree to which we can measure the same latent variable in each of the S populations. Essentially, we ask whether the 32 categories produced by all possible combinations of response patterns within each population can be represented by some lesser number of categories and whether these categories are the same (or similar) for each of the nations (groups). Using LCA, we test for model invariance by imposing across-sample equality restrictions on the conditional probabilities for each of the latent classes (e.g., $\pi_{i|11} = \pi_{i|12} = \cdots = \pi_{i|1S}$). Often, such equality constraints are overly restrictive, and the fit of the model to the original data is eroded well beyond the limits of chance variability. In

Table 1: Likelihood-ratio (G^2) and Pearson (χ^2) chi-squares for religious belief latent class models

Model	G^2	χ^2	df
Two class per country	1640.3	1839.7	140
Three class per country	287.8	439.5	94
Restricted five class per country	95.3	104.3	84
Final model	123.5	149.8	140

such instances, the researcher must decide how much invariance he or she is willing to accept before the latent classes in the S populations are no longer considered to represent similar latent types.

In Table 1 the likelihood ratio and Pearson goodness-of-fit chi-squared statistics for several SLCMs are reported. We first test whether a two-class model can adequately represent the 32 response patterns observed for the 6254 respondents in the seven nations. As we see from the chi-squares, we must reject the hypothesis that the observed response patterns can be represented by two classes ($G^2 = 1640.3$, $\chi^2 = 1839.7$, df $= 140$). Next, we test the hypothesis that the 32 patterns can be represented by three classes. Once again, we see that we must reject this hypothesis, although the chi-squares are substantially reduced and an inspection of the residuals suggests that, in each of the seven nations, the "extreme" patterns for the indicators (i.e., the YYYYY and NNNNN patterns) are underfitted with the three-class model. In the third line of Table 1, we report the chi-squares for a five-class model in which two of the classes are deterministically restricted: one responds "yes" to all five items, while the other responds "no" to all five items. The two deterministically restricted classes represent respondents who take what Duncan (1979) and Duncan *et al.* (1982) refer to as "ideological" responses; we refer to these classes as the "committed believers" and "committed nonbelievers," respectively. As these data indicate, this five-class model can be accepted. The final line of Table 1 reports the chi-squares for a five-class model in which ($140 - 84 =$) 56 across-nation equality restrictions have been imposed. As the data in line 4 of Table 1 indicate, the resulting model can be accepted within the limits of chance variability. Interestingly, we note in passing that the large difference between the G^2 and χ^2 is attributable to two respondents, one from Great Britain and one from Norway, who report belief in Hell and the Devil, but not in life after death, God, or Heaven.

The conditional probabilities of belief in each of the indicator items for each of the five classes across the seven nations are reported in Table 2. As the conditional probabilities for the first and fifth classes reflect, the probability of responding "yes" has been deterministically restricted to be 1.00 for the "committed believers" and 0.00 for the "committed nonbelievers." Consequently, classes 1 and 5 exhibit complete invariance among the seven nations. Classes 2 through 4, on the other hand, exhibit some between nation variance in the likelihood that respondents will respond "yes"

Table 2: Conditional probabilities of responding positively to religious beliefs in seven nations (Source: 1991 ISSP)

	God	Life after death	Heaven	Devil	Hell
1. Committed believers					
All seven nations	1.000	1.000	1.000	1.000	1.000
2. Believers					
Britain	.741	.694	.977	.778	.935
Germany	.811	.694	.977	.718	.935
Ireland	.957	.694	1.000	.718	1.000
Italy	.957	.694	1.000	.718	1.000
Netherlands	.811	.694	1.000	.612	.763
Norway	.811	.987	1.000	.939	1.000
United States	.957	.694	.977	.718	1.000
3. Positive believers					
Britain	1.000	.902	.981	.094	.124
Germany	.923	.761	.485	.094	.000
Ireland	.992	.999	1.000	.188	.208
Italy	.992	.902	1.000	.094	.482
Netherlands	1.000	.902	.872	.188	.000
Norway	.923	.902	1.000	.188	.124
United States	.992	.999	.954	.188	.000
4. Nonbelievers					
Britain	.556	.301	.147	.056	.010
Germany	.357	.140	.046	.000	.010
Ireland	.932	.425	.534	.056	.000
Italy	.855	.425	.046	.103	.000
Netherlands	.357	.425	.046	.024	.000
Norway	.357	.534	.046	.056	.010
United States	.932	.534	.534	.056	.000
5. Committed nonbelievers					
All seven nations	.000	.000	.000	.000	.000

to each of the five indicator items. For example, whereas class 2 respondents have a high probability of responding "yes" to all five of the religious belief indicator items, class 2 respondents in Ireland, Italy, and the United States are estimated to have a 0.957 likelihood of reporting belief in God and class 2 respondents in Britain are estimated to have a 0.741 likelihood of responding positively to this indicator item.

The issue of model invariance is well illustrated by the data in Table 2. Although there are many within-class, between-nation equality restrictions on the conditional probabilities, there remains some between-nation variance in these estimates. At this point, the comparative researcher must resolve for him- or herself how much invariance is to be tolerated before it is no longer plausible that the classes represent the same type of respondents in each of the nations (groups). As we have seen from the information in Table 1, the conditional probabilities of Table 2 are as restricted as possible; any additional between-nation equality restrictions on the conditional probabilities result in an unacceptably large erosion of the model fit.

Consider the case in which we accept the model fit presented in Tables 1 and 2. Except for the relatively high probability of reporting a belief in God among class 4 respondents in the United States, Ireland, and Italy, we might label the intermediate classes as follows. Class 2 respondents all have a relatively high likelihood of reporting belief in all five indicators; thus, we might label class 2 respondents "believers." Class 4 respondents all have a relatively low likelihood of reporting belief for any of the five indicators (given the exceptions noted earlier), so we might label this class "nonbelievers." And finally, class 3 respondents are likely to report a belief in life after death, God, and Heaven, but they are unlikely to report a belief in either the Devil or Hell; this class we will refer to as "positive believers," because they appear to be likely to believe in only the more positive, or rewarding, aspects of the afterlife, but not in the less rewarding aspects.

Once we have accepted a particular model as representing similar (or identical) underlying latent classes in each of the several groups, the next step is to compare the distribution of types among the multiple groups (McCutcheon, 1987a, 1987b). The SLCM latent class probabilities (i.e., π_{ts}) expressed in (2) may be presented as a latent joint distribution, such as that presented in Table 3. The "latent cross-tabulation" presented in Table 3 represents the maximum likelihood estimates of the relative proportion of the original seven-nation sample that is likely to be classified in each of the five classes.

The data reported in Table 3 illustrate a potentially important use of CA to complement the simultaneous latent class analysis of indicator items in cross-national research. The map presented in Figure 1 illustrates the results of a CA of the data

Table 3: Latent class probabilities from the restricted five-class model

Nation	Committed believers	Believers	Positive believers	Nonbelievers	Committed nonbelievers
Britain	.0280	.0111	.0284	.0334	.0236
Germany	.0268	.0105	.0387	.0640	.0048
Ireland	.0451	.0209	.0402	.0249	.0039
Italy	.0415	.0172	.0209	.0474	.0120
Netherlands	.0275	.0121	.0337	.0546	.0434
Norway	.0244	.0071	.0276	.0274	.0363
United States	.0816	.0419	.0101	.0231	.0061

presented in Table 3. CA of a latent joint distribution provides a singular value decomposition of the divergence from independence:

$$\Pi_{ts} = E_{ts} + D_s S \Lambda T^T D_t \tag{3}$$

where Π_{ts} is the two-way latent contingency table, D_s is a diagonal matrix with marginal row proportions Π_s (these will also equal the actual observed proportions p_s), D_t is a diagonal matrix with marginal column proportions Π_t, E_{ts} is the matrix with the independence proportions $E_{ts} = \Pi_t * \Pi_s^T$, S are the row scores normalized so that $S^T D_s S = I$, T are the column scores normalized so that $T^T D_t T = I$, and Λ is a diagonal matrix with the singular values.

As is clear from (3), CA focuses attention on the departures from independence in the latent cross-tabulation presented in Table 3. Consequently, we can think of (3) as dividing the joint distribution into two parts: the unstructured (independence) portion and the structured portion. It is the structured portion of the distribution that CA allows us to represent graphically in data maps such as that in Figure 1.

Inspection of the map in Figure 1 makes clear a set of relationships that is somewhat more difficult to discern from the tabled data presented in Table 3. Figure 1 indicates that the seven nations cluster differently with respect to these (latent) religious beliefs. The U.S. population appears at the right of the space, indicating that

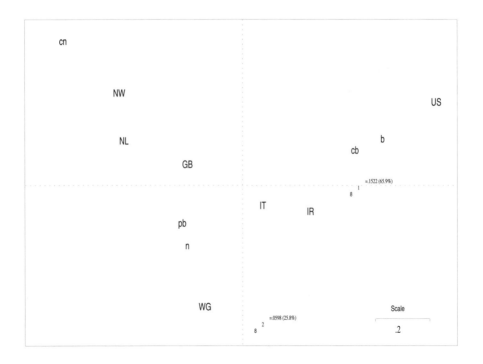

Figure 1: CA map for beliefs in seven nations.

there are relatively more believers and committed believers (b and cb, respectively) in the American sample, whereas the Dutch (NL) and Norwegian (NW) populations appear at the left of the space, indicating that there are relatively more committed nonbelievers (cn) among these populations. The Irish (IR) and Italian (IT) populations are mapped into the center and, although nearer the United States, appear to have relatively more positive believers and nonbelievers (pb and n, respectively), as do the British (GB) and Germans (WG). Thus, Figure 1 illustrates graphically the relative locations of the seven nations with respect to the latent dimensions of these religious beliefs.

3   CA Used Complementary to LCA with Panel Data

It is also possible that a researcher might have survey responses to an identical set of indicator items in two or more waves of a panel of respondents in one or more nations (see also van der Heijden *et al.*, 1994). For the following examples, we examine data from the Political Action Panel Study (Barnes and Kaase, 1979; Jennings *et al.*, 1991). In this panel study, independent panels of American ($N = 778$), German ($N = 846$), and Dutch ($N = 714$) respondents were interviewed in 1973 and again in 1981. In each of these years, respondents were asked what their response would be if their government proposed passing a law that the respondent clearly opposed. The respondents were asked (1) whether they would or would not be willing to sign a petition opposing the proposed law, (2) if they would or would not participate in a legal demonstration opposing the proposed law, and (3) if they would or would not participate in a sit-in opposing the proposed law.

We examine the patterns of responses to these three items in the 1973 wave (W_1) and the 1981 wave (W_2) as a set of latent classes. In the observed cross-tabulation, respondents can give one of eight possible response patterns (YYY to NNN) in each of the 2 years. Thus ($8^2 =$) 64 possible response patterns can be observed for each panel in the three nations. Consider the case in which the petition, demonstration, and sit-in questions (A, B, C, respectively) are asked in 1973, and the same three questions are asked again in 1981 (D, E, F). If we allow for one latent variable (X_t) at W_1 and another (Y_u) at W_2, we can express the latent class model with two latent variables as

$$\pi_{ijklmn} = \sum_t \sum_u \pi_{tu} \pi_{i|t} \pi_{j|t} \pi_{k|t} \pi_{l|u} \pi_{m|u} \pi_{n|u} \tag{4}$$

As in the multisample case, the first concern is whether the latent variable at W_1 (X) is invariant with respect to the latent variable at W_2 (Y). This model invariance may be tested by imposing a set of across-time equality constraints on the conditional probabilities relating each of the indicator variables to its respective latent variable. Thus, when there are equal numbers of classes at each panel wave (i.e., $T = U$) and a model with across-time equality constraints on the conditional probabilities (i.e.,

$\pi_{i|t} = \pi_{l|u}, \pi_{j|t} = \pi_{m|u}, \pi_{k|t} = \pi_{n|u}$) provides an acceptable fit to the observed data, we can represent the relationship between the latent variables X and Y as a latent turnover table.

In the Political Action Panel example, a four-class-per-year model fits the data well ($G^2 = 40.0$, df $= 45$), with the conditional probabilities restricted to fit a time-invariant, item-specific error rate model (.032, .020, and .011, respectively) with a Guttman ordering of the indicator variables (Clogg and Sawyer, 1981; McCutcheon, 1987a). Thus, as we would expect, it appears "easiest" for respondents to say they would sign a petition (A and D), somewhat more difficult for them to say they would participate in a legal demonstration, and most difficult for them to say that they would participate in a sit-in. As a consequence of the Guttman ordering, we expect respondents with the highest latent protest potential to have a high probability of agreeing that they would participate in all three forms of behavior, while those with the lowest latent protest potential would have a low expected probability of saying that they would engage in any of these forms of behavior.

The data in Table 4 represent the modeled latent class probabilities for the latent turnover table from the Dutch Political Action Panel. As these data show, the single largest latent class consists of respondents who were high on protest potential at both waves; approximately one quarter (.251) of the respondents had a high probability of reporting that they would engage in all three forms of protest in 1973 and again in 1981. Interestingly, however, although nearly half of the wave 1 Dutch population reported a high level of protest potential ($0.251 + 0.163 + 0.045 + 0.005 = 0.464$), a substantial segment of that number appears to have "migrated" to a lower level of protest potential by wave 2 ($0.163 + 0.045 + 0.005 = 0.213$).

The latent proportions in Table 4 can also be graphically displayed using CA; the one-dimensional map for these data is presented in Figure 2. Although the four latent classes could be displayed exactly in a three-dimensional space, a high-quality approximation (86.9%) can be displayed in a one-dimensional subspace. Moreover, the display in Figure 2 provides some interesting insights into the possible changes in latent protest potential among the Dutch respondents. Although we must be cautious concerning "overinterpretation" of the graphical display, it appears that Dutch

Table 4: Estimated latent turnover table for the Dutch sample

		Wave 2			
		T_1=high			T_4=low
Wave 1	U_1=high	.251	.163	.045	.005
		.064	.138	.080	.011
	U_4=low	.030	.059	.076	.012
		.011	.014	.031	.010

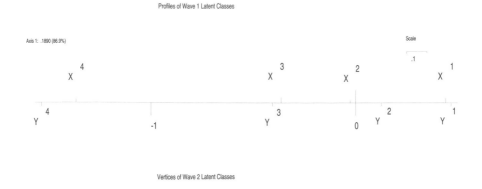

Figure 2: CA map for Dutch political protest potential.

respondents with "high" (X_1, Y_1) and "low" (X_4, Y_4) protest potential diverged somewhat over the 8 years between the two panel waves. Also, the two highest protest
potential classes of Dutch respondents at wave 1 (X_1 and X_2) appear to have become
more similar by wave 2 (Y_1 and Y_2).

A final example focuses on identical protest potential data collected on panels
of American and German samples in 1973 and 1981. In this example, we may ask a
set of questions similar to those posed with the Dutch panel data, although now with
a comparative perspective in mind: Does the protest potential of respondents in these
three democracies in 1973, a period of heightened activity, change by 1981 and does
it appear to change in a similar manner for respondents in each of the three nations?
We begin by modifying equation (4) as

$$\pi_{ijklmns} = \sum_t \sum_u \sum_s \pi_{tus} \pi_{i|ts} \pi_{j|ts} \pi_{k|ts} \pi_{l|us} \pi_{m|us} \pi_{n|us} \qquad (5)$$

Table 5 reports the wave 1 by wave 2 protest potential latent class proportions
for the German and American samples (π_{tus}/π_s). It is important to note that the
conditional probabilities are restricted to fit a nation- and time-invariant, item-specific
error rate model (0.032, 0.020, and 0.011, respectively) with a Guttman ordering of
the indicator variables. Thus, the latent classes in the three nations (as well as the
two panel waves) are invariant as represented by the relationship to the indicator
variables.

Unlike the case for latent protest potential for the Dutch sample (see Table 4), the
information in Table 5 indicates that among both German and American respondents
the highest likelihood is to be found among those who are at the second highest level
of protest potential at both panel waves (0.280 and 0.335, respectively). Although a
variety of approaches may be used to examine these data, we concatenate the matrices
presented in Tables 4 and 5 to obtain a 4 × 12 matrix, with rows representing the

Table 5: Estimated latent turnover table for the German and American samples

		Wave 2							
		German				American			
		T_1=high			T_4=low	T_1=high			T_4=low
	U_1=high	.039	.064	.029	.008	.144	.094	.004	.001
		.054	.280	.192	.040	.109	.335	.060	.006
Wave 1		.011	.058	.086	.034	.020	.108	.076	.009
	U_4=low	.003	.036	.038	.028	.002	.010	.013	.009

patterns of protest potential at wave 1 in the three nations and columns representing the nation-specific patterns of protest potential at wave 2.

The map in Figure 3 graphically displays the results of the CA for the 4 × 12 matrix of protest potential for the three nations. As this map indicates, a two-dimensional representation is required for a display of reasonable quality, although

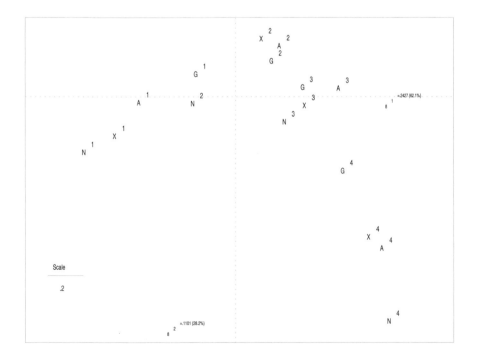

Figure 3: CA map for protest potential in three nations.

the curved pattern of the points clearly forms the common "horseshoe effect" (see, e.g., Greenacre, 1984, pp. 226–232; Greenacre, 1993a, p. 127).

Briefly, and again with caution against the overinterpretation of results, the patterns in the first dimension of Figure 3 indicate an interesting distribution of the latent classes at wave 1 (X_1 to X_4) with X_1 (high protest potential at wave 1) to the left of the space and the lower protest potential classes closely spaced in the right-hand portion of the space. Thus, the first axis appears to represent a "protest axis." The pattern for the nation-specific latent classes is also interesting; whereas the highest protest potential class for the wave 2 German (G_1) and American (A_1) samples are located in the left portion of the space, the *two* highest protest potential classes for the Dutch sample (N_1 and N_2) appear in this portion of the map. The remaining classes appear quite closely spaced in the right-hand portion of the space. Finally, the second dimension of Figure 3 also suggests an interesting difference between nations at wave 2; all of the instances of Dutch data (N_1 to N_4) from wave 2 are located in the bottom portion of the figure, while three instances of the German data (G_1 to G_3) are located in the upper portion and the American data are evenly divided between the upper and lower portion. Thus, although we must remain cautious, we may wish to interpret this as an "intensity axis," because the extreme responses (e.g., X_1 and X_4) are lower than the intermediate responses (e.g., X_2 and X_3).

4   Discussion

The number of survey data sets that are collected cross-nationally has grown dramatically in the past one to two decades. These survey data sets are often collected with the specific intent of facilitating comparative, cross-national survey research and typically include identical sets of indicator variables for latent variable analyses. Athough latent class analysis provides an attractive analytic technique for analyzing these data, the pattern of latent classes that results from such analyses may be difficult to represent. CA offers an attractive complement to simultaneous latent class analysis by providing a set of graphical displays that enables the visualization of the results of the simultaneous latent class analysis. Further, the results of latent class analysis of identical indicator variables in panel data, whether in a single panel or in multiple panels, can also be presented using CA. As the examples presented here indicate, the graphical displays provided by CA can be used as an attractive complement for presenting the results from complex latent class models.

Acknowledgments

The author acknowledges the generous support of the Fulbright Program, The Netherlands/America Commission for Educational Exchange, and the Work and Organization Research Centre of Tilburg University, The Netherlands.

Chapter 33

Graphical Display of Latent Budget Analysis and Latent Class Analysis, with Special Reference to Correspondence Analysis

L. Andries van der Ark and Peter G. M. van der Heijden

1 Introduction

Latent budget analysis (LBA) and latent class analysis (LCA) are methods for the analysis of contingency tables. They are equivalent techniques that lead to an identical visualization of the results of the data analyses. It is not widely known that LBA and LCA results can be visualized. Aided by two clarifying examples, we will illustrate these visualizations, and we will also show the relation between the graphical representation of LBA and LCA and that of correspondence analysis (CA), another method for the analysis of contingency tables.

The first set of data was originally published and analyzed by Guttman (1971). It is a two-way contingency table about the principal worries of Israeli adults (Table 1). The row variable is a combination of residence and father's residence, denoted by "residence," with $I = 5$ categories indexed by i. The column variable is the principal worry of the respondents, denoted by "worry," with $J = 8$ categories indexed by j.

The frequency of the cell corresponding to the ith row category and the jth column category is denoted by n_{ij}. The marginal row and column frequencies are denoted by $n_{i\cdot} = \sum_j n_{ij}$ and $n_{\cdot j} = \sum_i n_{ij}$ respectively. The total number of respondents is denoted by $n = \sum_i \sum_j n_{ij}$ (= 1554).

2   The Latent Budget Model

From the data matrix of Table 1 we can construct the matrix of proportions **P**, with elements p_{ij}, by dividing each element of the data by n: $p_{ij} = n_{ij}/n$. The marginal proportions of the rows and columns are denoted by $r_i \equiv p_{i\cdot} = \sum_j p_{ij}$ and $c_j \equiv p_{\cdot j} = \sum_i p_{ij}$, respectively. Since "residence" is an explanatory variable and "worry" is a response variable, we investigate the conditional proportions of "worry," given "residence," denoted by $p_{j|i} \equiv p_{ij}/p_{i\cdot} = n_{ij}/n_{i\cdot}$, rather than the unconditional proportions p_{ij}. This allows us to compare the categories of the variable "worry" between residence groups. If we collect the $p_{i\cdot}$ as entries of the $I \times I$ diagonal matrix $\mathbf{D}_r$ then the conditional proportions $p_{j|i}$ are found in the matrix $\mathbf{D}_r^{-1}\mathbf{P}$, which is presented in Table 2.

The rows of $\mathbf{D}_r^{-1}\mathbf{P}$ are vectors that contain only nonnegative elements and add up to one. We call such vectors *budgets*, in general, and the rows of $\mathbf{D}_r^{-1}\mathbf{P}$ *observed budgets* (in correspondence analysis these rows are referred to as *row profiles*). Normally, $\mathbf{D}_r^{-1}\mathbf{P}$ is of full rank, that is, rank$(\mathbf{D}_r^{-1}\mathbf{P}) = \min(I, J)$, equal to 5 in this example. In the latent budget model $\mathbf{D}_r^{-1}\mathbf{P}$ is approximated by $\mathbf{D}_r^{-1}\mathbf{\Pi}$, a matrix of conditional probabilities $\pi_{j|i}$, of rank K [$K \leq \min(I, J)$], such that $\pi_{j|i}$ is a mixture of K conditional probabilities $\pi_{j|k}$ ($k = 1, \ldots, K$). The mixing parameters are denoted

Table 1: Principal worries of Israeli adults (Guttman, 1971)[*]

Residence/ father's residence	Principal worries[a]							
	ENL	SAB	MIL	POL	ECO	OTH	MTO	PER	Total
Asia/Africa	61	70	97	32	4	81	20	104	469
Europe/America	104	117	218	118	11	128	42	48	786
Israel; father Asia/Africa	8	9	12	6	1	14	2	14	66
Israel; father Europe/America	22	24	28	28	2	52	6	16	178
Israel; father Israel	5	7	14	7	1	12	0	9	55
Total	200	227	369	191	19	287	70	191	1554

[a]ENL, enlisted relative; SAB, sabotage; MIL, military situation; POL, political situation; ECO, economic situation; OTH, other; MTO, more than one worry; PER, personal economics.
[*]Reprinted by permission of the Psychometric Society.

Table 2: Observed budgets

Residence/ father's residence	Principal worries[a]								
	ENL	SAB	MIL	POL	ECO	OTH	MTO	PER	Total
Asia/Africa	.130	.149	.207	.068	.009	.173	.043	.222	1.000
Europe/America	.132	.149	.277	.150	.014	.163	.053	.061	1.000
Israel; father Asia/Africa	.121	.136	.182	.091	.015	.212	.030	.212	1.000
Israel; father Europe/America	.124	.135	.157	.157	.011	.292	.034	.090	1.000
Israel; father Israel	.091	.127	.255	.127	.018	.218	.000	.164	1.000

[a]ENL, enlisted relative; SAB, sabotage; MIL, military situation; POL, political situation; ECO, economic situation; OTH, other; MTO, more than one worry; PER, personal economics.

by $\pi_{k|i}$. The latent budget model can be written as

$$\pi_{j|i} = \sum_{k=1}^{K} \pi_{k|i}\pi_{j|k} \tag{1}$$

The parameters in (1) are subject to the equality constraints

$$\sum_{j=1}^{J} \pi_{j|i} = \sum_{k=1}^{K} \pi_{k|i} = \sum_{j=1}^{J} \pi_{j|k} = 1 \tag{2}$$

and inequality constraints

$$0 \leq \pi_{j|i} \leq 1,\ 0 \leq \pi_{k|i} \leq 1,\ 0 \leq \pi_{j|k} \leq 1 \tag{3}$$

The idea for the latent budget model was introduced by Goodman (1974) and elaborated by Clogg (1981), de Leeuw *et al.* (1990), van der Heijden *et al.* (1992), and Siciliano and van der Heijden (1994). There are two ways to interpret the parameters of the latent budget model, which we will call the *mixture model interpretation* and the *MIMIC-model interpretation* (Multiple Indicator Multiple Cause model; Goodman, 1974). The mixture model interpretation is as follows. If we collect $\pi_{j|i}$ in an $I \times J$ matrix, then the rows of this matrix, denoted by $\boldsymbol{\pi}_i^{\mathsf{T}} = [\pi_{1|i}...\pi_{j|i}...\pi_{J|i}]$, are vectors with nonnegative elements that add up to one. These vectors are called *expected budgets*. The latent budget model writes these expected budgets as a mixture of the vectors $\tilde{\boldsymbol{\pi}}_k^{\mathsf{T}} = (\pi_{1|k}...\pi_{j|k}...\pi_{J|k})$ $(k = 1,\ldots,K)$, which are typical or *latent budgets*. We can write (1) as

$$\boldsymbol{\pi}_i^{\mathsf{T}} = \pi_{1|i}\tilde{\boldsymbol{\pi}}_1^{\mathsf{T}} + \cdots + \pi_{k|i}\tilde{\boldsymbol{\pi}}_k^{\mathsf{T}} + \cdots + \pi_{K|i}\tilde{\boldsymbol{\pi}}_K^{\mathsf{T}} \tag{4}$$

Hence, each expected budget is built up out of the K latent budgets, and the mixing parameters determine to what extent. If we interpret the latent budget model as a MIMIC model, then $\pi_{k|i}$ denote what proportion of row category i belongs to some latent class k, and $\pi_{j|k}$ denote how the subjects in each latent class k respond to the column categories j.

A schematic representation of a mixture model and a MIMIC model is given in Figure 1. For the mixture model the squares represent the expected budgets π_i and the circles the latent budgets $\tilde{\pi}_k$. The arrows represent $\pi_{k|i}$ and determine how each expected budget is built up in terms of the latent budgets. For the MIMIC model the squares on the left and right represent the row and column categories, repectively. The arrows on the left-hand side represent $\pi_{k|i}$ and the arrows on the right-hand side represent $\pi_{j|k}$. Hence the MIMIC model shows what proportion of each row category falls into each latent category and what proportion of each latent category responds to each column category.

In general, the latent budget model is not identifiable if $K > 1$ and no constraints other than (2) and (3) are imposed on the model. Therefore different sets of parameter estimates may be obtained for different starting values, but they provide the same estimates of the expected budgets. For a discussion of identifiability in the latent budget model we refer to de Leeuw *et al.* (1990) and van der Ark and van der Heijden (1996).

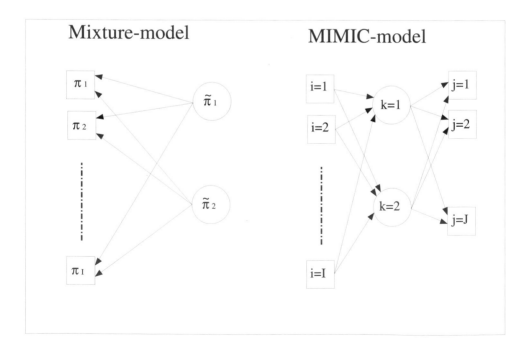

Figure 1: Graphical display of a mixture model and MIMIC model.

Table 3: $K = 1, K = 2$, and $K = 3$ latent budget solutions for the data of Table 1

	$K = 1$ latent budget solution	$K = 2$ latent budget solution		$K = 3$ latent budget solution		
	$k = 1$	$k = 1$	$k = 2$	$k = 1$	$k = 2$	$k = 3$
Mixing parameters						
Asia/Africa	1.000	.383	.617	.000	.477	.523
Europe/America	1.000	.832	.168	.235	.633	.133
Israel; father Asia/Africa	1.000	.424	.576	.116	.402	.482
Israel; father Europe/America	1.000	.721	.279	.436	.353	.210
Israel; father Israel	1.000	.576	.424	.205	.447	.348
Latent budgets						
Enlisted relative (ENL)	.129	.132	.123	.100	.149	.109
Sabotage (SAB)	146	.147	.145	.105	.170	.128
Military situation (MIL)	.238	.286	.145	.021	.429	.011
Political situation (POL)	.123	.187	.000	.250	.145	.000
Economic situation (ECO)	.012	.106	.006	.016	.015	.004
Other (OTH)	.185	.180	.194	.508	.000	.329
More than one worry (MTO)	.045	.054	.028	.000	.084	.000
Personal economics (PER)	.123	.000	.359	.000	.009	.420
Likelihood ratio χ^2	121.5	29.23		6.490		
Degrees of freedom	28	18		10		
Probability	.000	.050		.846		

The matrix $\mathbf{D}_r^{-1}\mathbf{P}$ in Table 2 was analyzed using the maximum likelihood estimation procedure of de Leeuw *et al.* (1990). The results of the latent budget analysis with $K = 1$, $K = 2$, and $K = 3$ latent budgets are presented in Table 3. We can see that the model with $K = 1$ latent budgets does not fit the data. In the model with $K = 2$ latent budgets, the goodness of fit has improved and now $100(121.5 - 29.2)/121.5 = 75.9\%$ of the dependence is modeled, but the fit is still not satisfactory. The model with $K = 3$ latent budgets fits the data very well, with 94.7% of the dependence modeled. We have transformed the parameter estimates such that $\hat{\pi}_{j=8|k=1} = \hat{\pi}_{j=4|k=2} = 0$ in the $K = 2$ latent budget model and $\hat{\pi}_{j=7|k=1} = \hat{\pi}_{j=8|k=1} = \hat{\pi}_{j=6|k=2} = \hat{\pi}_{j=7|k=3} = \hat{\pi}_{j=4|k=3} = \hat{\pi}_{k=1|i=1} = 0$ in the $K = 3$ latent budget model. These transformations were chosen so that as many parameter estimates as possible equal zero without altering the goodness of fit (see van der Ark and van der Heijden 1996). This facilitates the interpretation of the parameter estimates.

We will now interpret the parameter estimates for the model with $K = 3$ latent budgets to get insight into the data. One can characterize the latent budgets by the values of their categories presented in Table 3, but it is more appropriate to characterize these relative to the average. Therefore we interpret the latent budgets by comparing the estimates $\hat{\pi}_{j|k}$ $(k = 1, 2, 3)$ with the column marginals, $p_{\cdot j}$, which are also the elements of the latent budget in the $K = 1$ latent budget model, and attach a label to them. For example, the marginal proportion of MIL is 0.238. In the

$K = 3$ latent budget solution we can see that the estimated proportions of MIL are $0.021, 0.429$, and 0.011, respectively, for the first, second, and third latent budgets. Hence the second latent budget is characterized more than the other two budgets and more than average by people who feel the military situation is their principal worry. When we attach a label to the second latent budget, this feature should be considered. Besides MIL, the second latent budget is characterized by ENL and SAB, also larger than their respective marginals, which also deal with the endangerment of daily life by war and the undetermined category MTO ("more than one worry"). Hence this latent budget can be labeled "concerns for safety." In a similar way, we find that the first latent budget is characterized by POL and OTH, while personal and military concerns (PER, ENL, SAB, MIL) have very low existence or are absent. Hence we can label this latent budget "political and other worries." The third latent budget is dominated by PER, worries about personal economics, with OTH also present. Categories that denote nonegocentric concerns (MIL, POL, ECO) are almost absent in the third latent budget, hence this latent budget can be labeled "personal worries."

After the latent budgets are interpreted by the column categories we examine how the categories of the explanatory (row) variable are composed out of these latent budgets. For example, in the $K = 3$ latent budget solution, the category "residents from Europe and America" (EA) contributes 63.3% to the second latent budget. Hence EA can be described as a group whose principal worries are determined for the larger part by "concerns for safety."

3   Latent Class Model

The latent budget model is equivalent to the latent class model for two variables (see Clogg, 1981; van der Heijden *et al.*, 1992). The latent class model can be written as

$$\pi_{ij} = \sum_{k=1}^{K} \pi_k \pi_{i|k} \pi_{j|k} \tag{5}$$

For (5) we can write

$$\pi_{ij} = \sum_{k=1}^{K} \pi_k \frac{\pi_{ik}}{\pi_k} \frac{\pi_{jk}}{\pi_k} = \sum_{k=1}^{K} \pi_{ik} \frac{\pi_{jk}}{\pi_k} \Leftrightarrow \frac{\pi_{ij}}{\pi_i} = \sum_{k=1}^{K} \frac{\pi_{ik}}{\pi_i} \frac{\pi_{jk}}{\pi_k} \tag{6}$$

where the last expression is the equation of the latent budget model [see (1)]. Note that the latent budget model and the latent class model for two variables have the parameters $\pi_{j|k}$ in common. Equation (6) implies that, in the case of two variables, for each latent budget solution there is one corresponding latent class solution and vice versa. Therefore the estimation procedures and the unidentifiability of the model, mentioned in the previous section on LBA, apply to LCA as well (see van der Ark and van der Heijden, 1996). However, if we have an identified latent budget solution, such as presented in Table 3, we can get the corresponding latent class parameters

$\pi_{i|k}$ and π_k by using Bayes' theorem and the law of total probability

$$\pi_{i|k} = \frac{\pi_i \pi_{k|i}}{\sum_{k=1}^{K} \pi_i \pi_{k|i}} \quad \text{and} \quad \pi_k = \sum_{i=1}^{I} \pi_i \pi_{k|i} \tag{7}$$

The latent class parameter estimates for Table 1, corresponding to the mixing parameter estimates from Table 3 and reparameterized through (7), are presented in Table 4.

The reason for using either the latent class model or the latent budget model depends on the types of manifest variables. Since the latent class model studies the joint probabilities π_{ij}, the model is more appropriate if the row variable and the column variable are both response variables. The response variables are then independent given the latent class. The latent budget model is more appropriate if one of the variables is an explanatory variable and the other a response variable. Only if we regard "residence" as a response variable, then it is appropriate to interpret the latent class solution of Table 4. This might be considered if one accepts that a person can choose the country in which he or she lives. In this case we have a latent variable with three classes that determines the "principal worries" and "residence" of the respondents. We did not find, however, an appropriate way to label the classes in this way.

Since the latent class model comprises only response variables, the model can be extended easily to more than two variables. The latent class model with four variables, for example, is then

$$\pi_{ghij} = \sum_{k=1}^{K} \pi_k \pi_{g|k} \pi_{h|k} \pi_{i|k} \pi_{j|k} \tag{8}$$

From (8) we can see that the general latent class model is equivalent to the law of total probability where the response variables are independent conditional on the latent classes.

Table 4: $K = 1, K = 2$, and $K = 3$ latent class solutions for the data of Table 1

	$K = 1$ latent class solution	$K = 2$ latent class solution		$K = 3$ latent class solution		
	$k = 1$	$k = 1$	$k = 2$	$k = 1$	$k = 2$	$k = 3$
Residence						
Asia/Africa	.302	.176	.544	.000	.268	.560
Europe/America	.506	.640	.248	.656	.596	.238
Israel; father Asia/Africa	.042	.027	.071	.027	.032	.073
Israel; father Europe/America	.115	.126	.094	.276	.075	.086
Israel; father Israel	.035	.031	.043	.040	.029	.044
Worry		See latent budgets Table 3				
Latent class probabilities	1.000	.658	.324	.181	.537	.282

Table 5: Cross-classification of four manifest variables (McCutcheon, 1987)[*]

			Cooperation		
Purpose	Accuracy	Understanding	Interested	Cooperative	Impatient hostile
Good	Mostly true	Good	419	35	2
		Poor/fair	71	25	5
	Not true	Good	270	25	4
		Poor/fair	42	16	5
Depends	Mostly true	Good	23	4	1
		Poor/fair	6	2	0
	Not true	Good	43	9	2
		Poor/fair	9	3	2
Waste	Mostly true	Good	26	3	0
		Poor/fair	1	2	0
	Not true	Good	85	23	6
		Poor/fair	13	12	8

[*]Reprinted by permission of Sage Publications, Inc.

Table 5 contains the cross-classification of four response variables collected in the 1982 General Social Survey (see McCutcheon, 1987b, p. 31). The data comprise the evaluation of 1202 respondents in terms of the respondent's attitude toward the purpose of surveys and the accuracy of surveys in general and the respondent's cooperation and understanding of the survey. In Table 6 a latent class solution with three latent classes published by McCutcheon (1987b, p. 43) is presented. After performing latent class analysis, McCutcheon characterized the three latent classes by the type of respondent who belongs to each of them. Some of the parameters have been restricted post hoc to facilitate interpretation (see Table 6). The three respondent types (classes) are "Ideal," those who have a positive attitude toward surveys and understand the questions well; "Believers," those who have a positive attitude toward surveys but do not really grasp their content; and "Skeptics," those who mistrust surveys although they understand the questions rather well. For further discussion of LCA see McCutcheon (Chapter 32).

4   Visualization of the Latent Budget Model

Latent budgets are K vectors in the J-dimensional space of the response variable. For example, the $K = 3$ latent budgets of the latent budget model in Table 3 can be viewed as three vectors in an eight-dimensional space. The heads of these K vectors span a $(K - 1)$ dimensional subspace; that is, if $K = 1$ then the head of the latent budget is a point, if $K = 2$ the heads of the latent budgets can be connected by a one-dimensional line segment, and if $K = 3$ the heads of the latent budgets are

Table 6: $K = 3$ latent class solution of the data of Table 5 (McCutcheon, 1987)[*]

Manifest Variables	Respondent types		
	$k = 1$ (ideal)	$k = 2$ (believers)	$k = 3$ (skeptics)
Purpose			
Good	.887[a]	.887[a]	.110
Depends	.060[a]	.060[a]	.228
Waste	.053	.053	.661
Accuracy			
Mostly true	.617[a]	.617[a]	.000[b]
Not true	.383	.383	1.000
Cooperation			
Interested	.943	.683	.649
Cooperative	.057	.260	.248
Impatient/hostile	.000[b]	.058	.103
Understanding			
Good	1.000[b]	.338	.765
Poor/fair	.000	.662	.235
Latent class probabilities (π_k)	.619	.223	.158

[a] Equality constraints imposed.
[b] Exact indicator restriction imposed.
[*] Reprinted by permission of Sage Publications, Inc.

the vertices of a triangle (Figure 2). Because the expected budgets, π_i, are mixtures of the latent budgets, $\tilde{\pi}_k$ [see (4)], the expected budgets can be viewed as vectors whose heads lie in the space (point, line segment, triangle, ...) spanned by the latent budgets. The precise position of the expected budgets in this space is the weighted average of the latent budgets, where the weights are the mixing parameters, $\pi_{k|i}$. The mixing parameters serve as coordinates in a so-called barycentric coordinate system, which in the $K = 3$ case is also known as the "triangular coordinate system" (see e.g., Greenacre, 1993, p. 15). The models with $K = 2$ and $K = 3$ latent budgets can be visualized by depicting the space spanned by the latent budgets and plotting the expected budgets onto this space by means of their mixing parameters.

In Figure 3 we show the graphical display of the $K = 2$ latent budget model (for the parameter estimates, see Table 3). Here the line segment spanned by the two latent budgets is presented, with the head of the first latent budget on the right-hand side and the head of the second latent budget on the left-hand side. Now the first expected budget (AA), with mixing parameter estimates 0.383 and 0.617, is made up 38.3% by the first latent budget and 61.7% by the second latent budget. If we scale the line segment from 0 (the second latent budget) to 1 (the first latent budget), then the position of (AA) is .383, hence closer to the second latent budget than to the first latent budget. If the mixing parameter estimates were (1.000, .000), then the expected budget would be equal to the first latent budget and be positioned on the right end of

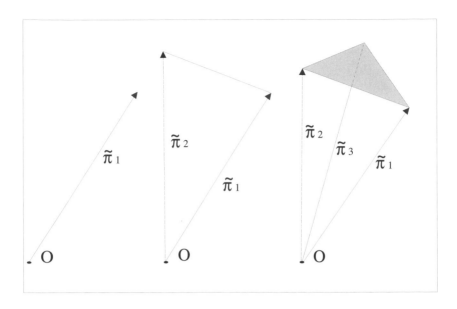

Figure 2: Visualization of $K = 1$, $K = 2$, and $K = 3$ latent budget model.

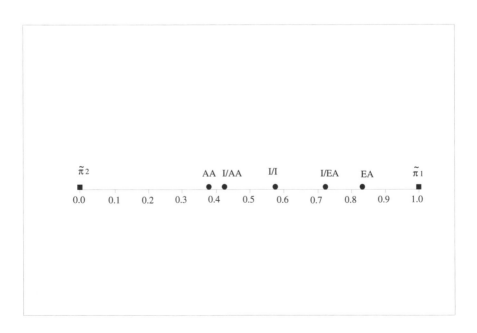

Figure 3: Graphical display of the $K = 2$ latent budget model.

the line segment. By depicting the $K = 2$ latent budget model in this way, we can see immediately that the expected budgets are composed of the latent budgets as their weighted average, where $\hat{\pi}_{1|i}$ and $\hat{\pi}_{2|i}$ ($i = 1, \ldots, 5$) denote the weights.

In Figure 4 a graphical representation is given of the $K = 3$ latent budget model (for the parameter estimates, see Table 3). By convention, the vertices are at equal distance and the upper vertex of the triangle represents the head of the first latent budget, the right-hand vertex represents the head of the second latent budget, and the left-hand vertex represents the head of the third latent budget. The side opposite a vertex is the area where the corresponding mixing parameters are zero; for example, the first expected budget (AA) with mixing parameter estimates (0.000, 0.477, 0.523) lies on the bottom side of the triangle, because the first parameter estimate is zero. The scales of this triangular coordinate system are drawn as dotted lines parallel to the three sides of the triangle. The second mixing parameter estimate $\hat{\pi}_{2|1} = 0.477$ of AA positions the point between the fourth and the fifth dotted line that parallels the left side of the triangle. The third mixing parameter estimate $\hat{\pi}_{3|1} = 0.523$ positions AA between the fifth and the sixth dotted line parallel to the right side of the triangle.

Figures 3 and 4 can be interpreted as a mixture model as well as a MIMIC model. The mixture model interpretation is a tool for understanding the composition of the expected budgets in terms of the latent budgets. The closer the expected budgets lie to the latent budgets, the greater the probability that a residence group will resemble the

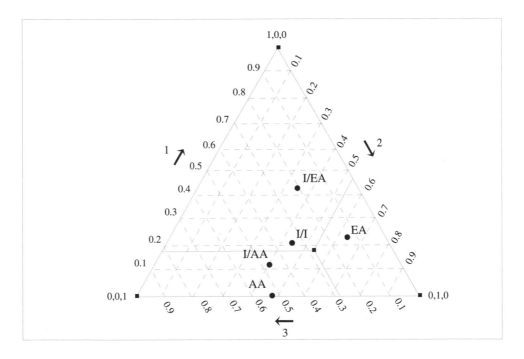

Figure 4: Graphical display of the $K = 3$ latent budget model.

latent budget, and the distance between two expected budgets determines the relative similarity among them. The vector of marginal column proportions with elements $\pi_{\cdot j} = \sum_i \pi_{ij}$ $(j = 1, \ldots, J)$ has also been plotted in Figure 4 as a solid square. This vector is the latent budget in the independence model and therefore represents the average budget. The vector of marginal proportions can serve as a reference vector for the expected budgets. Hence, if an expected budget is closer to a latent budget than the vector of marginal column proportions, then the expected budget resembles the particular latent budget more closely than average. The coordinates of the vector of marginal proportions are $\pi_k = \sum_i \pi_i \pi_{k|i}$, that is, 0.181, 0.537, 0.282. For example, from Table 3 we can see that I/I has mixing parameter estimates (0.205, 0.447, 0.348) and is closer to the vector of marginal proportions than any other row category. Hence the Israeli residents whose fathers also live in Israel display the most average pattern of worries.

The MIMIC model interpretation is a guide to an additional characterization of the latent budgets. We can consider Figure 4 such that the triangle displays the probability to enter the latent budgets; that is, the vertices denote the probability 1 that the subjects of a row category i belong to the corresponding latent budget ($\pi_{k|i} = 1$) and a probability 0 that they belong to the other latent budgets. In this interpretation the picture shows how the marginal row probabilities are distributed over the latent budgets, and we can label the budgets by this distribution. The point that denoted the vector of marginal column proportions now represents the average distribution of all subjects in the contingency table. If a row category is closer to a latent budget than the point representing the overall average, then the latent budget is characterized more than average by that row category. If the distance between two points in the figure is large, then the distribution of those two categories over the latent budgets is not similar; if the distance is small, then the two categories are distributed over the latent budgets in more or less the same way. We can see that the first latent budget $(1, 0, 0)$ is represented more than average by I/EA, EA, and I/I, the second latent budget $(0, 1, 0)$ can be interpreted as a budget typical for those who live in Europe or America, and the third latent budget $(0, 0, 1)$ is represented more than average by AA, I/AA, and I/I.

The categories of the column variable can also be represented graphically. This can be done if we rescale the elements of the latent budgets from $\pi_{j|k}$ into $\pi_{k|j}$ by

$$\pi_{k|j} = \frac{\pi_{j|k} \sum_{i=1}^{I} \pi_i \pi_{k|i}}{\pi_j} \tag{9}$$

[see (7)]. In Figure 5 a graphical representation of $\pi_{k|j}$ in the $K = 3$ latent budget solution is given for the data of Table 1 and the rescaled latent budget parameters [see (9)] are given in Table 7.

Figure 5 cannot be interpreted in terms of the mixture model for the rows, but must be interpreted according to the MIMIC model; that is, the vertices denote $\pi_{j|k} = 1$ ($k = 1, 2, 3$) and the squares in Figure 5 are the column categories. Their position in the triangle determined by $\pi_{k|j}$ denotes how the marginal probability of a particular observed category j is distributed over the three latent budgets. If one

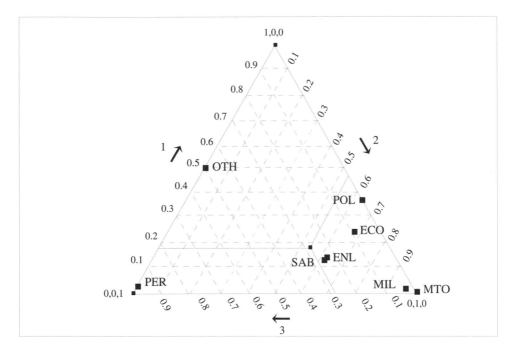

Figure 5: Graphical representation of the rescaled latent budgets in the $K = 3$ latent budget solution.

of the categories were positioned on a vertex, this category would be present only in the particular latent budget. Hence, MTO is present only in the second latent budget. If a category were positioned in the center of the figure [i.e., coordinates are $(\frac{1}{3}, \frac{1}{3}, \frac{1}{3})$], then the responses to that category would be equally distributed over the latent

Table 7: Rescaled latent budget elements of Table 3

Worries	$k = 1$	$k = 2$	$k = 3$
Enlisted relative (ENL)	.141	.620	.238
Sabotage (SAB)	.129	.625	.247
Military situation (MIL)	.017	.969	.014
Political situation (POL)	.368	.632	.000
Economic situation (ECO)	.239	.669	.092
Other (OTH)	.498	.000	.502
More than one worry (MTO)	.000	1.000	.000
Personal economics (PER)	.000	.038	.962
Latent class probabilities (π_k)	.181	.537	.282

budgets. If two points were plotted close together, for example ENL and SAB, then these categories have a similar distribution over the latent budgets. In this way we visualize the characterization of the latent budgets that has been given in Section 2. In Figure 5 the point that denoted the average distribution of all subjects over the latent budgets, with coordinates π_k ($k = 1, 2, 3$), is also plotted and now serves as a reference point for the column categories.

Notice that if we examine $\pi_{k|j}$ instead of $\pi_{j|k}$ the marginal column effects have disappeared. This means that if a marginal column proportion is very small, for example, the marginal proportion of ECO (.012), and we examine the actual proportions of the latent budgets with elements $\pi_{j|k}$, then ECO hardly plays a role in the interpretation of the latent budgets, because of its low marginal frequency. Categories with large marginal column proportions, on the other hand, tend to dominate, for example, MIL, which has a marginal proportion of 0.238. These differences disappear if we examine $\pi_{k|j}$, where we see how each category is distributed over the latent budgets.

Figures 4 and 5 can be overlaid. In this case the figure has to be interpreted according to the MIMIC model. Thus, the plot indicates to what extent the categories of the row and the column variables appear in a certain latent budget. This may help to interpret the latent budgets not only by means of the column categories but also by means of the row categories.

5   Visualization of the Latent Class Model

The idea of rescaling the parameters $\pi_{j|k}$ into $\pi_{k|j}$ can also be used to visualize the latent class parameters. If we have two response variables, we can depict $\pi_{k|i}$ ($i = 1, \ldots, I$) and $\pi_{k|j}$ ($j = 1, \ldots, J$) simultaneously. If we assume that the variables "residence" and "worry" from Table 1 are both response variables, visualization of the latent class model with three latent classes would be equivalent to Figures 4 and 5. The plot must be interpreted as a MIMIC model, however; that is, the picture reveals how the categories of the variables are distributed over the latent classes. In this way we can easily characterize the latent classes by the closeness of the category points to the corners of the triangle. If we overlay Figures 4 and 5, then we have a simultaneous representation of the row variable and the column variable.

As mentioned in Section 3, the latent class model can easily be extended to more than two variables. If we have more than two variables, say four, with corresponding latent parameters $\pi_{g|k}$ ($g = 1 \ldots G$), $\pi_{h|k}$ ($h = 1 \ldots H$), $\pi_{i|k}$ ($i = 1 \ldots I$), and $\pi_{j|k}$ ($j = 1 \ldots J$), [see (8)], they can be transformed into $\pi_{k|g}$ ($g = 1 \ldots G$), $\pi_{k|h}$ ($h = 1 \ldots H$), $\pi_{k|i}$ ($i = 1 \ldots I$), and $\pi_{k|j}$ ($j = 1 \ldots J$) by

$$\pi_{k|g} = \frac{\pi_k \pi_{g|k}}{\pi_g}; \; \pi_{k|h} = \frac{\pi_k \pi_{h|k}}{\pi_h}; \; \pi_{k|i} = \frac{\pi_k \pi_{i|k}}{\pi_i}; \; \pi_{k|j} = \frac{\pi_k \pi_{j|k}}{\pi_j} \qquad (10)$$

A graphical display of Table 6 is given in Figure 6. The manifest variables are depicted simultaneously. The rescaled parameter estimates obtained with (10) are given in Table 8.

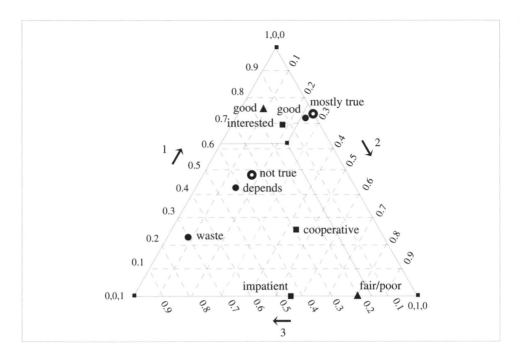

Figure 6: Display of McCutcheon solution.

Table 8: Reparameterized latent class solution of Table 6

	Respondent types		
Manifest Variables	$k = 1$ (ideal)	$k = 2$ (believers)	$k = 3$ (skeptics)
Purpose			
Good	.718	.259	.023
Depends	.429	.155	.416
Waste	.220	.079	.700
Accuracy			
Mostly true	.735	.265	.000
Not true	.493	.178	.329
Cooperation			
Interested	.696	.182	.122
Cooperative	.266	.438	.296
Impatient/hostile	.000	.443	.557
Understanding			
Good	.759	.093	.148
Poor/fair	.000	.799	.201
Latent class probabilities (π_k)	.619	.223	.158

In Figure 6 the solid circles denote the variable "purpose," the open circles denote "accuracy," the solid squares denote "cooperation," and the solid triangles denote "understanding." Figure 6 displays the characterization of the latent classes according to the results in Table 4. The first class (ideal respondents) is characterized more than average by all most positive categories of the variables. The second class (believers) is mostly characterized by a fair to poor understanding of surveys, but they are more cooperative than average. The third class (skeptics) is characterized by negative categories of all variables.

6 Relation of LBA and LCA to Correspondence Analysis

A problem with the display of LCA and LBA is that only the relative distances are visualized; Figures 4, 5, and 6 are equilateral triangles, while there are always two latent classes (budgets) that are more similar to each other than to the third one. We are able to solve this problem by using correspondence analysis (CA), and the solution follows from the relation of LBA and LCA to CA. Because LBA and LCA are equivalent, we will refer only to LCA in this section.

The relation between LCA and CA is rather close and has been studied before by, among others, Gilula (1979, 1983, 1984), Goodman (1987), de Leeuw and van der Heijden (1991), and van der Ark and van der Heijden (1996). Visualization of both models will give more insight into this relation. We recapitulate here first the analytic results of de Leeuw and van der Heijden (1991) and then illustrate the results by visualizing them.

Consider a two-way matrix with observed proportion p_{ij} of rank M. We define CA as

$$p_{ij} = p_{i \cdot} p_{\cdot j} \left(1 + \sum_{m=1}^{M} \alpha_m r_{im} c_{jm} \right) \tag{11}$$

where the scores r_{im} and c_{jm} are centered: $\sum_i p_{i \cdot} r_{im} = \sum_j p_{\cdot j} c_{jm} = 0$, and standardized: $\sum_i p_{i \cdot} r_{im}^2 = \sum_j p_{\cdot j} c_{jm}^2 = 1$. The parameters α_m are the singular values obtained from a singular value decomposition of the matrix with elements $(p_{ij} - p_{i \cdot} p_{\cdot j}) / \sqrt{(p_{i \cdot} p_{\cdot j})}$. When the matrix of proportions has full rank, then $M = \min(I - 1, J - 1)$. Decomposition (11) is also known as the canonical analysis of a contingency table (Gilula, 1984; Gilula and Haberman, 1986). In this context α_m is the mth canonical correlation between the quantified row and column variable, where the scores r_{im} are used as quantifications for the rows, and the scores c_{jm} are used as quantifications for the columns. These scores are often called the "standard coordinates" of a CA solution.

Suppose that we use only M^* ($1 \leq M^* < M$) dimensions of decomposition (11) to derive elements

$$p_{ij}^* = p_i.p_{.j} \left(1 + \sum_{m=1}^{M^*} \alpha_m r_{im} c_{jm} \right) \tag{12}$$

and we collect these approximations p_{ij}^* in a matrix $\mathbf{P}^*$. The matrix $\mathbf{P}^*$ is a reduced rank matrix of rank $M^* + 1$, and it provides an optimal approximation of the observed matrix in a least-squares sense (see, for example, Greenacre, 1984). Notice that $\mathbf{P}^*$ need not be a probability matrix, that is, a matrix with nonnegative elements adding up to one. Although it can be shown that $\sum_{ij} p_{ij}^* = 1$, some elements may be negative.

Both CA and LCA are reduced rank models. If a matrix can be decomposed by a K-class LCA, then it can also be decomposed by a $(K - 1)$-dimensional CA. However, contrary to what was stated by van der Heijden *et al.* (1989), the reverse does not hold in general. This can be seen from the fact that the factorization provided by LCA consists of nonnegative parameters only, whereas the parameters of CA may be negative. There is one special case, though. De Leeuw and van der Heijden (1991) prove that LCA and CA are equivalent in the two-class, one-dimensional case and then provide a counterexample to illustrate that this is not true in general for higher dimensions.

Let us now discuss the implications of these results for data analysis. Observed contingency tables that are of reduced rank seldom occur. If a matrix does not have a reduced rank, then we can still calculate the decomposition provided by (12). If we then consider only $M^* < \min(I - 1, J - 1)$ dimensions, then $\mathbf{P}^*$ need not be a probability matrix (see earlier). Therefore for CA estimated by least squares the above has limited practical relevance. It is relevant, however, for CA estimated by maximum likelihood, as proposed by Goodman (1985) and Gilula and Haberman (1986) (see also Siciliano *et al.*, 1993). Their model is

$$\pi_{ij}^* = a_i b_j \left(1 + \sum_{m=1}^{M^*} f_m u_{im} v_{jm} \right) \tag{13}$$

where the parameters have identification restrictions identical to those in (11) and (12): $\sum_i a_i u_{im} = \sum_j b_j v_{jm} = 0$ and $\sum_i a_i u_{im}^2 = \sum_j b_j v_{jm}^2 = 1$. A choice of M^* determines the rank of the matrix with elements π_{ij}^* and because (13) is estimated by maximum likelihood, this yields a probability matrix of reduced rank when $M^* < \min(I - 1, J - 1)$. This shows that for $K = 2$ the estimates of expected probabilities of both models will be equal, and therefore the fit of both models will be equal as well. For $K = 3$ it turns out that often, but not always, LCA and CA have identical estimates of expected probabilities (see van der Ark and van der Heijden, 1996, for more details). This is relevant for the visualization of LBA and LCA.

7 Simultaneous Visualization of the Correspondence Model and the Latent Budget Model

CA is usually employed to make graphical representations. The categories of the variables are plotted onto an M^*-dimensional space with M^* orthogonal axes.

An important concept in the visualization of CA is the chi-squared distance. The chi-squared distance $\delta^2_{i,i'}$ between rows i and i' of $\mathbf{\Pi}^*$ is defined as

$$\delta^2_{i,i'} = \sum_{j=1}^{J} \frac{\left((\pi^*_{ij}/\pi^*_i) - (\pi^*_{i'j}/\pi^*_{i'})\right)^2}{\pi^*_j} \tag{14}$$

From equation (14) we can see that the chi-squared distance is the squared difference between two expected row budgets π^*_{ij}/π^*_i weighted by the marginal proportion of the column π^*_j. Now we can plot each category of the row variable using $r_{im}\alpha_m$ ($m = 1 \ldots M^*$) as coordinates. Then the Euclidean distances between the rows in the plot are equal to chi-squared distances between the rows of $\mathbf{\Pi}^*$. We can also plot each category of the column variable using $c_{jm}\alpha_m$ ($m = 1 \ldots M^*$) as coordinates. Here the Euclidean distances between the columns in the plot are equal to chi-squared distances between the columns of $\mathbf{\Pi}^*$. These two graphical displays that are often produced for CA estimated by maximum likelihood could be enriched by supplementing them with points for latent budgets, when CA and LCA yield identical estimates of expected probabilities, and this would lead to an interpretation of the CA solution from a different perspective.

An example for the data in Table 1 is given in Figure 7. Let us denote the maximum likelihood estimates of the model by $\hat{\mathbf{\Pi}}^*$. Now in Figure 7 the row profiles of $\hat{\mathbf{\Pi}}^*$ are plotted onto the first two principal axes of the correspondence model, using $r_{im}\alpha_m$ ($i = 1 \ldots 5$; $m = 1, 2$) as coordinates. The columns of $\hat{\mathbf{\Pi}}^*$ have been plotted in the picture as well using standard coordinates c_{jm} ($j = 1 \ldots 8$; $m = 1, 2$). We can project the latent budgets onto Figure 7 to illustrate the relation between LBA and CA. We find the coordinates of the latent budgets by projecting them as supplementary points in the CA space, that is, $\sum_j \pi_{j|k} c_{jm}$ ($k = 1 \ldots 3$, $m = 1, 2$). The coordinates are given in Table 9. The horizontal axis differentiates basically on the origins of the respondents with (AA) and (I/AA) on the left-hand side, (EA) and (I/EA) on the right-hand side, and (I/I) in between, while the vertical axis differentiates on the actual residence of the respondents, residents of Israel on the upper side and citizens living abroad on the lower side. Notice that the triangle in Figure 7 is not the same as the triangle in Figure 4, because in Figure 4 by convention the distances between the latent budgets are unity, whereas in Figure 7 they are measured in the chi-squared metric. Also by convention, in Figure 4 the first latent budget is placed on top, whereas in Figure 7 the position of the latent budgets depends on the axes. Thus Figure 7 can

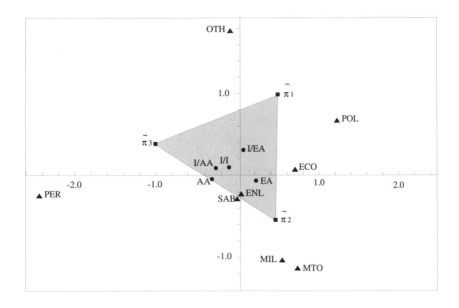

Figure 7: Latent budgets plotted in CA space.

Table 9: Coordinates of the expected budgets and latent budgets in Figure 7

	Dim. 1	Dim. 2
Row profiles (black dots)		
Asia/Africa	−.330	−.051
Europe/America	.212	−.058
Israel; father Asia/Africa	−.285	.089
Israel; father Europe/America	.069	.326
Israel; father Israel	−.101	.100
Column profiles (black triangles)		
Enlisted relative (ENL)	.057	−.238
Sabotage (SAB)	.101	−.288
Military situation (MIL)	.512	−1.069
Political situation (POL)	1.227	.669
Economic situation (ECO)	.687	.081
Other (OTH)	−.069	1.779
More than one worry (MTO)	.520	−1.160
Personal economics (PER)	−2.443	−.268
Latent budgets (black squares)		
First latent budget	.300	.997
Second latent budget	.440	−.545
Third latent budget	−1.032	.399

be viewed as a plot of the latent budget solution scaled in chi-squared distances, which allows a visual comparison of the similarities of the three latent budgets.

8 Discussion

We have shown how to visualize the results of LBA and LCA and how these visualizations are related to the visualizations of CA. A K-budget LBA, that is, a K-class LCA, is equivalent to a $(K - 1)$-dimensional CA when they yield the same estimated expected frequencies. In such a case, the visualization of the results of LBA and LCA can be plotted onto the CA map and vice versa.

LBA is a technique that can be used best when we have one explanatory and one response variable, and the question of interest is how the expected budgets can be composed of a smaller amount of typical or latent budgets. LCA can be used best when we want to study the relation between two or more discrete response variables. The question of interest is whether we can split up the sample into K latent classes such that the relation among the variables is satisfactorily explained by the classes.

On the other hand, CA visualizes how row profiles can be explained by continuous axes, which can be interpreted as latent traits. If the row profiles are equivalent to expected budgets, then the difference between LCA/LBA and CA could be summarized as the choice between a trait or state explanation of the latent budgets.

When the models have the same expected frequencies, plotting the latent budget solution or the latent class solution onto the CA map gives us the benefits of both models. On the one hand, we can see at a glance how the expected budgets are built up of prototypes, and on the other hand, we can assign latent trait scores to the expected budgets. An extra advantage is that the map allows a valid distance interpretation.

Chapter 34

Using New General Ordinal Logit Displays to Visualize the Effects in Categorical Outcome Data

Jay Magidson

1 Introduction

This chapter presents a way to visualize the effects (odds ratios) in the analysis of categorical outcome data through powerful graphical displays. Previously, results from such analyses consisted only of traditional listings of parameter estimates and related statistics that are difficult for the less technical user to interpret. No concise and informative display of the effects was available.

The choice of statistical model for analyzing a categorical response variable depends on whether the response consists of only two categories (dichotomous) or more than two categories (polytomous). In the former case, the model most used has been the multiplicative odds model, commonly expressed in additive form and referred to as the logit or log-odds model. In the latter case, the model choice depends further on whether or not the response categories are assumed to be ordered. Our focus here will be on situations in which the categorical outcome is either dichotomous or ordinal. Unless otherwise stated, we also assume that all predictor variables are categorical—either nominal or ordinal.

Ordinal outcomes occur naturally in many applications. Examples in survey analysis are a three-point rating scale (favorable, neutral, unfavorable), a five-point scale (strongly disagree, disagree, neutral, agree, strongly agree), and other ordered scales. In addition, any dichotomous (Yes, No) outcome may be expanded through the addition of a third response such as "Don't know/uncertain," which in some cases may represent a "middle response."

Despite the preponderance of ordinal outcomes, no single statistical model has emerged as the analysis standard. The two leading candidates are based on competing generalizations of simple odds—the cumulative logit model (McCullagh, 1980) and the adjacent category logit model (Goodman, 1979). Models based on these alternative generalizations can provide different interpretations of data, and the user must choose between them.

One advantage of the adjacent category model is that when scores are assigned to the outcome categories, the model is log-linear and hence current log-linear modeling software can be used to estimate the model parameters (Koch and Edwards, 1988). A second advantage of this model is that the maximum likelihood equations have a simple form, which permits various generalizations. Specifically, when scores are not known and hence cannot be assigned to all of the categories of the ordinal response, the model is log-bilinear and the maximum likelihood algorithm for log-linear modeling generalizes easily to enable estimation of the unknown scores simultaneously with the other parameters (Goodman, 1979). For further discussion and comparisons of these models, see Magidson (1996a), Agresti (1996), and Clogg and Shihadeh (1994).

The graphs presented here are suitable for displaying results from either the simple logit model or the adjacent category model. These graphical displays present a complete "picture" and therefore can reduce the possibility of error in interpretation. The ability to choose a simple graph as a way of specifying a model and viewing the resulting effect estimates in an intuitive graphical form is a powerful asset. Use of the displays represents a marked improvement over the traditional approach, which consists of inspection of parameter estimates and significance tests that are often difficult to integrate to produce a global insight.

Once the model is specified and the parameters estimated, the resulting graph can be examined visually along with traditional statistics that reflect the model fit and significance of the effects. If the user judges the model to be unsatisfactory, the current parameter settings may be altered through direct user manipulation of the graph and a new model can then be estimated that reflects the new settings. Thus, researchers can more readily and quickly implement the natural, interactive process that is prevalent in social science research.

2   Illustration of the Logit Display

To illustrate the benefits of visualizations of the effects in categorical data, we will first consider the data in Table 1 in which responses to the question "How much do you like your work?" are classified by two dichotomous personality characteristics. The EI variable categorizes respondents as either extroverts (E) or introverts (I) according to whether they indicate a preference for expressing ideas to others or for thinking

Table 1: Classification of job satisfaction by two personality characteristics

		How much do you like your work?			
EI	**JP**	**Not much**	**It's OK**	**A lot**	**No response**
Extrovert (E)	Judging (J)	255	2,698	10,215	1,265
Extrovert (E)	Perceiving (P)	268	2,031	4,693	839
Introvert (I)	Judging (J)	410	4,061	9,033	1,534
Introvert (I)	Perceiving (P)	366	2,281	3,684	795

From Martin and Macdaid (1995). Reprinted by permission of the author.

things out first before expressing ideas to others. The second characterization JP classifies persons as judging (J) or perceiving (P) depending on whether they indicate a preference for living a planned, decided, orderly way of life or one that is flexible and spontaneous. For more information on the EI and JP classifications see Myers and McCaulley (1985).

Figure 1 contains a graphical display of the results of applying the adjacent category logit model with unknown response scores. Figure 1 displays the "joint effects plot" in which separate effects lines are displayed for each joint (EI, JP)

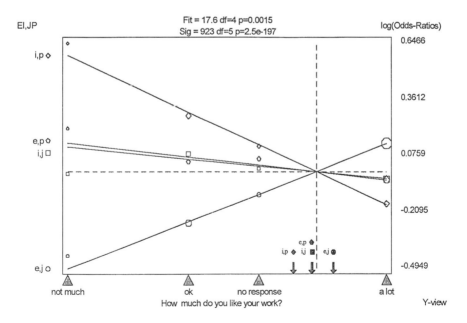

Figure 1: *Y*-view of ordinal logit model for personality characteristics data.

category. The slope of each of these lines represents the magnitude of the relationship between each of the four joint categories and job satisfaction. The extroverted judging (E,J) type is seen to be most likely and (I,P) least likely to be satisfied with their jobs by noting that the largest slope is associated with the "e,j" and smallest (most negative) slope with the "i,p" effects lines. The slope for the e,j effects line includes the main effects associated with extroversion and judging, as well as the extroversion × judging interaction effect.

The following information is embedded in the graphical display in Figure 1:

The EI effect: $\beta^{EI} = 0.11$

The JP effect: $\beta^{JP} = 0.10$

The EI*JP interaction effect: $\beta^{EI*JP} = 0.02$

The induced ordering for the response categories, which positions the no response category in proper proportion to the other categories

Arrows on the horizontal axis depicting the predicted response for each (EI, JP) personality type group

A comparison of the estimated expected odds ratios with the observed odds ratios to assess the model fit

The relative sample sizes associated with each cell of the table

The interpretation of each joint effect as a *baseline* odds ratio, to be described in the following

Moderate effects for both extroversion–introversion (EI) and judgment–perception (JP) variables can be ascertained from Figure 1 by noting the difference in slopes between two of the four effects lines in Figure 1. For example, the magnitude of the EI effect corresponds to the average of the distance between the "e,j" and "i,j" effects lines and the distance between the "e,p" and "i,p" effects lines. Formally, distance is defined as the difference between the slopes of the corresponding lines. The fact that the "e,j" and "i,j" lines are somewhat more distant than the "e,p" and "i,p" lines indicates the presence of a small EI*JP interaction effect.

The triangular markers along the bottom of the graph designate estimates for the response category scores including the "no response" category. Based on the estimated "no response" score, a missing response seems to reflect positive job satisfaction, somewhere between the responses of "OK" and "Like a lot."

Generalized odds ratios corresponding to the joint effects are calculated for each cell of the table. The 16 observed odds ratios for this example are represented by the circular, triangular, diamond, and square markers in Figure 1. The overall fit of the model is indicated by the relative closeness of these markers to the associated effects line, which represents the estimated effects expected under the model. The overall chi-squared lack-of-fit statistic assesses the extent to which the observed markers are distant from the corresponding lines. In this case, we have $X^2 = 17.6$ with 4 degrees of freedom, which reflects a statistically significant difference at the 0.01 level ($p = 0.0015$). However, through visual inspection we see that the observed markers are fairly close to the corresponding lines, which suggests that the significant difference is not substantively meaningful. Hence, we accept the fit of this model.

Upon further inspection, it can be seen that the markers farthest from the respective lines are those associated with the "not much" response category. These markers are the smallest in diameter, which is indicative of relatively small sample sizes. Because greater variability exists in those observed and estimated effects that are based on cells with smaller frequency counts, the distance between the "not much" category and the associated line could be interpreted as normal sampling variation rather than lack of model fit. The ability to ascertain a "look and feel" of a good fit for this example is an especially valuable feature when the overall sample size is large as in this example ($N = 44,428$). A large sample size contributes to a large chi-square statistic, which may often be used to judge a small effect estimate that is insignificant from a substantive perspective to be statistically significant.

3 Description of Methodology

For simplicity, we first describe the model associated with a two-way $I \times J$ table where a single predictor variable forms the I rows and an ordinal response forms the J columns. In this case, the adjacent category logit model can be presented in the following asymmetric form of *generalized baseline* logits (see Magidson, 1996a):

$$\Psi_{j,i} = \alpha_j + \beta_i(y_j - y_0) \quad i = 1, 2, \ldots, I; \; j = 1, 2, \ldots, J \tag{1}$$

where $\Psi_{j,i}$ represents the expected generalized logit associated with response category j given category i of the predictor variable, y_j denotes the score for the jth response category, and y_0 is the score assigned to a designated baseline response category designated by 0. Formally, $\Psi_{j,i} \equiv \ln(P_{ij}/P_{i0})$, where P_{ij} is the probability of response j and predictor level i and thus (P_{ij} / P_{i0}) is the odds in favor of response j relative to the baseline response category given predictor level i. More generally, the baseline response may be a weighted average of the response categories, in which case y_0 is the weighted average of the y-scores. For example, when the observed proportions p_j are used as the weights, y_0 is the mean of the y-scores. Figure 1, described further in Section 6, utilizes the observed proportions as weights in the definition of the baseline odds and odds ratios.

The intercept for the jth response category, α_j, is the baseline logit, representing the generalized logit associated with the designated baseline predictor category; formally, $\alpha_j = \ln(P_{0j}/P_{00})$. Since knowledge of the baseline categories is necessary to interpret both the odds of response and the associated effects, the baseline response and predictor categories should be selected on substantive grounds.

The β parameters are log-odds ratios; β_i represents the logarithm of the change in the odds of response associated with a change from the baseline predictor category to predictor category i. By "odds of response" is meant the odds of a unit change in the response variable—say from the baseline category assigned the score y_0 to a category assigned a score $y_0 + 1$. When we generalize the model later to the case of two or more predictors, each β parameter will correspond to a joint effect that can be decomposed into main effects and interaction effects.

Note that model (1) is invariant with respect to any linear transformation on y_j. That is, any scores $y_j' = a + by_j$, where $b \neq 0$, can be used in place of y_j without altering the statistical properties of the model. Although such a replacement will result in changes in α and β, the generalized logits and all statistical tests (including $H_0 : \beta_i = 0$) will be unchanged. Thus, it is the relative distance between the response categories as displayed in the graph rather than their actual quantitative values that is the essential part in these models.

4   Example 1: Clinical Trial Data

Our first detailed example uses a 2×5 cross-tabulation based on a clinical trial (DeJonge, 1983) involving two treatments (test drug, placebo) and five possible outcomes (marked improvement, moderate improvement, slight improvement, stationary, and worse). The "stationary" category is selected in this example as the baseline so that the odds of a positive change (i.e., improvement) will take on a value greater than 1 and the odds of a negative change will take on a value less than 1.

The outcome categories are assumed to be equally spaced through use of the equidistant y-scores -1, 0, 1, 2, and 3. Such a model is referred to as the equal adjacent odds ratio model by Koch and Edwards (1988).

It is instructive to show how the β_i effect estimates can be computed from the estimated expected frequency counts under this model. Table 2 displays the data and y-scores. Table 3 provides the estimated expected counts under the model.

Table 2: Observed counts for clinical trial data*

(X)Treatment	Worse	Stationary	Slight	Moderate	Marked
Test drug	1	13	16	15	7
Placebo	5	21	14	9	3
y-scores	-1	0	1	2	3

*Copyright John Wiley & Sons Limited. Reproduced with permission from *Statistics in Medicine*, H. DeJonge, 1983.

Table 3: Expected counts for clinical trial data

	Worse	**Stationary**	Slight	Moderate	Marked	Avg. scores
Test drug	1.58	**12.78**	15.10	15.12	7.41	1.3
Placebo	**4.42**	**21.22**	**14.90**	**8.88**	**2.59**	0.7
y-scores	-1	0	1	2	3	

There are four steps in performing the calculations:

1. Estimate the expected cell frequency counts under the baseline logit model; a computer program is required for this estimation, using maximum likelihood estimation.

2. Select the reference point to serve as the baseline for the odds ratio. For this example the "stationary" category of IMPROVEMENT and the "placebo" category for the TREATMENT are selected as the reference points that will be used to define the origin of the associated graph.

3. Calculate the expected odds by dividing each estimated expected count by the corresponding base count associated with the dependent variable reference category "stationary." Table 4 provides the results of calculating the odds.

4. Calculate the odds ratios by dividing each expected odds by the corresponding base odds associated with the predictor reference category "placebo." Table 5 provides the corresponding odds ratios.

Tables 4 and 5 illustrate how the odds and odds ratios are calculated from the expected counts. For example, for the "test drug" category of TREATMENT, the expected odds in favor of "marked" (vs. "stationary," the baseline reference category) IMPROVEMENT is $7.41/12.78 = 0.58$ (see Table 4). Note that by definition the expected odds in favor of "stationary" improvement equals 1. Then, the expected odds ratio associated with the "test drug" and "marked" improvement cell, for example, is computed by dividing the expected odds in favor of "marked" (vs. "stationary") improvement given the "test drug" by the expected odds in favor of a "marked" improvement given the "placebo"—that is, $0.58/0.12 = 4.75$ (Table 5).

Table 4: Expected (baseline) odds for clinical trial data; reference point: (improvement = stationary)

	Worse	Stationary	Slight	Moderate	Marked
Test drug	0.12	1.00	1.18	1.18	0.58
Placebo	0.21	1.00	0.70	0.42	0.12

Table 5: Expected odds ratios for clinical trial data; reference points: (improvement = stationary, treatment = placebo)

	Worse	Stationary	Slight	Moderate	Marked
Test drug	0.59	1.00	1.68	2.83	4.75
Placebo	1.00	1.00	1.00	1.00	1.00

Table 6: Expected log-odds ratios for clinical trial data; reference points: (improvement = stationary, treatment = placebo)

	Worse	**Stationary**	**Slight**	**Moderate**	**Marked**
Test drug	-0.52	0.00	0.52	1.04	1.56
Placebo	0.00	0.00	0.00	0.00	0.00

The equal adjacent category odds ratio model is one for which the differences between the log odds ratios in adjacent categories are identical. The logarithms of the odds ratios in Table 5 are listed in Table 6. From these log odds ratios it can be verified that the difference associated with adjacent categories is 0.52. For example, the difference in log odds ratios associated with "slight" and "stationary" is $0.52 - 0 = 0.52$; for "moderate" and "slight" it is $1.04 - 0.52 = 0.52$, and for "marked" and "moderate" it is $1.56 - 1.04 = 0.52$. The estimate of β_1 to four significant figures is 0.5197. Since there are $I - 1 = 1$ nonredundant effects for this example, by definition $\beta_2 = 0$. The odds of improvement, whether calculated from the adjacent categories "worse" to "stationary," "stationary" to "slight," "slight" to "moderate," or "moderate" to "marked," equals the odds ratio $\exp(0.5197) = 1.682$. As the odds ratio is the odds associated with the test drug divided by the corresponding odds associated with the placebo, the effect estimate states that the odds of improvement are 1.682 times as high (or 68.2% higher) for patients who received the test drug than patients who received the placebo.

A simple graphical display can be used to represent each of the different baseline logit models, clearly showing the relative distances between the response categories indicated by the response scores. The origin of the graph represents the selected baseline reference categories so that the interpretation of the odds ratios is straightforward. Hence, the slope of each line represents the change in the baseline odds associated with a change from the baseline predictor category.

The key to reexpressing model (1) in a form that permits such a meaningful graphical representation is to subtract the intercept from both sides of model (1) to yield

$$\Phi_{j.i} \equiv \beta_i(y_j - y_0) \tag{2}$$

where $\Phi_{j.i} \equiv \Psi_{j.i} - \alpha_j$ is the expected baseline log-odds ratio.

Figure 2 displays the effects as slopes of lines and marks the placement of the estimated expected outcome associated with each predictor level (i.e., $E[Y \mid X = x_i]$) as an arrow pointing onto the outcome score dimension. The baseline reference categories, "placebo" and "stationary," appear in the graph as the reference point (origin) for interpreting results. More generally, any contrast of the predictor and response categories may be used to define the baseline reference. For example, odds may be calculated relative to the weighted average of the y-scores.

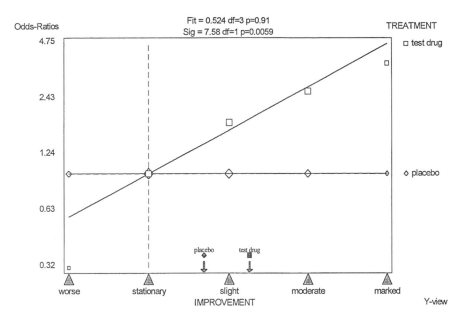

Figure 2: Y-view of the equal adjacent category odds ratio model for clinical trial data.

Separate "effects" lines are present for each of the two levels of the predictor variable. The slope of each of these "effects" lines represents the odds ratios in logarithmic units (i.e., the graph is a semilog chart). The slope of the test drug effects line is $\beta_1 = 0.5197$. The slope of the placebo effects line is $\beta_2 = 0$ because the placebo was selected as a reference category. Since odds ratios are more easily interpretable than log-odds ratios, in contrast to Figure 1, the units displayed along the y-axis in Figure 2 are odds ratios.

In summary, the plot of the expected log-odds ratio for the test drug ($i=1$) as a function of the IMPROVEMENT y-scores falls on a straight line having slope β_1, the "test drug effects line." Similarly, the plot of each expected log-odds ratio for the placebo ($i = 2$) falls on a straight line, the "placebo effects line." Because the placebo is selected as the baseline reference, the placebo effects line has a slope of zero.

Each plot contains the following components:

1. Vertical axis. The vertical axis represents a baseline response category or contrast used as the base in defining the generalized odds. In Figure 2, the vertical axis is aligned with the "stationary" marker, because this category is the selected reference category or base of the response variable.

2. Horizontal axis. The horizontal axis represents a baseline category or contrast used in defining the odds ratio. In Figure 2 it corresponds to the placebo effects line because the "placebo" category was selected as the prediction reference category.

3. Outcome scores. The scores assigned to or estimated for the response categories, referred to as y-scores, are signified by triangular markers at the bottom of the plot. For example, Figure 2 displays triangle markers associated with each level of IMPROVEMENT. The relative distances between these markers are a consequence of the y-scores assigned by the statistical model. In Figure 2, the distances between adjacent markers are equal, corresponding to the equal adjacent category odds-ratio model. More generally, if these distances are unknown they are estimated under the general log-bilinear model (see de Falguerolles, Chapter 35).

4. Effects lines. For each predictor category, the expected odds ratios are plotted on a logarithmic scale as a linear function of the y-scores. The resulting line is referred to as the effects line associated with that predictor level.

5. Origin. The origin, the point of intersection between the vertical and horizontal axes, denotes the reference point (or base cell) for interpreting the odds ratios. In Figure 2, the origin is associated with the (placebo, stationary) cell. A logarithmic scale in odds-ratio units is given for the vertical axis on the left-hand side of the display. For example, it can be seen from the test drug effects line in Figure 2 that the odds of having a "slight" improvement (vs. "stationary") is about 1.7 times as high for patients who received the test drug than for patients who received the placebo. The odds ratios are given more accurately in Table 5.

6. Observed generalized odds ratios. Observed odds ratios are calculated on the basis of the observed counts in Table 2 and appear in the plot as symbols, squares for the test drug and diamonds for the placebo. Larger symbols reflect cells that are based on larger sample sizes. The lack of fit of the model to the data is ascertained by examining how distant these symbols are from the corresponding effects lines. By construction, the baseline "stationary" category and the "placebo" points are on the effects lines.

7. Model fit and significance chi-squared statistics. The chi-squared statistic reported at the top of the graph along with the p value represents how well the model fits the data; the smaller the chi-squared value, the better the fit. In Figure 2 the fit is 0.524 with three degrees of freedom ($p = 0.91$), which indicates that the model provides a good fit to these data. The good fit is supported in the graph by the closeness of the observed log-odds ratio markers to the corresponding effect lines.

The significance chi-squared reported beneath the fit assesses the extent to which the effects are significantly different from the null effect (odds ratio of 1). In Figure 2, we see that the treatment effect is highly significant ($p = 0.0059$).

The sum of the fit and significance chi-square statistics always equals the independence chi-squared statistic. The sum of the corresponding degrees of freedom from these tests always equals the degrees of freedom associated with the test for independence. Hence, the significance test used by the ordinal logit models in general will have fewer degrees of freedom and therefore will be more powerful than the chi-squared test for independence in assessing the significance of an effect.

8. The predicted score associated with each predictor category is denoted by a vertical arrow pointing downward onto the horizontal axis at the particular point

(the "expected *y*-score") that corresponds to the maximum likelihood estimate for the expected value of the *y*-score conditional on that predictor level ($E[Y \mid X = x_i]$). Note that the predicted outcome score for "placebo" falls between "stationary" and "slight" improvement, while for "test drug" it falls between "slight" and "moderate" improvement. The distance between the arrows associated with the test drug and placebo represents the difference in the estimated expected *y*-scores and is significantly different from zero according to the test of significance displayed at the top of Figure 2, discussed earlier.

The estimated odds ratios provide relative measures of effect analogous to the correlation or regression coefficient. In addition, the predicted scores and the distance between them provide a visual measure of the absolute effect, an improvement from less than slight (the predicted score expected under the "placebo" condition) to somewhat more than slight (the predicted score expected under the "test drug" condition). Thus, the graphical display contains both relative and absolute measures of effect.

Our clinical trial example was selected to illustrate the simplest type of logit display, associated with $I = 2$ predictor categories. In this case, when one of these categories is chosen to be the baseline reference for calculation of the effect (odds ratio), there is only one nonredundant odds ratio, that associated with the other category. Hence, there is only one nontrivial effects line and the reference category in this case is represented by the horizontal axis, which serves as a baseline or null effects line.

5 Example 2: Nutrition Data

In this section, we will provide more general displays associated with data where $I > 2$ predictor categories are present, as well as a further elaboration for the situation of multiple predictor variables. We conclude by providing two alternative views of these displays, which we refer to as the *X*-view and the *XY*-view. For the *X*-view, scores associated with each predictor category are displayed on the horizontal axis. In this case, each outcome category corresponds to an effects line.

The *XY*-view differs from the view presented thus far (the *Y*-view) where scores associated with each outcome category are displayed on the horizontal axis and each predictor category corresponds to an effects line. The *XY*-view is analogous to the usual regression scatterplot, where scores are available for both *X* and *Y*. In this case, the regression curve $E(Y \mid X)$ is plotted as a function of the predictor scores.

Table 7 shows data from a national telephone survey of 1382 women conducted in the fall of 1980 and reported by Feick (1984). We consider the analysis of NUTRITION ("How much do you feel you know about nutrition?—almost nothing, not too much, some, quite a bit, a lot") as a function of READLABELS ("How often would you say you read nutrition and ingredient labels?—frequently, sometimes, never").

Unlike Example 1, where we restricted the outcome scores to be equidistant, this model involves no restrictions of any kind on the NUTRITION response scores. In

Table 7: Cross-classification of self-assessment of nutrition knowledge (NUTRITION) by reported frequency of reading labels and nutrition information (READLABELS)

	NUTRITION				
READLABELS	Almost nothing	Not too much	Some	Quite a bit	A lot
Frequently	6	57	302	243	70
Sometimes	23	115	251	106	17
Never	26	69	70	25	2

Reprinted with permission from *Journal of Marketing Research,* published by the American Marketing Association, Lawrence Feick, 1984.

this case, the scores are estimated simultaneously with the other model parameters. The results are displayed in Figure 3. Note that we now have separate effects lines for each of the three levels of the predictor variable. Each effects line is based on plotting the estimated expected odds ratios (see Table 10), which are calculated from the estimated expected counts (Table 8) using the "almost nothing" category of NUTRITION and the "never" category for READLABELS as the baseline references (Tables 9 and 10).

The utility of the odds-ratio effects lines as a measure of effect is straight-forward—the higher the slope of the line, the greater the nutrition knowledge. For example, the "frequently" effect line in Figure 3 shows that those who frequently read labels are 154 times as likely to have "a lot" of nutrition knowledge than those who "never" read labels. In other words, the odds of having "a lot" of nutrition knowledge (vs. "almost nothing") is 154 times as high for people who "frequently" read labels

Table 8: Expected counts for NUTRITION × READLABELS

	NUTRITION					
READLABELS	Almost nothing	Not too much	Some	Quite a bit	A lot	Average Scores
Frequently	5.50	56.79	304.69	240.95	70.07	2.90
Sometimes	24.19	115.51	244.63	110.84	16.83	2.36
Never	25.31	68.71	73.68	22.21	2.10	1.83
Estimated Scores	0.00	1.16	2.55	3.39	4.37	

Table 9: Expected odds for NUTRITION × READLABELS

	NUTRITION				
READLABELS	Almost nothing	Not too much	Some	Quite a bit	A lot
Frequently	1.00	10.33	55.43	43.83	12.75
Sometimes	1.00	4.77	10.11	4.58	0.70
Never	1.00	2.71	2.91	0.88	0.08

than for those who "never" read labels. Similarly, it can be seen in Figure 3 that those who "frequently" read labels are 50 times as likely to have "quite a bit" of nutrition knowledge than those who "never" read nutrition labels.

The triangular markers on the horizontal axis in Figure 3 depict the relative spacing between the NUTRITION outcome categories that is obtained using the estimated scores shown at the bottom of Table 8. First, we note that this relative spacing results in an ordering of the outcome categories that correctly reproduces the original ordering. That is, the score estimated for the category "a lot" is higher than that for "quite a bit," and so forth. Second, note that the spacing is not quite equidistant. For example, the "some" category is somewhat more distant from the "not too much" category (i.e., $2.55 - 1.16 = 1.39$) than from the "quite a bit" category ($3.39 - 2.55 = 0.84$).

The arrows on the horizontal axis in Figure 3 compare the predicted NUTRITION outcome for each of the three READLABELS levels. Note that the average READLABELS score for "never" and "sometimes" falls between "not too much"

Table 10: Expected odds ratios for NUTRITION × READLABELS

	NUTRITION				
READLABELS	Almost nothing	Not too much	Some	Quite a bit	A lot
Frequently	1.00	3.81	19.04	49.97	153.93
Sometimes	1.00	1.76	3.47	5.22	8.40
Never	1.00	1.00	1.00	1.00	1.00

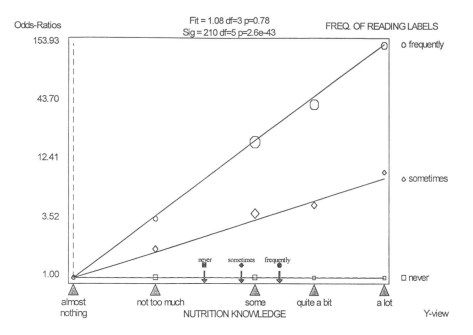

Figure 3: Y-view of ordinal logit model for nutrition data.

and "some" nutrition knowledge, while "frequently" falls between "some" and "quite a bit" of nutrition knowledge. The large difference between the lower average score for the base type "never" and the highest average score associated with "frequently" is displayed by the large distance between the two corresponding arrows in Figure 3. The predicted scores that are plotted are given in Table 8.

The lack-of-fit statistic is 1.08 with 3 degrees of freedom, indicating that this model, which makes no restrictions on the spacing between the response categories, provides an excellent fit to these data. A more parsimonious model that assumes equidistant spacing between the outcome categories also provides a good fit to the data—$X^2 = 5.46$ with 6 degrees of freedom ($p = 0.49$). The difference between these model fit chi-squares ($5.46 - 1.08 = 4.38$) with 3 degrees of freedom provides a test of the validity of the equidistant spacing restriction imposed by the more parsimonious model. As the difference between the chi-squares is not statistically significant, the hypothesis of equidistant spacing cannot be rejected.

Under either the unrestricted or equidistant spacing model, the effect of READ-LABELS on NUTRITION may be assessed by the chi-square significance statistic. For the equidistant spacing model in which scores are assumed to be equidistant, the significance $X^2 = 205$ with 2 degrees of freedom ($p < 0.0001$). Under the unrestricted model, the resulting significance level is similar, as shown above the graph in Figure 3.

Now once again consider the data in Table 1, where we have two predictor variables. Let i subscript the EI categories, j the JP categories, and k the response levels: $i = 1, 2; j = 1, 2; k = 1, 2, 3, 4$.

The model in this case becomes

$$\Phi_{k.ij} = \beta_i^{\text{EI}}(y_k - y_0) + \beta_j^{\text{JP}}(y_k - y_0) + \beta_{ij}^{\text{EI*JP}}(y_k - y_0)$$

$$= (\beta_i^{\text{EI}} + \beta_j^{\text{JP}} + \beta_{ij}^{\text{EI*JP}})(y_k - y_0)$$

where β^{EI} and β^{JP} are the main effects for EI and JP, respectively, and $\beta^{\text{EI*JP}}$ is the EI*JP interaction effect. The identifying conditions for this model, which yield the graphs in Figures 1 and 5, are

1. y_0 is the mean of the y scores so that $\Sigma_k p_k(y_k - y_0) = 0$
2. The β parameters are defined such that $\Sigma_i p_i \beta_i = \Sigma_j p_j \beta_j = \Sigma_i \Sigma_j p_{ij} \beta_{ij} = 0$

Hence, the log-odds ratios plotted in Figures 1 and 5 are defined with respect to the origin, which is identified by the "average" row condition and the "average" response score, the baseline references. In addition to these joint effects plots, partial effects plots are available to display each of the three effects separately (see, for example, Magidson, 1996b). For more general models of this type see Goodman (1983).

6   Alternative Views of the Model

In symmetric form, model (2) can be expressed as

$$\Phi_{ij}[= \Phi_{j.i} = \Phi_{i.j}] = \phi(x_i - x_0)(y_j - y_0) \tag{3}$$

For the asymmetric form of the model given earlier in (1) we set $\beta_i = \phi(x_i - x_0)$. The graphical displays based on this form of the model are called Y-views. An alternative form, known as the X-view, plots the odds ratios as a function of the x-scores: $\phi_{ij} = \gamma_j(x_i - x_0)$ where $\gamma_j \equiv \phi(y_j - y_0)$. Figure 4 shows the X-view for the clinical trial data, and Figure 5 provides the X-view for the personality characteristics data of Table 1. One advantage of the X-view is that the predictor variable is plotted along the horizontal or x-axis as is usually done in regression analysis. A disadvantage is that arrows representing the predicted values associated with each predictor level are not available in this view.

Another useful display is similar to the traditional regression view where the predicted value $E(Y \mid X)$ is plotted as a function of X. Plotted together with a scatterplot of quantitative (x, y) observations, we refer to this view as the XY-view. Since the curve $E(Y \mid X)$ has many favorable properties, it is called the "universal" regression by Magidson (1996b). For example, the change in the predicted value of Y associated with a unit increase in X equals the product of two quantities—the association parameter ϕ, which ranges from $-\infty$ to ∞, and the conditional variance of Y given X.

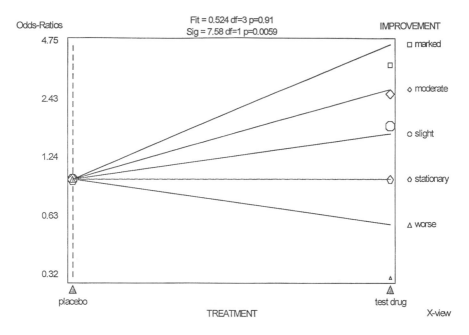

Figure 4: *X*-view of the equal adjacent category odds ratio model for clinical trial data.

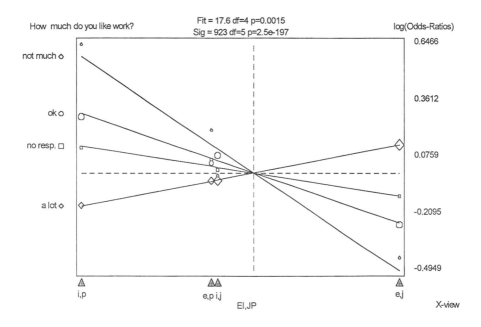

Figure 5: *X*-view of ordinal logit model for personality characteristics data.

Hence, regardless of the value of X, the change in the expected value for Y is either always positive or always negative depending on whether ϕ is positive or negative, a result that justifies calling the model a "monotonic regression" model:

$$\frac{\partial E(Y \mid X)}{\partial X} = \phi V(Y \mid X)$$

Since any given sample contains only a finite number of observations, even two continuous variables can be viewed as categorical, because at most they take on only a finite number of values. Figure 6 contrasts the universal/monotonic regression curve with the traditional linear regression line estimated by ordinary least squares for real continuous data so that both x-scores and y-scores are known. Is the true relationship linear or not?

The model used in Figure 6 is the form appropriate for continuous (and discrete quantitative) variables where scores are available for both X and Y. In this case, the scores are set equal to the observed quantitative values. Magidson (1996b) generalized odds ratios further to apply to continuous variables and showed that model (3) holds true under the bivariate normal distribution itself.

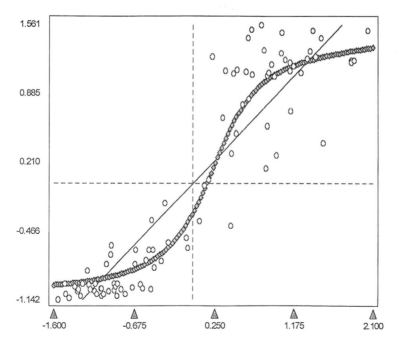

Figure 6: Comparison of universal regression model and OLS linear regression model for two continuous variables.

7 Conclusion

Recently, much attention in the literature has been paid to adjacent category logit models of both the log-linear and the log-bilinear variety for analyzing dichotomous or ordinal response variables—for example, see Ishii-Kuntz (1994) and Clogg and Shihadeh (1994). In this chapter, we have presented some new graphical representations for this important class of models and illustrated these graphs using several different data sets.

Traditional tabular results from estimation of these models are often complex and the parameter estimates may be very difficult to interpret. Proper interpretation requires both knowledge of the choice of coding for the variables plus knowledge of the scores associated with the categories of the variables. By providing an integrated picture of the model results, the graphs provide valuable information to the researcher, which saves time and reduces the likelihood of errors in interpretation.

The graphs can also be used as the basis of a graphical user interface to provide great simplification in the specification as well as the interpretation of results from such models. If the user judges the model to be unsatisfactory, the current parameter settings may be altered through direct user manipulation of the graph and a new model can then be estimated that reflects the new settings. Thus, researchers can more readily and quickly implement the natural, interactive process that is prevalent in social science research, in an active, participatory manner. Moreover, overlaying fit and significant statistics can supplement the powerful graph with important summary statistics.

Acknowledgments

The author wishes to acknowledge the important contributions of Frank Derrick and Alexander Ahlstrom to the preparation of the manuscript. The author is also indebted to Keith Potts of Beneficial Management Corp. for his assistance regarding the *XY* form of the graph.

Software Notes

The software and related computer technology for producing the X-view and Y-view of the graphical displays are the subject of issued and pending patents, but the graph itself is not patented. Persons interested in computer programs that produce this graph should write to GOLDminer, P.O. Box 1, Belmont, MA 02178 USA.

Chapter 35

Log-Bilinear Biplots in Action

Antoine de Falguerolles

1 Introduction

Generalized bilinear models aim at analyzing data arrays where one of the interactions can be described by a bilinear term. This leads to the ability to display the corresponding interaction visually by means of biplots (Gabriel, 1971; Gower and Hand, 1996).

Bilinear (also called "biadditive") models have long been known in the area of analysis of variance (Tukey, 1949; Mandel, 1971; Dorkenoo and Mathieu, 1993; Denis and Gower, 1996). Factor analysis is referred to as a bilinear model by Kruskal (1978). Obviously, bilinear structures also encompass methods such as principal component analysis, correspondence analysis, and nonsymmetrical correspondence analysis (see Balbi, Chapter 21). All these methods can be formulated as statistical models defined by the reconstitution formula, where a row $\times$ column effect is modeled through a bilinear term or equivalently through an inner product in a low-dimensional subspace. The fixed effect model, discussed by Caussinus (1986) in the context of principal component analysis and of some extensions of the analysis of count data in two-way tables, provides further examples of bilinear models.

It can be shown that all these statistical analyses implicitly assume a Gaussian distribution for the response variable. However, the association and correlation models considered by Goodman (1986) are interesting and useful examples of this modeling approach where Poisson or multinomial or product multinomial distributions are assumed for the counts of two-way contingency tables. The general bilinear model can be readily extended into other areas of application involving the more elaborate statistical settings discussed by Nelder and Wedderburn (1972) and McCullagh and Nelder (1989). Examples of such extensions can be found in Choulakian (1996), de Falguerolles and Francis (1992, 1994, 1995), and van Eeuwijk (1995).

2   Generalized Bilinear Models

2.1   Data Structure

It is assumed that the data of interest may be cross-classified by two factors, which, for simplicity, will be called the row and the column factors, denoted by R and C. The row and column factors have levels i ($i = 1,\ldots,I$) and j ($j = 1,\ldots,J$). The row (respectively, column) factor level for the sth observation is given by $i(s)$ [respectively, by $j(s)$]. Obvious examples of such data structures are the counts $y_s = y_{ij}$ of a simple two-way contingency table, or the counts of a more elaborate two-way table where the cross-classifying factors are themselves obtained by interactive coding of several polytomous variables.

In accordance with generalized linear modeling, the response variable values, y_s ($s = 1,\ldots,n$), are assumed to be the observed values of independent random variables Y_s with known distribution, expected values denoted by μ_s, and prior weights w_s. In practice, a variance function of the mean, possibly involving a scale factor, suffices to take into account the distributional assumptions. The expected values μ_s are related to predictors η_s by a link function $g(\mu_s) = \eta_s$.

2.2   Bilinear Model

The model formula for the predictor consists of a linear model (possibly null) and an additional bilinear term of reduced rank K that models the interaction between R and C:

$$g(\mu_s) = \eta_s = \text{linear model}_s + \sum_{k=1}^{K} \sigma_k \alpha_{i(s),k} \beta_{j(s),k}$$

The α_k form the *row score vectors* of order k and the β_k the *column score vectors* of order k, which are the analogues of the left and right singular vectors in a singular value decomposition. The σ_k are the *generalized singular values*, which can be taken strictly positive and arranged in decreasing order. Once the model is fitted, estimates for the parameters and for the expected values (the so-called fitted values), as well as the usual goodness-of-fit statistics (G^2 and χ^2), are obtained.

2.3   Identification

Identification constraints are introduced in order to identify the scores; these may be centered and orthonormalized with respect to given metrics (usually defined by a diagonal matrix of weights). Centering can be specified for either the row or column scores, for both the row and column scores, or not at all. Note that, for identification purposes, if a row effect (respectively, a column effect) is included in the linear part of the model, the column score vectors (respectively, the row score vectors) are to be centered.

If diagonal weights (w_i' for the rows and w_j'' for the columns) are assumed, centering provides the following constraints:

$$\sum_{i=1}^{I} w_i' \alpha_{i,k} = 0, \quad \sum_{j=1}^{J} w_j'' \beta_{j,k} = 0, \quad k = 1, \ldots, K$$

whereas orthonormalization provides the additional constraints:

$$\sum_{i=1}^{I} w_i' \alpha_{i,k} \alpha_{i,k'} = \begin{cases} 0, & k \neq k' \\ 1, & k = k' \end{cases} \qquad \sum_{j=1}^{J} w_j'' \beta_{j,k} \beta_{j,k'} = \begin{cases} 0, & k \neq k' \\ 1, & k = k' \end{cases}$$

It must be emphasized that the choice of identification constraints affects the parameter values in the model and therefore may possibly distort the patterns in the associated graphical displays. However, this choice does not affect the fitted values for the data.

In some instances, the score vectors are further restricted to belong to some linear subspaces generated by exogenous variables, the latter being sometimes referred to as "instrumental variables." Examples can be found in ter Braak (1988), Gilula and Haberman (1988), Böckenholt and Böckenholt (1990b), and Böckenholt and Takane (1994).

2.4   Biplots

The bilinear structure of the predictor allows several biplots to be constructed: the rank K restricted interaction between level i of the row factor and level j of the column factor is equal to the inner product of the K-dimensional vectors $[\sigma_1^\gamma \alpha_{i,1}, \ldots, \sigma_K^\gamma \alpha_{i,K}]$ and $[\sigma_1^{1-\gamma} \beta_{j,1}, \ldots, \sigma_K^{1-\gamma} \beta_{j,K}]$ where γ is any fixed value in the open interval $(0, 1)$. These vectors provide the coordinates for plotting the row level i and the column level j, respectively, in the associated K-dimensional biplot. Note that a value $\gamma = 0.5$ reflects the equal treatment of the row and column factor in defining the interaction. It is the only one considered in the sequel.

3   A Three-Way Table

To illustrate the flexibility of generalized bilinear models, a three-way contingency table of suicide rates in West Germany is considered. It is shown how generalized bilinear models can be inserted in the hierarchy of possible linear models while allowing the visualization of the interesting interaction terms in the form of biplots.

Table 1: Suicide behavior: age by sex by cause of death

	Men Cause of death[a]								
Age	c1	c2	c3	c4	c5	c6	c7	c8	c9
10–14	4	0	0	247	1	17	1	6	9
15–19	348	7	67	578	22	179	11	74	175
20–24	808	32	229	699	44	316	35	109	289
25–29	789	26	243	648	52	268	38	109	226
30–34	916	17	257	825	74	291	52	123	281
35–39	1118	27	313	1278	87	293	49	134	268
40–44	926	13	250	1273	89	299	53	78	198
45–49	855	9	203	1381	71	347	68	103	190
50–54	684	14	136	1282	87	229	62	63	146
55–59	502	6	77	972	49	151	46	66	77
60–64	516	5	74	1249	83	162	52	92	122
65–69	513	8	31	1360	75	164	56	115	95
70–74	425	5	21	1268	90	121	44	119	82
75–79	266	4	9	866	63	78	30	79	34
80–84	159	2	2	479	39	18	18	46	19
85–89	70	1	0	259	16	10	9	18	10
90+	18	0	1	76	4	2	4	6	2

Women
Cause of death

Age	c1	c2	c3	c4	c5	c6	c7	c8	c9
10–14	28	0	3	20	0	1	0	10	6
15–19	353	2	11	81	6	15	2	43	47
20–24	540	4	20	111	24	9	9	78	67
25–29	454	6	27	125	33	26	7	86	75
30–34	530	2	29	178	42	14	20	92	78
35–39	688	5	44	272	64	24	14	98	110
40–44	566	4	24	343	76	18	22	103	86
45–49	716	6	24	447	94	13	21	95	88
50–54	942	7	26	691	184	21	37	129	131
55–69	723	3	14	527	163	14	30	92	92
60–64	820	8	8	702	245	11	35	140	114
65–69	740	8	4	785	271	4	38	156	90
70–74	624	6	4	610	244	1	27	129	46
75–79	495	8	1	420	161	1	29	129	35
80–84	292	3	2	223	78	0	10	84	23
85–89	113	4	0	83	14	0	6	34	2
90+	24	1	0	19	4	0	2	7	0

[a]c1, suicide by solid or liquid matter; c2, suicide by toxification of gas at home; c3, suicide by toxification of other gas; c4, suicide by hanging, strangling, suffocating; c5, suicide by drowning; c6, suicide with guns and explosives; c7, suicide with knives . . .; c8, suicide by jumping; c9, suicide by other methods.

531

3.1   The Suicide Data

The data, shown in Table 1, are frequencies of suicide classified by sex, method of suicide, and age group. Originally from Heuer (1979, Table 1), these data have been quite extensively used in statistical work (see van der Heijden and de Leeuw, 1985; van der Heijden and Worsley, 1988; Friendly, 1994a). The corresponding data are reproduced in Table 1. S, M, and A, respectively, represent the factor sex (2 levels), method of suicide (9 levels), and age group (17 levels).

3.2   Correspondence Analysis Used Complementary to Log-Linear Analyses

The entries in the three-way table can be assumed to be independent, Poisson distributed with saturated model [AMS]:

$$\log(\mu_{ams}^{AMS}) = v + v_a^A + v_m^M + v_s^S + v_{am}^{AM} + v_{as}^{AS} + v_{ms}^{MS} + v_{ams}^{AMS}$$

This model involves as many independent parameters as cell counts. Hence, unsaturated models obtained by removing interactions terms in a hierarchical manner are of special interest.

The all-two-way-interaction model is

$$\log(\mu_{ams}^{AMS}) = v + v_a^A + v_m^M + v_s^S + v_{am}^{AM} + v_{as}^{AS} + v_{ms}^{MS}$$

where the three-way interaction term v_{ams}^{AMS} has been removed. Note that, in the absence of missing data, the maximum likelihood fitted values for this model reproduce all the two-way margins of the observed table. As a consequence, all the one-way margins are also reproduced. It then follows that the fitted values from an all-two-way-interaction model have the same Burt table as the observed data. Therefore multiple correspondence analysis of the Burt matrix cannot reveal more structure in the data than this model. Accordingly, the all-two-way-interaction model and the value of its associated deviance can be used as benchmarks in modeling.

A further restricted model is [AS][M]:

$$\log(\mu_{ams}^{AMS}) = v + v_a^A + v_m^M + v_s^S + v_{as}^{AS}$$

where the maximum likelihood fitted values reproduce the two-way margin [AS] and all one-way margins [A], [M], and [S].

Noting that none of the restricted log-linear models has an acceptable fit (see Table 2), the strategy of analyzing the residuals from a log-linear model by the correspondence analysis (CA) of an ad hoc two-way table can be considered (for further details see van der Heijden *et al.*, 1989). Along that line, van der Heijden and de Leeuw (1985) perform the CA of a two-way table $R \times C$ in which the column factor C is the method of suicide (M) and the factor R is obtained by interactively coding the factors age (A) and sex (S), thus creating an R factor with $17 \times 2 = 34$ levels. As a result, van der Heijden and de Leeuw (1985) get reduced rank approximations of

Table 2: Log-linear models, associated degrees of freedom (df), and chi-squared goodness-of-fit statistics (G^2 and χ^2) for the suicide data (although not necessary in our framework, we have added the value 0.1 to each level, following van der Heijden and de Leeuw, 1995)

Model	df	G^2	χ^2
$[A][M][S]$	280	12337.14	12304.05
$[A][MS]$	272	6857.76	6522.38
$[AM][S]$	152	7779.68	7198.33
$[AS][M]$	264	10313.80	9995.32
$[AM][AS]$	136	5756.34	5369.09
$[AM][MS]$	144	2300.30	2255.64
$[AS][MS]$	256	4834.42	4518.90
$[AM][AS][MS]$	128	429.19	435.63

the residuals from model $[AS][M]$. They retain a rank $K = 2$ approximation, and the associated biplot is reproduced in Figure 1.

3.3 A Log-Bilinear Analysis

In the spirit of the two-stage analysis just outlined, the following log-bilinear models are considered:

$$\log(\mu_{ams}^{AMS}) = v + v_a^A + v_m^M + v_s^S + v_{as}^{AS} + \sum_{k=1}^{K} \sigma_k \alpha_{as,k}^{AS} \beta_{m,k}^M.$$

All these models include model ($[AS][M]$) as a baseline model, and Table 3 gives the chi-squared statistics corresponding to inclusion of a restricted bilinear interaction $[ASM]$ of rank $K = 1, 2, 3$. It appears that $K = 3$ is needed to obtain a better fit than that of the all two-way interaction model (see Table 3 and Table 2).

To compare the scores in the bilinear term with those of CA, marginal proportions are used as weights in the identification constraints (see Section 2.3). The corresponding estimated generalized singular values are given in Table 4. The variation of the generalized singular values reflects the dramatic changes in the chi-squared statistics when successive bilinear terms are introduced (see Table 3 and Table 4).

3.4 Biplot Interpretation

It appears that the biplot (see Figure 1) provided by the first two scores in a log-bilinear model is somewhat similar to that obtained in the first two dimensions when using CA to analyze the residuals from the baseline log-linear model $[AS][M]$ (see Figure 2). Roughly, both biplots stress the differences in methods of suicide between men and women and the different use of methods as age varies. In this example, the approach based on CA compares favorably with the log-bilinear modeling approach.

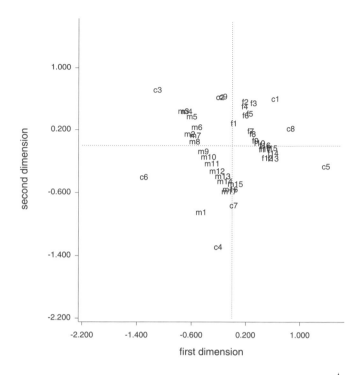

Figure 1: Biplot in the first two dimensions obtained by log-bilinear modeling with $K = 3$. Males are labelled m1 to m17 and females f1 to f17, where the number refers to the age group.

Table 3: Adding restricted [ASM] interactions to the baseline model [AS][M] where $\log(\mu_{ams}^{AMS}) = v + v_a^A + v_m^M + v_s^S + v_{as}^{AS}$

Model	df	G^2	χ^2
$v + v_a^A + v_m^M + v_s^S + v_{as}^{AS}$	264	10313.80	9995.33
$v + v_a^A + v_m^M + v_s^S + v_{as}^{AS} + \sum_{k=1}^{1} \sigma_k \alpha_{as,k}^{AS} \beta_{m,k}^{M}$	224	4534.92	4313.62
$v + v_a^A + v_m^M + v_s^S + v_{as}^{AS} + \sum_{k=1}^{2} \sigma_k \alpha_{as,k}^{AS} \beta_{m,k}^{M}$	186	622.20	624.58
$v + v_a^A + v_m^M + v_s^S + v_{as}^{AS} + \sum_{k=1}^{3} \sigma_k \alpha_{as,k}^{AS} \beta_{m,k}^{M}$	150	321.12	319.66

Table 4: Generalized singular values for $K = 3$

Generalized singular values	Squared generalized singular values
10.40	108.23
6.66	44.35
2.96	8.78

4 A Square Table

The data (Clogg and Shihadeh, 1994, Table 4.1) derive from the National Survey of Black Families and consist of a 7×7 cross-classification of religion at age 16 and actual religion (Sherkat, 1992) (Table 5). The data define a square contingency table where there is a one-to-one correspondence between the levels of the row factor and the levels of the column factors. Clogg and Shihadeh (1994) note that the cells on the main diagonal account for most of the structure of the table. Concerning the

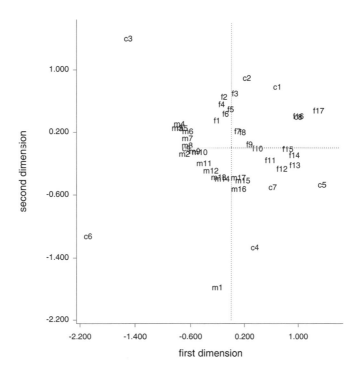

Figure 2: Biplot in the first two dimensions obtained by correspondence analysis.

Table 5: Cross-classification of religion by religion at age 16

		Religion at age 16					
Religion	Li	Me	Ba	Co	Ot	Ca	No
Liberal	41	8	14	1	0	1	1
Methodist	0	212	23	5	3	2	2
Baptist	4	58	980	15	9	7	10
Conservative	2	20	103	120	6	5	6
Other	2	5	42	4	19	8	5
Catholic	5	7	20	0	1	97	2
None	9	28	95	17	8	14	61

off-diagonal cells, they show that there is some evidence for a symmetric row and column interaction. It follows that quasi-symmetric models (Becker, 1990) also allow the biplot visualization of the symmetrical interactions.

Note that, like generalized linear models, generalized bilinear models can cope with missing entries and structural values: an observation can be excluded from a fit either by setting its prior weight to zero or by introducing a specific parameter in the linear term. This possibility is illustrated on the diagonal cells of the square table, which, from now on, are excluded in all fits.

4.1   Log-Linear Modeling of Symmetry

A quasi-symmetry model with Poisson distribution and log link can be fitted:

$$\log(\mu_{ij}) = v + v_i^R + v_j^C + v_{ij}^{RC}$$

where $v_{ij}^{RC} = v_{ji}^{RC}$ (Caussinus, 1965). The model, which exactly fits the diagonal cells (or equivalently excludes them from the analysis), provides a reasonable fit to the data: the corresponding chi-squared statistics are $G^2 = 25.97$ and $\chi^2 = 25.82$ with 15 degrees of freedom. This implies that the symmetry in this data set is worth looking at further.

4.2   Goodman RC-Association Model

A bilinear approach that does not imply symmetrical interactions for the off-diagonal cells can be considered. The Goodman RC-association model (Goodman, 1986, 1991) assumes a Poisson distribution for the cell counts with the following structure for their means:

$$\log(\mu_{ij}) = v + v_i^R + v_j^C + \sum_{k=1}^{K} \sigma_k \alpha_{i,k} \beta_{j,k}$$

Table 6: Log-bilinear models, degrees of freedom, and chi-squared statistics for $K = 0, 1, 2, 3$, for the off-diagonal cells

Model	df	G^2	χ^2
$v + v_i^R + v_j^C$	29	57.01	59.47
$v + v_i^R + v_j^C + \sum_{k=1}^{K=1} \sigma_k \alpha_{i,k} \beta_{j,k}$	18	26.60	23.03
$v + v_i^R + v_j^C + \sum_{k=1}^{K=2} \sigma_k \alpha_{i,k} \beta_{j,k}$	9	8.05	6.41
$v + v_i^R + v_j^C + \sum_{k=1}^{K=3} \sigma_k \alpha_{i,k} \beta_{j,k}$	2	.88	.80

The chi-squared statistics corresponding to increasing values of K are given in Table 6. It appears that the model with $K = 2$ provides a very good fit to the data.

4.3 Quasi-Symmetric Bilinear Models

Quasi-symmetry and symmetry models are obviously connected to the class of quasi-symmetric models considered by Becker (1990), and it turns out that this class of models is itself related to the class of bilinear models. For a log link, a quasi-symmetric model assumes that

$$\log(\mu_{ij}) = \text{linear model}_{ij} + \sum_{k=1}^{K} \sigma_k \alpha_{i,k} \alpha_{j,k}$$

where the $\alpha_{i,k}$ are the common row and column scores of order k, usually centered and orthonormalized. In this formula, the σ_k cannot always be taken to be strictly positive. This is known as the problem of "inverse factors" (Benzécri, 1973a). In this case, the biplot interpretation is preserved by constraining the corresponding row and column score vectors to have opposite signs:

$$\sum_{k=1}^{K} \sigma_k \alpha_{i,k} \alpha_{j,k} = \sum_{k=1}^{K} |\sigma_k| \alpha_{i,k} \left(\frac{\sigma_k}{|\sigma_k|} \alpha_{j,k} \right)$$

Typical linear model formulas in a quasi-symmetric setting are $v + v_i^R + v_j^C$, $v + v_i^R + v_j^C + \mu \delta_{ij}$, and $v + v_i^R + v_j^C + \mu_i \delta_{ij}$ (where $\delta_{ij} = 1$ if $i = j$, otherwise 0), depending on the assumptions pertaining to the diagonal cells: no effect, constant effect, or specific effect, respectively. Note that the last formula is equivalent to the first with diagonal cells excluded from the fit. It is thus retained in following analyses.

By using an adaptation of the three-dimensional representation of quasi-symmetry (Bishop *et al.*, 1975), quasi-symmetric models for the off-diagonal cells can be fitted as a regular Goodman RC-association model. The data are replicated

Table 7: Quasi-symmetric models, degrees of freedom, and chi-squared statistics in the modeling of off-diagonal cells

Model	df	G^2	χ^2
$v + v_i^R + v_j^C$	29	57.01	59.47
$v + v_i^R + v_j^C + \sum_{k=1}^{K=1} \|\sigma_k\|\alpha_{i,k} \left(\frac{\sigma_k}{\|\sigma_k\|}\,\alpha_{j,k}\right)$	23	36.66	33.54
$v + v_i^R + v_j^C + \sum_{k=1}^{K=2} \|\sigma_k\|\,\alpha_{i,k} \left(\frac{\sigma_k}{\|\sigma_k\|}\,\alpha_{j,k}\right)$	18	28.59	28.54
$v + v_i^R + v_j^C + v_{ij}^{RC} \,(v_{ij}^{RC} = v_{ji}^{RC})$	15	25.97	25.82

twice with exchanged row and column indices in the restricted interaction term:

$$\log(\mu_{ij1}) = \log(\mu_{ij}) = \text{linear model}_{ij} + \sum_{k=1}^{K} \sigma_k \alpha_{i,k} \beta_{j,k}$$

$$\log(\mu_{ij2}) = \log(\mu_{ij}) = \text{linear model}_{ij} + \sum_{k=1}^{K} \sigma_k \alpha_{j,k} \beta_{i,k}$$

Indeed, the fit coerces the row score vectors (α_k) and the corresponding column score vectors (β_k) to be equal (possibly up to a sign) and thus preserves the positivity of the associated generalized singular values (σ_k).

4.4   Biplot Interpretation

Table 7 reports the statistics of fit obtained for $K = 1, 2$. The quasi-symmetric model of order two provides an acceptable fit. However, there is an inversion in the second dimension. The biplot of the corresponding symmetrical interactions is reproduced in Figure 3. It stresses the lack of mobility between conservative (4) and liberal (1), or conservative (4) and catholic (6), and the mobility between methodist (2) and baptist (3), or liberal (1) and catholic (6). It should be emphasized that this biplot visualizes the religious mobility as captured by the off-diagonal cells. Therefore the proximity between the row and column markers corresponding to a same category has no meaning in this context.

5   Discussion

The flexibility of bilinear models makes it easy to unify two methodological streams: the stream of exploratory data analysis methods (e.g., principal component analysis, correspondence analysis, multiple correspondence analysis, biplot decomposition of matrices) and the stream of data modeling in which probabilistic models are formu-

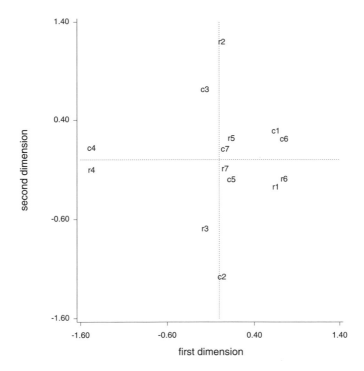

Figure 3: Biplot in the first two dimensions provided by the quasi-symmetric model of rank $K = 2$.

lated, fitted, and tested. The class of bilinear models allows multivariate descriptive techniques to be reformulated as models (see Gower, 1989b) and modeling can be performed in a descriptive way. A serious danger of this flexibility is that of overfitting the data. This can be partly alleviated by implementing model-selection methods and cross-validation procedures in this context. But this is another story.

Computational Note

Using an extension of the non-linear iterative partial least-squares (NIPALS) procedure introduced by Wold (1966), generalized bilinear models can be fitted in most software programs which can fit general linear models (de Falguerolles and Francis, 1992). The overall measure of fit is the (quasi-)deviance, and standard procedures for assessing the adequacy of these models are currently available in this context (McCullagh and Nelder, 1989). The risk of reaching local optima in those fits is dealt with by considering several random initializations for the scores.

References

Adam, G., Bon, F., Capdevielle, J. and R. Mouriaux. 1970. *L'ouvrier français en 1970*. Paris: FNSP.

Adorno, T., Frenkel-Brunswik, E., Levinson, D. and R. Sanford. 1950. *The Authoritarian Personality*. New York: Harper & Row.

Advanced Visual Systems. 1992. *AVS Users Guide*. Waltham, MA: AVS Inc.

Agresti, A. 1990. *Categorical Data Analysis*. New York: Wiley.

Agresti, A. 1996. *An Introduction to Categorical Data Analysis*. New York: Wiley.

Ahn, H. 1991. *Effects of missing responses in multiple-choice data on dual scaling results*. Unpublished doctoral dissertation. The University of Toronto.

Akaike, H. 1974. A new look at the statistical model identification. *IEEE Transactions on Automatic Control*, **19**, 716–723.

Andersen, E. B. 1994. *The Statistical Analysis of Categorical Data* (3rd ed.). Berlin: Springer.

Andersen, E. B. 1995. Graphic diagnostics in correspondence analysis. Research report 108. Copenhagen: University of Copenhagen.

Andrews, D. F. 1972. Plots of high dimensional data. *Biometrics*, **28**, 125–136.

Andrich, D. 1988. The application of an unfolding model of the PIRT type to the measurement of attitude. *Applied Psychological Measurement*, **12**, 33–51.

Anheier, H. K., Gerhards, J. and F. P. Romo. 1995. Forms of capital and social structure in cultural fields: Examining Bourdieu's social topography. *American Journal of Sociology*, **100**, 859–903.

Arabie, Ph., Boorman S. A. and P. R. Levitt. 1978. Constructing blockmodels: How and why. *Journal of Mathematical Psychology*, **17**, 21–63.

Bahadur, R. R. 1961. A representation of the joint distribution of responses to *n* dichotomous items, in H. Solomon (ed.), *Studies in Item Analysis and Prediction*, pp. 158–176. Stanford: Stanford University Press.

Balbi, S. 1992. On stability in nonsymmetrical correspondence analysis using bootstrap. *Statistica Applicata*, **4**, 543–552.

Balbi, S. 1994. Influence and stability in non symmetrical correspondence analysis. *Metron*, **52**, 111–128.

Balbi, S. and R. Siciliano. 1994. *Analisi longitudinale non simmetrica di tabelle di contingenza a tre vie*. Atti della XXXVII Riunione della Società Italiana di Statistica, vol. 1, pp. 345–356. San Remo.

Balbi, S., D'Esposito, M. R., Golia, F. and A. Pane. 1995. Le adozioni in Campania. Un'indagine campionaria, in M. Cavallo (ed.), *Adozioni dietro le quinte*, pp. 165–204. Milano: FrancoAngeli.

Barnes, S. H. and M. Kaase. 1979. *Political Action*. Beverly Hills, CA: Sage.

Barrere-Maurisson, M., Battagliola, F. and A. Daune-Richard. 1985. The course of women's careers and family life, in B. Roberts, R. Finnegan and D. Gallie (eds.), *New Approaches to Family Life*, pp. 431–458. Manchester: Manchester University Press.

Barry, J. T., Walby, S. and B. Francis. 1990. Graphical exploration of work history data. *Quaderni di Statistica e Matematica Applicata alle Scienze Economico-Sociali*, **12**, 65–74.

Bartlett, M. S. 1935. Contingency table interactions. *Journal of the Royal Statistical Society, Supplement 2*, 248–252.

Baudelot, C. 1990. Aimez–vous les maths?, in M. Bécue, L. Lebart, and N. Rajadell (eds.), *Proceedings of 1ᵉ Journées d'Analyse Statistique des Données Textuelles*, pp. 13–27. Barcelona: Servei de Publicaciones de la UPC.

Becker, H. 1986. Distributing modern art, in H. Becker (ed.), *Doing Things Together*, pp. 67–79. Evanston, IL: Northwestern University Press.

Becker, M. 1990. Quasisymmetric models for the analysis of square contingency tables. *Journal of the Royal Statistical Society, Series B*, **52**, 369–378.

Becker, R. A., Chambers, J. M. and A. R. Wilks. 1988. *The new S language. A programming environment for data analysis and graphics*. Pacific Grove, CA: Wadsworth.

Bécue, M. 1991. *Análisis de Datos Textuales. Métodos Estadísticos y Algoritmos*. Paris: CISIA.

Bécue, M. and R. Peiro. 1993. Les quasi-segments pour une classification automatique des questions ouvertes, in Anastex (ed.), *Proceedings of 2º Journées d'Analyse Statistique de Données Textuelles,* pp. 411–423. Paris: Editions de Télécom.

Bell, D. 1976. *The Cultural Contradictions of Capitalism.* New York: Basic Books.

Bellah, R. N., Madsen, R. and W. M. Sullivan. 1985. *Habits of the Heart. Individualism and Commitment in American Life.* Berkeley: University of California Press.

Belson, W. A. and J. A. Duncan. 1962. A comparison of the check-list and the open response questioning system. *Applied Statistics,* **2**, 120–132.

Benzécri, J-P. 1969. Statistical analysis as a tool to make patterns emerge from data, in S. Watanabe (ed.), *Methodologies of Pattern Recognition,* pp. 35–74. New York: Academic Press.

Benzécri, J-P. 1973a. *L'Analyse des Données.* Vol. 1: *La Taxinomie.* Vol. 2: *L'Analyse des Correspondances.* Paris: Dunod.

Benzécri, J-P. 1973b. La place de l'a priori, in *Encyclopedia Universalis,* Vol. 17, pp. 11–24.

Benzécri J-P. *et al.* 1981. *Pratique de l'analyse des données,* Vol. 3: *Linguistique & Lexicologie.* Paris: Dunod.

Benzécri, J-P. 1992. *Correspondence Analysis Handbook.* New York: Marcel Dekker.

Bernard, J. M., Le Roux, B., Rouanet, H. and M. A. Schiltz. 1989. Analyse des données multidimensionelles par le langage d'interrogation de données (LID): Au delà de l'analyse des correspondances. *Bulletin de Méthodologie Sociologique,* **23**, 3–46.

Bernard, M. Cl. and Ch. Lavit. 1985. Statistique. *Cahiers de l'Analyses des Données,* **10**, 11–15.

Bertin, J. 1967. *Sémiologie graphique.* Paris: Mouton.

Bertin, J. 1981. *Graphics and Graphic Information Processing.* New York: de Gruyter.

Bertin, J. 1983. *Semiology of Graphics.* Madison, WI: University of Wisconsin Press.

Bickel, P. J., Hammel, J. W. and J. W. O'Connell. 1975. Sex bias in graduate admissions: Data from Berkeley. *Science,* **187**, 398–403.

Billiet, J. and H. de Witte. 1995. Attitudinal dispositions to vote for a 'nex' extreme right-wing party. The case of 'Vlaams Blok.' *European Journal of Political Research,* **27**(2), 181–202.

Birch, M. W. 1963. Maximum likelihood in three-way contingency tables. *Journal of the Royal Statistical Society, Series B*, **25**, 220–233.

Bishop, Y. V. V., Fienberg, S. E. and P. W. Holland. 1975. *Discrete Multivariate Analysis.* Cambridge: MIT Press.

Blau, P. M. and O. D. Duncan. 1967. *The American Occupational Structure.* New York: Free Press.

Blasius, J. 1994. Correspondence analysis in social science research, in M. J. Greenacre and J. Blasius (eds.), *Correspondence Analysis in the Social Sciences*, pp. 23–52. London: Academic Press.

Blasius, J. and M. J. Greenacre. 1994. Computation of correspondence analysis, in M. J. Greenacre and J. Blasius (eds.), *Correspondence Analysis in the Social Sciences*, pp. 53–78. London: Academic Press.

Blossfeld, H.–P., Hamerle, A. and K. U. Mayer. 1989. *Event History Analysis.* Hillsdale, NJ: Lawrence Erlbaum.

Böckenholt, U. 1997. Concomitant variables in latent change models, in J. Rost and R. Langeheine (eds.), *Applications of Latent Trait and Latent Class Models in the Social Sciences*, pp. 361–369. Waxman.

Böckenholt, U. and I. Böckenholt. 1990a. Modeling individual differences in unfolding preference data. *Applied Psychological Measurement*, **14**, 257–269.

Böckenholt, U. and I. Böckenholt. 1990b. Canonical analysis of contingency tables with linear constraints. *Psychometrika*, **55**, 633–639.

Böckenholt, U. and I. Böckenholt. 1991. Constrained latent class analysis: Simultaneous classification and scaling of discrete choice data. *Psychometrika*, **56,** 699–716.

Böckenholt, U. and R. Langeheine. 1996. Latent change in recurrent choice data. *Psychometrika*, **61**, 285–302

Böckenholt, U. and Y. Takane. 1994. Linear Constraints in Correspondence, in M. J. Greenacre and J. Blasius (eds) *Correspondence Analysis in the Social Sciences*, pp. 112–127. London: Academic Press.

Bogart, L. 1967. No opinion, don't know, and maybe no answer. *Public Opinion Quarterly*, **31**, 311–345.

Bollen, K. A. 1989. *Structural Equations with Latent Variables.* New York: Wiley.

Bonnet, P., Le Roux, B. and G. Lemaine. 1996. Analyse géometrique des données: une enquête sur le racisme. *Mathématiques, Informatique et Sciences Humaines*, **136**, 5–24.

Borg, I. 1977. Geometric representation of individual differences, in J. C. Lingoes, E. E. Roskam, and I. Borg (eds.), *Geometric Representations of Relational Data: Readings in Multidimensional Scaling,* pp. 609–656. Ann Arbor, MI: Mathesis.

Borg, I. and P. J. F. Groenen. 1997. *Modern Multidimensional Scaling.* New York: Springer.

Borg, I. and J. Lingoes. 1987. *Multidimensional Similarity Structure Analysis.* New York: Springer.

Borg, I. and S. Shye. 1995. *Facet Theory: Form and Content.* Newbury Park, CA: Sage.

Borg, I. and T. Staufenbiel. 1993. Facet theory and design for attitude measurement and its application, in D. Krebs and P. Schmidt (eds.), *New Directions in Attitude Measurement,* pp. 206–237. New York: de Gruyter.

Bossuyt, P. 1990. *A Comparison of Probabilistic Unfolding Theories for Paired Comparison Data.* New York: Springer.

Bourdieu, P. 1979. *La Distinction. Critique Sociale du Jugement.* Paris: Les Editions de Minuit.

Bourdieu, P. 1980. Quelques propriétés des champs, in P. Bourdieu (ed.), *Questions de sociologie.* Paris: Les Editions de Minuit.

Bourdieu, P. 1981. La représentation politique, éléments pour une théorie du champ politique. *Actes de la recherche en sciences sociales,* **36–37,** 3–22.

Bourdieu, P. 1984. *Distinction.* Boston, MA: Harvard University Press.

Bourdieu, P. 1991. Inzwischen kenne ich alle Krankheiten der soziologischen Vernunft. Pierre Bourdieu im Gespräch mit Beate Krais, in P. Bourdieu, J. C. Chamboredon and J. C. Passeron (eds.), *Soziologie als Beruf,* pp. 269–283. Berlin: de Gruyter.

Bourdieu, P. 1993. *The Field of Cultural Production.* Oxford: Polity Press.

Bourdieu, P. 1996. *The Rules of Art.* Cambridge: Polity Press.

Bradley, R. A., Katti, S. K. and I. J. Coons. 1962. Optimal scaling for ordered categories. *Psychometrika,* **27,** 355–374.

Bradu, D. and K. R. Gabriel. 1978. The biplot as a diagnostic tool for models of two-way tables. *Technometrics,* **20,** 47–68.

Breiman, L., Friedman, J. H., Olshen, R.A. and C. J. Stone. 1984. *Classification and Regression Trees.* Belmont, CA: Wadsworth.

Brody, C. J. 1986. Things are rarely black and white: Admitting gray into the Converse model of attitude stability. *American Journal of Sociology*, **92**, 657–677.

Budge, I. and D. J. Farlie. 1983. *Explaining and Predicting Elections: Issue Effects and Party Strategies in Twenty-Three Democracies*. London: Allen & Unwin.

Buerklin, W. and D. Roth (eds.). 1994. *Das Superwahljahr*. Köln: Bund Verlag.

Campbell, L., Gurin, G. and W. E. Miller. 1954. *The Voter Decides*. Evanston, IL: Row & Peterson.

Caraux, G. 1984. Réorganisation et représentation visuelle d'une matrice de données numériques: un algorithme itératif. *Revue de Statistique appliquée*, **32**, 5–23.

Carlier, A. and G. Ewing. 1992. Longitudinal analysis of contingency tables with applications to demographic data. *Computational Statistics*, **7**, 329–353.

Carlier, A. and P. M. Kroonenberg. 1996. Decompositions and biplots in three-way correspondence analysis. *Psychometrika*, **61**, 355–374.

Carroll, J. D. 1972. Individual differences and multidimensional scaling, in R. N. Shepard, A. K. Romney, and S. B. Nerlove (eds.), *Multidimensional Scaling: Theory and Applications in the Behavioral Sciences*, Vol. 1. New York: Seminar Press.

Carroll, J. D. and J. J. Chang. 1970. Analysis of individual differences in multidimensional scaling via an *N*-way generalization of "Eckart–Young" decomposition. *Psychometrika*, **35**, 283–320.

Carroll, J. D., Kumbasar, E. and A. K. Romney. 1997. An equivalence relation between correspondence analysis and classical metric multidimensional scaling for the recovery of Euclidean distances. *British Journal of Mathematical and Statistical Psychology*, **50**, 81–92.

Carroll, J. D. and S. Pruzansky. 1980. Discrete and hybrid scaling models, in E. D. Lantermann and H. Feger (eds.), *Similarity and Choice*, pp. 108–139. Vienna: Huber.

Carton, A., Swyngedouw, M., Billiet, J. and R. Beerten. 1993. *Source Book of the Voters' Study in Connection with the 1991 General Election*. Leuven: I.S.P.O./S.O.I.

Caussinus, H. 1965. Contribution à l'analyse statistique des tableaux carrés. *Annales de la Faculté des Sciences de l'Université de Toulouse*, **29**, 77–182.

Caussinus, H. 1986. Models and Uses of Principal Components Analysis, in J. de Leeuw et al. (eds.), *Multidimensional Data Analysis*, pp.149–170. Leiden: DSWO.

Celeux, G. and Y. Lechevallier. 1982. Méthodes de segmentation non paramétriques. *Revue de Statistique Appliquée, 30*, 39–53.

Charniak, E. 1993. *Statistical Language Learning.* Cambridge: MIT Press.

Choulakian, V. 1988. Exploratory analysis of contingency tables by loglinear formulation and generalizations of correspondence analysis. *Psychometrika, 53*, 235–250.

Choulakian, V. 1996. Generalized bilinear models. *Psychometrika, 61*, 271–283.

Choulakian, V., Lockhart, R. and M. Stephens. 1994. Cramer–von Mises statistics for discrete distributions. *Canadian Journal of Statistics, 22*, 125–137.

Christensen, R. 1990. *Log-Linear Models.* New York: Springer.

Ciampi, A. 1991. Generalized regression trees. *Computational Statistics and Data Analysis, 12*, 57–78.

Ciampi, A. and J. Thiffault. 1987. Recursive partition and amalgamation (RECPAM) for censored survival data: Criteria for tree selector. *Statistical Software Newsletter, 14*, 78–81.

Cleveland, W. S. 1985. *The Elements of Graphing Data.* Monterey: Wadsworth.

Cleveland, W. S. 1993. *Visualizing Data.* Summit: Hobart Press.

Cleveland, W. S. 1994. *The Elements of Graphing Data.* New Jersey: Hobart Press.

Cleveland, W. S. and M. E. McGill (eds.). 1988. *Dynamic Graphics for Data Analysis.* Belmont, CA: Wadsworth.

Clogg, C. C. 1981. Latent structure models of mobility. *American Journal of Sociology, 86*, 836–868.

Clogg, C. C., Rudas, T. and L. Xi. 1995. A new index of structure for the analysis of models for mobility tables and other cross-classifications, in P. V. Marsden (ed.), *Sociological Methodology 1995,* pp. 197–222. Oxford: Blackwell.

Clogg, C. C. and D. O. Sawyer. 1981. A comparison of alternative models for analyzing the scalability of response patterns, in S. Leinhart (ed.), *Sociological Methodology 1981,* pp. 240–280. San Francisco: Jossey-Bass.

Clogg, C. C. and E. S. Shihadeh. 1994. *Statistical Models for Ordinal Variables.* Thousand Oaks, CA: Sage.

Converse, P. E. 1964. The nature of belief systems in mass politics, in D. E. Apter (ed.), *Ideology and Discontent,* pp. 206–261. Glencoe: Free Press.

Converse, P. E. 1970. Attitudes and non-attitudes: Continuation of a dialogue, in E. R. Tufte (ed.), *The Quantitative Analysis of Social Problems,* pp. 168–189. Reading, MA: Addison-Wesley.

Converse, P. E. 1976/77. Predicting no opinion in the polls. *Public Opinion Quarterly,* **40**, 515–530.

Coombs, C. H. 1950. Psychological scaling without a unit of measurement. *Psychological Review,* **57**, 148–158.

Coombs, C. H. 1964. *A Theory of Data.* New York: Wiley.

Coombs, C. and L. Coombs. 1976. "Don't know": Item ambiguity or respondent uncertainty. *Public Opinion Quarterly,* **40**, 497–514.

Coombs, C. H. and R. C. Kao. 1960. On a connection between factor analysis and multidimensional unfolding. *Psychometrika,* **25**, 219–231.

Cox, C. and K. R. Gabriel. 1982. Some comparisons of biplot displays and pencil-and-paper exploratory data analysis method, in R. L. Launer and A. F. Siegel (eds.), *Modern Data Analysis,* pp. 45–82. New York: Academic Press.

Cox, T. F. and M. A. A. Cox. 1994. *Multidimensional Scaling.* London: Chapman & Hall.

Crane, D. 1987. *The Transformation of the Avant Garde.* Chicago: University of Chicago Press.

CRISP. 1989. Les élections au Grand-Duché de Luxembourg. Rapport sur les élections législatives et européennes de juin 1989, étude réalisée pour la Chambre des Députés du Grand-Duché de Luxembourg.

Croon, M. A. 1990. Latent class analysis with ordered latent classes. *British Journal of Mathematical and Statistical Psychology,* **43,** 171–192.

Cuadras, C. M. 1992. Probability distributions with given multivariate marginals and given dependence structure. *Journal of Multivariate Analysis,* **42**, 51–66.

Cuadras, C. M. and J. Fortiana. 1995. A continuous metric scaling solution for a random variable. *Journal of Multivariate Analysis,* **52**, 1–14.

D'Alton, M. E. and A. H. DeCherney. 1993. Prenatal diagnosis. *New England Journal of Medicine,* **328**, 114–120

D'Ambra, L. and N. C. Lauro. 1989. Non symmetrical analysis of three-way contingency tables, in R. Coppi and S. Bolasco (eds.), *Multiway Data Analysis,* pp. 301–315. Amsterdam: North Holland.

Darroch, J. N. 1962. Interaction in multi-factor contingency tables. *Journal of the Royal Statistical Society, Series B,* **24,** 251–263.

Darroch, J. N. 1974. Multiplicative and additive interaction in contingency tables. *Biometrika,* **61**, 207–214.

Davey, B. A. and H. A. Priestley. 1990. *Introduction to Lattices and Order.* Cambridge: Cambridge University Press.

Davies, R. B., Elias, P. and R. D. Penn. 1992. The relationship between a husband's unemployment and his wife's participation in the labour force. *Oxford Bulletin of Statistics,* **54**, 145–171.

Davison, M. L. 1983. *Multidimensional Scaling.* New York: Wiley.

Dawid, A. P. 1979. Conditional independence in statistical theory (with discussion). *Journal of the Royal Statistical Society, Series B,* **41**, 1–31.

de Falguerolles, A. and B. Francis. 1992. Algorithmic approaches for fitting bilinear models, in Y. Dodge and J. Whittaker (eds.), *COMPSTAT 92, Computational Statistics,* Vol. 1, pp. 77–82. Heidelberg: Physica.

de Falguerolles, A. and B. Francis. 1994. An algorithmic approach to bilinear models for two-way contingency tables, in E. Diday *et al.* (eds.), *New Approaches in Classification and Data Analysis,* pp. 518–524. Berlin: Springer.

de Falguerolles, A. and B. Francis. 1995. Fitting bilinear models in GLIM. *GLIM Newsletter,* **25**, 9–20.

DeJonge, H. 1983. Deficiencies in clinical reports for registration of drugs. *Statistics in Medicine,* **2**, 155–166.

de Leeuw, J. 1973. *Canonical Analysis of Categorical Data.* Psychological Institute, Leiden University, The Netherlands, Report RN 007-68 (reprinted as a book and published in 1984 from DSWO Press, Leiden).

de Leeuw, J. 1984. Models of data. *Kwantitatieve Methoden,* **5**, 17–30.

de Leeuw, J. 1988. Models and techniques. *Statistica Neerlandica,* **42**, 91–98.

de Leeuw, J. 1990. Data modeling and theory construction, in J. Hox and J. D. Jong–Gierveld (eds.), *Operationalization and Research Strategy,* pp. 229–244. Amsterdam: Swets and Zeitlinger.

de Leeuw, J. 1994. Statistics and the sciences, in I. Borg and P. Mohler (eds.), *Trends and Perspectives in Empirical Social Research,* pp. 139–148. Berlin: de Gruyter.

de Leeuw, J. and W. J. Heiser. 1980. Multidimensional scaling with restrictions on the configuration, in P. R. Krishnaiah (ed.), *Multivariate analysis,* Vol. V, pp. 501–522. Amsterdam: North Holland.

de Leeuw, J. and P. G. M. van der Heijden. 1991. Reduced rank models for contingency tables. *Biometrika,* **78**, 229–232.

de Leeuw, J., van der Heijden, P. G. M. and P. Verboon. 1990. A latent time budget model. *Statistica Neerlandica,* **44**, 1–21.

Denis, J. G. and J. Gower. 1996. Asymptotic confidence regions for biadditive models: Interpreting genotype–environment interactions. *Applied Statistics,* **45**, 479–493.

DeSarbo, W. S., Manrai A. K. and L. A. Manrai. 1994. Latent class multidimensional scaling: A review of recent developments in the marketing and psychometric literature, in R. Bagozzi (ed.), *The Handbook of Marketing Research,* pp. 190–222. London: Blackwell.

Diday, E. 1992. From data to knowledge: Probabilist objects for a symbolic data analysis, in Y. Dodge and J. Whittaker (eds.), *COMPSTAT 92, Computational Statistics,* pp. 193–214. Heidelberg: Physica.

Dilnot, A. and M. Kell. 1987. Male unemployment and women's work. *Fiscal Studies,* **8**, 1–16.

DiMaggio, P. 1993. On metropolitan dominance: New York in the urban network, in M. Shefter (ed.), *Capital of the American Century. The National and International Influence of New York City,* pp. 193–218. New York: Russel Sage Foundation.

DiMaggio, P. 1996. Are art-museum visitors different from other people? The relationship between attendance and social and political attitudes in the United States. *Poetics,* **24**, 161–180.

Dorkenoo, K. M. M. and J.-R. Mathieu. 1993. Étude d'un modèle factoriel d'analyse de la variance comme modèle linéaire généralisé. *Revue de Statistique Appliquée,* **41**, 43–57.

Doosje, B. J. and F. W. Siero. 1991. Invloed van discrepantie tussen boodschap en eigen standpunt, kwaliteit van argumenten en motivatie tot elaboratie op de attitude t.o.v. de auto-milieu-problematiek, in R. W. Meertens, A. P. Buunk and R. van der Vlist (eds.), *SocialePsychologie, Voorlichting en Maatschappelijke Problemen.* Den Haag: Vuga.

Duncan, O. D. 1979. Indicators of sex-typing: Traditional and egalitarian, situational and ideological responses. *American Journal of Sociology,* **85**, 251–260.

Duncan, O. D., Sloane, D. M. and C. Brody. 1982. Latent classes inferred from response-consistency effects, in K. G. Jöreskog and H. Wold (eds.), *Systems Under Indirect Observation* Part I, pp. 19–64. Amsterdam: North Holland.

Duncan, O. D., Stenbeck, M. and C. J. Brody. 1988. Discovering heterogeneity: Continuous versus discrete latent variables. *American Journal of Sociology,* **93**, 1305–1321.

Durkheim, E. 1961. *Moral Education*. New York: Free Press.

Eckart, C. and G. Young. 1936. The approximation of one matrix by another of lower rank. *Psychometrika,* **1**, 211–218.

Elchardus, M. 1994. "Verschillende werelden. Over de ontdubbeling van links en rechts", *Samenleving en Politiek,* vol. 1 (7), pp. 5–18.

Elizur, D., Borg, I., Hunt, R. and I. Magyari-Beck. 1991. The structure of work values: A cross-cultural comparison. *Journal of Occupational Behavior,* **12**, 21–38.

Establet, R. and G. Felouzis. 1993. Mis en oeuvre comparative de l'analyse de contenu et lexicométrique sur un corpus de rédactions scolaires, in Anastex (ed.), *Proceedings of 2$\underline{0}$ Journées d'Analyse Statistique des Données Textuelles,* pp. 283–303. Paris: Editions de Télécom.

Everitt, B. S. 1978. *Graphical Techniques for Multivariate Data*. London: Heinemann.

Falter, J. W. 1995. Weniger Überraschungen als erwartet. Das "Superwahljahr" 1994, in R. Alterhof and E. Jesse (eds.), *Das wiedervereinigte Deutschland,* pp. 25–44. Düsseldorf: Droste.

Farebrother, R. W. 1987. Mechanical representation of the L_1 and L_2 estimation problem, in Y. Dodge (ed.), *Statistical Data Analysis Based on the L1 Norm and Related Methods,* pp. 455–464.

Farebrother, R. W. 1988. On an analogy between classical mechanics and maximum likelihood estimation. *Österreichische Zeitschrift für Statistik und Informatik,* **18**, 303–305.

Faulkenberry, G. D. and R. Mason. 1978. Characteristics of nonopinion and no opinion response groups. *Public Opinion Quarterly,* **40**, 497–514.

Fehlen, F. 1993. Le panachage, un héritage lourd de conséquences, in: *Forum,* **147**, Luxembourg.

Feick, L. 1984. Analyzing marketing research data with association models. *Journal of Marketing Research,* **21**, 376–386.

Ferber, R. 1966. Item nonresponse in a consumer survey. *Public Opinion Quarterly,* **30**, 399–415.

Ferligoj, A., Ule, M. and T. Rener. 1991. Sex differences in "don't know" rate: The case of Slovenia. *Wisdom,* **1/2**, 1–21.

Fienberg, S. E. 1980. *The Analysis of Cross-Classified Categorical Data* (2nd ed.). Cambridge, MA: MIT Press.

Fienberg, S. E. and J. P. Gilbert. 1970. The geometry of a 2 × 2 contingency table. *Journal of the American Statistical Association,* **65**, 694–701.

Fisher, R. A. 1940. The precision of discriminant functions. *Annals of Eugenics,* **10**, 422–429.

Fleck, R. 1996. Frankreich 1981 bis 1996. Idealtypischer Verlauf staatlicher Kunstinnovation. *Kursiv,* **3**, Special Issue: Kultur-Politik, 22–26.

Formann, A. K. 1992. Linear logistic latent class analysis for polytomous data. *Journal of the American Statistical Association,* **87**, 476–486.

Francis, B. and M. Fuller. 1996. Visualisation of event histories. *Journal of the Royal Statistical Society Series A,* **159**, 301–308.

Francis, J. and L. Busch. 1975. What we don't know about 'I don't knows.' *Public Opinion Quarterly,* **39**, 207–218.

Frey, B. S. and W. E. Pommerehne. 1989. *Muses and Markets. Explorations in the Economics of the Arts.* Oxford: Basil Blackwell.

Frick, U., Laschat, M. and J. Rehm. 1995. Neonatologische Versorgung in Wien. Scientific report to the Vienna's Hospitals Association. Vienna.

Friendly, M. 1992a. Graphical methods for categorical data. *Proceedings of the SAS Users' Group International Conference,* **17**, 1367–1373.

Friendly, M. 1992b. Mosaic displays for loglinear models. *Journal of the American Statistical Association,* **87**, 61–68.

Friendly, M. 1994a. Mosaic displays for multi-way contingency tables. *Journal of the American Statistical Association,* **89**, 190–200.

Friendly, M. 1994b. *A fourfold display for 2 by 2 by k tables.* York University, Department of Psychology Reports, No. 217.

Friendly, M. 1994c. SAS/IML graphics for fourfold displays. *Observations,* **3**, 47–56.

Friendly, M. 1995. Conceptual and visual models for categorical data. *The American Statistician,* **49**, 153–160.

Froeschl, K. A. 1992. Semantic metadata: query processing and data aggregation, in Y. Dodge and J. Whittaker (eds.), *Computational Statistics,* pp. 357–362. Heidelberg: Physica.

Gablik, S. 1985. *Has Modernism Failed?* New York: Thames & Hudson.

Gabriel, K. R. 1971. The biplot-graphic display of matrices with applications to principal component analysis. *Biometrika,* **58**, 453–467.

Gabriel, K. R. 1981a. Biplot display of multivariate matrices for inspection of data and diagnosis, in V. Barnett (ed.), *Interpreting Multivariate Data,* pp. 147–174. Chichester: Wiley.

Gabriel, K. R. 1981b. Biplot, in S. Kotz, N. L. Johnson and C. Read (eds.), *Encyclopedia of Statistical Sciences,* Vol. I, pp. 262–265. New York: Wiley.

Gabriel, K. R. 1995. MANOVA biplots for two-way contingency tables, in W. J. Krzanowski (ed.), *Recent Advances in Descriptive Multivariate Analysis,* pp. 227–268. Oxford: Clarendon.

Gabriel, K. R. and C. L. Odoroff. 1990. Biplots in biomedical research. *Statistics in Medicine,* **9,** 469–485.

Gabriel, K. R. and S. Zamir. 1979. Lower rank approximation of matrices by least squares with any choice of weights. *Technometrics,* **21,** 489–498.

Gale, W. A., Church, K. W., and D. A. Yarowski. 1992. A method for disambiguating word senses in a large corpus. *Computer and the Humanities,* **26,** 415–439.

Galinat, W. and I. Borg. 1987. On symbolic temporal information: Beliefs about the experience of duration. *Memory and Cognition,* **15,** 308–317.

Ganter, B. and R. Wille. 1989. Conceptual Scaling, in F. Roberts (ed.), *Applications of Combinatorics and Graph Theory to the Biological and Social Sciences,* pp. 139–167. New York: Springer.

Ganter, B. and R. Wille. 1996. *Formale Begriffsanalyse—Mathematische Grundlagen.* Berlin: Springer.

Giegler, H. and H. Klein. 1994. Correspondence analysis of textual data from personal advertisements, in M. Greenacre and J. Blasius (eds.), *Correspondence Analysis in the Social Sciences,* pp. 282–301. London: Academic Press.

Gifi, A. 1990. *Nonlinear Multivariate Analysis.* New York: Wiley.

Gilbert, G. N. 1981. *Modelling Society.* London: Allen and Unwin.

Gilula, Z. 1979. Singular value decomposition of probability matrices: Probability aspects of latent dichotomous variables. *Biometrics,* **66,** 339–344.

Gilula, Z. 1983. Latent conditional independence in two-way contingency tables: A diagnostic approach. *British Journal of Mathematical and Statistical Psychology,* **36,** 114–122.

Gilula, Z. 1984. On some similarities between canonical correlation models and latent class models for two-way contingency tables. *Biometrika,* **71,** 523–529.

Gilula, Z. and S. J. Haberman. 1986. Canonical analysis of contingency tables by maximum likelihood. *Journal of the American Statistical Association,* **81**, 780–788.

Gilula, Z. and S. J. Haberman. 1988. The analysis of multivariate contingency tables by restricted canonical and restricted association models. *Journal of the American Statistical Association,* **83**, 760–771.

Gilula, Z. and S. J. Haberman. 1994. Conditional log-linear models for analyzing categorical panel data. *Journal of the American Statistical Association*, **89**, 645–656.

Gini, C. 1912. *Variabilit e mutabilit, contributo allo studio delle distribuzioni e relazioni statistiche.* Studi Economico-Giuridici della R. Universit à di Cagliari.

Gittins, R. 1985. *Canonical Analysis: A Review with Applications in Ecology.* Berlin: Springer.

Gold, E. M. 1973. Metric unfolding: Data requirements for unique solution and clarification of Schönemann's algorithm. *Psychometrika,* **38**, 555–569.

Goodman, L. A. 1974. The analysis of systems of qualitative variables when some of the variables are unobservable. A modified latent structure approach. *American Journal of Sociology,* **79**, 1179–1259.

Goodman, L. A. 1979. Simple models for the analysis of associations in cross-classifications having ordered categories. *Journal of the American Statistical Association,* **76**, 320–334.

Goodman, L. A. 1981. Association models and the bivariate normal for contingency tables with ordered categories. *Biometrika,* **68**, 347–355.

Goodman, L. A. 1983. The analysis of dependence in cross-classifications having ordered categories, using log-linear models for frequencies and log-linear models for odds. *Biometrics,* **39**, 149–160.

Goodman, L. A. 1984. *The Analysis of Cross-Classifications Having Ordered Categories.* Cambridge, MA: Harvard University Press.

Goodman, L. A. 1985. The analysis of cross-classified data having ordered and/or unordered categories: Association models, correlation models, and asymmetry models for contingengy tables with or without missing entries. *Annals of Statistics,* **13**, 10–69.

Goodman, L. A. 1986. Some useful extensions of the usual correspondence analysis approach in the analysis of contingency tables (with discussion). *International Statistical Review,* **54**, 243–270.

Goodman, L. A. 1987. New methods for analyzing the intrinsic character of qualitative variables using cross-classified data. *American Journal of Sociology,* **93**, 529–583.

Goodman, L. A. 1991. Measures, models, and graphical displays in the analysis of cross-classified data (with discussion). *Journal of the American Statistical Association,* **86**, 1085–1138.

Goodman, L. A. and C. C. Clogg. 1992. New methods for the analysis of occupational mobility tables and other kinds of cross-classifications. *Contemporary Sociology,* **21**, 609–622.

Goodman, L. A. and W. H. Kruskal. 1954. Measures of association for cross-classifications. *Journal of the American Statistical Association,* **49**, 732–764.

Gower, J. C. 1966. Some distance properties of latent-root and vector methods in multivariate analysis. *Biometrika,* **53**, 325–338.

Gower, J. C. 1975. Generalized Procrustes analysis. *Psychometrika,* **40**, 33–51.

Gower, J. C. 1989a. Generalized canonical analysis, in R. Coppi and S. Bolasco (eds.), *Multiway Data Analysis*, pp. 221–232. Amsterdam: Elsevier.

Gower, J. C. 1989b. Discussion of "A combined approach to contingency table analysis using correspondence analysis and log-linear analysis." *Applied Statistics,* **38**, 249–292.

Gower, J. C. 1993. The construction of neighbour-regions in two dimensions for prediction with multi-level categorical variables, in O. Opitz, B. Lausen and R. Klar (eds.), *Information and Classification: Concepts–Methods–Applications,* pp.174–189. Berlin: Springer.

Gower, J. C. and D. J. Hand. 1996. *Biplots.* London: Chapman & Hall.

Gower, J. C. and S. Harding. 1988. Non-linear biplots. *Biometrika,* **73**, 445–455.

Gray, L. N. and J. S. Williams. 1981. Goodman and Kruskal's τ_b: Multiple and partial analogs. *Sociological Methods & Research,* **10**, 50–62.

Green, P. E. 1978. *Analyzing Multivariate Data.* Hinsdale, IL: Dryden.

Green, P. E. and V. Rao. 1972. *Applied Multidimensional Scaling.* Hinsdale, IL: Dryden.

Greenacre, M. J. 1984. *Theory and Applications of Correspondence Analysis.* London: Academic Press.

Greenacre, M. J. 1988a. Clustering the rows and columns of a contingency table. *Journal of Classification,* **5**, 39–51.

Greenacre, M. J. 1988b. Correspondence analysis of multivariate categorical data by weighted least squares. *Biometrika,* **75,** 457–467.

Greenacre, M. J. 1989. The Carroll–Green–Schaffer scaling in correspondence analysis: A theoretical and empirical appraisal. *Journal of Marketing Research,* **26,** 358–365.

Greenacre, M. J. 1990. Some limitations of multiple correspondence analysis. *Computational Statistics Quarterly,* **3,** 249–256.

Greenacre, M. J. 1991. Interpreting multiple correspondence analysis. *Applied Stochastic Models and Data Analysis,* **7,** 195–210.

Greenacre, M. J. 1992. Biplots in correspondence analysis. *Journal of Applied Statistics,* **20,** 251–269.

Greenacre, M. J. 1993.*Correspondence Analysis in Practice.* London: Academic Press.

Greenacre, M. J. 1994. Multiple and joint correspondence analysis, in M. J. Greenacre and J. Blasius (eds.), *Correspondence Analysis in the Social Sciences,* pp. 141–161. London: Academic Press.

Greenacre, M. J. and J. Blasius (eds.). 1994a. *Correspondence Analysis in the Social Sciences.* London: Academic Press.

Greenacre, M. J. and J. Blasius. 1994b. Preface, in M. J. Greenacre and J. Blasius (eds.), *Correspondence Analysis in the Social Sciences,* pp. vii–xv. London: Academic Press.

Greenacre, M. J. and Browne, M. W. 1986. An efficient alternating least-squares algorithm to perform multidimensional unfolding. *Psychometrika,* **51,** 241–250.

Greenacre, M. J. and T. Hastie. 1987. The geometric interpretation of correspondence analysis. *Journal of the American Statistical Association,* **82,** 437–446.

Gueguen, A. and J. P. Nakache. 1988. Méthode de discrimination basée sur la construction d'un arbre de décision binaire. *Revue de Statistique Appliquée,* **36,** 19–38.

Guilbaut, S. 1983. *How New York Stole the Idea of Modern Art. Abstract Expressionism, Freedom, and the Cold War.* Chicago: University of Chicago Press.

Guttman, L. 1941. The quantification of a class of attributes: A theory and method of scale construction, in P. Horst (ed.), *The Prediction of Personal Adjustment,* pp. 319–348. New York: Social Science Research Council.

Guttman, L. 1944. A basis for scaling qualitative data. *American Sociological Review,* **91,** 139–150.

Guttman, L. 1946. An approach for quantifying paired comparison and rank order. *Annals of Mathematical Statistics,* **17,** 144–163.

Guttman, L. 1954. A new approach to factor analysis: The radex, in P. F. Lazarsfeld (ed.), *Mathematical Thinking in the Social Sciences,* pp. 258–348. New York: Free Press.

Guttman, L. 1971. Measurement as structural theory. *Psychometrika,* **36,** 329–347.

Guttman, L. 1977. What is not what in statistics. *The Statistician,* **26,** 81–107.

Guttman, L. and S. Levy. 1991. Two structural laws for intelligence tests. *Intelligence,* **15,** 79–103.

Haberman, S. J. 1995. Computation of maximum likelihood estimates in association models. *Journal of the American Statistical Association,* **90,** 1438–1446.

Haeusler, L. 1984. Analyse lexicale de réponses libres. Le coût de l'électricité. *Rapport Crédoc.* Paris: Crédoc.

Hagenaars, J. 1990. *Categorical Longitudinal Data.* Newbury Park, CA: Sage.

Hand, D. J. 1992. Microdata, macrodata, and metadata, in Y. Dodge and J. Whittaker (eds.), *Computational Statistics,* pp. 325–340. Heidelberg: Physica.

Hartigan, J. A. and B. Kleiner. 1981. Mosaics for contingency tables, in W. F. Eddy (ed.), *Computer Science and Statistics: Proceedings of the 13th Symposium on the Interface.* New York: Springer.

Hayashi, C., Suzuki, T. and M. Sasaki. 1992. *Data Analysis for Social Comparative Research: International Perspective.* Amsterdam: North Holland.

Hearnshaw, H. M. and D. J. Unwin (eds.). 1994. *Visualisation in Geographical Information Systems.* New York: Wiley.

Heiser, W. J. 1981. *Unfolding Analysis of Proximity Data.* University of Leiden, The Netherlands.

Heiser, W. J. 1989. Order invariant unfolding analysis under smoothness restriction, in G. DeSoete, H. Feger, and K. C. Klauer (eds.), *New Developments in Psychological Choice Modeling,* pp. 3–31, Amsterdam: Elsevier.

Heuer, J. 1979. *Selbstmord bei Kindern und Jugendlichen.* München: Klett.

Hoerl, A. E. and R. W. Kennard. 1970. Ridge regression. *Technometrics,* **12,** 55–67 and 69–82.

Hoffmann, D. and J. de Leeuw. 1992. Interpreting multiple correspondence analysis as a multidimensional scaling method. *Marketing Letters,* **3,** 259–272.

Hoijtink, H. 1990. A latent trait model for dichotomous choice data. *Psychometrika,* **55,** 641–656.

Hojo, H. 1994. A new method for multidimensional unfolding. *Behaviormetrika,* **21,** 131–147.

Horstman, I. and A. Slivinski. 1985. Location models as models of product choice. *Journal of Economic Theory,* **36,** 367–386.

Householder, A. S. and G. Young. 1938. Matrix approximation and latent roots. *American Mathematical Monthly,* **45,** 165–171.

Hubert, L. and P. Busk. 1976. Normative location theory: Placement in continuous space. *Journal of Mathematical Psychology,* **14,** 187–210.

Inglehart, R. 1987. Value change in industrial societies. *American Political Science Review,* **81,** 1189–1203.

Inglehart, R. 1990. *Culture Shifts in Advanced Industrial Society.* Princeton: Princeton University Press.

Ishii-Kuntz, M. 1994. *Ordinal Log-Linear Models.* Thousand Oaks, CA: Sage.

Iyengar, S. and D. R. Kinder. 1987. *News That Matter.* Chicago: University of Chicago.

Jambu, M. 1989. *Exploration informatique et statistique de données.* Paris: Dunod.

Jennings, M. K., van Deth, J., *et al.* 1991. *Continuities in Political Action: A Longitudinal Study of Political Orientations in Three Western Democracies.* New York: de Gruyter.

Jobson, J. D. 1992. *Applied Multivariate Data Analysis.* Springer Texts in Statistics. New York: Springer.

Johnson, E. J. and A. Tversky. 1984. Representations of perceptions of risk. *Journal of Experimental Psychology,* **113,** 55–70.

Kamakura, W. A. and R. K. Srivastava. 1986. An ideal-point probabilistic choice model for heterogeneous preferences. *Marketing Science,* **5,** 199–218.

Karr, A. 1993. *Probability.* New York: Springer.

Kass, G. V. 1980. An exploratory technique for investigating large quantities of categorical data. *Applied Statistics,* **29,** 119–127.

Keiding, N. 1990. Statistical inference in the Lexis diagram. *Royal Society of London Philosophical Transactions: Series A,* **332,** 487–509.

Kelly, E. and P. J. Stone. 1975. *Computer Recognition of English Words Senses.* Amsterdam: North Holland.

Kendall, M. G. 1948. *Rank Correlation Methods.* London: Griffin.

Klein, F. 1872. Vergleichende Betrachtungen über neuere geometrische Forschungen, in R. Fricke and A. Ostrowski (eds.) (1921), *Felix Klein: Gesammelte Mathematische Abhandlungen*, Vol.1. Berlin: Springer.

Knoke, D. and P. J. Burke. 1980. *Log-Linear Models.* Newbury Park, CA: Sage.

Knutsen, O. and E. Scarbrough. 1995. Cleavage politics, in J. W. van Deth and E. Scarbrough (eds.), *The Impact of Values. Beliefs in Government Volume Four,* pp. 492–523. Oxford: Oxford University Press.

Koch, G. G. and S. Edwards. 1988. Clinical efficacy trials with categorical data, in K. Peace (ed.), *Biopharmaceutical Statistics for Drug Development,* pp. 403–457. New York: Marcel Dekker.

Koh, E.-K. 1993. Visualising the stock market crash. *IEEE Computer Graphics and Applications,* **13**, 14–16.

Konner, M. and C. Worthman. 1980. Nursing frequency, godadal function and birth spacing among !Kung hunter-gatherers. *Science,* **207**, 788–791.

Krantz, D. H., Luce, R. D., Suppes, P. and A. Tversky. 1971. *Foundations of Measurement,* Vol I. New York: Academic Press.

Kroonenberg, P. M. 1983. *Three-Mode Principal Component Analysis: Theory and Applications.* Leiden: DSWO.

Kroonenberg, P. M. 1989. Singular value decompositions of interactions in three-way contingency tables, in R. Coppi and S. Bolasco (eds.), *Multiway Data Analysis,* pp. 169–184. Amsterdam: North Holland.

Kroonenberg, P. M. 1996. *3WAYPACK User's Manual.* Leiden: Department of Education, Leiden University.

Kruskal, J. B. 1964a. Multidimensional scaling by optimizing goodness of fit to a nonmetric hypothesis. *Psychometrika,* **29**, 1–27.

Kruskal, J. B. 1964b. Nonmetric multidimensional scaling: A numerical method. *Psychometrika,* **29**, 28–42.

Kruskal, J. B. 1978. Factor analysis: Bilinear methods, in W. H. Kruskal and J. M. Tanur (eds.), *International Encyclopedia of Statistics*, 307–330. New York: Macmillan.

Kruskal, J. B. and M. Wish. 1978. *Multidimensional Scaling.* Beverly Hills, CA: Sage.

Krzanowski, W. J. 1988. *Principles of Multivariate Analysis.* Oxford: University Press.

Krzanowski, W. J. and F. H. C. Marriott. 1994. *Multivariate Analysis.* Part I, *Distributions, Ordination and Inference.* London: Edward Arnold.

Kumbasar, E., Romney, A. K. and W. H. Batchelder. 1994. Systematic biases in social perception. *American Journal of Sociology,* **100**, 477–505.

Lafon, P. 1981. *Dépouillement et statistique en lexicométrie.* Paris: Slatkine– Champion.

Lamont, M. and M. Fournier (eds.). 1992. *Cultivating Differences. Symbolic Boundaries and the Making of Inequality.* Chicago: University of Chicago Press.

Lancaster, H. O. 1951. Complex contingency tables treated by the partition of the chi-square. *Journal of Royal Statistical Society Series B,* **13**, 242–249.

Lang, G. E. and K. Lang. 1981. Watergate: An exploration of the agenda-building process, in *Mass Communication Review Yearbook 2,* pp. 447–468.

Langeheine, R. and F. van de Pol. 1990. A unifying framework for Markov modeling in discrete space and discrete time. *Sociological Methods and Research,* **18**, 416–441.

Lauro, N. C. and S. Balbi. 1995. The analysis of structured qualitative data, in A. Rizzi (ed.), *Some Relations between Matrices and Structures of Multidimensional Data Analysis,* pp. 53–92. Pisa: Applied Mathematics Monographs, CNR.

Lauro, N. C. and L. D'Ambra. 1984. L'analyse non symétrique des correspondances, in E. Diday *et al.* (eds.), *Data Analysis and Informatics, III,* pp. 433–446. Amsterdam: North Holland.

Lauro, N. C. and R. Siciliano. 1989. Exploratory methods and modelling for contingency tables analysis: An integrated approach. *Statistica Applicata,* **1**, 5–32.

Lazarsfeld, P. F. 1944. The controversy over detailed interviews—an offer for negotiation. *Public Opinion Quarterly,* **8**, 38–60.

Lazarsfeld, P. F. 1950. The logical and mathematical foundation of latent structure analysis, in S. A. Stouffer, L. Guttman, E. A. Suchman, P. F. Lazarsfeld, S. A. Star, and J. A. Clausen (eds.), *Studies in social psychology in World War II.* Vol IV: *Measurement and Prediction,* pp. 362–412. Princeton: Princeton University Press.

Lazarsfeld, P. F. and N. W. Henry. 1968. *Latent Structure Analysis.* New York: Houghton-Mifflin.

Lebart, L. 1982a. Exploratory analysis of large sparse matrices, with application to textual data, *COMPSTAT 82*, pp. 67–76. Heidelberg: Physica.

Lebart, L. 1982b. L'Analyse statistique des réponses libres dans les enquêtes socio-économiques. *Consommation,* **1**, 39–62.

Lebart, L. 1987. Conditions de vie et aspirations des fran ais, évolution et structure des opinions de 1978 à 1986. *Futuribles,* **Sept 1987**, 25–56.

Lebart, L. and A. Salem. 1994. *Statistique Textuelle.* Paris: Dunod.

Lebart, L., Morineau, A. and M. Bécue (with the collaboration of L. Haeusler and P. Pleuvret). 1989. *SPAD.T, Manuel de l'utilisateur.* Paris: CISIA.

Lebart, L., Morineau, A. and K. M. Warwick. 1984. *Multivariate Descriptive Statistical Analysis.* Chichester, UK: Wiley.

Lebart, L., Morineau, A. and M. Piron. 1995. *Statistique exploratoire multidimensionelle.* Paris: Dunod.

Lebart, L., Salem, A. and E. Berry E. 1991. Recent development in the statistical processing of textual data. *Applied Stochastic Models and Data Analysis,* **7**, 47–62.

LeFevre, M., Scanner, L., Anderson, S. and R. Tsutakawa. 1992. The relationship between neonatal mortality and hospital level. *Journal of Family Practice,* **35**, 259–264.

Le Roux, B. 1991. Sur la construction d'un protocole additif de référence. *Mathématiques, Informatique et Sciences Humaines,* **114**, 57–62.

Le Roux, B. and H. Rouanet. 1984. L'analyse multidimensionnelle des données structurées, *Mathématiques, Informatique et Sciences Humaines,* **85**, 5–18.

Lexis, W. 1875. *Einleitung in die Theorie der Bevölkerungsstatistik.* Strassbourg: Trübner.

Light, R. J. and B. H. Margolin. 1971. An analysis of variance for categorical data. *Journal of the American Statistical Association,* **66**, 534–544.

Lindsay, B., Clogg, C. and J. Grego. 1991. Semiparametric estimation in the Rasch model and related exponential response models, including a simple model for item analysis. *Journal of the American Statistical Association,* **86**, 96–107.

Lingoes, J. C. 1981. Testing regional hypotheses in multidimensional scaling, in I. Borg (ed.), *Multidimensional Data Representations: When and Why,* pp. 280–310. Ann Arbor, MI: Mathesis.

Lingoes, J. C. 1987. *Guttman-Lingoes Nonmetric PC Series Manual.* Ann Arbor, MI: Mathesis.

Lipset, S. M. and S. Rokkan. 1967. Cleavage structure, party stystems, and voter alignments: An introduction, in S. M. Lipset and S. Rokkan (eds.), *Party Systems and Voter Alignments,* chap. 1. New York: Free Press.

Loewenstein, G. F. and J. Elster. 1992. *Choice over Time.* New York: Russell Sage Foundation.

Lombardo, R. and P. M. Kroonenberg. 1993. Non-symmetrical correspondence analysis—Some examples. *Bulletin of the International Statistical Institute, Contributed Papers 49th Session,* book 2, 127–128. Florence.

Lorwin, V. 1971. Segmented pluralism: Ideological cleavages and political cohesion in the smaller European democracies. *Comparative Politics,* **3**, 141–175.

Luce, R. D. 1959. *Individual Choice Behaviour.* New York: Wiley.

Luhmann, N. 1983. Öffentliche Meinung, in N. Luhmann (ed.), *Politische Planung,* pp. 9–34. Opladen: Westdeutscher Verlag.

Luijkx, R. 1994. *Comparative Log-Linear Analyses of Social Mobility and Heterogamy.* Tilburg, The Netherlands: Tilburg University Press.

Luytens, K., Symons, F. and M. Vuylsteke-Wauters. 1994. Linear and non-linear canonical correlation analysis: An exploratory tool for the analysis of group-structured data. *Journal of Applied Statistics,* **21**, 43–6.

Madsen, M. 1976. Statistical analysis of multiple contingency tables: Two examples. *Scandinavian Journal of Statistics,* **3**, 97–106.

Magidson, J. 1995. Introducing a new graphical model for the analysis of an ordered categorical response—Part I. *Journal of Targeting, Measurement and Analysis for Marketing (UK),* **2**, 133–148.

Magidson, J. 1996a. Maximum likelihood assessment of clinical trials based on an ordered categorical response. *Drug Information Journal,* **30**, 143–170.

Magidson, J. 1996b. Introducing a new graphical model for the analysis of an ordered categorical response—Part II. *Journal of Targeting, Measurement and Analysis for Marketing (UK),* **3**, 214–227.

Mandel, J. 1971. A new analysis of variance model for non-additive data. *Technometrics,* **13**, 1–13.

Marcotorchino, F. 1987. Block seriation problems: A unified approach. *Applied Stochastic Models and Data Analysis,* **3**, 73–91.

Mardia, K. V., Kent, J. T. and J. M. Bibby. 1979. *Multivariate Analysis.* London: Academic Press.

Margolin, B. H. and R. J. Light. 1974. An analysis of variance for categorical data, II: Small sample comparisons with chi square and other competitors. *Journal of the American Statistical Association,* **69**, 755–764.

Martin, Ch. and G. Macdaid. 1995. Looking at type and career exploration. Presented at APT XI, the Eleventh Biennial International Conference of the Association of Psychological Type (July 11–16, 1995), Kansas City, MO.

Mathes, R. and U. Freisens. 1990. Kommunikationsstrategien der Parteien und ihr Erfolg, in M. Kaase and H.–D. Klingemann (eds.), *Wahlen und Wähler,* pp. 531–568. Opladen: Westdeutscher Verlag.

Mathworks. 1995. *The Student Edition of MATLAB.* Englewood Cliffs, NJ: Prentice Hall.

Maxwell, A. E. 1961. Canonical variate analysis when the variables are dichotomous. *Educational and Psychological Measurement,* **21**, 259–271.

McCormick, B. H., DeFanti, T. A. and M. D. Brown (eds.). 1987. Visualisation in scientific computing. (Special issue ACM SIGGRAPH). *Computer Graphics,* **21**.

McCullagh, P. 1980. Regression models for ordinal data (with discussion). *Journal of the Royal Statistical Society Series B,* **42**, 109–142.

McCullagh, P. and J. A. Nelder. 1989. *Generalized Linear Models* (2nd ed.). London: Chapman & Hall.

McCutcheon, A. L. 1987a. Sexual morality, pro-life values and attitudes towards abortion: A simultaneous latent structure analysis for 1978–1983. *Sociological Methods and Research,* **16**, 256–275.

McCutcheon, A. L. 1987b. *Latent Class Analysis.* Beverly Hills, CA: Sage.

Meloen, J., van der Linden, G. and H. de Witte. 1994. Authoritarianism and political racism in Belgian Flanders. A test of the approaches of Adorno *et al.,* Lederer and Altemeyser, in R. Holly (ed.), *Political Consciousness and Civic Education during the Transformation of the System,* pp. 72–108. Warsaw: Institute of Political Studies, Polish Academy of Sciences.

Menger, K. 1954. On variables in mathematics and natural science. *British Journal of the Philosophy of Science,* **5**, 134–142.

Menger, K. 1955. Random variables from the point of view of a general theory of variables, in J. Neyman (ed.), *Third Berkeley Symposium on Mathematical Statistics and Probability.* Berkeley, CA: University of California Press.

Menger, K. 1961. Variables, constants, fluents, in H. Feigl and G. Maxwell (eds.), *Current Issues in the Philosophy of Science*. New York: Holt, Rinehart & Winston.

Meulman, J. 1986. *A Distance Approach to Nonlinear Multivariate Analysis*. Leiden: DSWO.

Meyer, R. 1992. Multidimensional scaling as a framework for correspondence analysis and its extensions, in M. Schader (ed.), *Analyzing and Modeling Data and Knowledge*, pp. 63–72. Heidelberg: Springer.

Middendorp, C. 1991. *Ideology in Dutch Politics. The Democratic System Reconsidered 1970–1985*. Assen/Maastricht: van Gorcum.

Miller, T. C., Densberger, M. and J. Krogman. 1983. Maternal transport and the perinatal denominator. *American Journal of Obstetrics and Gynecology,* **147**, 19–24.

Modanlou, H. D., Dorchester, W., Freeman, R. K. and C. Rommal. 1980. Perinatal transport to a regional perinatal center in a metropolitan area: Maternal versus neonatal transport. *American Journal of Obstetrics and Gynecology,* **138**, 1157–1164.

Mola, F. and R. Siciliano. 1994. Alternative strategies and CATANOVA testing in two-stage binary segmentation, in E. Diday *et al.* (eds), *New Approaches in Classification and Data Analysis,* pp. 316–323. Berlin: Springer.

Mola, F. and R. Siciliano. 1996. Visualizing data in tree-structured classification. *Proceedings of the Conference of the International Federation of Classification Society,* Vol. 1, pp. 65–68.

Mola, F. and R. Siciliano. 1997. A fast splitting procedure for classification trees, to appear in D. Hand (ed.), *Statistics and Computing*. London: Chapman & Hall.

Mola, F., Klaschka, J. and R. Siciliano. 1996. Logistic classification trees, in A. Prat (ed.), *COMPSTAT 96. Proceedings in Computational Statistics,* pp. 373–378. Heidelberg: Physica.

Moulin, R. 1987. *The French Art Market*. New Brunswick, NJ: Rutgers University Press.

Moulin, R. 1992. *L'artiste, l'institution et le marché*. Paris: Flammarion.

Moulin, R. 1996. Heirs. The sociological identity of artists, in B. von Bismarck, D. Stoller, and U. Wuggenig (eds.), *Games, Fights, Collaborations,* pp. 160–162. Ostfildern b. Stuttgart: Cantz.

Myers, I. B. and M. H. McCaulley. 1985. *Manual: A Guide to the Development and Use of the Myers-Briggs Type Indicator.* Palo Alto, CA: Consulting Psychologists Press.

Nakao, K. and A. K. Romney. 1993. Longitudinal approach to subgroup formation: Re-analysis of Newcomb's fraternity data. *Social Networks,* **15**, 109–131.

Nelder, J. A. and R. W. M. Wedderburn. 1972. Generalized linear models. *Journal of the Royal Statistical Society, Series A,* **135**, 370–384.

Newcomb, T. M. 1956. The prediction of interpersonal attraction. *The American Psychologist,* **11**, 575–586.

Newcomb, T. M. 1961. *The Acquaintance Process.* New York: Holt, Rinehart & Winston.

Nishisato, S. 1978. Optimal scaling of paired comparison and rank order data: An alternative to Guttman's formulation. *Psychometrika,* **43**, 263–271.

Nishisato, S. 1980. *Analysis of Categorical Data: Dual Scaling and Its Applications.* Toronto, Canada: University of Toronto Press.

Nishisato, S. 1988. Assessing quality of joint graphical displays in correspondence analysis and dual scaling, in E. Diday, Y. Escoufier, L. Lebart, J. Pagés, Y. Schektman, and R. Tomassone (eds.), *Data Analysis and Informatics,* V, pp. 409–416. Amsterdam: North Holland.

Nishisato, S. 1990. Dual scaling of designed experiments, in M. Schader and W. Gaul (eds.), *Knowledge, Data and Computer-Assisted Decisions,* pp. 115–125. Berlin: Springer.

Nishisato, S. 1993. On quantifying different types of categorical data. *Psychometrika,* **58,** 617–629.

Nishisato, S. 1994. *Elements of Dual Scaling: An Introduction to Practical Data Analysis.* Hillsdale, NJ: Lawrence Erlbaum.

Nishisato, S. 1996. Gleaning in the field of dual scaling. *Psychometrika,* **61**, 559–600.

Nishisato, S. and H. Ahn. 1994. When not to analyze data: Decision-making on missing responses in dual scaling. *Annals of Operations Research,* **55**, 361–378.

Nishisato, S. and M. Kolić. 1992. *From dual scaling to clustering: An alternative to joint graphical display.* A paper presented at the annual meeting of the Psychometric Society, Columbus, OH.

Nishisato, S. and I. Nishisato. 1994. *Dual Scaling in a Nutshell.* Toronto: MicroStats.

Nishisato, S. and W. J. Sheu. 1984. A note on dual scaling of successive categories data. *Psychometrika,* **49**, 493–500.

Nordlie, P. G. 1958. *A Longitudinal Study of Interpersonal Attraction in a Natural Group Setting.* Ph.D. dissertation, University of Michigan.

Obladen, M., Luttkus, A., Rey, M., Metze, B., Hopfenmüller, W. and J. W. Dudenhausen. 1994. Differences in morbidity and mortality according to type of referral of very low birthweight infants. *Journal of Perinatal Medicine,* **22**, 53–64.

Okamoto, Y. 1995. Unfolding by the criterion of the fourth quantification method. *Behaviormetrika,* **22**, 126–134 (in Japanese, with an English abstract).

Partchev, I. 1995. Using multivariate analysis to visualize Bulgarian political space, in I. Partchev (ed.), *Multivariate Analysis in the Behavioral Sciences: Philosophic to Technical,* pp. 164–170. Sofia: Marin Drinor.

Poulsen, C. S. 1990. Mixed Markov and latent Markov modeling applied to brand choice data. *International Journal of Marketing,* **7**, 5–19.

Pressat, R. 1961. *L'Analyse Démographique.* Paris: Presses Universitaires de France.

Radunski, P. 1980. *Wahlkämpfe: Moderne Wahlkampfführung als politische Kommunikation.* München: Olzog.

Rao, C. R. 1964. The use and interpretation of principal component analysis in applied research. *Sankhyā. The Indian Journal of Statistics, Series A,* **26**, 329–358.

Rao, C. R. 1973. *Linear Statistical Inference and its Applicatons.* New York: Wiley.

Rao, C. R. 1995. A review of canonical coordinates and an alternative to correspondence analysis using Hellinger distance. *Qüestiió* **19**, 23–63.

Rapoport, R. B. 1982. Sex differences in attitude expression: A generational explanation. *Public Opinion Quarterly,* **46**, 86–96.

Rapoport, R. B. 1985. Like mother, like daughter, intergenerational transmission of DK response rates. *Public Opinion Quarterly,* **49**, 198–208.

Reinert, M. 1983. Une méthode de classification descendante hiérarchique: Application à l'analyse lexicale par contexte. *Cahiers de l'Analyse des Données,* **8**, 187–198.

Renwick, M. 1992. Foetal abnormality: An audit of its recognition and management. *Archives of Diseases in Childhood,* **67**, 770–774.

Riedwyl, H. and M. Schüpbach. 1983. *Siebdiagramme: Graphische Darstellung von Kontingenztafeln.* Technical Report No. 12, University of Bern: Institute for Mathematical Statistics.

Riedwyl, H. and M. Schüpbach. 1994. Parquet diagram to plot contingency tables, in F. Faulbaum (ed.), *Softstat '93: Advances in Statistical Software,* pp. 293–299. New York: Fischer.

Risson, A., Rolland, P., Bertin J. and J. H. Chauchat. 1994. *AMADO: Analyse graphique d'une MAtrice de DOnnées, guide de l'utilisateur du logiciel (AMADO, User Guide)*. Saint-Mandé, France: CISIA.

Romney, A. K., Batchelder, W. H. and T. J. Brazill. 1995. Scaling semantic domains, in R. D. Luce *et al.* (eds.), *Geometric Representations of Perceptual Phenomena: Papers in Honor of Tarow Indow's 70th Birthday,* pp. 267–294. Mahwah, NJ: Lawrence Erlbaum.

Romney, A. K., Boyd, J. P., Moore, C. C., Batchelder, W. H. and T. J. Brazill. 1996. Culture as shared cognitive representations. *Proceedings of the National Academy of Sciences*, **93**, 4699–4705.

Romney, A. K., Shepard, R. N. and S. B. Nerlove. 1972. *Multidimensional Scaling: Theory and Applications in the Behavioral Sciences,* Vol. II, *Applications*. New York: Seminar Press.

Rost, J. 1988a. Test theory with qualitative and quantitative latent variables, in R. Langeheine and J. Rost (eds.), *Latent Trait and Latent Class Models,* pp. 147–171. New York: Plenum.

Rost, J. 1988b. Measuring attitudes with a threshold model drawing on a traditional scaling concept. *Applied Psychological Measurement,* **12**, 397–409.

Rost, J. 1990. *LACORD. Latent Class Analysis for Ordinal Variables. A FORTRAN Program.* (2nd ed.) Kiel: IPN, University of Kiel.

Rouanet, H. and B. Le Roux. 1993. *Analyse des données multidimensionelles*. Paris: Dunod.

Rouget, B., Sagot-Duvauroux, D. and S. Pflieger (eds.). 1991. *Le marché de l'art contemporain en France.* Paris: La Documentation Française.

Roy, S. N. and S. K. Mitra. 1956. An introduction to some nonparametric generalizations of analysis of variance and multivariate analysis. *Biometrika,* **43**, 361–376.

Rudas, T., Clogg, C. C. and B. G. Lindsay. 1994. A new index of fit based on mixture methods for the analysis of contingency tables. *Journal of the Royal Statistical Society, Series B,* **56**, 623–639.

Rudas, T. and R. Zwick. 1997. Estimating the importance of differential item functioning. *Journal of Educational and Behavioral Statistics,* **22**, 31–45.

Rudolph, C. S. and S. R. Borker. 1987. *Regionalization. Issues in Intensive Care for High Risk Newborns and Their Families.* New York: Praeger.

Salem, A. 1984. La typologie des segments répétés dans un corpus, fondée sur l'analyse d'un tableau croisant mots et textes. *Les Cahiers de l'Analyse des Donnés,* **9**, 489–500.

Salem, A. 1987. *Pratique des segments répétés. Essai de statistique textuelle.* Paris: Klincksieck.

Salem, A. 1995. Les unités lexicométriques, in S. Bolasco, L. Lebart and A. Salem (eds.), *Analisis Statistica dei Dati Testuali (Proceedings of the JADT95),* pp. 19–26. Rome: CISU.

Sall, J. 1991a. The conceptual model behind the picture. *ASA Statistical Computing and Statistical Graphics Newsletter,* **2**, 5–8.

Sall, J. 1991b. The conceptual model for categorical responses. *ASA Statistical Computing and Statistical Graphics Newsletter,* **3**, 33–36.

SAS. 1994. *JMP: Statistics for the Macintosh from the SAS Institute.* Cary, NC: SAS Institute.

Sasaki, M. and T. Suzuki. 1989. New directions in the study of general social attitudes: Trends and cross-national perspectives. *Behaviormetrika,* **26**, 9–30.

Schrott, P. R. and D. J. Lanoue. 1994. Trends and perspectives in content analysis, in I. Borg and P. P. Mohler (eds.), *Trends and Perspectives in Empirical Social Research,* pp. 327–345. Berlin: de Gruyter.

Schoenberg, I. J. 1935. Remarks to Maurice Fréchet's article "Sur la définition axiomatique d'une classe d'espaces distanciés vectoriellement applicables sur l'espace de Hilbert." *Annals of Mathematics,* **36**, 724–732.

Schönemann, P. H. 1970. On metric multidimensional unfolding. *Psychometrika,* **35**, 167–176.

Schönemann, P. H. and M. M. Wang. 1972. An individual difference model for the multidimensional analysis of preference data. *Psychometrika,* **37**, 275–309.

Schuman, H. 1966. The random probe: A technique for evaluating the validity of closed questions. *American Sociological Review,* **21**, 218–222.

Schuman, H. and F. Presser. 1981. *Question and Answers in Attitude Surveys.* New York: Academic Press.

Schwarz, S. H. and W. Bilsky. 1987. Toward a unified psychological structure of human values. *Journal of Personality and Social Psychology,* **53**, 550–562.

Seber, G. A. F. 1984. *Multivariate Observations.* New York: Wiley.

Shenai, J. P. 1993. Neonatal transport—outreach educational program. *Pediatric Clinics of North America,* **40**, 275–285.

Shepard, R. N. 1962a. The analysis of proximities: Multidimensional scaling with an unknown distance function (I). *Psychometrika,* **27**, 125–139.

Shepard, R. N. 1962b. The analysis of proximities: Multidimensional scaling with an unknown distance function (II). *Psychometrika*, **27**, 219–246.

Shepard, R. N., Romney, A. K. and S. B. Nerlove. 1972. *Multidimensional Scaling: Theory and Applications in the Behavioral Sciences,* Vol. I, *Theory.* New York: Seminar Press.

Sherkat, D. E. 1992. Theory and methods in religious mobility research. *Social Science Research,* **22**, 208–227.

Shiffman, S. S., Reynolds, M. L. and F. W. Young. 1981. *Introduction to Multidimensional Scaling: Theory, Methods and Applications.* New York: Academic Press.

Shulman, J., Edmonds, L. D., McClearn, A. B., Jensvold, N. and G. M. Shaw. 1993. Surveillance for and comparison of birth defect prevalences in two geographic areas—United States, 1983–88. *Mortality and Morbidity Weekly Reports,* **42**, 1–7.

Shye, S. and D. Elizur. 1994. *Introduction to Facet Theory and Intrinsic Data Analysis.* Newbury Park, CA: Sage.

Siciliano, R. 1990. Asymptotic distribution of eigenvalues and statistical tests in non symmetric correspondence analysis. *Statistica Applicata,* **3**, 259–276.

Siciliano, R., Mooijaart, A. and P. G. M. van der Heijden. 1993. A probabilistic model for nonsymmetric correspondence analysis and prediction in contingency tables. *Journal of the Italian Statistical Society,* **1**, 85–106.

Siciliano, R. and P. G. M. van der Heijden. 1994. Simultaneous latent budget analysis of a set of two-way tables with constant-row-sum-data. *Metron,* **52**, 155–180.

Siegel, S. and N. J. Castellan. 1988. *Nonparametric Statistics for the Behavioral Sciences* (2nd ed.) New York: McGraw–Hill.

Silicon Graphics. 1993. *IRIS Explorer User's Guide.* Mountain View, CA: Silicon Graphics.

Slater, P. 1960. The analysis of personal preferences. *British Journal of Statistical Psychology,* **3**, 119–135.

Snee, R. D. 1974. Graphical display of two-way contingency tables. *American Statistician,* **28**, 9–12.

Snow, R. E., Kyllonen, P. C. and B. Marshalek. 1984. The topography of ability and learning correlations, in R. J. Sternberg (ed.), *Advances in the Psychology of Human Intelligence,* (Vol. 3), pp. 47–103. Hillsdale, NJ: Lawrence Erlbaum.

Spangenberg, N. 1990. *Familienkonflikte eßgestörter Patientinnen: Eine empirische Untersuchung mit Hilfe der Repertory-Grid-Technik.* Habilitationsschrift am Fachbereich Humanmedizin der Justus-Liebig-Universität Gießen.

Spangenberg, N. and K. E. Wolff. 1991. Comparison of biplot analysis and formal concept analysis in the case of a repertory grid, in H. H. Bock and P. Ihm (eds.), *Data Analysis and Knowledge Organization,* pp. 104–112. Berlin: Springer.

Srole, L. 1956. Social integration and certain corollaries: An exploratory study, *American Sociological Review,* **21**, 709–716.

Stevens, S. S. 1968. Measurement, statistics and the schemapiric view. *Science,* **161**, 849–856.

Stram, D. O., Wei, L. J. and J. H. Ware. 1988. Analysis of repeated ordered categorical outcomes with possibly missing observations and time-dependent covariates. *Journal of the American Statistical Association,* **83**, 631–637.

Strobino, D. M., Frank, R., Oberdorf, A. M., Shachtman, R., Kim, Y. J., Callan, N. and D. Nagey. 1993. Development of an index of maternal transport. *Medical Decision-Making,* **13**, 64–73.

Sonquist, J. A. and J. N. Morgan. 1964. *The Detection of Interaction Effects.* Ann Arbor Institute for Social Research. University of Michigan.

Symons, F., Deknopper, E., Rymenans, J. and M. Vuylsteke-Wauters. 1983. De geometrische voorstelling van multidimensionele gegevens: theoretische inleiding. *Biometrie-Praximetrie,* **23**, 121–148.

Swyngedouw, M. 1992. National elections in Belgium: The breakthrough of extreme right in Flanders. *Regional Politics and Policy,* **2**, 62–75.

Takane, Y. 1986. Multiple discriminant analysis for predictor variables measured at various scale levels, in W. Gaul and M. Schader (eds.), *Classification as a Tool of Research,* pp. 439–445. Amsterdam: Elsevier.

Takane, Y. 1987. Analysis of contingency tables by ideal point discriminant analysis. *Psychometrika,* **52**, 493–513.

Takane, Y. 1989a. Ideal point discriminant analysis and ordered response categories. *Behaviormetrika,* **26**, 31–46.

Takane, Y. 1989b. Ideal point discriminant Analysis: Implications for multiway data analysis, in R. Coppi and S. Bolasco (eds.), *Multiway Data Analysis,* pp. 287–299. Amsterdam: Elsevier.

Takane, Y., Bozdogan, H. and T. Shibayama. 1987. Ideal point discriminant analysis. *Psychometrika,* **52**, 371–392.

Tenenhaus, M. and F. W. Young. 1985. An analysis and synthesis of multiple correspondence analysis, optimal scaling, dual scaling, homogeneity analysis and other methods for quantifiying categorical data. *Psychometrika,* **50**, 91–119.

ter Braak, C. J. F. 1986. Canonical correspondence analysis: A new eigenvector technique for multivariate direct gradient analysis. *Ecology,* **67**, 1167–1179.

ter Braak, C. J. F. 1988. Partial canonical correspondence analysis, in H. H. Bock (ed.), *Classification and Related Methods of Data Analysis,* pp. 551–558. Amsterdam: North Holland.

ter Braak, C. J. F. 1990. Interpreting canonical correlation analysis through biplots of structure and weights. *Psychometrika.* **55**, 519–532.

The BMS (K. M. van Meter, Schiltz, M.-A., Cibois P. and L. Mounier). 1994. Correspondence analysis. A history and French sociological perspective, in M. J. Greenacre and J. Blasius (eds.), *Correspondence Analysis in the Social Sciences,* pp. 128–137. London: Academic Press.

Thisted, R. A. (ed.). 1994. *CIS Extended Database.* American Statistical Association (Alexandria, VA), and the Institute of Mathematical Statistics (Hayward, CA).

Thornes, B. and J. Collard. 1979. *Who Divorces?* London: Routledge & Kegan Paul.

Torgerson, W. S. 1952. Multidimensional scaling: I. Theory and method. *Psychometrika,* **17**, 401–419.

Torgerson, W. S. 1958. *Theory and Methods of Scaling.* New York: Wiley.

Travis, D. T. 1991. *Effective Color Displays: Theory and Practice.* New York: Academic Press.

Tucker, L. R. 1960. Intra-individual and inter-individual multidimensionality, in H. Gulliksen and S. Messick (eds.), *Psychometric Scaling: Theory and Applications,* pp. 155–167. New York: Wiley.

Tucker, L. R. 1966. Some mathematical notes on three-mode factor analysis. *Psychometrika,* **31**, 279–311.

Tukey, J. W. 1949. One degree of freedom for non-additivity. *Biometrics,* **5**, 232–242.

Tukey, J. W. 1959. A quick, compact, two sample test to Duckworth's specifications. *Technometrics,* **1**, 31–48.

Tukey, J. W. 1977. *Exploratory Data Analysis.* Reading, MA: Addison Wesley.

van Buuren, S. and J. L. A. van Rijckevorsel. 1992. Imputation of missing categorical data by maximizing internal consistency. *Psychometrika,* **57**, 567–580.

van de Geer, J. P. 1993. *Analysis of Categorical Data* (2 vols.). Newbury Park, CA: Sage.

van der Ark, L. A. and P. G. M. van der Heijden. 1996. *Geometry and Identifiability in the Latent Budget Model.* Methods Series MS-96-3. Utrecht: ISOR.

van der Heijden, P. G. M. 1987. *Correspondence Analysis of Longitudinal Categorical Data.* Leiden: DSWO.

van der Heijden, P. G. M., de Falguerolles, A. and J. de Leeuw. 1989. A combined approach to contingency table analysis and log-linear analysis (with discussion). *Applied Statistics,* **38**, 249–292.

van der Heijden, P. G. M. and J. de Leeuw. 1985. Correspondence analysis used complementary to loglinear analysis. *Psychometrika,* **50,** 429–447.

van der Heijden, P. G. M., Mooijaart, A. and J. de Leeuw. 1989. Latent budget analysis, in A. Decarli, B. J. Francis, R. Gilchrist and G. U. H. Seeber (eds.), *Statistical Modelling. Lecture Notes in Statistics 57,* pp. 301–313. Berlin: Springer.

van der Heijden, P. G. M., Mooijaart, A. and J. de Leeuw. 1992. Constrained latent budget analysis, in Marsden, P. V. (ed.), *Sociological Methodology, 22,* pp. 279–320. Cambridge: Blackwell.

van der Heijden, P. G. M., Mooijaart, A. and Y. Takane. 1994. Correspondence analysis and contingency table models, in M. J. Greenacre and J. Blasius (eds.), *Correspondence Analysis in the Social Sciences,* pp. 79–111. London: Academic Press.

van der Heijden, P. G. M. and K. J. Worsley. 1988. Comment on "Correspondence analysis used complementary to loglinear analysis." *Psychometrika,* **53**, 287–291.

van Deth, J. W. 1995. Introduction: The impact of values, in J. W. van Deth and E. Scarbrough (eds.), *The Impact of Values. Beliefs in Government Volume Four,* pp. 1–18. Oxford: Oxford University Press.

van Deth, J. W. and E. Scarbrough. 1995. The concept of values, in J. W. van Deth and E. Scarbrough (eds.), *The Impact of Values. Beliefs in Government Volume Four,* pp. 21–47. Oxford: Oxford University Press.

van Eeuwijk, F. A. 1995. Multiplicative interaction in generalized linear models. *Biometrics,* **51**, 1017–1032.

van Rijckevorsel, J. L. A. and J. de Leeuw. 1988. *Component and Correspondence Analysis: Dimension Reduction by Functional Approximation.* New York: Wiley.

van Schuur, W. 1984. *Structure in Political Beliefs, a New Model for Stochastic Unfolding with Application to European Party Activists.* Amsterdam: CT-Press.

Velleman, P. and L. Wilkinson. 1993. Nominal, ordinal, interval, and ratio typologies are misleading. *American Statistician*, **47**, 65–72.

Verger, A. 1991. Le champ des avantgardes. *Actes de la recherche en sciences sociales,* **88**, 2–40.

von Bismarck, B., Stoller, D. and U. Wuggenig (eds.). 1996. *Games, Fights, Collaborations. The Game of Boundary and Transgression.* Ostfildern b. Stuttgart: Cantz.

Vuylsteke-Wauters, M. (1994). *Projectie en rotatie: paradigma's in statistische technieken*, Universitair Rekencentrum/Universitair Centrum voor Statistiek, KU Leuven.

Ward, J. H. 1963. Hierarchical grouping to optimize an objective function. *Journal of the American Statistical Association,* **58**, 236–244.

Weibel, P. (ed.). 1994. *Kontext Kunst.* Köln: DuMont.

Weischedel, R., Meteer M., Schwartz R., Ramshaw L. and J. Palmucci. 1993. Coping with ambiguity and unknown words through probabilistic models. *Computational Linguistics*, **19**, 361–382.

Weller, S. C. and A. K. Romney. 1990. *Metric Scaling: Correspondence Analysis.* Newbury Park, CA: Sage.

White, H. and C. White. 1993. *Canvasses and Careers.* Chicago: University of Chicago Press.

Whittaker, J. 1990. *Graphical Models in Multivariate Statistics.* New York: Wiley.

Wiggins, L. M. 1973. *Panel Analysis: Latent Probability Models for Attitude and Behaviour Processes.* Amsterdam: Elsevier.

Wilkinson, L. 1989. *SYGRAPH: The System for Graphics.* Evanston, IL: SYSTAT.

Wille, R. 1982. Restructuring lattice theory: an approach based on hierarchies of concepts, in I. Rival (ed.), *Ordered Sets,* pp. 445–470. Dordrecht-Boston: Reidel.

Wille, R. 1987. Bedeutungen von Begriffsverbänden, in B. Ganter, R. Wille, und K. E. Wolff (Hrsg.), *Beiträge zur Begriffsanalyse,* pp. 161–211. Mannheim: B.I.-Wissenschaftsverlag.

Wille, R. 1992. Concept lattices and conceptual knowledge systems. *Computers & Mathematics with Applications,* **23**, 493–515.

Wold, H. 1966. Nonlinear estimation by iterative least squares procedures, in H. David (ed.), *Research Papers in Statistics, Festschrift for J. Neyman,* pp. 411–444. New York: Wiley.

Wolff, K. E. 1994. A first course in formal concept analysis—how to understand line diagrams, in F. Faulbaum (ed.), *SoftStat '93, Advances in Statistical Software 4*, pp. 429–438. Stuttgart: Fischer.

Wolff, K. E. 1996. Comparison of graphical data analysis methods, in F. Faulbaum and W. Bandilla (eds.), *Softstat '95, Advances in Statistical Software 5*, pp. 139–151. Stuttgart: Lucius & Lucius.

Wolff, K. E., Gabler, S. and I. Borg. 1994. Formale Begriffsanalyse von Arbeitswerten in Ost-und Westdeutschland. *ZUMA-Nachrichten*, **34**, 69–82.

Wottawa, H. 1980. *Grundriß der Testtheorie*. München: Juventa.

Wuggenig, U. and P. Mnich. 1994. Explorations in social spaces: Gender, age, class fractions and photographical choices of objects, in M. Greenacre and J. Blasius (eds.), *Correspondence Analysis in the Social Sciences*, pp. 302–323. London: Academic Press.

Xi, L. 1996. *The Mixture Index of Fit*. Ph.D. Thesis. Department of Statistics, Pennsylvania State University.

Xi, L. and B. G. Lindsay. 1997. A note on calculating the $\pi*$ index of fit for the analysis of contingency tables. *Sociological Methods and Research*, **25**, 248–259.

Yates, J. F. and E. R. Stone. 1992. The risk construct, in J. Fank Yates (ed.), *Risk-Taking Behavior*, pp. 1–25. Chichester: Wiley.

Young, F. W. and R. M. Hamer. 1987. *Multidimensional Scaling: History, Theory and Applications*. Hillsdale, NJ: Lawrence Erlbaum.

Young, G. and A. S. Householder. 1938. Discussion of a set of points in terms of their mutual distances. *Psychometrika*, **3**, 19–22.

Yule, G. U. 1912. On the methods of measuring association between two attributes. *Journal of theRoyal Statistical Society*, **75**, 579–642.

About the Authors

Jacques Allard is Professor in the Department of Mathematics and Statistics, Université de Moncton, Canada. Most of his work is in applications of statistics, particularly to fisheries management and other fields related to the environment. He is consultant to both government and private industry. He is the author of *Concepts Fondamentaux de la Statistique* (Addison-Wesley 1992), an introductory statistics textbook in the social sciences used in several institutions in French Canada.
Address: Department of Mathematics and Statistics, Université de Moncton, Moncton, N.B. Canada E1A 3E9. E-mail: allardj@umoncton.ca

Tomàs Aluja-Banet is Professor of Statistics and Data Analysis at the Universitat Politècnica de Catalunya, Barcelona, Spain. His interests are mainly in multivariate data analysis, descriptive factorial analysis, and classification techniques and their application to social sciences.
Address: Departament d'Estadística i Investigació Operativa, Universitat Politècnica de Catalunya, C/Pau Gargallo 5, E-08028 Barcelona, Spain. E-mail: aluja@eio.upc.es

Simona Balbi is Researcher in the Dipartimento di Matematica e Statistica, Università "Federico II," Napoli, Italy. Her research is in multidimensional data analysis techniques and their applications in socioeconomic surveys, with special attention to repeated surveys, quality control, and measuring customer satisfaction.
Address: Dipartimento di Matematica e Statistica, Università "Federico II," Via Cintia, Monte S. Angelo, I-80126 Napoli, Italy. E-mail: sb@unina.it

Mónica Bécue Bertaut is Associate Professor in the Department of Statistics, Universitat Politècnica de Catalunya, Barcelona, Spain. Her interests are primarily in statistical methods for textual data, analysis of open responses, and software for textual data design. She is coauthor (with Ludovic Lebart, Alain Morineau, and Laurence Haeusler) of the statistical package SPAD.T for statistical analysis of texts.
Address: Departament d'Estadística i Investigació Operativa, Universitat Politècnica de Catalunya, C/Pau Gargallo 5, E-08028 Barcelona, Spain.
E-mail: monica@eio.upc.es

Jaak Billiet is Professor of Social Methodology in the Department of Sociology, Katholieke Universiteit Leuven, Belgium. He is also the project leader of the Inter-University Centre for Political Opinion Research, which organizes the General Election Studies in Flanders (Belgium). His main research interests are measurement error in social surveys and the methodology of longitudinal and comparative research on social and cultural change, in particular in the fields of ethnic prejudice and political values.

Address: Department of Sociology, KU Leuven, E. Van Evenstraat 2B, B 3000 Leuven, Belgium. E-mail: jaak.billiet@soc.kuleuven.ac.be

Jörg Blasius is Researcher at the Zentralarchiv für Empirische Sozialforschung at the University of Cologne. His main research interests focus on multivariate explorative analysis, urban research, lifestyles, and methods of empirical social research. He is coeditor (with Michael Greenacre) of *Correspondence Analysis in the Social Sciences* (Academic Press, UK, 1994) and he gives courses on correspondence analysis at the Essex Summer School.

Address: University of Cologne, Zentralarchiv für Empirische Sozialforschung, Bachemer Str. 40, D-50931 Köln. E-mail: blasius@za.uni-koeln.de

Ulf Böckenholt is Associate Professor of Psychology at the University of Illinois at Urbana-Champaign, USA. His interests are primarily in the analysis of multivariate categorical data and mathematical models of judgment and choice.

Address: Department of Psychology, University of Illinois at Urbana-Champaign, Champaign, IL 61820. E-mail: ubockenh@s.psych.uiuc.edu

Ingwer Borg is Scientific Director at the Center for Survey Methodology (ZUMA) in Mannheim, Germany, and Professor of Psychology at the University of Giessen, Germany. He is author or editor of 10 books and numerous articles on data analysis, facet theory, and various substantive topics of psychology. He recently finished a book on multidimensional scaling with Patrick Groenen.

Address: ZUMA, P.O. Box 122155, D-68072 Mannheim, Germany. E-mail: borg@zuma-mannheim.de

Timothy J. Brazill is a Ph.D. candidate in the Social Network Analysis Program at the University of California, Irvine. His professional interests include the cognition of social structure and its behavioral consequences, the measurement of network properties such as acquaintanceship volume and social proximity, and methodology and statistics.

Address: School of Social Sciences, Social Science Plaza 5157, University of California, Irvine, CA 92697-5100, USA. E-mail: tbrazill@uci.edu

André Carlier is Senior Researcher at the Laboratoire de Statistique et Probabilités, Université Paul Sabatier, Toulouse, France. His interests focus on multivariate analysis in general and in particular the analysis of three-way contingency tables. He is a coauthor of the book *Analyse Discriminante sur Variables Qualitatives* (Polytechnica, Paris, 1995) and has written a set of S-plus functions for multidimensional data analysis.

Address: Laboratoire de Statistique et Probabilités, Université Paul Sabatier, 118 route de Narbonne, F-31400 Toulouse, France. E-mail: carlier@cict.fr

Jean Hugues Chauchat is Professor of Applied Statistics at the Université Lumière of Lyon, France. His present research interest lies in data analysis and surveys and he is directing a postgraduate program on the gathering, organization, and processing of data with a view to improving the decision-making process in the firm. He has published a number of articles on optimal coding, the measurement of attitudinal differences, and various applications of statistics to political sciences, medicine, social psychology, and marketing.

Address: Faculté des Sciences Economiques et de Gestion, Université Lumière, 16 quai Claude Bernard, F-69007 Lyon, France. E-mail: chauchat@univ-lyon2.fr

Vartan Choulakian is Professor in the Department of Mathematics and Statistics, Universitè de Moncton, Canada. His interests are primarily in generalized multilinear models, measures of vector correlations, and goodness of fit using Cramer–van Mises statistics.

Address: Department of Mathematics and Statistics, Université de Moncton, Moncton, N.B.,Canada E1A 3E9. E-mail: choulav@umoncton.ca

Clifford C. Clogg was Distinguished Professor of Sociology and Professor of Statistics at the Pennsylvania State University and he was also affiliated with the Population Research Institute of Penn State. He was an outstanding statistician and sociologist and made important contributions to the methodology of demography and sociology and to the methods and theory of categorical data analysis, including extensions of the log-linear model and latent class analysis. In addition to papers in several scholarly journals, he published a book on measuring unemployment, was coauthor of a book on statistical models for ordinal variables, and was coeditor of a volume on latent variable analysis and of the *Handbook of Statistical Modeling in the Social and Behavioral Sciences.* Furthermore, he had been applications and coordinating editor of the *Journal of the American Statistical Association* and editor of *Sociological Methodology* for several years. His contribution to this book is one of the last papers he worked on.

Carles M. Cuadras is Professor of Statistics in the Departament d'Estadística, Universitat de Barcelona, Spain. His main interests are in biostatistics, construction of distributions with given marginals, and methodological applications of multidimensional scaling in regression and classification. In 1981 he published the first Spanish book on multivariate analysis.

Address: Departament d'Estadística, Universitat de Barcelona, Diagonal 645, E 08028 Barcelona, Spain. E-mail: carlesm@porthos.bio.ub.es

Antoine de Falguerolles is Senior Lecturer in the Laboratoire de Statistique et Probabilités at the Université Paul Sabatier, Toulouse, France. His current work concerns the integration of techniques of "Analyse des Données" and of modeling in the analysis of multivariate data, with special interest in square tables.

Address: Laboratoire de Statistique et Probabilités, Université Paul Sabatier, 118 route de Narbonne, F-31062 Toulouse Cedex, France. E-mail: falguero@cict.fr

Jan de Leeuw is Professor and Director, UCLA Statistics Program. He is a former president of the Psychometric Society and the editor of the *Journal of Educational and Behavioral Statistics*, the *Journal of Multivariate Analysis*, and the Advanced Quantitative Methods book series. He is also the World Wide Web editor for the Institute of Mathematical Statistics.

Address: UCLA Statistics Program, 8142 Math Sciences Building, Box 951554, Los Angeles, CA 90095-1554, USA. E-mail: deleeuw@stat.ucla.edu

Hans de Witte is Head of the Sector Labour at the Higher Institute of Labour Studies, located at the Katholieke Universiteit Leuven. He is also a social psychologist with interests in social and cultural attitudes of the working class, on which he recently published a book. He is also involved in research on extreme right wing ideologies and on ethnocentrism. He supervises a number of research projects about the cultural differences between the social classes.
Address: Higher Institute of Labour Studies, E. Van Evenstraat 2E, B-3000 Leuven, Belgium. E-mail: hans.dewitte@hiva.kuleuven.ac.be

Fernand Fehlen is Senior Researcher in the "Cellule Statistique et Decision" of the Centre de Recherche Public, Centre Universitaire, Luxembourg. He teaches statistics and sociology at the Centre Universitaire and the Institut d'Etudes Educatives et Sociales in Luxembourg. His main interest is the description of the Luxembourg social space, and particularly the political and cultural fields, in terms of Bourdieu's sociology.
Address: Centre de Recherche Public, Centre Universitaire, 13 rue de Bragance, L-1255 Luxembourg, Luxembourg. E-mail: fehlen@crpcu.lu

Josep Fortiana is Associate Professor, Departament d'Estadística, Universitat de Barcelona, Spain. His interest is in computational statistics and methodological applications of multidimensional scaling in regression and classification.
Address: Departament d'Estadística, Universitat de Barcelona, Gran Via, 585, E-08007 Barcelona, Spain. E-mail: fortiana@cerber.mat.ub.es

Brian Francis is Senior Lecturer and Assistant Director in the Centre for Applied Statistics at Lancaster University, United Kingdom. As well as having an interest in data visualization techniques, he has other interests in generalized linear and bilinear models, statistical models for criminological data, and statistical software design. He is a co-developer of GLIM4 and has coauthored several books on GLIM and statistical modelling.
Address: Centre for Applied Statistics, Fylde College, Lancaster University, LA1 4YF, United Kingdom. E-mail: b.francis@lancaster.ac.uk

Ulrich Frick is Head of Biometrics at estimate GmbH, a consulting company for health planning projects and clinical trials. He studied psychology and jazz composition and has been working in applied statistics with specialization in biometrics and public health. His main research interests are medical decision making, health planning, and risk perception.
Address: estimate GmbH, Schmiedberg 6, D-86152 Augsburg, Germany. E-mail: estimate@t-online.de

Michael Friendly is Associate Professor of Psychology and Coordinator of the Statistical Consulting Service at York University, Canada. He is an associate editor of the *Journal of Computational and Graphical Statistics* and has been working on the development of graphical methods for categorical data.
Address: Psychology Department, York University, Toronto, Canada M3J 1P3. E-mail: friendly@yorku.ca

Mark Fuller is a Graphics Specialist with LightWork Design Ltd. and a former research associate in the Centre for Applied Statistics at Lancaster University,

United Kingdom. His interests are in scientific visualization, computer graphics, and perception.
Address: LightWork Design, 60 Clarkehouse Road, Sheffield, S10 2LH, United Kingdom. E-mail: markf@lightwork.co.uk

Siegfried Gabler is Director of the Department of Statistics at the Centre for Survey Research and Methodology (ZUMA) in Mannheim, Germany. His interests are primarily correspondence analysis and sampling from finite populations. He recently published a book on minimax solutions in sampling from finite populations.
Address: ZUMA, Department of Statistics, P.O. Box 122155, D-68072 Mannheim, Germany. E-mail: gabler@zuma-mannheim.de

K. Ruben Gabriel is Professor in the Department of Statistics, University of Rochester, NY. He was formerly professor and chair of the Department of Statistics at the Hebrew University of Jerusalem, Israel. His principal interests are in lower rank approximation and visual display of multivariate data, for which he proposed the biplot in 1971, and in the design and analysis of weather modification experiments.
Address: Department of Statistics, University of Rochester, Rochester, NY 14627, USA. E-mail: krg1@db1.cc.rochester.edu

M. Purificación Galindo Villardón is Professor and Head of the Departamento de Estadística y Matemática Aplicadas, Universidad de Salamanca, Spain. Her main research interest is in multivariate data analysis, particularly the application of biplot and correspondence analysis to clinical and environmental data. Recent work includes contributions to automatic interaction detection models.
Address: Departamento de Estadística y Matemática Aplicadas, C/Espejo 2, E-37007 Salamanca, Spain. E-mail: pgalindo@gugu.usal.es

John Gower is Senior Researcher in the Department of Statistics, Open University, England. He was formerly head of the Biomathematics Division and Statistics Departments at Rothamsted Experimental Station. His interests are primarily in biometrical methods and applications, multidimensional scaling, classification, and biplots. He has recently published a book on biplots.
Address: Department of Statistics, The Open University, Walton Hall, Milton Keynes, MK7 6AA, United Kingdom. E-mail: j.c.gower@open.ac.uk

Michael Greenacre, formerly at the University of South Africa, is Professor of Statistics in the Department of Economics and Business, Pompeu Fabra University, Barcelona, Spain. His main interests are in the visualization of high-dimensional data, particularly correspondence analysis. He has published or edited three books on correspondence analysis and has given short courses in the United States, United Kingdom, Germany, Norway, Finland, and South Africa.
Address: Departament d'Economía i Empresa, Universitat Pompeu Fabra, Ramon Trias Fargas 25-27, E-08005 Barcelona, Spain. E-mail: michael@upf.es

Patrick J. F. Groenen is Researcher in the Department of Data Theory, Leiden University, The Netherlands. He has written several articles in the area of multidimensional scaling, optimization, and data analysis. He recently finished a book on multidimensional scaling with Ingwer Borg.

Address: Department of Data Theory, Leiden University, P.O. Box 9555, NL-2300 RB Leiden, The Netherlands. E-mail: groenen@rulfsw.fsw.leidenuniv.nl

Simon J. Harding is a member of the Statistics Department, Rothamsted Experimental Station, United Kingdom. His interests include statistical programming, multivariate analysis, and graphics. He is also a developer of the Genstat statistical package, which was used to generate the analyses and color figures for the chapter on generalized biplots in this book.

Address: Statistics Department, IACR Rothamsted, Harpenden, Herts AL5 2JQ, United Kingdom. E-mail: simon.harding@bbsrc.ac.uk

Willem J. Heiser is Professor in the Department of Data Theory, Leiden University, The Netherlands. His main research areas are multidimensional scaling, clustering, and graph models. He is co-author and editor of the "Albert Gifi" (1990) book on nonlinear multivariate analysis. Currently, he is editor of the journal *Psychometrika*.

Address: Department of Data Theory, Leiden University, P.O. Box 9555, NL-2300 RB Leiden, The Netherlands. E-mail: heiser@rulfsw.fsw.leidenuniv.nl

Jörg Kastl is Assistant Professor in the Department of Sociology, University of Tübingen, Germany. His research areas are theoretical and methodological problems of sociology, especially in the field of communication theory and sociology of mass media.

Address: Institut für Soziologie, Universität Tübingen, Wilhelmstr. 36, D-72074 Tübingen, Germany. E-mail: joerg michael.kastl@uni-tuebingen.de

Pieter M. Kroonenberg is Senior Lecturer in Applied Statistics in the Department of Education, Leiden University, The Netherlands. His interests are primarily in both theoretical and applied three-way data analysis. He is the author of a book on three-mode principal component analysis and a commercially available package for three-way data analysis.

Address: Vakgroep Pedagogiek, Rijksuniversiteit Leiden, Wassenaarseweg 52, NL-2333 AK Leiden, The Netherlands. E-mail: kroonenb@rulfsw.fsw.leidenuniv.nl

Michael Laschat is a medical doctor with specialization in child anesthesiology. He is working as Senior Physician at the Municipal Children's Hospital, Cologne, Germany. He has been working as a medical consultant to various planning projects, such as the forecasting of future medical requirements for children in Vienna, potential substitution of inpatient by outpatient treatment, and evaluation of neonatal care.

Address: Städtische Kinderklinik, Anästhesie, Hamburger Straße, D-50668 Köln, Germany.

Ludovic Lebart is Senior Researcher at the Centre National de la Recherche Scientifique in Paris. His research interests are the exploratory analysis of qualitative and textual data, with emphasis on large data sets. He has coauthored several books on descriptive multivariate statistics and survey methodology.

Address: Ecole Nationale Supérieure des Télécommunications, 46 rue Barrault, F-75013 Paris, France. E-mail: lebart@eco.enst.fr

Brigitte Le Roux is Professor in the Département de Mathématiques et Informatique, Université René Descartes, Paris. Her main interests are in geometric data analysis, with emphasis on large-size tables, structured data, stability problems, and applications to behavioral and social sciences. She has recently coauthored several books about descriptive and inductive data analysis.

Address: UFR Math-Info, Université René Descartes, 45 Rue des Saints Pères, F-75270 Paris Cedex 06, France. E-mail: lerb@math info.univ-paris5.fr

Jay Magidson is founder and President of Statistical Innovations Inc., a firm that specializes in software development, educational training, and consulting in statistical modeling. He is the developer of the SPSS CHAID computer software. He has taught at the university level, served on government advisory boards, and consulted with businesses in marketing research/direct marketing. His current research interests include Jungian psychological types.

Address: Statistical Innovations Inc., 375 Concord Avenue, Suite 007, Belmont, MA 02178, USA.

Bernd Martens is Assistant Professor at the University of Tübingen, Department of Sociology. He has a particular interest in sociological data analysis and his recent work is connected with sociology of technology and environmental issues.

Address: Institut für Soziologie, Universität Tübingen, Wilhelmstr. 36, D-72074 Tübingen, Germany. E-mail: bernd.martens@uni-tuebingen.de

Stephen A. Matthews is Adjunct Assistant Professor of Geography at Penn State University. He is the Director of the Computer Core and Co-Director of the Geographic Information Analysis Core at the Population Research Institute, Penn State University. His research interests focus on medical geography and relations between population and environmental processes.

Address: Population Research Institute, The Pennsylvania State University, 601 Oswald Tower, University Park, PA 16802-6211. E-mail: matthews@pop.psu.edu

Allan L. McCutcheon is the Donald O. Clifton Distinguished Professor of Sociology, Survey Research and Methodology, and Director of the Gallup Research Center, at the University of Nebraska-Lincoln. His interests are primarily in methods for modeling categorical survey data.

Address: Gallup Research Center, University of Nebraska-Lincoln, 200 North 11th Street, Lincoln, NE 68588-0241, USA. E-mail: amccutch@unlinfo.unl.edu

Jacqueline J. Meulman is Associate Professor in the Department of Data Theory, Leiden University, The Netherlands, and she holds an adjunct position as Professor of Psychology at the University of Illinois at Urbana-Champaign. Her main research areas are multivariate analysis with optimal scoring and multidimensional scaling. She is coauthor and editor of the "Albert Gifi" (1990) book on nonlinear multivariate analysis.

Address: Department of Data Theory, Leiden University, P.O. Box 9555, NL-2300 RB Leiden, The Netherlands. E-mail: meulman@rulfsw.fsw.leidenuniv.nl

Francesco Mola is Associate Professor of Statistics in the Department of Mathematics and Statistics, University of Naples, Italy. He received the Research Doctorate degree in Computational Statistics and Applications in 1993. His research interests are in computational statistics, multidimensional data analysis, and artificial intelligence.

Address: Dipartimento di Matematica e Statistica, Universitá di Napoli Federico II, Via Cintia, Monte S. Angelo, I-80126 Naples, Italy.
E-mail: f.mola@dmsna.dms.unina.it

Carmella C. Moore is Lecturer in Anthropology and Assistant Researcher in the Department of Family Medicine, University of California, Irvine. She is coprincipal investigator with Arthur J. Rubel of an NSF funded study of cultural knowledge of tuberculosis in poor Mexican patients. Her research interests include medical and psychological anthropology, neuropsychiatry, and quantitative methods.
Address: Department of Anthropology, University of California, Irvine, CA 92697-5100. E-mail: ccmoore@orion.oac.uci.edu

Eduard Nafría is a part-time lecturer in statistics at the Universitat Politècnica de Catalunya, Barcelona, Spain. He has a computer science degree and is Statistics Director of Sofres España S.A. His main interest is the application of multivariate techniques for decision making.
Address: SOFRES AM, Cami de Can Calders 4, E-08190 Sant Cugat del Valles, Barcelona, Spain.

Shizuhiko Nishisato is Professor in the Graduate Department of Education, University of Toronto, Canada. He was formerly the editor of *Psychometrika* and president of the Psychometric Society. His research interest is in dual scaling. He has published five books on dual scaling.
Address: OISE/UT, University of Toronto, 252 Bloor Street West, Toronto, Ontario, Canada M5S 1V6. E-mail: snishisato@oise.utoronto.ca

Ivailo Partchev is Senior Lecturer in Statistics and Survey Methods at the Department of Sociology, Sofia University, Bulgaria. His interests are primarily in multivariate statistical analysis, sampling and variance estimation, and public opinion research. Since 1989, he has done much electoral research in Bulgaria and other Balkan countries.
Address: Department of Sociology, Sofia University, Zarigradsko shose 125, block 4, 1113 Sofia, Bulgaria. E-mail: ivailo@sclg.uni-sofia.bg

John Pritchard is a Research Associate in the Centre for Applied Statistics and graphics consultant in the Information System Services department at Lancaster University, United Kingdom. His interests are in scientific visualization systems and their application to complex real life data and in computer intensive numerical computing.
Address: Centre for Applied Statistics, Fylde College, Lancaster University, LA1 4YF, United Kingdom. E-mail: j.pritchard@lancaster.ac.uk

Jürgen Rehm is Associate Professor of Biostatistics and Preventive Medicine at the University of Toronto and director of the social and evaluation research department of the Addiction Research Foundation, Toronto, Canada. He has published books on bank robbery from the offender's perspective, and intuitive predictions and professional forecasts and numerous articles within the field of epidemiology and public health.
Address: Addiction Research Foundation, 33 Russel Street, Toronto, Ontario, M5S 2S1, Canada.

Alban Risson is Researcher in Statistics and Computer Sciences at the Université Claude Bernard, Lyon, France. His doctoral thesis deals with applications of statistics, data processing, and graphic information processing, in quality management.
Address: PRISMa, Université Claude Bernard, F-69622 Villeurbanne, France. E-mail: risson@univ.lyon1.fr

A. Kimball Romney is Research Professor in the School of Social Sciences at the University of California, Irvine. He formerly held positions at Chicago, Stanford, and Harvard. His research interests include measurement and scaling, cognitive anthropology, and social networks.
Address: School of Social Sciences, Social Science Plaza 2139, University of California, Irvine, CA 92697-5100, USA. E-mail: akromney@uci.edu

Henry Rouanet is Director of Research at the Centre National de la Recherche Scientifique, in the Département de Mathématiques et Informatique, Université René Descartes, Paris. Since 1979, he has been the head of the multidisciplinary *Groupe Mathématiques et Psychologie*. His main research interests in statistics are analysis of variance and Bayesian inference for complex designs. He has coauthored several books about statistics in social sciences.
Address: UFR Math Info, Université René Descartes, 45 Rue des Saints Pères, F-75270 Paris Cedex 06, France. E-mail: rouanet@math info.univ-paris5.fr

Tamás Rudas is head of the Centre for Applied Statistics at the Central European University. He is also affiliated with the Social Science Informatics Center (TÁRKI) in Budapest, Hungary. Formerly he was head of the Statistics Group of the Institute of Sociology at the Eötvös University. His main interests are categorical data analysis, model selection, and statistical algorithms.
Address: Centre for Applied Statistics, The Central European University, Nádor u. 9, H-1051 Budapest, Hungary. E-mail: rudas@tarki.hu

Roberta Siciliano is Associate Professor of Statistics in the Department of Mathematics and Statistics, University of Naples, Italy. Her research interests include modeling categorical data, latent variable models for longitudinal compositional data, classification, and regression trees. She is a member of the board of directors of the European Regional Section of the International Association for Statistical Computing (IASC-ERS).
Address: Dipartimento di Matematica e Statistica, Università di Napoli Federico II, Via Cintia, Monte S. Angelo, I-80126 Naples, Italy. E-mail: r.sic@dmsna.dms.unina.it

Frans Symons was Professor of Biometrics and Plant Ecology at the Department of Biology, Catholic University of Leuven. He was also affiliated with the University Center of Statistics, where he was a member of the steering committee and taught advanced data analysis in the masters program. As a biologist and statistician, he was a pioneer in introducing statistical data analysis in the domains of plant ecology and biology in Leuven. He was the author of a number of articles in scholarly journals. His matrix formulation of the different multivariate techniques allowed him to give a very concise overview of these methods. Every year he added a new level of insight to this overview. His approach to linear and nonlinear canonical correlation as an exploratory tool for the analysis of grouped data, which he published in the *Journal* *of Applied Statistics*, was applied in several disciplines, including the social sciences. His contribution to this book was made in collaboration with social scientists and a mathematician, and was one of his last papers.

Yoshio Takane is Professor of Psychology, McGill University, Montréal, Canada. His recent interests are primarily in constrained principal component analysis/canonical correlation analysis, statistical expert systems, and statistical bases of artificial neural network models.
Address: Department of Psychology, McGill University, 1205 Dr. Penfield Avenue, Montréal, Québec H3A 1B1, Canada. E-mail: takane@takane2.psych.mcgill.ca

Christian Tarnai is Associate Professor in the Department of Education, University of Münster, Germany. His main research interests are in statistical methodology in the social sciences and research on higher education.
Address: Department of Education, Universität Münster, D-48143 Münster, Germany. E-mail: tarnai@uni-muenster.de

Victor Thiessen is Professor of Sociology and Associate Dean of the Faculty of Arts and Social Sciences at the Dalhousie University, Halifax, Canada. His main research interest is in interpersonal dynamics within families. At the present time he is the principal investigator for the Nova Scotia School-to-Work Transition project, which documents youths' efforts to prepare themselves for the world of work.
Address: Faculty of Arts and Social Sciences, Dalhousie University, Halifax, Nova Scotia, Canada. E-mail: thiessen@is.dal.ca

L. Andries van der Ark is a Ph.D. student in the Department of Methodology and Statistics, Utrecht University, The Netherlands. His thesis is about models for compositional data.
Address: Department of Methodology and Statistics, Utrecht University, P.O. Box 80140, NL-3508 TC Utrecht, The Netherlands. E-mail: a.vanderark@fsw.ruu.nl

Peter G. M. van der Heijden is Professor in the Department of Methodology and Statistics, Utrecht University, The Netherlands. He is interested in relations between categorical data analysis methods such as correspondence analysis, latent class analysis, and log-linear analysis; in capture–recapture applications in demography and in the social sciences; and in randomized response as a tool for measuring sensitive topics.
Address: Department of Methodology and Statistics, Faculty of Social Sciences, Utrecht University, Postbus 80.140, NL-3508 TC Utrecht, The Netherlands. E-mail: p.vanderheijden@fsw.ruu.nl

José Luis Vicente-Villardón is Professor in the Departamento de Estadística y Matemática Aplicadas, Universidad de Salamanca, Spain. His main research interest is in multivariate data analysis, especially biplot and unfolding. Recent work is related to the integration of these two multidimensional data analysis techniques in gradient analysis of ecological data.
Address: Departamento de Estadística y Matemática Aplicadas, C/Espejo 2, E-37007 Salamanca, Spain. E-mail: villardon@gugu.usal.es

Magda Vuylsteke-Wauters is Professor of Statistics in the International Study Program in Statistics, Katholieke Universiteit Leuven, Belgium. She is also head of

the user group in the university's computing center and responsible for the statistical consultancy and courses in applied statistics in this center. Her experience and research are mainly in statistical computing, statistical graphics, and multivariate analysis.
Address: Computing Centre KU Leuven, De Croylaan 52a, B-3001 Heverlee, Belgium. E-mail: magda.vuylsteke@cc.kuleuven.ac.be

Karl Erich Wolff is Professor of Mathematics in the department Mathematik und Naturwissenschaften, Fachhochschule Darmstadt, Germany. He is deputy chairman of the *ErnstSchröderZentrum für Begriffliche Wissensverarbeitung e.V.* at the Technische Hochschule Darmstadt. His main research field is conceptual knowledge processing, especially formal concept analysis and its applications in psychology, sociology, and industry.
Address: Fachbereich Mathematik und Naturwissenschaften, Fachhochschule Darmstadt, Schöfferstr. 3, D-64295 Darmstadt, Germany. E-mail: wolff@mathematik.thdarmstadt.de

Ulf Wuggenig is Associate Professor of Cultural Sociology and Sociology of Arts at the University of Lüneburg, Germany. His main research areas include methods of social research, sociology of inequality and lifestyles, research on higher education, and the study of visual arts and media from a sociological point of view.
Address: Department of Cultural Sciences, Universität Lüneburg, D-21332 Lüneburg, Germany. E-mail: wuggenig@uni-lueneburg.de

Index